2nd International Conference on Structural Nano Composites

(NANOSTRUC 2014)

IOP Conference Series: Materials Science and Engineering Volume 64

Madrid, Spain
20 – 21 May 2014

ISBN: 978-1-63439-875-6
ISSN: 1757-8981

Printed from e-media with permission by:

Curran Associates, Inc.
57 Morehouse Lane
Red Hook, NY 12571

Some format issues inherent in the e-media version may also appear in this print version.

Printed by Curran Associates, Inc. (2015)

Institute of Physics
Dirac House, Temple Back
Bristol BS1 6BE UK

Phone: 44 1 17 929 7481
Fax: 44 1 17 920 0979

techtracking@iop.org

Additional copies of this publication are available from:

Curran Associates, Inc.
57 Morehouse Lane
Red Hook, NY 12571 USA
Phone: 845-758-0400
Fax: 845-758-2634
Email: curran@proceedings.com
Web: www.proceedings.com

TABLE OF CONTENTS

IOP Conf. Series: Materials Science and Engineering **64** (2014) 011001 doi:10.1088/1757-899X/64/1/011001

Preface: 2nd International Conference on Structural Nano Composites (NANOSTRUC2014)

J Njuguna[1] and R Verdejo
Institute for Innovation, Design and Sustainability, School of Engineering, Robert Gordon University, Aberdeen, AB10 7GJ, UK
Institute of Polymer Science and Technology, Consejo Superior de Investigaciones Científicas (ICTP-CSIC), C/ Juan de la Cierva,3, 28006, Madrid, Spain

E-mail: j.njuguna@rgu.ac.uk

Abstract. The NANOSTRUC 2014 took place at CSIC, Madrid, Spain. The conference theme on 'Nanosciences and Nanotechnologies – Recent Advances towards Nanoproducts and Applications'. The conference aimed to promote activities in various areas of materials and structures by providing a forum for exchange of ideas, presentation of technical achievements and discussion of future directions. NANOSTRUC conferences brings together an international community of experts to discuss the state-of-the-art, new research results, perspectives of future developments, and innovative applications relevant to structural materials, engineering structures, nanocomposites, modelling and simulations, and their related application areas.

1. Introduction

There has been explosive growth in the production of man-made nanostructures. Consequently, nanotechnology and nanosciences has been the greatest impetus to technological and industrial development in the 21st Century and has been recognized as the resource for the next industrial revolution [1]. The evolving technology has already influenced a large number of industrial segments, and the economic activity generated from it has been high in magnitude and wide in scope.

The nanotechnology-based products, which have had a huge impact on almost each industrial sector, have now a considerable consumer market share. In their latest research study, analysts have identified that the global nanotechnology industry has been growing at a rapid pace with rising applications in sectors primarily electronic, energy and healthcare sector. Hence, the global nanotechnology market is anticipated to grow at a CAGR of around 19% during 2013-2017 [1].These nanostructures, with one or more dimensions in the 1–100 nm size range, when engineered appropriately exhibit a variety of unique and tunable chemical and physical properties. With ease of manufacture in bulk, nanostructures are finding increasing practical application in many areas including in the development of nanocomposite materials for future use in technological applications.

2. NANOSTRUC 2014

The 2nd International Conference on Structural Nano Composites (NANOSTRUC 2014) is a follow up of NANOSTRUC 2012 that took place at Cranfield in the UK [2,3]. The conference theme on 'Nanosciences and Nanotechnologies – Recent Advances towards Nanoproducts and Applications'

[1] To whom any correspondence should be addressed.

IOP Conf. Series: Materials Science and Engineering **64** (2014) 011001 doi:10.1088/1757-899X/64/1/011001

was very well explored in this conference that attracted participants from 30 Countries across the 5 Continents.

In additional to excellent Keynote Lectures [4], NANOSTRUC continue to support and champion young researchers. The NANOSTRUC 2014 Posters Awards jointly supported by Wiley VCH and Nanotechnology Industries Association (NIA) were won by (1st Award) Mehmet Tutgun poster on 'Proton Conducting Composite Membranes of Crosslinked Poly(vinyl alcohol)–Sulfosuccinic Acid Ester and Hexagonal Boron Nitride for High Temperature Fuel Cells'; 2nd Award was won by Beatriz Lucio who presented a poster on 'Characterization of phase structures of novel metallo-polyurethanes; and the 3rd Award was won by Massimiliano Bartolomei poster presentation titled ' Water through Graphynes' Pores: First-Principles Penetration Barrier and Force Field Optimization'.

The conference aimed to promote activities in various areas of materials and structures by providing a forum for exchange of ideas, presentation of technical achievements and discussion of future directions. NANOSTRUC brings together an international community of experts to discuss the state-of-the-art, new research results, perspectives of future developments, and innovative applications relevant to structural materials, engineering structures, nanocomposites, modelling and simulations, and their related application areas.

The nanoparticles and nanocomposites have a wide range of applications in various fields such as medicine, textiles, cosmetics, agriculture, optics, food packaging, optoelectronic devices, semiconductor devices, aerospace, automotive, construction and catalysis. Advancements in the nanotechnology industry promise to offer improvements in capabilities across a spectrum of applications. This is of immense strategic importance to the high performance sector, which has historically leveraged technological advances in materials. It was great pleasure to welcome many delegates to NANOSTRUC 2014 at CSIC [4], Madrid, Spain. The next NANOSTRUC conference will be in 2016, venue to be confirmed.

Acknowledgements

We would like to acknowledge the hard work, professional skills and efficiency of the team, which ensured the general organisation, in particular the NANOSTRUC 2014 Scientific and Organising Committees, Session Chairs and all assistants. We would also like to acknowledge support from our industrial and Media partners - Wiley VCH; Nanotechnology Industries Association (NIA); UK's Knowledge Transfer Network- Nanotechlogy; Innovation (Greece); The Minerals, Metals & Materials Society (TMS), Nafigate; Benthan Books; European Commission FP7 & LIFE 11 programmes through ECOSHELL and SIRENA projects. Special thanks to our Host Consejo Superior de Investigaciones Científicas (CSIC) in Madrid and other academic partners including Robert Gordon University, Cranfield University and University of Johannesburg.

References

[1] Nanotechnology Market Outlook 2017, *RNCOS* Research Reports, 2013.
[2] Njuguna J 2012 *IOP Conf. Ser.: Mater. Sci. Eng.* **40** 011001
[3] Njuguna J and Sachse S, 2012 NANOSTRUC 2012 Book of Abstracts. 2012
[4] Njuguna, J, Verdejo R and Starost K, 2014, NANOSTRUC Book of Abstracts, 2014.

IOP Conf. Series: Materials Science and Engineering **64** (2014) 011002 doi:10.1088/1757-899X/64/1/011002

Peer review statement

All papers published in this volume of *IOP Conference Series: Materials Science and Engineering* have been peer reviewed through processes administered by the proceedings Editors. Reviews were conducted by expert referees to the professional and scientific standards expected of a proceedings journal published by IOP Publishing.

Mechanical and electrical properties of low density polyethylene filled with carbon nanotubes

Maziyar Sabet[1*], Hassan Soleimani[2]

[1*]Faculty of Engineering, Department of Chemical Engineering, Universiti Teknologi PETRONAS (UTP), Bandar Seri Iskandar, 31750, Tronoh, Perak, Malaysia

[2]Faculty of Science and Information Technology, Department of Fundamental and Applied Sciences, Universiti Teknologi PETRONAS (UTP), Bandar Seri Iskandar, 31750, Tronoh, Perak, Malaysia

*Corresponding author. Tel: +60-177496501. E-mail: Maziyar.sabet@petronas.com.my

Abstract. Carbon nanotubes (CNTs) reveal outstanding electrical and mechanical properties in addition to nanometer scale diameter and high aspect ratio, consequently, making it an ideal reinforcing agent for high strength polymer composites. Low density polyethylene (LDPE)/CNT composites were prepared via melt compounding. Mechanical and electrical properties of (LDPE)/CNT composites with different CNT contents were studied in this research.

Keywords: Low density polyethylene, Carbon nanotube, Composites, Properties

1.Introduction

Carbon nanotube (CNT) have gained attention as reinforcing fillers due to its outstanding properties [1], but it is also regarded as ultimate fillers for some advanced applications [2]. CNTs hold an extremely high elastic modulus about 1 TPa, comparable to that of diamond (1.2 TPa) and describe strengths of 10-100 times that of strongest steel [3]. In addition, they exhibit electrical conductivity as high as 10^5-10^7 S/m [4] and can convert an insulating polymer into a conducting composite at very low content due to their extraordinarily high aspect ratio [5]. The introduction of nanofillers like CNTs in polymer fibers can direct multifunctional high-performance materials which bring together high strength with electrical conductivity [6]. The keys to achieve maximum performances from composite are to obtain a homogeneous distribution of nanoparticles within polymer matrix, and to have best nanoparticle polymer adhesion, critical for load transfer from matrix to particle [7, 8]. The present research looked into both dependence of CNT content on mechanical and electrical properties in low density polyethylene (LDPE) contain CNT compounds.

Experimental

Materials and Sample Preparation

Low density polyethylene (LDPE) as granules with grade LH0075, density of 0.921 g. cm^{-3} and a MFI of 0.89 g.$(10\ min)^{-1}$, was obtained from the Bandar Imam Petrochemical Company (Iran). Sample preparation LDPE-CNT powder (CNT) composites were prepared by melt-blending commercial LDPE with CNT powder in a Brabender mixer. CNT powder was supplied by Nanostructrued and Amorphous Materials Inc. Mixing time was set to 15 min at 120°C to disperse CNT powder into LDPE matrix more uniformly and to avoid thermal degradation of LDPE, LDPE/CNT compounds were pre-heated for 5 min, and then compression molded according to ASTM D-1928 into sheets (with dimensions of 150×150×2 mm) at 150 ºC under a pressure of 10MPa for 3 min.

Tensile strength (TS) and Elongation at Break (EB)

TS and EB were determined according to ASTM D-638-03-2003 on an Instron model 4411 testing machine (UK). The test procedure was carried out at a cross head speed of 50 mm/min and room temperature conditions.

Electrical Properties

Samples for dielectric and conductivity measurements were coated with silver paint prior to measurements. Two metallic electrodes were then connected to samples using silver wires. The dielectric constant and resistivity of samples were measured by employing an impedance analyzer (Agilent model 4294) in frequency range of 40–10^7 Hz at room temperature. The dielectric constant was determined using an AndoTRS-10T and 6303 dielectric loss measurement system (ANDO Electric Co., Japan and FRIBORG Co. Sweden) in accordance with ASTM D-149 and 150 using an Ando insulation resistance set at 23°C and a frequency of 1 MHz.

Results and Discussion

Mechanical Properties

The stress/strain curves of CNT/LDPE composounds with varying tube contents are shown in Figure 1. It is displayed that CNT addition (0-10 wt%) to LDPE affected on increasing chain rigidity and enhanced stress values of compounds. Whereas, CNT addition (0-10 wt%) to LDPE decreased strain values and compound with 10 wt% CNT content broke at 17% strain values. It is understandable that, due to curving and coiling nature of CNTs, strong mechanical interlocking among nanotubes appeared and promoted reinforcement in polymer matrix which restricted chain mobility and made compounds less flexible against stress.

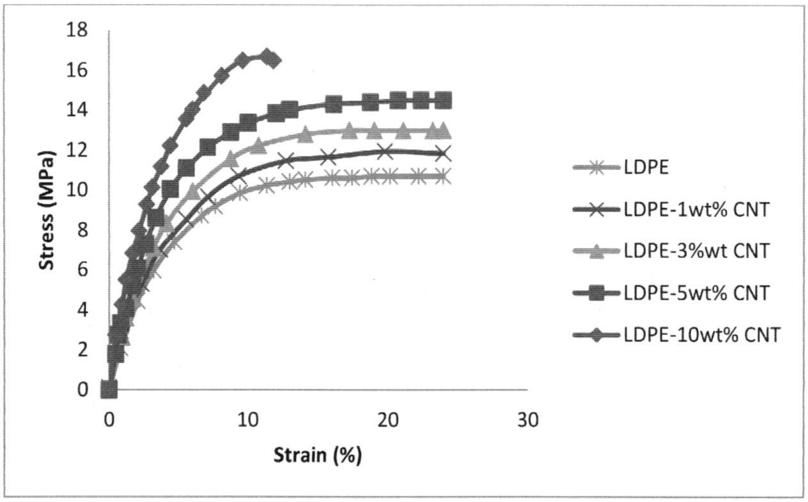

Figure1. LDPE and LDPE/CNTs Stress/strain of composites with varying CNT content

Figure 2 shows elastic moduli of LDPE, as function of CNT percentages. The augmentation of mechanical properties of composites demands a high degree of load transfer between matrix and nanotubes. If interfacial adhesion between phases is weak, nanotubes behave as holes or nanostructured flaws, introducing local stress concentrations, and benefits of CNT properties are lost [9-11]. The nanotubes must be well dispersed. In case of poor dispersion, they will fail by separation of bundle rather than by failure of nanotube itself, resulting in significantly decreased strength [12-15]. Strength in composites is a complicated issue involving load transfer, stress concentrations and defect distribution, especially in case of fibers. The data suggest that significant property enhancement, at low CNT percentages (0-10 wt%), could be attributed to highly dispersed and well aligned nanotubes and/or good adhesion between nanotubes and polymer matrix.

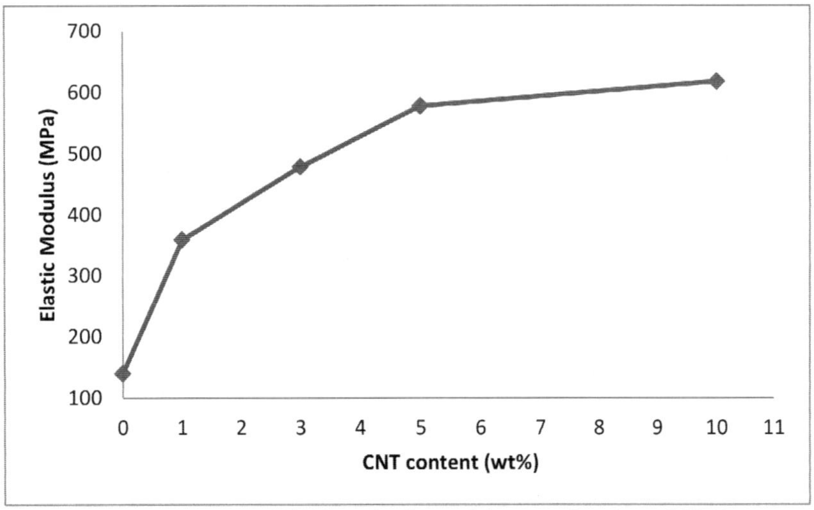

Figure 2. LDPE and LDPE/CNTs elastic modulus as a function of CNT content

Figure 3 presents storage modulus of LDPE and LDPE/CNT composites as a function of temperature. LDPE storage modulus and its composites reduced with temperature increase. Storage modulus of LDPE/CNT composites tended to augment with increasing CNT content. This implied effectiveness of CNTs reinforcing effect. The storage modulus maximum value was gained with 10 wt% CNT content. Figure 3 displays LDPE and its composites tangent delta versus temperature. A broad peak located at -20°C for LDPE, and -10°C for LDPE/CNT composites that can be attributed to LDPE glass transition temperature (*T*g). LDPE molecular chains are frozen below glass transition temperature and LDPE chains mobility increases above *T*g.

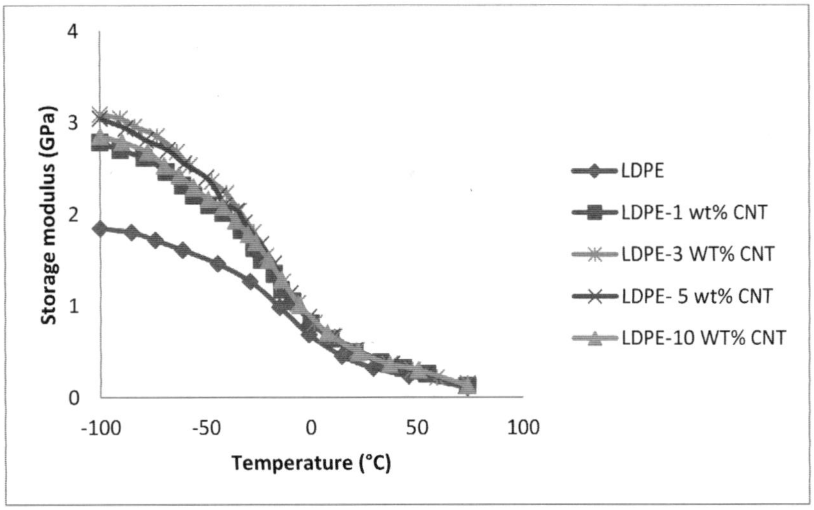

Figure 3. LDPE and LDPE/CNTs Storage modulus as a function of temperature

It was studied that there is a moderate enhancement of Young's modulus when CNT content changes from 0 to 10 wt%, as shown in Figure 4. The structural change of composites with increasing CNT content is responsible for this, because when CNT content was increased to 5 wt%, a continuous CNT network is formed throughout polymer matrix, as confirmed by rheological tests as discussed earlier [15-17]. It is understandable that, due to curving and coiling nature of CNTs, strong mechanical interlocking among nanotubes appears and promotes reinforcement.

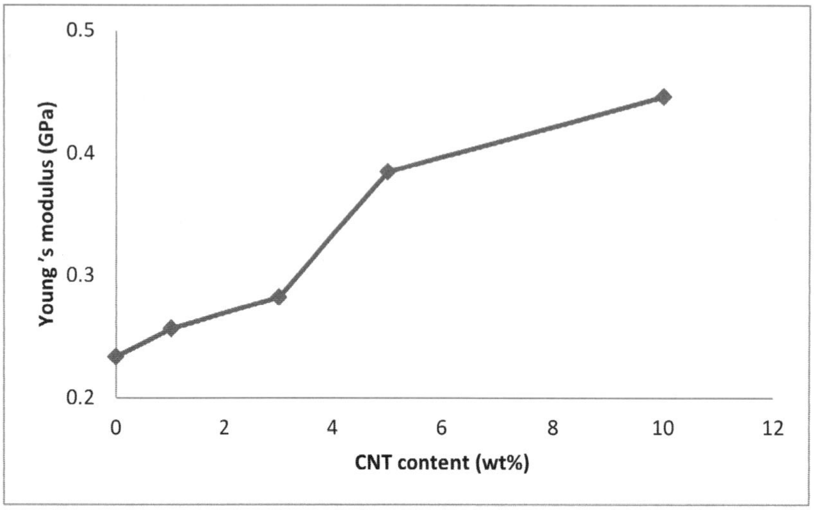

Figure 4. LDPE and LDPE/CNTs Young's moduli as a function of CNTs content

Electrical Properties

Figure 5 presents dielectric constant changes with various CNT contents for LDPE/CNT composites. The dielectric constant of LDPE/CNT composites augmented slightly with increasing CNT content up to 3 wt%. Above this content, dielectric constant of composites enhanced sharply. The dielectric constant of LDPE/5 wt%CNT composite was four orders of magnitude higher than that of LDPE/3 wt%CNT. No sharp increase in dielectric constant was observed even CNT content reached 10 wt%. In most polymer/CNT composites, bundle structure is formed resulting from agglomeration of filler particles at vicinity of percolation threshold. Such a structure is favorable at relatively lower filler volume fraction [18-20]. In this respect, it becomes apparent that fillers of LDPE/CNT composites containing lower CNT weight content are dispersed as isolated agglomerates in the matrix rather than a network of linked particles.

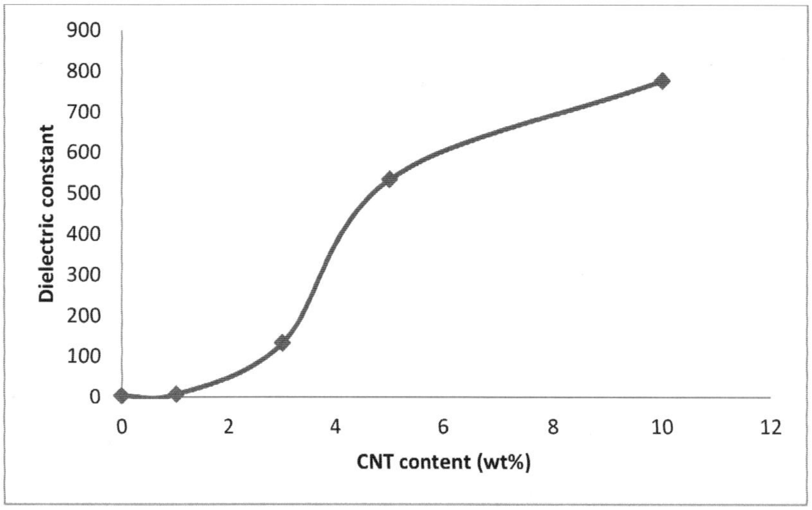

Figure 5. LDPE/CNT composites dielectric constant as a function of CNT content

Figure 6 shows dielectric constant dependence on LDPE/CNT composites frequency in range of 10^2–10^7 Hz. The pure LDPE dielectric constant and its composites with low CNT content (1 wt%) was independent of frequency. However, as CNT content reached 3 wt%, a noticeable dielectric constant dependence on frequency was observed. The frequency dependence of dielectric constant of materials having percolation threshold [21] which can be expressed as below:

$$\text{Dielectric constant} \propto \omega^{-\mu} \quad (1)$$

Where $\omega = 2\pi f$ is the radial frequency and μ are critical exponent. The μ for LDPE/ 10 wt% CNT composite was determined to be 0.26 [10].

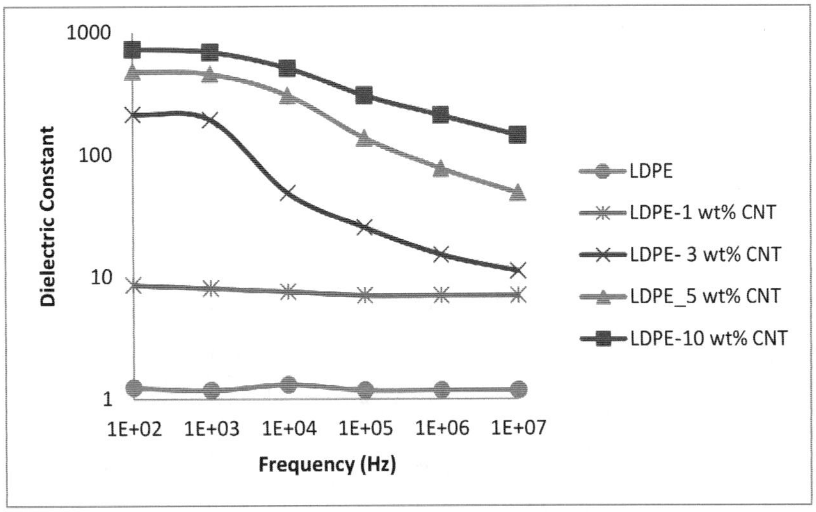

Figure 6. LDPE and LDPE/CNT composites as a function of frequency

Figure 7 shows conductivity dependence of LDPE/CNT composites and frequency at room temperature. Figure 7 reveals that conductivity increased with CNT content (0-10 wt%) and frequency range of $1\text{–}10^7$ Hz. The LDPE conductivity and its composites augmented almost linearly with frequency. However, conductivity of LDPE/10 wt% CNT composite was frequency independent below a critical value (i.e. < 20 Hz) which was characteristic of a non-dielectric material with a pronounced direct current [21].

The dependence of conductivity, conductivity, on frequency of dielectric materials can be described by:

$$\text{conductivity} \propto \omega^{v} \quad (2)$$

where ω is $\omega = 2\pi f$ and v value for LDPE/10 wt% CNT composite was determined to be ~0.66 at higher frequency regime. The low v value is associated with polarization effect between CNT clusters [22]. This condition was almost obeyed for LDPE/10 wt% CNT composite, indicating that this composite was confined in percolation regime, facilitating formation of CNT network within LDPE matrix. The conductivity did not have a significant increment with increasing $f < 20$ HZ, which suggested that electrical conducting networks were not formed in LDPE/CNT composites. Specifically, at low CNT concentration, CNT are individually dispersed within polymeric matrix. This leads to composites having a low electrical conductivity. Above $f_{CNT} > 20$, increase in conductivity is very remarkable. Such a marked increase in conductivity values proposed there was a critical concentration, known also as percolation threshold of fillers [10]. Definitely, when f_{CNT} was above the f_c, electrons could easily transfer between CNT through a tunnel effect and an Ohmic contact because distance between CNT was very short and electrical conducting networks were formed in LDPE/CNT composites.

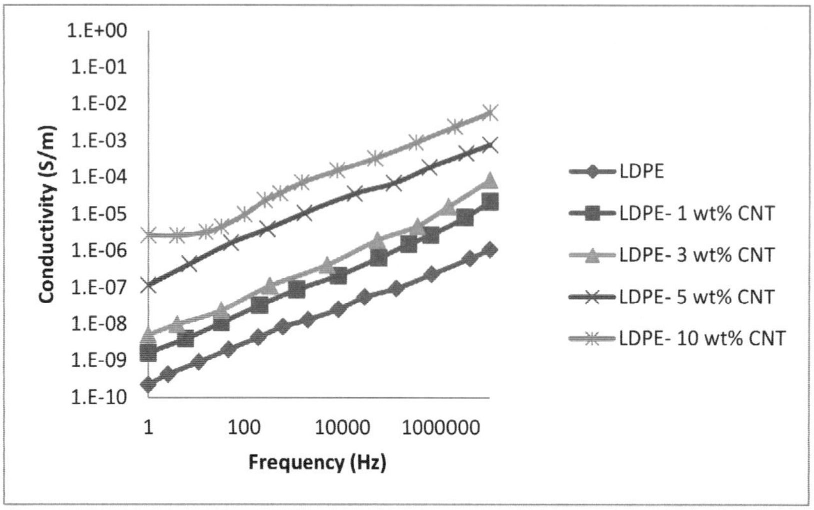

Figure 7. LDPE and LDPE/CNT composites conductivity as a function of frequency

Figure 8 expresses conductivity values of LDPE/CNTs, as a function of CNT content. Percolation theory deals with effect of varying in a random system and number of interconnections present [9, 10]. In this case filler bundles began to connect each other to form a filler network throughout entire composite, resulting in several orders of magnitude increased in electrical conductivity of composites. The transition from isolated bundle to connected network of conducting filler was referred to as percolation transition [10-17]. However, It was emphasized that electrical conductivity of LDPE, soon after percolation, resulted increase of about nine orders of magnitude at nanotube loadings (0-10 wt%).

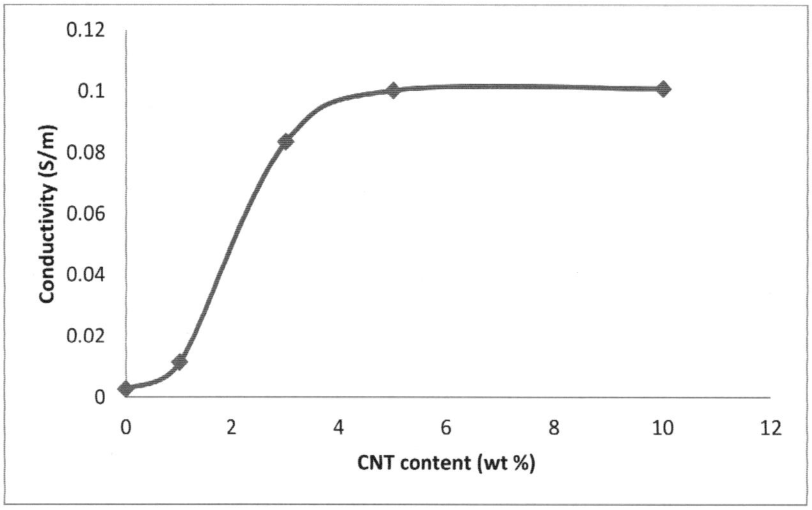

Figure 8. LDPE electrical conductivity as function of CNT content

Conclusion

It was revealed that 10 wt% CNTs addition to LDPE increased young's modulus and TS values of composites by 96% and 60%, respectively. Mechanical properties of pure LDPE augmented with CNTs addition. Electrical tests presented that DC value of LDPE content 10 wt% CNT composite was much bigger than pure LDPE and it is augmented slightly with further CNT addition up to 3 wt%. Above 3 wt% CNT content, DC of composites increased sharply. The frequency dependence of electrical properties of LDPE content 10 wt% CNT was well described with percolation theory. Conductivity and DC properties of LDPE/CNT composites first increased gently with addition of CNT content, and raised significantly as CNT weight fraction was above 3 wt%. Electrical properties also showed that electrical conductivity of LDPE matrix, resulted a significant increase, with quite low nanotube loadings (above 1 wt%).

References

1. Han, Z., Fina, A., (2011). Thermal conductivity of carbon nanotubes and their polymer nanocomposites: A review. Progress in Polymer Science; 36: 914–944.
2. Kuila, T., Bose, S., Hong, C.E., Uddin, M.E., Khanra, P., Kim, N.M., Lee, J.H., (2011). Preparation of functionalized graphene/linear low density polyethylene composites by a solution mixing method; 49: 1033–1037.
3. Guadagno, L., Vertuccio, L., Sorrentino, A., Raimondo, M., Naddeo, C., Vittoria, V., et al. (2009). Mechanical and barrier properties of epoxy resin filled with multi-walled carbon nanotubes. Journal of Carbon; 47: 10 2419-2430.
4. Giuliana, G., Roberta, D., Lieto, G., Patimo et. al. (2011). Structure-property relationships on uniaxially oriented carbon nanotube/polyethylene composites. Journal of Polymer; 52 1124-1132.
5. Du, J., Zhao, L., Zeng, Y., Zhang, L., Li, F., Liu, P., Liu, C., (2011). Comparison of electrical properties between multi-walled carbon nanotube and graphene nanosheet/high density polyethylene composites with a segregated network structure. Carbon; 49: 1094–1100.
6. Bauhofer, W., Kovacs, J.Z (2009). A review and analysis of electrical percolation in carbon nanotube polymer composites. Journal of Composites Science and Technology; 69: 10 1486-1498.
7. Xiao, K.Q., Zhang, L.C. Zarudi, I. (2007), Mechanical and rheological properties of carbon nanotube-reinforced polyethylene composites. Journal of Composites Science and Technology; 67 177–182.
8. Cha, S.I.; Kim, K.T.; Arshad, S.N.; Mo, C.B.; Hong, S.H. (2005). Extraordinary strengthening effect of carbon nanotubes in metal-matrix composites processed by molecular-level mixing. Journal of Advanced Materials; 17: 11, 1377–1381.
9. Cha, S.I.; Kim, K.T.; Lee, K.H.; Mo, C.B.; Hong, S.H. (2005) Strengthening and toughening of carbon nanotube reinforced alumina composite fabricated by molecular level mixing process. Journal of Scripta Materials; 53: 7, 793–797.
10. Liang, G.D., Tjong, S.C. (2006), Electrical properties of low-density polyethylene/multiwalled carbon nanotube composites. Journal of Materials Chemistry and Physics; 100 132–137.
11. Singh, B. P., Prabha, Parveen, S., et. al., (2011). Designing of multiwalled carbon nanotubes reinforced low density polyethylene composites for suppression of electromagnetic radiation. Journal of Nanoparticle Research; 13:7065–7074.
12. Dorigato, A., Pegoretti, A.P., (2010). Linear low-density polyethylene/silica micro- and nanocomposites: dynamic rheological measurements and modeling. Journal of eXPRESS Polymer Letters; 4: 115–129.
13. Liang, G.D., Tjong, S.C., (2006). Electrical properties of low-density polyethylene/multiwalled carbon nanotube nanocomposites. Journal of Materials Chemistry and Physics; 100: 132–137.
14. Chen-Feng, K., Hsu-Chiang, K., Chen-Chi, M.M., Chia-Hsun, C., Han-Lang, W., (2007). The preparation of carbon nanotube/linear low density polyethylene composites by a water-crosslinking reaction, Journal of Materials Letters; 61:2744–2748.
15. Singh, B., Saini, P., Gupta, T., Garg, P., Kumar, G., Panda, I., (2011). Designing of multiwalled carbon nanotubes reinforced low density polyethylene nanocomposites for suppression of electromagnetic radiation. Journal of Nanoparticle Research; 13: 7065-7074.
16. Qinghua, Z., Sanjay, R., Dajun, C., Dirk, L., Piet, J.L., (2006). Low percolation threshold in single-walled carbon nanotube/high density polyethylene composites prepared by melt processing technique. Journal of Carbon; 44: 778–785.
17. Tapas, K., Saswata, B., Chang, E. Hong, M.E.U, Partha, K., Nam, H.K., Joong, H.L., (2011). Preparation of functionalized graphene/linear low density polyethylene composites by a solution mixing method. Journal of Carbon; 49: 1033– 1051.
18. Petra, P., Tschke, M.A.G., Ingo, A., Sergej, D., Dirk, L., (2004). Rheological and dielectrical characterization of melt mixed polycarbonate-multiwalled carbon nanotube composites. Journal of Polymer; 45: 8863–8870.

19. Tony, M., Petra, P., tschke, P., Halley, M.M., Steven, E.J.B., Gerard, P., Brennan, D.B., Patrick, L., Darren, M., John, P.Q., (2005). Polyethylene multiwalled carbon nanotube composites. Journal of Polymer; 46: 8222–8232.
20. Marius, C.C., Matthew, J.H.E., Manias, G.C., Alberto, F., Gunter, B., Rakesh, K.G., Charles, A.W., (2007). The influence of carbon nanotubes, organically modified montmorillonites and layered double hydroxides on the thermal degradation and fire retardancy of polyethylene, ethylene vinyl acetate copolymer and polystyrene. Journal of Polymer; 48: 6532-6545.
21. Changchun, W., Zhi-Xin, G., Shoukuan, F., Wei, W., Daoben, Z., (2004). Polymers containing fullerene or carbon nanotube structures. Journal of Progress Polymer Science; 29 1079–1141.
22. Zdenko, S., Dimitrios, T., Konstantinos, P., Costas, G., (2010). Carbon nanotube–polymer composites: Chemistry, processing, mechanical and electrical properties. Journal of Progress in Polymer Science; 35: 357–401.

IOP Conf. Series: Materials Science and Engineering **64** (2014) 012002 doi:10.1088/1757-899X/64/1/012002

Zero Bias Anomaly in an Individual Suspended Electrospun Nanofiber

Asaf Avnon[1], Vitaliy Datsyuk[1], Svitlana Trotsenko[1], Bei Wang[2], Shuyao Zhou[2], Niels Grabbert[3] and Ha-Duong Ngo[3]

[1] Institut für Experimentalphysik, Freie Universität Berlin, Arnimallee 14, 14195 Berlin, Germany

[2] Research Center of Microperipheric Technologies, Technische Universität Berlin, TiB4/2-1, Gustav-Meyer-Allee 25, 13355 Berlin, Germany

[3] Microsystem Engineering (FB I), University of Applied Sciences, Wilhelminenhofstr. 74 (C 525), 12459 Berlin, Germany

E-mail: `Avnon@physik.fu-berlin.de`

Abstract. We report observing a double broad Kondo-like zero bias conductance peak at low temperatures in individual suspended electrospun nanofibers Poly(methyl methacrylate)-multiwalled carbon nanotubes. This anomalous behavior is suppressed at higher temperatures. We attribute this to the existence of correlated double impurity system inside the nanofiber. From the results we calculate a Kondo-like temperature for the nanofiber to be ~31.7-34K.

1. Introduction

Mesoscopic devices offer a venue for observing fundamental many-body phenomenas. One of the many phenomenas encountered is the zero bias anomaly at low temperature where the devices presents unusual tunneling behavior. While Coulomb blockade is a first order tunneling process, and does not allow tunneling when states are not aligned, higher order processes allow tunneling through virtual intermediate states. One such higher order process is the Kondo effect[1]. The Kondo effect is realized in quantum dot device which consists of a small electronic island connected by tunnel barriers to extended leads. For this kind of device with a localized unpaired spin embedded between metal leads, it is possible to observe signatures of Kondo effect in its conduction properties. Kondo physics incorporates the growth of correlated state comprising the local spin interacting with spins of conduction electrons of the leads. In devices as noted above, the result is enhanced transmission, manifested by a zero bias peak in the conductance. Kondo effect was observed in carbon nanotubes[2; 3], fullerenes[4], graphene[5], single molecule transistors containing metal ions[6] and semiconductor quantum dots[7] to name a few and has been on the focus of many papers[2; 4; 8–15]. So far this was not observed in nanofibers.

In this Letter, we present an early experimental data of a double peak zero bias anomaly resembling Kondo behavior in an individual nanofiber. This is done through transport spectroscopy over ultra long (~ 100μm) individual suspended electrospun nanofiber Poly(methyl methacrylate)-multiwalled carbon nanotubes. We discuss the adaptability of the Kondo model and calculate the Kondo-like temperature associated with this kind of behavior[16]. This system is of importance since it presents an unusual correlated two impurity electron system observed only few times albeit in much lower temperatures.

2. Experimental

For the fabrication of a substrate with trenches, we used 100mm⟨100⟩ wafers with one polished side. A cleaning process was performed using H_2SO_4+H_2O_2. This was followed by a photolithography process with a resist thickness of 1.5μm. The trenches were etched using deep silicon etching followed by resist stripping and another cycle of H_2SO_4+H_2O_2 cleaning. For the insulation of the substrate, a 600nm thick SiO_x layer was deposited on the wafer using wet oxidation at 1000°C. For the production of the nanofibers a 0.6g of MWCNTs (From Bayer-Baytubes C150[17]) was dispersed in 60g of dimethyl acetamide (DMAC, Carl Roth, GMbH) with 1g of hydroxypropyl cellulose (HPC) (from Aldrich Mw 100.000) by ultra-sonicator for 30 minutes. 3g of PMMA (from Aldrich Mw 996.000) were dissolved in 30g of acetone. DMAC and acetone were used without any additional purification. For dispersion tip sonicator Sonopuls HD3100, Bandelin GmbH, equipped with a cup horn operating at 10kHz and 100W output power was used. Components were mixed and dispersed by ultra-sonicator for additional 30 minutes. Nanofibers were electrospun over a substrate Si/SiO_x using Yflow®2.2.D-300 lab - scale electrospinning unit with single-nozzle electrospinning (Yflow Sistemas y Desarrollos S.L) with an injector-collector distance of 15 cm, an applied voltage of +9.5/-8.0kV, and a flow rate of 2ml/h. Candidate nanofibers suspended over trenches were located using an optical microscope and afterward were connected through direct bonding of Aluminum wires onto the nanofibers. We determined the diameter of the nanofibers with atomic force microscopy imaging was performed with Park XE150 in tapping mode. We performed electrical measurements with a Keithley 4200 − SCS and an Oxford closed cycle cryostat with 2-probe in Helium ambiance.

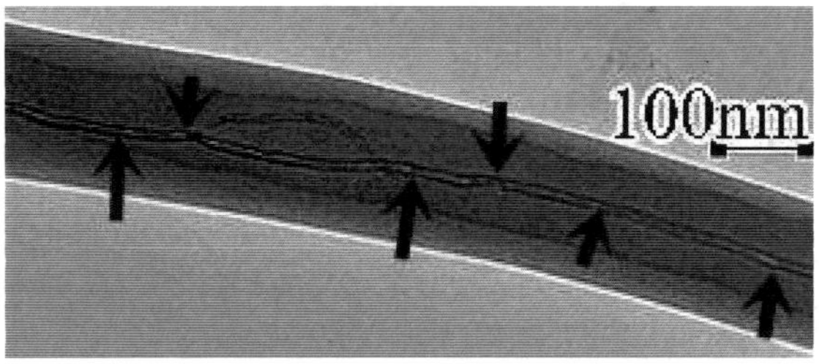

Figure 1. A typical transmission electron microscopy image of the nanofiber. The MWCNTs are aligned as a chain inside the nanofiber. Along this chain defects and impurities, marked with black arrows accumulate creating quantum dots.

3. Results

As Fig. 2(a) shows, the conductance drops gradually in dependence to the temperature. This type of behavior was recorded in various devices[18; 19] and also shows up in all of our devices. For highly resistive devices as the temperatures drop the conductance is suppressed in the zero bias regime and a Kondo-like peaks begin to rise sharply until saturation at lower temperatures[20]. It is noteworthy that we notice a resolved double peak as Fig. 2(a) shows. This kind of behavior has also been reported in the past[8; 12]. As Fig. 2(b) shows, it is roughly Lorentzian in shape and becomes narrower and higher in lower temperatures. The extracted maxima and full width at half maximum (FWHM) values from Fig. 2(b) through a fit for Lorentzian are presented at Fig. 3 in dependence with temperature. Fig. 3(a) and Fig. 3(b)

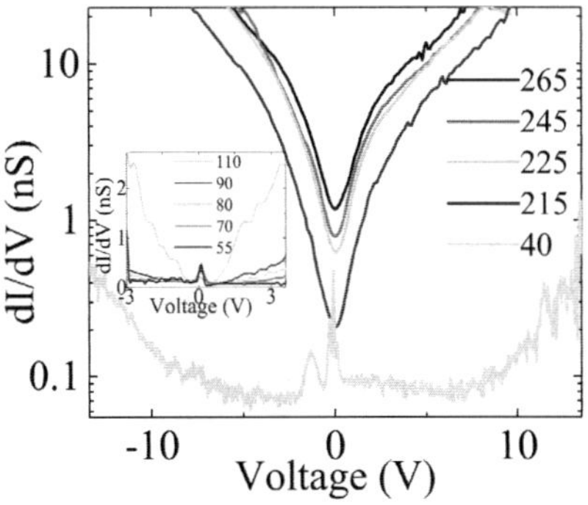

(a)

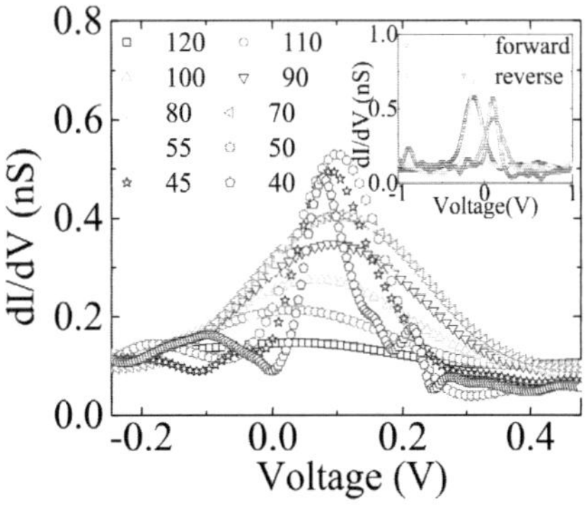

(b)

Figure 2. 2(a) Temperature dependence of the conductance in log scale. The conductance is suppressed close to zero bias while in low temperature a Kondo-like double peak appears. In the inset it is possible to see the interplay between the normal conduction and the emergent zero bias anomaly. 2(b) A close-up of the temperature dependence of the main Kondo-like feature. It can be noticed that normal conductance in higher voltages is suppressed as the temperature drops while a zero bias sharp peaks appear. Under 40K the Kondo-like peaks cannot be traced due to suppression of conduction. In the inset, in low temperatures, the Zero bias anomaly is shifted between forward and reverse sweeps but does not change its location in different temperatures. The secondary lower peak is absorbed by noise and cannot be noticed until reaching lower temperature. The differential conductance of all graphs was obtained through numerical differentiation and using Savitzky-Golay smoothing in an effort to reduce the noise at the cost of the bias voltage resolution.

are a fit of the Kondo model (continuous line) according to Eq. 1 and Eq. 2 respectively to the data (Black points).

4. Discussion

Crystallographic defects affect multiwalled carbon nonotubes (MWCNTs) electrical and thermal properties. Usually the result is lowered electrical[21] conductivity which is related with reduced mean free path and thermal conductivity[22] which is related with increased relaxation rate of the phonons through the defective region of the MWCNTs. Ultrasonification through the electrospinning process of the nanofiber creates defects over the MWCNTs[23; 24] as Fig. 1 shows. As our nanofiber is rich with impurities and defects- it is reasonable to believe that for highly resistive devices, at low temperatures, along the conduction nanochain we can get a quantum dot. In MWCNTs and carbon based materials in general, defects of single monoatomic vacancies induce magnetic properties[25]. This is reflected in our results through a Kondo-like zero bias peak we observe in Fig. 2(b). This Zero-bias peak is Lorentzian in shape, grows narrower and higher as we lower the temperature. Therefore, we assign this to the Kondo effect which is associated to logarithmic increase in the peak maximum conductance toward lower temperatures. We analyze the results in a similar manner. We adapt Lorentzians to the zero-bias peaks and plotted the extracted values for maximum conductance G_{max}[6; 15; 16; 26; 27] and peak width -full width at half maximum (FWHM)[6; 15; 16; 26; 27] as a function of the temperature according to these equations, respectively,

$$G_{max}(T) = G_C + G_0(1 + (2^{1/s} - 1)(T/T_K)^2)^{-s}, \tag{1}$$

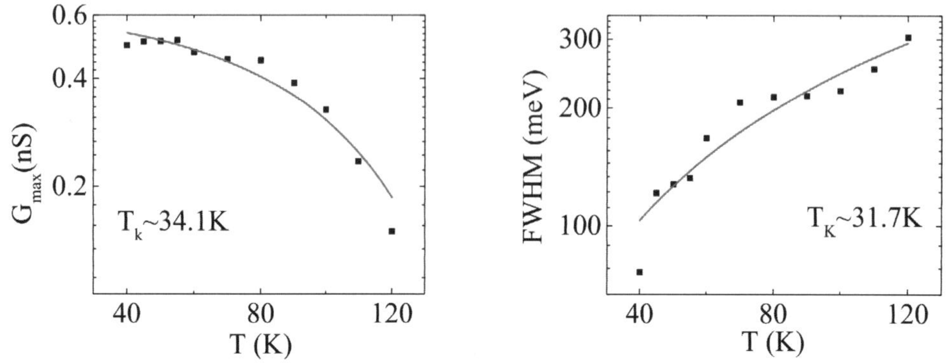

(a) (b)

Figure 3. 3(a) Temperature dependence maximum conductance at Zero bias with G_{max} collected from Fig. 2(b). The data (Black points) is fitted to the Kondo model (continuous line) according to Eq. 1 resulting with $T_K = 34.1$K. 3(b) Temperature dependence with FWHM collected from Fig. 2(b) through FWHM Lorentzian fits performed on each of the zero-bias peaks. The data (Black points) is fitted with the Kondo model (continuous line) according to Eq. 2 resulting with $T_K = 31.7$K

$$\Lambda_{FWHM} = g_0\sqrt{(\pi k_B T)^2 + 2(k_B T)^2}, \tag{2}$$

where T_K is Kondo temperature, G_0 is the conductance range, $s = 0.22$ is the spin parameter for carbon nanotubes (0.22 for spin 1/2, the simplest case[28]) G_C is an offset value and g_0 is a dimensionless parameter. Both fits yield a similar result for the critical temperature which indicates the validity of the Kondo model in this case.

In Kondo phenomena the width of the localized state, Γ is the sum of the level widths due to the couplings Γ_S,Γ_D of the localized state to the source and drain, respectively. An asymmetric source-drain coupling ($\Gamma_S \gg \Gamma_D$) results in high Kondo temperature (set by Γ) and a low total conductance (limited by Γ_D). This is identical to the case we observe in this system -low conductance with high Kondo-like temperature. From the inset of Fig. 2(a) it is possible to notice that there is an apparent interplay between the emergent zero bias anomaly and the conductance staircase[29]. As the first order process of conductance staircase trails off, the zero bias peaks rise. The conductance staircase is not symmetrical which again is a signature of asymmetrical coupling to the leads. This is also exemplified from the inset to 2(b) where we notice a shift between the forward and the reverse sweeps.

We find that the saturation value of the conductance is lower than the expected conductance quantum. This has been reported in the past[6]. We attribute this to poor transmission in the leads and over the nanochain of conduction($\Gamma_S \gg \Gamma_D$ or vice versa), insufficient local thermalization or insufficient Kondo screening. This seems plausible since the zero bias peaks and the conductance in high voltage are of the same magnitude. At temperatures 40K and lower, it is impossible to spot the zero bias peak due to noise. It is conceivable that in lower temperatures the nanochain of conduction is hampered by insufficient local thermalization. This again may be the result of lowered total conductance. As can be noticed, the peaks are broadened. FWHM is in the size of ~330meV at 110K. While normally the energy gaps in MWCNTs are relatively small ($<$ 100meV), a substantial amount of defects can raise it. We also notice a split in the zero bias peak or a broad side band. This was observed recently for single walled carbon nanotubes[30], quantum point contacts[31] as well as in other devices. It was explained as two-impurity Kondo effect with coupled spins. We believe that this is also a demonstration of similar behavior as Fig. 2 shows. Two impurity sites with unequal coupling strength Γ to the adjacent leads results with asymmetric double-peak zero bias anomaly. with the minimum conductance between the two peaks roughly at zero bias voltage. This result was observed earlier albeit in microkelvin temperatures[31]. The coupling strength remains to be tested in future experiments.

5. Summary

To summarize, we have shown a double Kondo-like effect in an individual nanofiber performing as quantum dot. This situation is associated with a two-impurity Kondo effect. We calculated a Kondo-like temperature of $\sim 31.7 - 34$K for the nanofiber for the main feature. Connecting a nanofiber consisting multiwalled carbon nanotubes aligned in series widens the scope of possibilities to investigate different regimes in the mesoscopic electron transport. Further experimental work will be needed to unambiguously support this picture.

Acknowledgments

The authors would like to acknowledge T. Schurig and C. Assman from Physikalisch-Technische Bundesanstalt (PTB) for allowing access to PTB facilities. Parts of this work were supported by the European Research Council under grant number 210642.

References

[1] Kondo J 1964 *Prog, Theor. Phys.* **32** 37–49 URL http://dx.doi.org/10.1143/PTP.32.37
[2] Nygard J, Cobden D H and Lindelof P E 2000 *Nature* **408** 342–346 URL http://dx.doi.org/10.1038/35042545
[3] Liang W, Bockrath M and Park H 2002 *Phys. Rev. Lett.* **88**(12) 126801 URL http://link.aps.org/doi/10.1103/PhysRevLett.88.126801
[4] Yu L H and Natelson D 2004 *Nano Letters* **4** 79–83 (*Preprint* http://pubs.acs.org/doi/pdf/10.1021/nl034893f) URL http://pubs.acs.org/doi/abs/10.1021/nl034893f
[5] Popinciuc M, Calado V E, Liu X L, Akhmerov A R, Klapwijk T M and Vandersypen L M K 2012 *Phys. Rev. B* **85**(20) 205404 URL http://link.aps.org/doi/10.1103/PhysRevB.85.205404
[6] Wagner S, Kisslinger F, Ballmann S, Schramm F, Chandrasekar R, Bodenstein T, Fuhr O, Secker D, Fink K, Ruben M and Weber H B 2013 *Nat Nano* **8** 575–579 ISSN 1748-3387 URL http://dx.doi.org/10.1038/nnano.2013.133
[7] Grobis M, Rau I G, Potok R M and Goldhaber-Gordon D 2007 URL http://dx.doi.org/10.1002/9780470022184.hmm523
[8] Cronenwett S M, Oosterkamp T H and Kouwenhoven L P 1998 *Science* **281** 540–544 (*Preprint* http://www.sciencemag.org/content/281/5376/540.full.pdf) URL http://www.sciencemag.org/content/281/5376/540.abstract
[9] Parks J J, Champagne A R, Hutchison G R, Flores-Torres S, Abruña H D and Ralph D C 2007 *Phys. Rev. Lett.* **99**(2) 026601 URL http://link.aps.org/doi/10.1103/PhysRevLett.99.026601
[10] Ngo Dinh S, Bagrets D A and Mirlin A D 2010 *Phys. Rev. B* **81**(8) 081306 URL http://link.aps.org/doi/10.1103/PhysRevB.81.081306
[11] Vidan A, Stopa M, Westervelt R M, Hanson M and Gossard A C 2006 *Phys. Rev. Lett.* **96**(15) 156802 URL http://link.aps.org/doi/10.1103/PhysRevLett.96.156802

[12] De Franceschi S, Hanson R, van der Wiel W G, Elzerman J M, Wijpkema J J, Fujisawa T, Tarucha S and Kouwenhoven L P 2002 *Phys. Rev. Lett.* **89**(15) 156801 URL http://link.aps.org/doi/10.1103/PhysRevLett.89.156801

[13] Goldhaber-Gordon D 1998 *Phys. Rev. Lett.* **81** 5225–5228 URL http://dx.doi.org/10.1103/PhysRevLett.81.5225

[14] Choi D J, Rastei M V, Simon P and Limot L 2012 *Phys. Rev. Lett.* **108**(26) 266803 URL http://link.aps.org/doi/10.1103/PhysRevLett.108.266803

[15] Hanson R, Kouwenhoven L P, Petta J R, Tarucha S and Vandersypen L M K 2007 *Rev. Mod. Phys.* **79**(4) 1217–1265 URL http://link.aps.org/doi/10.1103/RevModPhys.79.1217

[16] Hewson A 1997 *The Kondo Problem to Heavy Fermions* Cambridge Studies in Magnetism (Cambridge University Press) ISBN 9780521599474 URL http://books.google.com/books?id=fPzgHneNFDAC

[17] wall Carbon Nanotubes B M, Processing N and Product D 2010 Baytubes C 150 P

[18] Aleshin A N 2007 *Physics of the Solid State* **49** 2015–2033 ISSN 1063-7834 URL http://www.springerlink.com/index/10.1134/S1063783407110017

[19] Bockrath M, Cobden D H, Lu J, Rinzler A G, Smalley R E, Balents L and McEuen P L 1999 *Nature* **397** 598–601 ISSN 0028-0836 URL http://dx.doi.org/10.1038/17569

[20] Avnon A, Datsyuk V, Trotsenko S, Wang B, Zhuo S, Grabbert N and Ngo H D 2013 Quasi one dimensional transport in individual composite electrospun nanofibers manuscript under submission

[21] Beuneu F, l'Huillier C, Salvetat J P, Bonard J M and Forró L 1999 *Phys. Rev. B* **59**(8) 5945–5949 URL http://link.aps.org/doi/10.1103/PhysRevB.59.5945

[22] Mingo N, Stewart D A, Broido D A and Srivastava D 2008 *Phys. Rev. B* **77**(3) 033418 URL http://link.aps.org/doi/10.1103/PhysRevB.77.033418

[23] Koshio A, Yudasaka M, Zhang M and Iijima S 2001 *Nano Letters* **1** 361–363 URL http://pubs.acs.org/doi/abs/10.1021/nl0155431

[24] Lu K, Lago R, Chen Y, Green M, Harris P and Tsang S 1996 *Carbon* **34** 814 – 816 ISSN 0008-6223 URL http://www.sciencedirect.com/science/article/pii/000862239689470X

[25] Makarova T and Palacio F 2006 *Carbon Based Magnetism: An Overview of the Magnetism of Metal Free Carbon-based Compounds and Materials* (Elsevier Science) ISBN 9780080460376 URL http://books.google.com/books?id=Qj9R7Tr45tUC

[26] Costi T A, Hewson A C and Zlatic V 1994 *Journal of Physics: Condensed Matter* **6** 2519 URL http://stacks.iop.org/0953-8984/6/i=13/a=013

[27] Sasaki S 2000 *Nature* **405** 764–767 URL http://dx.doi.org/10.1038/35015509

[28] Vecino E, Buitelaar M, Martín-Rodero A, Schönenberger C and Yeyati A L 2004 *Solid State Communications* **131** 625 – 630 ISSN 0038-1098 new advances on collective phenomena in one-dimensional systems URL http://www.sciencedirect.com/science/article/pii/S0038109804004181

[29] Balberg I, Azulay D, Goldstein Y, Jedrzejewski J, Ravid G and Savir E 2013 *The European Physical Journal B* **86** 1–17 ISSN 1434-6028 URL http://dx.doi.org/10.1140/epjb/e2013-40200-7

[30] Eichler A, Weiss M and Schönenberger C 2011 *Nanotechnology* **22** 265204 URL http://stacks.iop.org/0957-4484/22/i=26/a=265204

[31] Iqbal M J, Levy R, Koop E J, Dekker J B, de Jong J P, van der Velde J H M, Reuter D, Wieck A D, Aguado R, Meir Y and van der Wal C H 2013 *Nature* **501** 79–83 ISSN 0028-0836 URL http://dx.doi.org/10.1038/nature12491

IOP Conf. Series: Materials Science and Engineering **64** (2014) 012003 doi:10.1088/1757-899X/64/1/012003

BROADBAND DIELECTRIC SPECTROSCOPY OF NANOCOMPOSITES BASED ON PVDF AND EXPANDED GRAPHITE

A Linares[1], JC Canalda[1], A Sanz[1], A Szymczyk[2], Z Rosłaniec[2] and TA Ezquerra[1]
[1]Instituto de Estructura de la Materia (IEM-CSIC), C/ Serrano 121, 28006 Madrid, Spain
[2]West Pomeranian University of Technology, Institute of Material Science and Engineering, Piastow Av. 19, PL-70310 Szczecin, Poland

E-mail: alinares@iem.cfmac.csic.es

Abstract. Nanocomposites based on poly (vinylidene fluoride) (PVDF) and expanded graphite (EG) were prepared by non-solvent precipitation from solution with different EG concentrations. Films were obtained by compression molding and their structural and dielectric properties studied. From Wide Angle X-ray Scattering (WAXS) experiments, it can be assessed that for all EG concentrations the α-crystalline phase of PVDF is the predominant crystalline form. However, for composites with high nanoadditive content, higher than 3 wt.%, the β-crystalline phase is also detected. Dielectric spectroscopy results showed that the nanocomposites present both high dielectric constant and electrical conductivity at low percolation threshold.

1. Introduction

Nanoadditives based on carbon are receiving increasing attention because of the unique properties of their nanocomposites in comparison with those of traditional composites [1]. Recently graphene and expanded graphite (EG) [2] have attracted considerable interest due to its extraordinary characteristics. Poly(vinilydene fluoride) (PVDF) is a technologically important polymer because of its high dielectric constant, mechanical strength, and resistance to solvents, acids, and bases [3]. Moreover, due to its piezo- and pyroelectric activity this polymer has found use in electromechanical and thermoelectrical transducers. PVDF having a repeat unit of [- CH_2 – CF_2 -] is known to exhibit at least four crystalline phases, known as α, β, γ and δ [4]. All these phases differ structurally and the conditions under which a specific conformation can be obtained depend strongly on the processing and thermal or mechanical treatments that the polymer undergoes. It is known that improved electrical performances such as high dielectric constant, piezoelectricity and pyroelectricity are obtained if PVDF presents β-phase and, to a lesser extent, γ-phase. In this work, we prepare nanocomposites based on PVDF and expanded graphite with different compositions. Providing a comprehensive study of this type of composites can be of great importance because of their broad applications in electrochemical systems such as sensors, electromagnetic interference absorbers and high-charge storage capacitors among others.

2. Experimental

Commercial PVDF powder (Aldrich, Spain), with M_w ~ 534.000 g/mol, was used as polymer matrix. The nanoadditive was expanded graphite (SGL Carbon SE, Germany), which, as shown in Fig. 1, forms big stacks of densely packed EG particles, producing worm-like aggregates with an average thickness between 450 and 560 nm and graphene platelets size ranged from 16 μm to 46 μm (99%) [2].

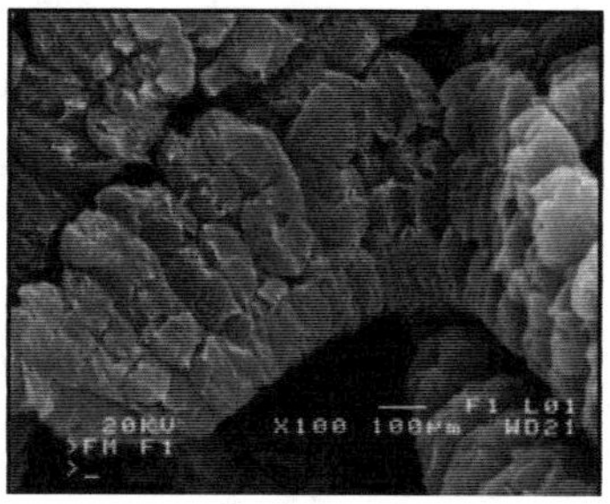

Fig.1. Scanning electron micrograph (SEM) of expanded graphite

N,N-dimethylacetamide (Sigma-Aldrich) was used as solvent and milli-Q quality deionized water as non-solvent. In short, PVDF was dissolved in DMA and different amounts (0.5, 1, 1.5 1.75, 2, 3 and 4 wt.%) of expanded graphite were also dispersed in DMA. Then, the dispersion of nanoadditive/DMA was added to the PVDF/DMA solution and subsequently the mixture was precipitated into ice distilled water. Then, films of about 1 mm thick were obtained by compression molding. Wide Angle X-ray Scattering (WAXS) measurements were performed by means of a Seifert XRD 3000 h/h diffractometer using Ni-filtered Cu Kα radiation (λ = 0.154 nm) at a scanning speed of 0.02 °/s. For dielectric experiments, a sandwich geometry was used and circular gold electrodes were deposited onto the surfaces of the film sample. Measurements were performed, at room temperature, over a frequency window of 10^{-2}–10^{7} Hz, using a Novocontrol spectrometer integrating an ALPHA dielectric interface.

3. Results and discussion

3.1. Wide Angle X-ray Scattering (WAXS) results

Fig. 2 shows representative diffraction patterns corresponding to the homopolymer, expanded graphite and some of their nanocomposites. As shown, the diffractogram of PVDF exhibits the characteristic Bragg maxima associated to the (100), (020), (110) and (021) reflections of α-crystalline phase [4]. The pattern corresponding to EG shows the diffraction peak associated to the (002) reflection of the graphitic layer structure.

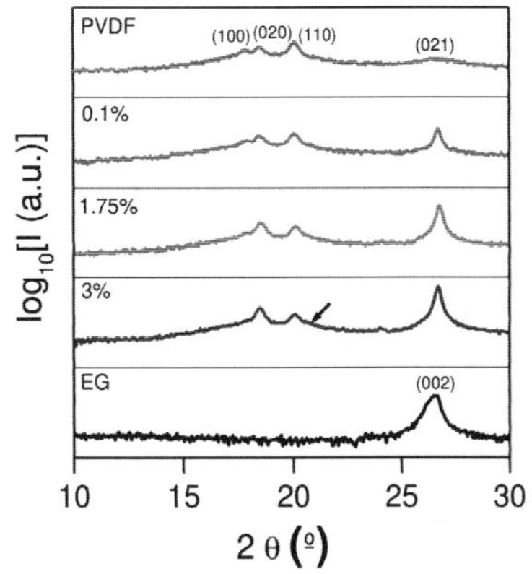

Fig.2. Diffraction patterns of expanded graphite (EG) and of the PVDF/EG nanocomposites for different EG content as labeled

Nanocomposites exhibit mostly the characteristic diffraction maxima corresponding to the PVDF α-crystal phase, being the (021) reflection overlapped by the (002) of the nanoadditive. However, as indicated in Fig. 2 by an arrow, at high nanoadditive content, a shoulder at $2\theta = 20.7°$, which is associated to the (200)/(110) reflections of the β-crystal phase, is also observed. Moreover, it is noteworthy that the (100) reflection decreases and an interchange between the relative intensities associated to (020) and (110) reflections can be observed. This phenomenon has been associated to the nucleation effect of additives in polymer crystallization [5].

3.2. Broadband electrical conductivity

Fig. 3 shows the electrical conductivity as a function of frequency. At low nanoadditive content, σ(F) follows a linear dependence with frequency with a slope close to 1, similar to that followed by the insulating PVDF matrix. However, for higher concentrations, σ(F) adopts a characteristic behavior which can be formally depicted by the so called universal dynamic response [6] described by a law of the type:

$$\sigma(F) = \sigma_{dc} + \sigma_{ac} = \sigma_{dc} + A\,F^S \tag{1}$$

where σ_{dc} is the frequency independent direct current conductivity and $0 < S < 1$.

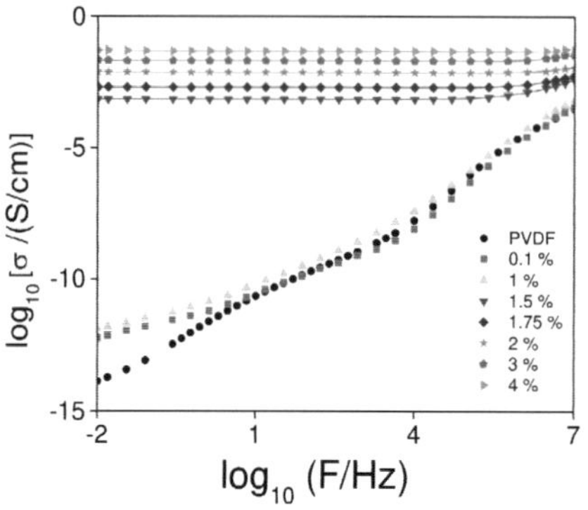

Fig. 3. Broadband electrical conductivity of PVDF/EG nanocomposites for different EG content as labeled

This law introduces a critical frequency, F_c, above which $\sigma(F) = \sigma_{ac} \propto F^S$. The continuous lines in Fig. 3 correspond to fits of eq. 1 to the experimental data. From these fits σ_{dc} values can be extracted. The value of the conductivity at the lowest measured frequency (10^{-2} Hz) for the insulating samples has been considered as σ_{dc} for comparative purpose.

3.2.1 Direct current electrical conductivity

Fig. 4 represents σ_{dc} data as a function of filler concentration. As shown, a characteristic percolating behavior is observed. For low nanoadditive content, the conductivity corresponds to an insulating material. With increasing filler content, at a critical concentration, ϕ_c, the conductivity starts a sudden increase and, according to percolation theory, a continuous network of physically connected particles appears and the insulator–conductor transition occurs. The *dc* conductivity above the critical concentration can be analyzed in terms of the percolation theory [7] by means of:

$$\sigma_{dc} \propto (\emptyset - \emptyset_c)^t \tag{2}$$

where t is a critical exponent. Although the critical concentration, ϕ_c, depends on the lattice in which particles are accommodated, t depends primarily on the dimensionality of the percolating system and not on the details of the geometric structures or the interactions [7]. Theoretical calculations [8] propose values of t between 1.6 and 2 for three-dimensional systems. The dashed line in Fig. 4 shows the fittings of eq. 2 to experimental data with t value around 1.4 and $\phi_c \sim 1.31$wt.%. Below ϕ_c, electrical conductivities higher than that of the insulating matrix are expected provided that a non-physically connected particles conduction mechanism is present.

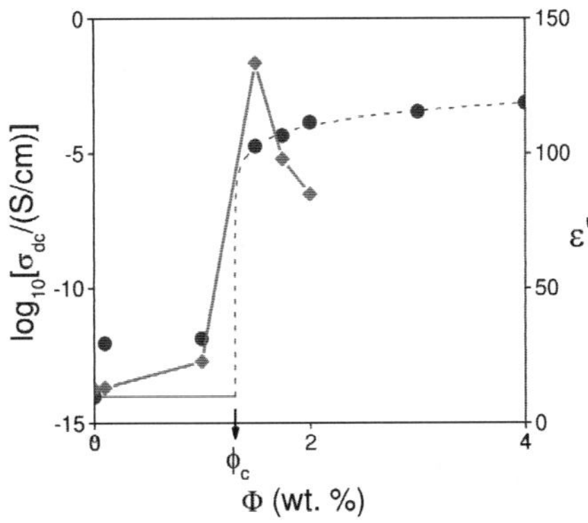

Fig.4. DC electrical conductivity (•) and dielectric constant (♦), at 1000 Hz, for PVDF/EG nanocomposites as a function of the weight concentration of EG

3.2.2 Alternating electrical conductivity

As far as the *ac* conductivity is concerned, Fig. 3 shows that for frequencies above F_c, the frequency dependent conductivity component is observed and the conductivity increases approximately as described by the power law: σ_{ac} (F) $\propto$ F^S. For these nanocomposites, the *S*-exponent, which is calculated as the slope of the high frequency region in Fig. 3, starts from $S = 1$ (corresponding to insulating specimens) and decreases with increasing nanoadditive concentration until $S \approx 0.8$. Based on the works of Bergman and Imry [9] and Straley [10], the so-called inter-cluster polarization model has been proposed to describe the frequency dependence of conductivity of percolating systems near percolation threshold, ϕ_c. According to this model a value for *S*-exponent $\approx$ 0.72 can be expected, which is very close to our results. This fact suggests that in this type of nanoadditives, with low aspect ratio, there are a significant amount of polymer interfaces that can contribute to *ac* conductivity through inter-cluster polarization. Likewise, this type of conduction mechanism could also explain the *dc* conductivity behavior observed at nanoadditive concentration below ϕ_c.

3.3 Dielectric constant

Relating to dielectric constant, Fig. 4 shows ε' values, taken at 1000 Hz, as a function of the nanoadditive concentration. As can be seen, ε' increases with nanoadditive loading. Moreover, near the percolation threshold ε' undergoes a remarkable increase. This effect is compatible with the increment of interfaces as nanoadditive concentration increases. In Fig. 4 a clear maximum is observed, which indicates the concentration at which the amount of interfaces starts to decrease most likely due to the presence of physically connected clusters. At room temperature and 1000 Hz, PVDF presents a value of ε' ≈ 12, while for the nanocomposite with 1.5 wt.% of EG the ε' value is higher than 130.

4. Conclusions

Nanocomposites films based on PVDF and expanded graphite (EG), with different EG concentrations, have been prepared by non-solvent precipitation from solution, and subsequent compression molding. Then, structural and electrical properties have been studied. WAXS studies showed that α-crystalline phase of PVDF is the predominant crystal form in all composites. However, at high content of nanoadditive, the β-crystal phase is also present. Moreover, with increasing EG the intensity of the peak assigned to the (100) reflection decreases, and an interchange between the relative intensities associated to (020) and (110) reflections can be observed. From dielectric spectroscopy analysis, it was concluded that these nanocomposites present high dielectric constant and electrical conductivity at low percolation threshold (ϕ_c ~1.31 wt. %)

5. References

[1] Linares A *et al* 2008 *Macromolecules* **41** 7090

[2] Paszkiewicz S *et al* 2012 *J. Polym. Sci. Part B-Polym. Phys.* **50** 1645

[3] Kerbow D L and Sperati C A 1999 *Physical Constants of Fluoropolymers, Polymer Handbook 4th ed.* (New York, Wiley)

[4] Lovinger A J 1982 *Developments in Crystalline Polymers* (London, Applied Science)

[5] García-Gutiérrez M C *et al* 2008 *Macromolecules* **41** 844

[6] Kremer F and Schönhals A 2002 *A Broad band Dielectric Spectroscopy* (Berlin, Springer)

[7] Stauffer D 1985 *Introduction to Percolation Theory* (London, Taylor&Francis)

[8] Connor M T *et al* 1998 *J. Phys. Rev. B* **57** 2286

[9] Bergman D J and Imry Y 1977 *Phys. Rev. Lett.* **39** 1222

[10] Straley J P 1977 *Phys. Rev. B* **15** 5733

Acknowledgements

The authors thank the financial support from the Spanish Ministry of Science and Innovation (MICINN), Grant MAT2012-33517 and from the Polish National Science Centre and the Slovak Academy of Sciences in the frame of ERA-NET project APGRAPHEL.

2nd International Conference on Structural Nano Composites (NANOSTRUC 2014) IOP Publishing
IOP Conf. Series: Materials Science and Engineering **64** (2014) 012004 doi:10.1088/1757-899X/64/1/012004

Spinning, structure and properties of PP/CNTs and PP/carbon black composite fibers

A Marcincin, M Hricova and A Ujhelyiova

Slovak University of Technology in Bratislava, Faculty of Chemical and Food Technology, Department of Fibres and Textile Chemistry, Radlinskeho 9, 812 37 Bratislava, Slovak Republic

anton.marcincin@stuba.sk

Abstract. In this paper, the effect of the compatibilisers-dispersants and other nanofillers on melt spinning of the polypropylene (PP) composites, containing carbon nanotubes (CNTs), and carbon black pigment (CBP) has been investigated. Further, the structure and selected properties of composite fibers, such as mechanical and electrical have been studied. The results revealed, that percolation threshold for PP/CBP composite fibres was situated within the concentration of 15 - 20 wt%, what is several times higher than for PP/CNTs fibers.

1. Introduction

The polypropylene (PP) fibers containing inorganic or organic fillers represent a large part of the synthetic fibre production. The knowledge and experience from the developments in mass pigmented, delustered and filled fibers have been widely involved in practice. The new formulations of PP concentrate dispersions (concentrates, masterbatches) and modified spinning methods have been developed to guarantee an acceptable spinning conditions and new or improved properties of fibres such as high colour strength and colour uniformity [1], enhanced mechanical properties [2], thermal and UV stability [3] and electrical conductivity [4].

The developments in inorganic and organic nanoparticles with one size below 100 nm have opened up a new ways for investigation and use of PP (nano) composite fibers. [5, 6]. Specially carbon nanotubes and other kinds of carbon particles are very attractive for new materials based particularly on PP/CNTs composites and composite fibers. A large number of original papers and numerous overview articles, focussed on developments in the polymer/carbon nanotubes (CNTs) (nano)composites [6-8] and (nano)composite fibers [9-11] have been published in the last two decades. In structural composites CNTs acts as effective reinforcing nanofibrous filler and enhances mechanical properties of polymers such as tensile strength and elastic modulus [12-14]. Polymer/CNTs functional composites improve the electrical and thermal conductivity [15, 16] and other properties such as electromagnetic and microvawe absorption.

In this paper, the effect of the the MWCNTs functionalisation, spinning equipments, conditions and compatibilisers-dispersants on rheological properties, and melt spinning of the polypropylene (PP), containing carbon nanotubes (MWCNTs), and CBP has been investigated. Further, the structure and selected properties of composite fibers, such as mechanical and electrical properties are presented and discussed.

2. Experimental

2.1. Materials used

Polymers: PP HF 500R (PP500R), MFR 25 g/10 min, flakes; PP Moplen HP 561N (PP561N), MFR 11 g/10 min, produced by Basell Polyolefins, Italia.

PP Tatren TH 1810 (PP1810), MFR 20.8 g/10 min, produced by Slovnaft Plc., Slovakia.

Nanofillers: Multi-Wall Carbon Nanotubes - Nanocyl® 7000 (MWCNT), average diameter 10 nm, length 0.1-10 μm, surface area 250-300 m^2/g, carbon purity 90%, metal oxide impurity 10%, produced by Nanocyl S.A., Belgium

Carbon Black Pigments (CBP): Vulcan XC 605 (V605) and Printex L6 (PL6), produced by Degussa GmbH, Germany.

Compatibilisers: Slovacid 44P (S44P), ester of stearic acid and polypropylene glycol, Sasol GmbH, Germany; Tegopren 6875 (TEG), alkyl ester polydimethylsiloxane, Degussa GmbH, Germany; Novanik 3010 (NOV), copolymer of ethylene oxide and propylene oxide, Sasol GmbH, Germany; PP-g-MA – polypropylene grafted by maleic anhydride, Clariant GmbH, Germany; Polypropylene glycol 2000 (PEG), Aldrich. Oxazoline derivative, (ALOXA), synthesised in laboratory.

2.2. Preparation of PP/MWCNT composite fibres

Two-step technology was used for preparation of PP/MWCNT nanocomposite fibres:

1. Preparation of the dispersion of MWCNT in PP500R: Powdered PP500R, MWCNT (and compatibilisers) were mixed in high rpm mixer for defined time. Powder mixture was melted and kneaded using twin screw corotating extruder ϕ 28 mm. Temperature of extruder zones were 80, 150, 220, 225, 225, 225, 232°C, Temperature of extrudated melt was 229°C. Extrudate was cooled and cut.
2. The PP/MWCNT chips were dried at 80°C and then they were melted and spun using extruders ϕ 16 mm and ϕ 32 mm. The temperatures of extruder zones were 80, 240, 250, 250°C and spinning temperature was 250°C. Metering of the melt 16g/min; spinning speed 150m/min.

Fibres were drawn using laboratory drawing machine at 120°C.

PP/CBP composite fibres were using the same two-step process. Masterbatch, containing of 30 wt% of V605 or PL6 prepared in the first step, was mixed with PP1810 on final concentration of the pigment and simultaneously spun at 270°C in the second step. Fibres were drawn at 120°C.

2.3.Methods used

The capillary rheoviscosimeter Göttfert (with extruder ϕ=20 mm) for measurement of the rheological properties of concentrates was used. The Newton equation $\tau = \eta.\gamma^{\cdot}$ and Ostwald de Waele power law $\tau = K.\gamma^n$, where K – coefficient, n – deviation from the Newtonian flow (power law exponent), for evaluation of the basic rheological characteristics of concentrates were used.

DSC 7 (Perkin Elmer) was used for evaluation of thermal properties of the PP/CNT fibres, for estimation of their supermolecular structure and interactions of components at the interphase. The heating and cooling rate was 10°C/min, temperature range of 50-280°C.

Mechanical properties: An Instron (Type 3343) was used for measurements of the tensile strength and elongation at break according to ISO 2062:1993 as well as Young's modulus.

Electrical properties:a) Direct current electrical conductivity of the PP composites (strings) and fibres was measured by the standard four-contact method (FCM). b) The resistivity of the linear textiles method (RLT) according to accredited test (CSN 800059) is suitable for measurement of the electrostatic and electroconductive properties of the fibres. Resistivity was expressed in [Ω.m].

3.Results and discussion

3.1.PP/MWCNT fibres with improved mechanical-physical properties

The favourable conditions at preparation of PP/MWCNT composites (masterbatches), optimal spinning and drawing conditions with suitable compatibilisers and selected nanofillers in addition, contribute to high draw ratio of LOY PP composite fibres, to their higher tenacity and Young's modulus and to very low breaking elongation (Table 1).
The results reveal enhanced tenacity and Young's modulus of selected composite fibres about 25-30 % compared with unmodified fibres. The similar effect of MWCNT wit other additive was achieved for dynamical-physical properties e.g. dynamical modulus of fibres.
DSC analysis (results are not in the short paper) of composite fibres indicates, that effect of solid particles and dispersants within the investigated concentration on thermal properties (on phase structure) of fibres is negligible. MWCNT fibrous particles act as medium nucleating agents.

Table 1 The tenacity σ, elongation at break ε, Young's modulus E and their coefficients of variation for PP561N/ MWCNT (0.1%) composite fibres

Fibre composition	Draw ratio λ	T_t [tex]	σ [cN/tex]	$CV_σ$ [%]	ε [%]	$CV_ε$ [%]	E [N/tex]	CV_E [%]
PP561N+PP500R	7.0	10.9	64.8	2.9	22.6	7.6	7.2	2.8
PP+0.1% CNT	7.0	10.8	65.1	7.6	22.6	8.1	7.3	5.3
PP+0.1% CNT+0.1% D40	7.2	9.9	76.1	6.8	20.7	6.9	8.7	7.2
PP+0.1% CNT+0.1% C15A	7.5	9.9	72.8	5.3	21.6	5.5	7.9	5.3
PP+0.1% CNT+0.1% C15A+S44P	7.5	10.1	71.7	8.3	20.3	5.0	8.2	8.8
PP+0.1% CNT+0.1% C15A+TEG	7.5	9.7	81.2	2.1	21.7	6.2	9.5	4.5
PP+0.1% CNT+0.1% C15A+S44P+TEG	7.5	8.9	77.0	2.3	21.7	4.4	9.1	5.7
PP+0.1% CNT+0.1% PP-g-MA+TEG	7.2	10.3	76.6	6.2	22.0	9.3	8.9	6.0
PP+0.1% CNT+0.05% ALOXA+S44P	7.5	9.3	74.0	5.8	20.0	4.4	8.8	6.4

The following technology and equipment is necessary to keep, to ensure acceptable spinning of PP/MWCNT composite fibres with enhanced mechanical properties:

- Two step process: a) preparation of polymer composite (concentrate) and b) melt spinning of polymer composite with final concentration of MWCNT and additive in chips form is recommended.
- The powder form of PP for mixing with MWCNT and other additives before melt processing of concentrate (composite) is required
- The application of equipment with high kneading and homogenization effect (lower temperature) is crucial for preparation of PP/MWCNT composites (masterbatch), suitable for melt sinning. The twin screw extruder with diameter $\phi = 28$ mm and higher is needed.
- The low spinning speed process producing of LOY composite fibres is appropriated.
- The higher draw ratio in cold drawing at optimal conditions, the higher both tenacity and Young's modulus can be achieved.

The knowledge and experiences obtained were utilized at spinning of PP/MWCNTs composite fibers with high content of fibrous nanofiller. The results are in the next chapter.

3.2. Electrically conductive PP/MWCNT composite fibres

3.2.1.Rheological properties and spinning of PP/MWCNT composites

The melt viscosity of the PP/MWCNT composites increases gradually with content of the MWCNT in PP matrix. Already 1.0 wt% of nanofiller in PP causes the first break on the dependences of both the power law exponent and viscosity on content of MWCNT. The second break is observed between 6 wt% and 8 wt% of MWCNT in the PP. This corresponds with qualitative change in number of contacts of fibrous particles, their mutual interactions and higher electrical conductivity (Table 2).

Table 2 Dependence of viscosity η, power law exponent n and electrical conductivity σ (four contact method) on concentration of MWCNT in PP500R/MWCNT composites.

Concentrate composition	n	η [Pa.s]		σ [S.cm^{-1}]
		γ=100 s^{-1}	γ=300 s^{-1}	
PP561R+PP500R	0.41	329	173	-
1% CNT	0.40	302	156	$<1.9.10^{-9}$
4% CNT	0.39	378	194	$3.2.10^{-5}$
6% CNT	0.48	416	235	$5.6.10^{-2}$
8% CNT	0.29	777	357	$5.1.10^{-1}$

3.2.2.Mechanical properties of PP/MWCNT composite fibres

Tenacity and Young's modulus of PP/MWCNT composite fibres decrease gradually with content of the fibrous particles in PP fibres. The low elongation at break (about 15%) and high non-uniformity of the mechanical properties (CV) are characteristic for the fibres (Table 3). The compatibilisers-dispersants (PEG and NOV) have significantly improved spinability and deformation of the fibres in drawing. In addition, they have improved the non-uniformity expressed by lover CVσ and CVε in comparison with the PP/MWCNT fibres without compatibiliser-dispersant.

The decrease of the tensile strength and elastic modulus of the PP/MWCNT fibres containing higher content of the fibrous nanofiller (MWCNT) is in contrary with the mechanical properties of the PP composite fibres with low content of nanofiller (Table 1).

Table 3 The tenacity σ, elongation at break ε, Young's modulus E and their coefficients of variation of PP500R/MWCNT composite fibres, drawn at low gradient of deformation

Content of MWCNT [%]	Draw ratio λ	T_t [tex]	σ [cN/tex]	$CV_σ$ [%]	ε [%]	$CV_ε$ [%]	E [N/tex]	CV_E [%]
1.0	7.6	63.2	47.5	10.7	15.7	21.6	4.4	8.0
4.0	6.8	112.6	38.4	10.1	14.8	26.3	4.4	8.4
6.0	9.3	93.0	35.0	15.8	13.1	44.8	4.3	6.7
8.0	7.0	195.8	33.3	15.0	14.3	8.5	4.1	2.2

The microscopic observation of the PP/MWCNT fibres reveals a significant difference in the morphology of the fibres without compatibiliser-dispersant and that with dispersant. The MWCNT is uniformly distributed in the fine structure of the PP matrix without dispersant. On the contrary, the floccules of the MWCNT are formed in the abundance of the polyglycol based dispersant, immiscible with PP. The floccules are easy deformed at drawing and form of the oriented dispersed phase in the PP matrix.

3.2.3.Electrical properties of the PP composite fibres

Electrical properties of the PP/MWCNT composite fibres were measured using two methods: standard four contact method (FCM) and resistivity of the linear textile method (RLT). Results are presented in the Table 4. The percolation threshold is dependent on shape of sample measured and shifted from the PP composites to PP composite drawn fibres within the nanofiller concentration of 4.0–8.0 wt%. The electrically conductive composites was obtained already at 4.0 wt% of conductive particles, but 8.0 wt% was unsufficient for conductive drawn fibres (Table 4). Electrical conductivity of PP/MWCNT fibres is compared with conductivity of PP/CBP V605 (Table 5). The results reveal that percolation threshold for PP/CBP composite fibres is situated within the concentration of 15 - 20 wt%, what is two times higher than for PP/MWCNT fibres (Table 4). The decrease of the conductivity of drawn fibres was confirmed also for PP/CBP fibres.

Table 4 Electrical resistivity ρ of the PP500R/MWCNT composite fibres evaluated by method of resistivity of linear textiles (RLT)

Content of MWCNT [%]	**ρ [Ω.m]**		
	composites	**undrawn fibres**	**drawn fibres**
0	$1.1.10^{13}$	$3.4.10^{13}$	$9.2.10^{13}$
1.0	$1.3.10^{13}$	$1.1.10^{13}$	$8.6.10^{13}$
4.0	$2.5.10^{5}$	$1.3.10^{13}$	$3.1.10^{13}$
6.0	$5.4.10^{3}$	$4.3.10^{6}$	$6.7.10^{9}$
8.0	$1.4.10^{3}$	$3.4.10^{4}$	$2.2.10^{5}$

Table 5 Electrical resistivity ρ of the PP1810/V605 composite fibres evaluated by method of resistivity of linear textiles (RLT)

Content of V605 pigment [%]	**ρ [Ω.m]**	
	undrawn fibres	**drawn fibres**
0	$\sim 10^{13}$	$\sim 10^{13}$
5	$4.0.10^{13}$	$5.1.10^{12}$
10	$1.9.10^{13}$	$1.1.10^{13}$
15	$2.6.10^{13}$	$5.7.10^{12}$
20	$2.6.10^{4}$	$6.3.10^{4}$
25	$6.8.10^{2}$	$1.2.10^{3}$
30	$8.0.10^{1}$	$6.8.10^{2}$

Improvement of the rheological properties and spinability of the PP/MWCNT/CBP L6 composite fibres by polar additives (PEG and NOV) led to decrease their electrical conductivity compared to PP/MWCNT fibres. This phenomena can be explained by separation and flocculation of the fibrous nanofiller in the PP matrix , what shifts the percolation threshold to higher concentration of electrically conductive additives.

4. Conclusions

On the basis of experimental results the following conclusions can be drawn:

- Two step process, preparation of polymer composite and melt spinning of polymer composite with final concentration of MWCNT and additives in chips form is recommended for acceptable spinning and properties of fibres.
- The application of equipment with high kneading and homogenization effect is crucial for preparation of PP/MWCNT composites suitable for melt sinning. The twin screw extruder with minimum diameter $\phi = 28$ mm is needed.
- The low spinning speed process producing of LOY composite fibres is appropriate, providing the high deformability of fibres in cold drawing.
- The MWCNT particles with high aspect ratio contribute in PP matrix to creating the electrically conductive net at essentially lower concentration than carbon black pigment. The percolation threshold corresponds with filler concentration at which the viscosity began to increase rapidly.
- The electrical conductivity of the composite fibres decreases with increase of their deformation and orientation.

Acknowledgement: This work was supported by the Slovak Research and Development Agency under the contract No: VMSP-P-0007-09.

References

[1] Marcincin A 2002 Modification of fire-forming polymers by additives *Progr. in Pol. Sci.* **27** pp 853-913
[2] Marcincin A *et al* 2013 Structure, thermal and mechanical properties of PP/organoclay composite fibres *Fibr. & Text. In EE* **21** pp 18-23
[3] Marcinčin A. *et al* 2009 Effect of inorganic (nano)fillers on the UV barrier properties, photo and thermal degradation of polypropylene fibres *Fibr. & Text. In EE* **17** pp 29-35
[4] Vogel R., Jenschke W and Beyrenther R. 2000 Generation of an antistatic polyolefin fiber by adding a carbon black filled component *Chem. Fib. Int.* **50** pp 584-586
[5] Lee GW *et al* 2008 Carbon nanotube dispersion and exfoliation in polypropylene and structure and properties of the resulting composites *Polymer* **49** pp 1831-1840
[6] 7Bikiaris D 2010 Microstructure and properties of polypropylene/carbon nanotube nanocomposites *Mat.* **3** pp 2884-2946
[7] Spitalsky Z *et al* 2010 Carbon nanotube-polymer composites: Chemistry, processing, mechanical and electrical properties *Progr. in Pol. Sci.* **35** pp 357-401
[8] Rahmat M and Hubert P 2011 Carbon nanotube–polymer interactions in nanocomposites: A review *Comp. Sci. and Techn.* **72** pp 72-84
[9] Du J *et al* 2007 The present status and key problems of carbon nanotube based polymer composites *eXPRESS Pol. Lett.* **1** pp 253-273
[10] Min B G *et al* 2009 *Polymer/carbon nanotube composite fibers - An overview* (Functional Composites of Carbon Nanotubes and Applications) ed K-P Lee, A L Gopalan and F DS Marquis (Kerala: Transworld Research Network) pp 43-73
[11] Chou T-W *et al* 2010 An assessment of the science and technology of carbon nanotube-based fibers and composites *Comp. Sci. and Techn.* **70** pp 1-19
[12] Xushan G *et al* 2005 Application of multi-wall carbon nanotubes in polymer fibers *Chem. Fib. Int.* **55** pp 170-172
[13] Katiyar P *et al* 2008 Comparison of mechanical properties of various types of carbon nanotubes in polypropylene fiber *RJTA* **12** pp80-86
[14] Szentes A *et al* 2010 Mechanical properties of polypropylene/multi walled carbon nanotube composites *Hung. J. Ind. Chem.* **38** pp 67-70
[15] Wen M *et al* 2012 The electrical conductivity of carbon nanotube/carbon black/polypropylene composites prepared through multistage stretching extrusion *Polymer* **53** pp 1602-1610
[16] Chien AN *et al* 2013 Functional polymer-polymer/carbon nanotube bi-component fibers *Polymer* **54** pp 6210-6217

IOP Conf. Series: Materials Science and Engineering **64** (2014) 012005 doi:10.1088/1757-899X/64/1/012005

Mechanical Properties of Individual Composite Poly(methyl-methacrylate) -Multiwalled Carbon Nanotubes Nanofibers

Niels Grabbert[1,4,*], Bei Wang[3,*], Asaf Avnon[2,*], Shuyao Zhuo[3], Vitaliy Datsyuk[2], Svitlana Trotsenko[2], Piotr Mackowiak[3], Katrin Kaletta[4], Klaus-Dieter Lang[3,4] and Ha-Duong Ngo[1,4]

[1] Microsystem Engineering (Department I), University of Applied Sciences, Wilhelminenhofstr. 74, 12459 Berlin, Germany
[2] Institut für Experimentalphysik, Freie Universität Berlin, Arnimallee 14, 14195 Berlin, Germany
[3] Research Center of Microperipheric Technologies, Technische Universität Berlin, TiB4/2-1, Gustav-Meyer-Allee 25, 13355 Berlin, Germany
[4] Fraunhofer Institute for Reliability and Microintegration, Gustav-Meyer-Allee 25, 13355 Berlin, Germany
[*] These Authors contributed equally

E-mail: `Niels.Grabbert@izm.fraunhofer.de`, `Wangbei325@hotmail.de`, `Avnon@physik.fu-berlin.de`

Abstract. Multiwalled carbon nanotubes with their superb mechanical properties are an unique filler material for polymer composites. Here, we present an investigation of mechanical properties of electrospun Poly-(methyl-methacrylate) multiwalled carbon nanotubes composite nanofibers. The method of electrospinning was used to fabricate suspended individual Poly-(methyl-methacrylate) multiwalled carbon nanotubes nanofibers. In order to reinforce the nanofibers, different high concentration of multiwalled carbon nanotubes were used. Transmission electron microscopy measurements reveal a successful filling of the nanofibers. The different types of nanofibers were deposited at SiO_2 substrates. Which were previously etched, to create trenches for bend tests. Followed by fixing the nanofiber with a focus ion beam platinum deposition at the trench edges. An atomic force microscopy was used to perform the mechanical nanofiber bending tests over trenches. The results were compared with pristine Poly-(methyl-methacrylate) nanofibers to nanofibers with 15 weight% and 20 weight% multiwalled carbon nanotubes composite fibers. We observed that pristine nanofibers have Young's modulus of 136 MPa, while for composite nanofibers with 15 weight% have 2.65 GPa and with 20 weight% have 6.06 GPa (at room temperature and air ambiance). This corresponds to an increase of Young's modulus of 19 fold between the pristine nanofibers and the 15 weight% of mutliwalled carbon nanotubes filled nanofibers. Therefore the increase of the Young's modulus compared between the pristine and the 20 weight% MWCNT filled nanofibers corresponds to 45 fold.

1. Introduction

Nanocomposites combine a few nanomaterial components in order to introduce new functionalities in a matrix or fiber to harness some of their characteristics. Such components as carbon nanotubes composite materials have attracted great attention from academia and industry alike, due to their novel mechanical, electrical and thermal characteristics [1–6]. The application range of nanofibers is vast and can be extended with the use of novel nanofillers such as carbon nanotubes, TiO_2 [3]. Among the various application fields for nanofibers is the field of sensors for chemical detection, optical sensors [7–9] or sensors for biomedical uses. Notable works in the biomedical fields are for virus removal filters [10] or tissue engineering [11]. These examples provide a strong evidence for the future wide field of applications. Other applications like air filter systems sold by United Air Specialists Inc. [12] have matured into the industrial level. We present an investigation of mechanical properties of individual composite Poly(methyl-methacrylate) (PMMA) -Multiwalled Carbon Nanotubes (MWCNT) nanofibers. We used a concentration of 15 weight% and 20 weight% MWCNT embedded in PMMA nanofiber. Atomic force bending test with pristine PMMA nanofibers, with 15 weight% and 20 weight% MWCNT reinforced composite nanofibers were performed. We present different Young's moduli measurement methods of pristine polymer nanofibers and compare the results with the composite MWCNT-PMMA nanofibers. Finally we discuss the strong mechanical reinforcement with MWCNTs nanofillers. We observe a 45 fold increase of the Young's modulus with 20 weight% MWCNT-PMMA composite nanofiber compared to the pristine nanofibers.

2. Experiment

A single-side polished 4 inch standard silicon wafers were used as basic material. A cleaning step with H_2SO_4+H_2O_2 was realized and followed by H_2O dehydration baking. A primer (Hexamethyldisilazan) was used to improve the adhesion between the wafer surface and the later posistive photo resist (AZ1514H). The resist was spun on with a height of 1.5 µm and followed by prebake, lithography and postbake. The resist was developed afterwards with MIF 326. Deep Silicon Etching (BOSCH-process) was used to create high aspect ratio trenches. After etching the resist stripping and H_2SO_4+H_2O_2 cleaning was performed again. A 670 nm thick SiO_2 layer was deposited on the wafer and was realized with a wet oxidation process at 1000 ° C. Different concentrations of MWCNTs (From Bayer-Baytubes C150 [13]) were dispersed in dimethyl acetamide (DMAC, Carl Roth, GMbH) and with hydroxypropyl cellulose (HPC) (from Aldrich). The solution was sonicated for 30 minutes. 3 g of PMMA (from Aldrich Mw) were dissolved in acetone. A Sonopuls HD3100 sonicator (from Bandelin GmbH) were used. The dispersions were performed at 10 kHz and 100 W output power. Both solutions were mixed together and dispersed by sonication for additional 30 minutes. Electrospinning was processed to create the different types of nanofibers over Si/SiO_2 substrates and carbon layerd copper Transmission Electron Microscope (TEM) grids. That was realized by using Yflow R2.2.D-300 lab - scale electrospinning unit with single-nozzle electrospinning (Yflow Sistemas y Desarrollos S.L). With injector-collector distances for the various fiber types of 15-16 cm, an applied voltages of +6 / -8,5 KV, +9.5/-8.0kV, and a flow rate of 2ml/h. Before operating the bending tests, a TEM imaging with already manufactured grids were operated, to analyze the CNT distribution and to check the successfully filling of PMMA-matrices with MWCNTs. Fig. 1 illustrates TEM images of a created 15 weight% MWCNT-PMMA composite nanofiber.

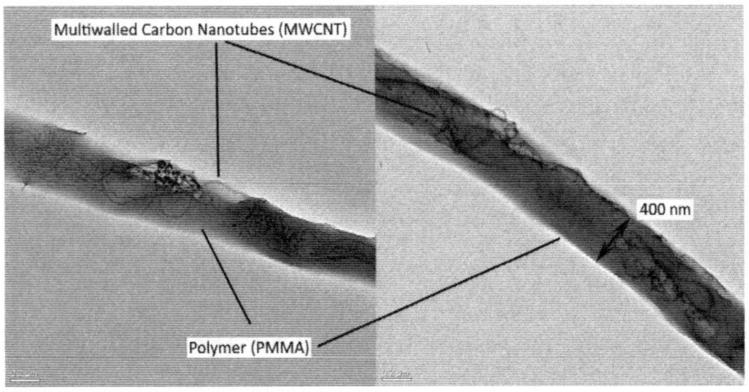

Figure 1. Transmission electron microscopy images of 15 weight% MWCNT-PMMA compsite nanofiber

To clamp the nanofibers permanently at the trench edges, a Focus Ion Beam (FIB) platinum deposition was operated. A Zeiss FIB Leo 1540XB with a Gemini column was used. Pt pads of 3 by 3 µm were deposited with current of 100 pA. The pad height of ≈ 315 nm were realized and was determined afterwards using atomic force imaging with an XE150 AFM (from Park Systems). Figure 2 show the fix composite nanofiber over an etched trench.

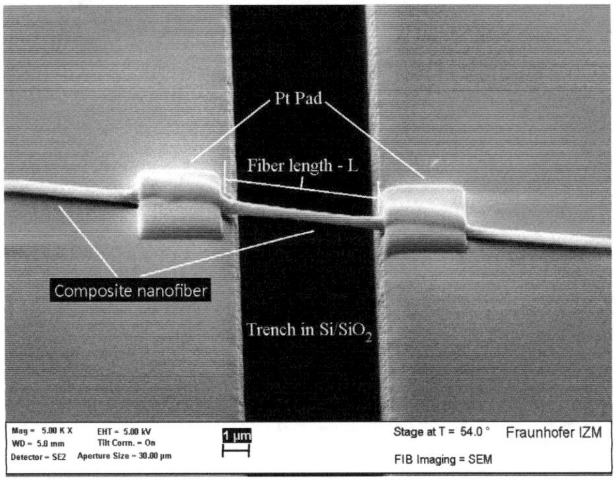

Figure 2. An electron microscope image of a 15 weight% MWCNT-PMMA composite nanofiber fixed with Pt pads on Si/SiO_2 trench structure

The AFM was equipped with a suitable contact-mode PNP-TR tip from Nanoworld [14]. All AFM bending test were performed at room temperature and air ambiance. Each nanofiber was investigated at three different positions along its length. In total two samples of pristine PMMA, 3x 15 weight%-MWCNT-PMMA and three samples of 20 weight%-MWCNT-PMMA were measured.

Pristine PMMA Nanofiber	1	2
average Young's modulus [GPa]	**1,21E-01**	**1,87E-01**
average Young's modulus +Δ[GPa]	8,92E-02	5,88E-02
average Young's modulus -Δ[GPa]	5,74E-02	5,69E-02

Table 1. Average Young's modulus of each pristine PMMA nanofiber at different positions

15 weight%-MWCNT-PMMA Nanofiber	1	2	3
average Young's modulus [GPa]	**3,54E+00**	**1,90E+00**	**2,08E+00**
average Young's modulus +Δ[GPa]	2,30E+00	4,20E-01	8,11E-01
average Young's modulus -Δ[GPa]	1,22E+00	4,28E-01	9,46E-01

20 weight%-MWCNT-PMMA Nanofiber	1	2	3
average Young's modulus [GPa]	**8,96E+00**	**4,48E+00**	**6,20E+00**
average Young's modulus +Δ[GPa]	2,46E+00	5,90E+00	3,27E+00
average Young's modulus -Δ[GPa]	3,49E+00	2,59E+00	3,27E+00

Table 2. Average Young's moduli of each 15 and 20 weight%-MWCNT-PMMA Nanofiber at different positions

3. Results

Fig. 3 illustrates the determined Young's moduli of separated nanofiber types. The results were calculated with Eq. 2, which was derived from the sample construction using differential beam theory [15–18]. Where E is the effective Young's modulus, a - force position, F - applied force, L - fiber length, I - moment of inertia, R- fiber radius and ω as the fiber displacement.

$$I = \frac{\pi R^4}{4} \tag{1}$$

$$E = \frac{Fa^3[L-a]^3}{3I\omega L^3} \tag{2}$$

It was observed, that pristine nanofibers have average Young's modulus of $136^{+109}_{-72.4}$ MPa, while filled nanofibers, with 15 weight% have $2.65^{+3.2}_{-1.51}$ GPa and with 20 weight% have $6.06^{+5.36}_{-4.18}$ GPa. As fig 3 shows, a substantial of the Young's modulus was observed, compared with pristine PMMA nanofiber to the 15 weight% and 20 weight% MWCNTs composite nanofibers. The average diameters with 436^{+9}_{-9} nm for pristine PMMA nanofibers, 330^{+24}_{-59} nm for 15 weight%-MWCNT-PMMA nanofibers and 284^{+7}_{-9} nm for the 20 weight%-MWCNT-PMMA nanofibers were determined.

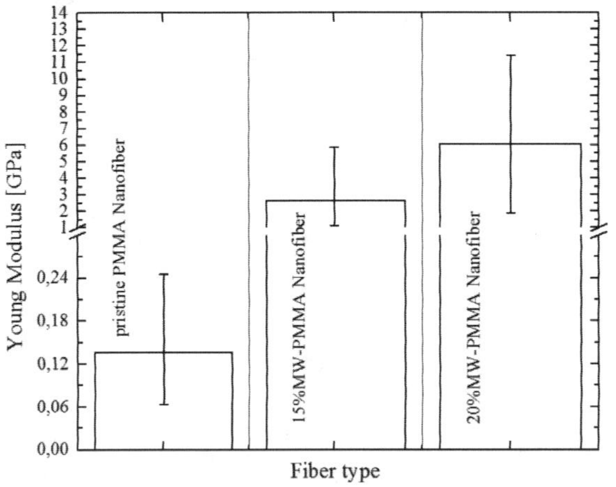

Figure 3. Total average Young's moduli from each measured fiber type

4. Discussion

Clearly, it can be seen that the Young's moduli increases significantly with the use of MWCNTs as filler material. The mean reinforcement from pristine PMMA nanofiber to 15 weight%-MWCNT-PMMA nanofiber was about 19.4 fold. Moreover, the reinforcement from pristine PMMA nanofiber to 20 weight%-MWCNT-PMMA nanofiber was even greater. The Young Modulus increase was here about the 44.5 fold, than without MWCNT as filler. This is a huge gain of the mechanical properties and highlights again MWCNTs as unique filler material. Furthermore the deviation values of the Young's moduli increases even with the use of MWCNTs. There are many reasons for that deviance. One reason could be the randomized distribution of the MWCNTs in the polymer matrix, observed in TEM images. It was recognized, that the MWCNTs were not well aligned in the nanofiber. This causes different defects likewise, empty polymer nanofiber, short and long MWCNTs influenced by sonication [19], (compare fig. 1). Another reason is, that the nanofibers have non uniform diameters, due to MWCNT bundles or empty polymer matrices. The morphology of the pristine nanofibers were much more uniform compared to the composite nanofibers. During all steps humidity plays a significant role for stiffness of polymers. Water absorption results in swelling of polymers, detaching molecular chain apart leading to lower strength [20]. Another failure source could be, that the AFM tip might be not hit the fiber always exactly in the middle of its diameter. A slip or a tangential strike of the AFM tip during the bending tests is also a possible scenario affecting our results. The Young's modulus of the pristine PMMA nanofiber with $136^{+109}_{-72.4}$ MPa and with a diameter of 436^{+9}_{-9} nm is quite small compared the Young's modulus of bulk PMMA, which is in average 3 GPa [21]. This result was determined in past literature values [22]. Lu-Qi Liu et al. used electrospun PMMA nanofibers as well with diameters of 380^{+61}_{-61} nm and using a nano tensile tester [1, 23] and determined a Young's modulus with $72.8^{+33.2}_{-33.2}$ MPa. At the tensile test the nanofibers were glued with liquid epoxy to an AFM tip and loaded in a SEM chamber with a micro manipulation system. The free end of the nanofiber were glued afterwards to the chamber stage. Then the tension test was realized by pulling the nanofiber end. Meanwhile the AFM cantilever was video-recorded and analyzed. The cantilever deflection was recorded and used to determine the force [24] using the thermal-noise calibration method of Sader et al. [25] and results in the Young's modulus calculation. The small deviation of both Young's modulus results, may be due to in the different measurement methods. The nano tensile Lu-Qi Liu et al. [22] used is more precise but also complicated, then the nano bending test. At the nano tensile test, the specimen only will be stressed with pull stresses. This eliminates the rise of shear forces. At the bending test performed here, shear forces cannot be completely excluded. It is assumed that the results with the tensile test are more accurate. But nevertheless the results of the Young's modulus, which were determined in this work, are comparable to prior results in literature [22]. The smaller Young's modulus of pristine PMMA nanofibers compared to bulk PMMA is justified with different reasons. Shearing forces during electrospinning decrease the molecular density and the rapid solidification leads to a mechanical stressed fiber [22]. These reasons bring about to a decrease of the nanofiber stiffness. Moreover the use of FIB induces strong thermal effects, which has extreme weakening effect on polymer stiffness. It has been demonstrated, that the Young's modulus and tensile strength of electrospun nanofibers are generally lower, than at commercial scaled fibers. This is caused the unoptimized molecular orientation of the polymer matrix [26, 27].

5. Summary

In summary, we observed that MWCNT reinforced PMMA nanofibers present a higher Young's moduli, compared with a pristine PMMA nanofiber. The reinforcement rises with the increasing weight percentage of MWCNT per nanofiber. Even the elastic behavior increases demonstrably in the composite nanofibers. The low Young's modulus of the pristine PMMA nanofiber, compared to the Young's modulus of bulk PMMA, was indicated and compared to other research group.

6. Acknowledgments

The authors would like to acknowledge T. Schurig and C. Assman from Physikalisch-Technische Bundesanstalt (PTB) for allowing access to PTB facilities. Parts of this work were supported by the European Research Council under grant number 210642. Moreover we are grateful to Mr. Sören Selve, from the ZELMI of the Technische Universität Berlin, who operate the TEM imaging. The authors would like to acknowledge Mr. Thomas Fritzsch and Mr. Oswin Ehrmann, from the Fraunhofer Institute for Reliability and Microintegration, for their financial support and possibility to utilize the FIB.

References

[1] E.P.S. Tan and C.T.Lim, Mechanical characterization of nanofibers A review, Composites Science and Technology 66 (2006) 1102 1111
[2] A. L. Andrady, SCIENCE AND TECHNOLOGY OF POLYMER NANOFIBERS, Wiley
[3] Zheng-Ming Huanga, Y.-Z. Zhangb, M. Kotakic, S. Ramakrishna, A review on polymer nanofibers by electrospinning and their applications in nanocomposites, Composites Science and Technology 63 (2003) 22232253
[4] V. Datsyuk, S. Trotsenko, S. Reich, Carbon-nanotubepolymer nanofibers with high thermal conductivity, Carbon 52 (2013) 605 620
[5] A. Avnon, B. Wang, S. Zhou, V. Datsyuk, S. Trotsenko, N. Grabbert, and H.-D. Ngo, Quasi one dimensional transport in individual electrospun composite nanobers, AIP ADVANCES 4, 017110 (2014)
[6] A. Avnon, V. Datsyuk, S. Trotsenko, B. Wang, S. Zhuo, N. Grabbert, and H.-D. Ngo, Zero Bias anomaly in an individual suspended electrospun nanofiber (2013), manuscript under submission.
[7] D. Aussawasathien, J.-H. Dong, L. Dai, electrospun polymer nanofiber sensors, Synthetic Metals 154 (2005) 3740
[8] B. Ding, M. Wang, J. Yu and G. Sun, Gas Sensors Based on Electrospun Nanofibers, Sensors 2009, 9, 1609-1624; doi:10.3390/s90301609
[9] L. Zhang, J. Lou, and L. Tong, Micro/Nanofiber Optical Sensors, Photonic Sensors (2011) Vol. 1, No. 1: 31-42
[10] Bingyu Bai, Electrospun chitosan nanofibers for virus removal - Master Thesis, Michigan Technological University
[11] C. P. Barnes, S. A. Sell, E. D. Boland, D. G. Simpson, G. L. Bowlin, Nanofiber technology: Designing the next generation of tissue engineering scaffolds, Advanced Drug Delivery Reviews 59 (2007) 14131433
[12] L. Li, M. W. Frey, T. B. Green, Modification Of Air Filter Media With Nylon-6 Nanofibers
[13] B. M. wall Carbon Nanotubes, N. Processing, and D. Product, Baytubes C 150 P, (2010)
[14] Nanoworld AfM Cantilever (PNP-TR)
[15] Schnell, Gross, Hauger, Technische Mechanik 1 Statik, Springer Verlag
[16] Schnell, Gross, Hauger, Technische Mechanik 2 Elastostatik, Springer Verlag
[17] James M. Gere, Barry J. Goodno, Mechanics of Materials, Cengage Learning
[18] Russel C Hibbeler, Technische Mechanik 2 Festigkeitslehre, Pearson
[19] X.-L. Xiea, Y.-W. Maia, X.-P. Zhoub, Dispersion and alignment of carbon nanotubes in polymer matrix: A review, Materials Science and Engineering R 49 (2005) 89112
[20] A.M.S. Hamouda, The influence of humidity on the deformation and fracture behaviour of PMMA, Journal of Materials Processing Technology 124 (2002)
[21] MilenaKoleva, Poly(methylmethacrylate) (PMMA), CAEDS Injection Moulding Materials
[22] Lu-Qi Liu, Dimitrios Tasis, Maurizio Prato, and Hanoch Daniel Wagner, Tensile Mechanics of Electrospun Multiwalled Nanotube/Poly(methyl methacrylate) Nanofibers, Adv. Mater. 2007, 19, 12281233
[23] E. P. S. Tan, S. Y. Ng, C. T. Lim, Tensile testing of a single ultrafine polymeric fiber, Biomaterials 2005, 26
[24] I. Kaplan-Ashiri, S. R. Cohen, K. Gartsman, V. Ivanovskaya, T. Heine, G. Seifert, I. Wiesel, H. D. Wagner, and R. Tenne, On the mechanical behavior of WS2 nanotubes under axial tension and compression, PNAS
[25] J. E. Sader, J. W. M. Chon, P. Mulvaney, Calibration of rectangular atomic force microscope cantilevers, Rev. Sci. Instrum. 1999, 70, 3967
[26] E. Zussman, M. Burman, A. L. Yarin, R. Khalfin, Y. Cohen, Tensile deformation of electrospun nylon-6,6 nanofibers, Journal of Polymer Science Part B. 2006, 44, 1482
[27] R. Inai, M. Kotaki, S. Ramakrishna, Structure and properties of electrospun PLLA single nanofibres, Nanotechnology 2005, 16, 208

IOP Conf. Series: Materials Science and Engineering **64** (2014) 012006 doi:10.1088/1757-899X/64/1/012006

Influence of carbon nanotubes on the properties of epoxy based composites reinforced with a semicrystalline thermoplastic

A Díez-Pascual, P Shuttleworth, E Gónzalez-Castillo, C Marco, M Gómez-Fatou and G Ellis

Institute of Polymer Science and Technology (ICTP-CSIC), Juan de la Cierva 3, 28006 Madrid, Spain

E-mail: adiez@ictp.csic.es

Abstract. Novel ternary nanocomposites based on a thermoset (TS) system composed of triglycidyl p-aminophenol (TGAP) epoxy resin and 4,4'-diaminodiphenylsulfone (DDS) curing agent incorporating 5 wt% of a semicrystalline thermoplastic (TP), an ethylene/1-octene copolymer, and 0.5 or 1.0 wt% multi-walled carbon nanotubes (MWCNTs) have been prepared via physical blending and curing. The influence of the TP and the MWCNTs on the curing process, morphology, thermal and mechanical properties of the hybrid nanocomposites has been analyzed. Different morphologies evolved depending on the CNT content: the material with 0.5 wt% MWCNTs showed a matrix-dispersed droplet-like morphology with well-dispersed nanofiller that selectively located at the TS/TP interphase, while that with 1.0 wt% MWCNTs exhibited coarse dendritic TP areas containing agglomerated MWCNTs. Although the cure reaction was accelerated in its early stage by the nanofillers, curing occurred at a lower rate since these obstructed chain crosslinking. The nanocomposite with lower nanotube content displayed two crystallization peaks at lower temperature than that of pure TP, while a single peak appearing at similar temperature to that of TP was observed for the blend with higher nanotube loading. The highest thermal stability was found for TS/TP (5.0 wt%)/MWCNTs (0.5 wt%), due to a synergistic barrier effect of both TP and the nanofiller. Moreover, this nanocomposite displayed the best mechanical properties, with an optimal combination of stiffness, strength and toughness. However, poorer performance was found for TS/TP (5.0 wt%)/MWCNTs (1.0 wt%) due to the less effective reinforcement of the agglomerated nanotubes and the coalescence of the TP particles into large areas. Therefore, finely tuned morphologies and properties can be obtained by adjusting the nanotube content in the TS/TP blends, leading to high-performance hybrid nanocomposites suitable for structural and high-temperature applications.

1. Introduction

Epoxy resins are versatile thermosetting polymers widely used in structural applications because of their high stiffness, chemical and heat resistance derived from the cross-linked structure, and easy processability. However, they are inherently brittle and have low wear resistance, which hinder their use for certain structural, adhesive, or coating applications. To address this problem, thermoplastic modifiers [1], rubber particles [2] or hyperbranched polymers [3] have been blended with epoxy resins. On the other hand, multi-walled carbon nanotubes (MWNTs) have outstanding mechanical, thermal and electrical properties and low density, thus are ideal candidates for reinforcing polymer matrices. Poor nanofiller-matrix interfacial adhesion and MWCNT agglomeration are generally the main challenges when dealing with nanotube-reinforced polymer composites. However, it has been recently demonstrated [4,5] that the uneven or asymmetric distribution of nanofillers within the matrix can lead to significant improvements in key properties. In the present work, we have followed this approach in order to achieve property improvements in an aerospace-grade high performance epoxy system. To attain this goal, the resin has been modified with a semicrystalline thermoplastic, an ethylene/1-octene copolymer and MWCNTs. An extensive characterization has been carried out in order to analyze the effect of MWCNT content, location and level of dispersion on the morphology, thermal and mechanical properties of the resulting hybrid composites.

2. Experimental

The epoxy (TS) system consisted of a high-performance trifunctional epoxy, triglycidyl p-aminophenol (TGAP) with a 4,4′-diaminodiphenyl sulfone (DDS) curing agent, supplied by Huntsman Advanced Materials, LLC. CVD MWCNTs were obtained from Cheap Tubes Inc (VT, USA). The thermoplastic (TP), an ethylene/1-octene copolymer with commercial name DOWLEX SC2107, was provided by Dow Chemical. The TP particle size was reduced using a cryogenic impact mill.

The neat TS was prepared by blending TGAP and DDS (100/49 resin/hardener wt% ratio) in a planetary centrifugal mixer for 5 min at 300 rpm and room temperature. The mixture was then poured into silicone moulds and cured in an oven at 130 °C for 1.5 h, followed by postcuring at 177 °C for 2 h. Blends of TS/TP (5 wt%) were prepared in a two-stage process: The TP powder was initially mixed with the TGAP in the planetary centrifuge for 10 min at 500 rpm. After a few minutes of stirring at 130 °C, DDS was added stepwise under continuous stirring until a homogeneous mixture was obtained, and the blend was then subjected to the above indicated curing cycle. Regarding TS/TP/MWCNT composites, the TP powder was initially blended with TGAP in the planetary mixer for 10 min at 500 rpm; afterward, the MWCNTs were added and the mixing continued for another 10 min. Subsequently, the mixture was stirred at 130 °C and the curing agent was added, and finally it was cured under the same conditions.

3. Results and discussion

The effect of the TP and MWCNTs on the curing behaviour of TS was investigated by differential scanning calorimetry (DSC), and the calorimetric parameters derived from the non-isothermal thermograms are collected in Table 1: curing enthalpy (ΔH), curing onset (T_o) and the slope of the curve in the region of the curing peak (*m*). The neat resin begins the curing at 130 °C, and shows a single step process with a slope of -0.55 W/g °C. Similar behaviour is observed for the TS/TP binary mixture, with the curing occurring at a slightly lower rate. However, composites incorporating MWCNTs start the curing at a lower temperature, and exhibit a two-stage process, the first step taking place at a significantly lower rate and the second at similar rate to that of the neat TS. These results hint that the MWCNTs have a dual effect: firstly, due to their higher thermal conductivity, they accelerate the onset of the curing process, which is initiated in those regions where CNTs are preferentially located. Secondly, they appear to impose a certain hindrance in the epoxy crosslinking that takes place at a lower rate. Afterwards, since curing is a dynamic process, it is extended to the entire composite, occurring at similar rate and the same temperature for all the samples compared to the neat matrix. Interestingly, as the CNT content increases, the aforementioned effects are stronger: T_o shifts to lower temperature and the slope of the first stage decreases. This is related to the morphology of the composites, as discussed later. Further, ΔH decreases in the presence of the MWCNTs, also pointing towards a hampering of the epoxy crosslinking. In fact, it has been reported [6] that well-dispersed CNTs can act as a physical hindrance to the monomer mobility and thus lower the heat of reaction.

Table 1. Thermal data derived from DSC experiments

SAMPLE	ΔH^a (J/g)	T_o^a (°C)	m^a (W/g °C)	T_m (°C)	T_c (°C)	X_c (%)
TS	203	130	-0.55			
TP				121	99	39
TS/TP (5 wt%)	177	130	-0.46	115	72/16	41
TS/TP (5wt%)/MWCNT (0.5 wt%)	175	128	-0.023/-0.45	123	82/23	48
TS/TP (5wt%)/MWCNT (1.0 wt%)	189	124	-0.015/-0.48	112	93	36

[a]Determined from non-isothermal curing thermograms. ΔH: curing enthalpy; T_o: curing onset; *m*: slope of the non-isothermal curve in the region of the curing peak. T_c and T_m: crystallization and melting temperatures, respectively. X_c: degree of crystallinity.

The morphology of the samples was analyzed by scanning electron microscope (SEM), and typical *images* of ternary composites with 0.5 and 1.0 wt% are shown in Fig. 1. In TS/TP (5.0 wt%)/MWCNT (0.5 wt%) nanocomposite (Fig. 1a and 1b), the TP particles are randomly and homogenously dispersed within the TS, showing matrix-dispersed droplet type morphology. The MWCNTs appear to be preferentially located at the TS/TP interphase (see Fig. 1b), probably due to their higher affinity with the TP owing to their hydrophobic nature. This phenomenon has been regarded as a mechanism to reduce the size of the minor polymer phase. Further, no MWCNT aggregates were observed in the whole examined regions. The well-dispersed nanofillers act as a solid barrier around the TP particles, thereby preventing the coalescence. However, SEM micrographs of TS/TP (5.0 wt%)/MWCNTs (1.0 wt%) show a very different morphology (Fig. 1c and 1d). The size of the TP domains is considerably larger, and they appear interconnected due to the coalescence of several particles, leading to a gross dendritic morphology. The MWCNTs are located within the TP phase (Fig. 1d), and appear quite aggregated. A plausible explanation could be that the higher CNT concentration resulted in higher viscosity of the phases, hindering their diffusion and migration over the composite.

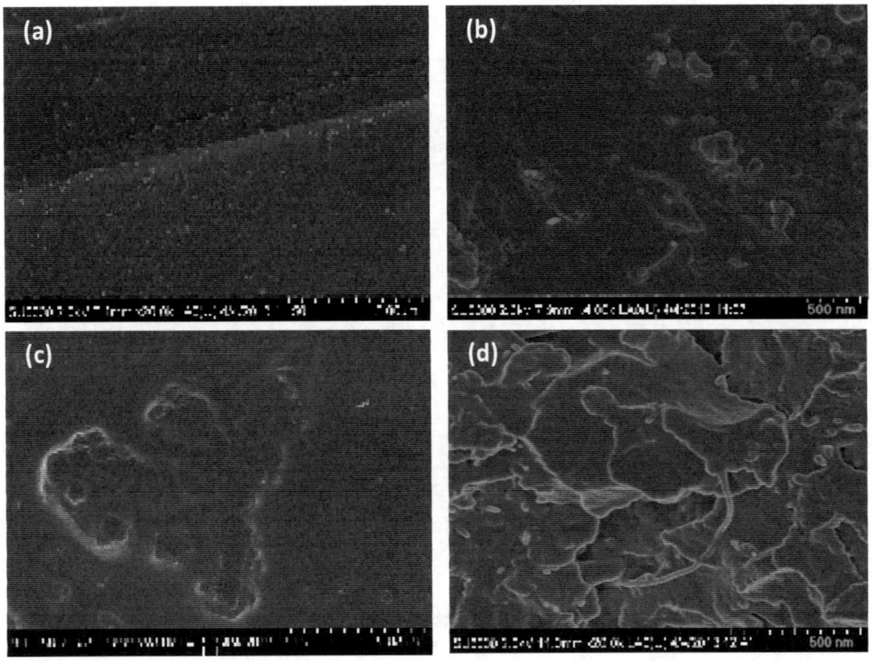

Fig. 1 SEM micrographs at different magnifications of: (a) and (b) TS/TP (5 wt%)/MWCNT (0.5 wt%); (c) and (d) TS/TP (5 wt%)/MWCNT (1.0 wt%).

The crystallization behaviour of the composites was investigated by DSC (Fig. 2). Neat TP shows a sharp crystallization peak T_c at about 100 ºC, with a slight tail corresponding to smaller crystal sizes at the lower temperature side. On the contrary, TS/TP (5.0 wt%) displays two small, broad exothermic peaks at much lower temperatures than that of the neat TP, around 72 and 16 ºC (Table 1). These are probably related to two populations of crystals with different average crystal sizes. According to the SEM images, the TP phase forms isolated and well-dispersed micro domains confined within the epoxy phase. Each domain can have a different nucleating activity depending on the size, degree of dispersion, presence of impurities, etc. An analogous behaviour has been reported for immiscible polymer blends where the crystalline polymer is finely dispersed in small domains (at the micro- or nanoscale), and is denominated "confined or fractionated crystallization" [7]. TS/TP (5 wt%)/MWCNT (0.5 wt%) also shows two broad crystallization peaks related with TP particles with different crystal sizes, hence different degrees of confinement. Besides, the T_c is shifted to higher values, probably due to the nucleating effect of the nanofillers. However, TS/TP (5 wt%)/MWCNT (1.0 wt%) only exhibits a single crystallization. The disappearance of the lower temperature exotherms on increasing the MWCNT content is indicative of the elimination of the confinement effect in this sample. From the SEM images, both TP and the MWCNTs are agglomerated, with considerably large TP domains, hence the polymer crystallizes in a similar manner to neat TP. Further, the T_c is slightly higher than that of TS/TP(5.0 wt%)/MWCNTs(0.5 wt%), which could be related to the dispersion of the nanofillers inside the TP domains.

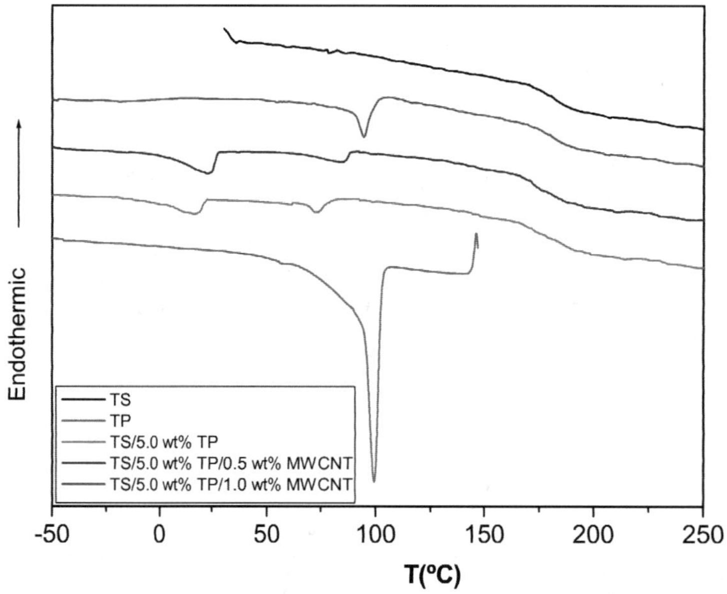

Fig. 2 Dynamic DSC cooling (bottom) scans for the different samples

The thermal stability of the samples was investigated under nitrogen atmosphere and the results are presented in Fig. 3. The initial degradation temperature of TS increases upon addition of both TP and MWCNTs, due to the higher thermal stability of these materials compared to the neat resin. The hybrid containing 0.5 wt% MWCNTs shows around 25 C increment compared to that of neat TS. This outstanding increase is attributed to a synergetic stabilizing effect of the isolated TP domains and the well-dipersed MWCNTs that can act as a barrier and effectively hinder the diffusion of the degradation products from the bulk of the matrix to the gas phase. Further, the increase in thermal conductivity due to the presence of the MWCNTs facilitates heat dissipation within the nanocomposite. However, for the hybrid with 1.0 wt% MWCNTs the improvement is smaller (~15 ºC), since both the TP and the MWCNTs are quite agglomerated, hence exert a less efficient barrier effect. Qualitatively similar trends are found for the temperatures corresponding to 10% weight loss and the maximum rate of weight loss.

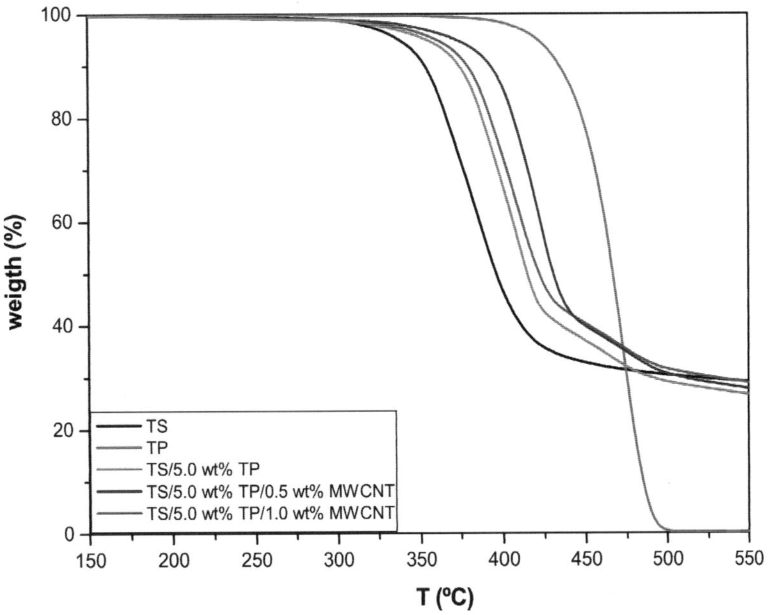

Fig. 3 TGA thermograms under nitrogen atmosphere at a rate of 10 °C/min

The mechanical properties of the hybrids were characterized by tensile tests (Fig. 4). The Young's modulus (E) of TS is ~2.8 GPa. The TP acts as a plasticizer and causes a 16% decrease in the matrix modulus. In contrast, the incorporation of MWCNTs imparts the high stiffness of the fillers to the matrix, and the TS/TP (5 wt%)/MWCNT (0.5 wt%) has ~12% higher E than TS. In this case, the nanotubes are predominantly located at the TS/TP interphase and might improve the matrix-TP interfacial adhesion. However, for the nanocomposite with 1.0 wt% MWCNTs, E is slightly lower than that of the matrix due to the less effective reinforcement of the agglomerated nanotubes that are mainly located inside the TP domains. Regarding the tensile strength (σ_y), the trends observed are similar to those found for the modulus, albeit the percentage increments attained are higher, i.e. ~24% for TS/TP (0.5 wt%)/MWCNT (0.5 wt%) compared to neat TS, suggesting that the nanotubes are more efficient in enhancing the matrix strength.

Focusing on the elongation at break (ε_b), a noticeable improvement is attained upon addition of the TP to TS due to its higher ductility. Nevertheless, TS/TP/MWCNT samples display lower ε_b than TS/TP (0.5 wt%), the reduction being more pronounced for that with higher nanotube loading since agglomerates strongly obstruct the plastic deformation of the matrix. The area under the tensile curve is a measure of the toughness of the material. As expected, TS/TP (0.5 wt%) shows higher toughness than the neat matrix (about 45%); the incorporation of 1.0 wt% MWCNTs to TS or TS/TP results in moderate toughness reductions, since the presence of agglomerates nucleates cracks, resulting in premature failure hence lower impact resistance. However, TS/TP (5.0 wt%)/MWCNT (0.5 wt%) shows ~52% higher value than that of TS. It seems that the toughening effect is boosted in this sample due to a synergistic action between the TP and the MWCNTs. Moreover, this superior performance could be related to the matrix-dispersed droplet type morphology observed for this composite, where small TP domains were found to be homogeneously dispersed within the matrix.

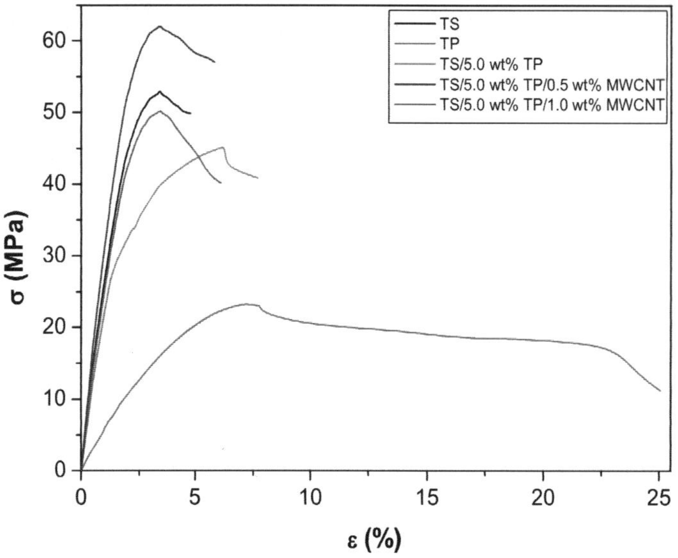

Fig. 4 Room temperature tensile curves for the different samples.

4. Conclusions

Novel ternary nanocomposites based on a thermoset epoxy matrix toughened with a semicrystalline thermoplastic copolymer and reinforced with MWCNTs have been developed. A significant change in the phase morphology of TS/TP blends from droplet-dispersed type to agglomerated TP domains with dendritic structure was found when the CNT concentration was increased from 0.5 to 1.0 wt%. The MWCNTs have a dual effect on the epoxy curing process: they accelerated the cure onset albeit hindered its progress since they impeded chain crosslinking. The nanocomposite with the lowest nanotube content showed two crystallization peaks at lower temperature than that of neat TP, related to confined crystallization of isolated domains with different average crystal sizes. However, a single peak appearing at similar temperature to that of TP was observed for that with 1.0 wt% MWCNTs, since the coalescence of the TP domains eliminated the confinement effect. The highest thermal stability and best mechanical properties were found for TS/TP (5.0 wt%)/MWCNTs (0.5 wt%) nanocomposite, due to synergistic effects of both TP and the nanofiller.

Acknowledgments

Financial support from the Ministerio de Ciencia e Innovación (MICINN, Project MAT2010-21070-C02-01) is gratefully acknowledged. AD thanks to the Consejo Superior de Investigaciones Cientificas (CSIC) for a JAE-doc postdoctoral contract cofinanced by the EU. PS thanks to the MICINN for a Juan de la Cierva postdoctoral contract (JCI-2011-10836). EIGC acknowledges CONACYT of the Mexican Government for a postgraduate fellowship.

References

[1] Hourston D J and Lane J M 1992 *Polymer* **33** 1379
[2] Frigione M, Acierno D and Mascia L 1999 *J. Appl. Polym. Sci.* **73** 1457
[3] Ratna D, Varley R, and Simon G P 2003 *J. Appl. Polym. Sci.* **89** 2339
[4] Li Y and Shimizu H 2008 *Macromolecules* **41** 5339
[5] Liu L *et al* 2009 *Polymer* **50** 3072
[6] Gónzalez-Dominguez J M *et al* 2011 *J. Phys. Chem. C* **115** 7238
[7] Salmeron-Sanchez M *et al* 2006 *J. Polym. Sci. Part B: Polym. Phys* **44** 815

Influence of carbon nanotubes on mechanical properties and structure of rigid polyurethane foam

E. Ciecierska[1], M. Jurczyk-Kowalska[1], P. Bazarnik[1], M. Kulesza[1], M. Lewandowska[1], M. Kowalski[2], S. Krauze[2]

1. Warsaw University of Technology, Faculty of Materials Science and Engineering, Woloska 141, 02-507 Warsaw
2. Transport Vehicles Upholstery Manufacturer TAPS - Maciej Kowalski, Solec 3/5, 94-247 Łódź, Poland

Corresponding author: eciecierska@meil.pw.edu.pl

Abstract. In this work, the influence of carbon nanotubes addition on foam structure and mechanical properties of rigid polyurethane foam/nanotube composites was investigated. Scanning electron microscopy was performed to reveal the foam porous structure and distribution of carbon nanotubes. To determine the mechanical properties, three point bending tests were carried out.

1. Introduction

Carbon nanotubes (CNTs) are considered to be one of the best candidates for reinforcement of polymer matrix composites because of the unique combination of high strength, electrical and thermal conductivity and high temperature action resistance [1]. However, CNTs exhibit high tendency for agglomeration, which implies that it is difficult to obtain composites with their optimal dispersion [2, 3, 4]. Also, it is well known that their addition to solid polymers increases significantly strength but its impact on the properties of rigid foam is less known and described only in a small number of publication [5, 6, 7]. We report here the recent results on the influence of CNTs addition on structure and mechanical properties of rigid polyurethane foam.

2. Experimental

In this work, high purity (95%) multi wall CNTs supplied by Nanocyl and Polyurethane system supplied by BASF were used to produce foams with 3 different CNTs concentrations (namely 0.01, 0.05 and 0.1%). Pure polyurethane foam was used as a reference sample. Nanotubes were dispersed in polyol by using three roll milling. The mixture of CNTs and polyol was mixed several times by the rollers. In each subsequent pass the gap between the rolls was reduced in order to deaglomerate CNTs. The mixture of poliol and CNTs prepared in such a way was mixed by a mechanical stirrer with isocyanate and put into mold.
The structure of foams and dispersion of CNTs was examined by using scanning electron microscope. Mechanical properties of these composites were determined in three point bending test from which bending strength and modulus were evaluated.

3. Results and discussion

Pure polyurethane foam is presented in Fig. 1. This foam features two populations of pores: large with size of about 100μm and small with sizes of about 0.2μm The addition of CNTs changes significantly the viscosity of the polymer, which influences the foam growth time. As a result, the pores' structure changed, as illustrated in Fig. 2. In particular, one can observe an increase in the number and size of small pores whereas the structure of large pores remains unchanged. The spatial distribution of small pores is inhomogeneous. On the other hand, the dispersion of CNTs seems to be nearly uniform, although some places with higher nanotubes concentration are also visible in the microstructure.

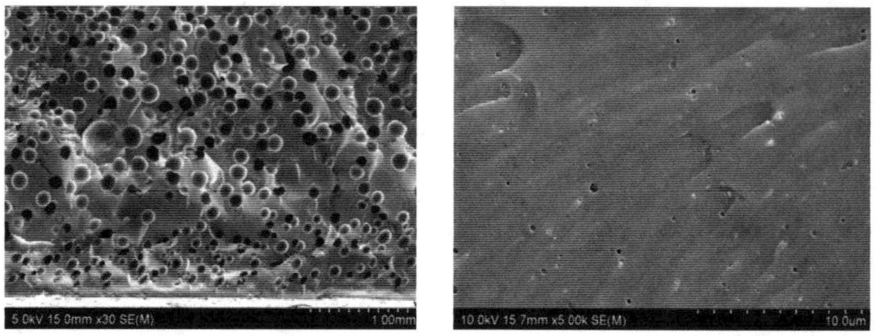

Fig. 1. Structure of pure polyurethane foam

Fig. 2. Structure of polyurethane foam/carbon nanotubes composites

The results of bending tests of pure polyurethane foam and its composites with a different addition of CNTs are presented in Fig. 3. The results show that only small addition of CNTs (0.01%) results in strength and Young modulus improvement. Higher concentration of CNTs decreases both values, which can be attributed to higher porosity of the composite samples.

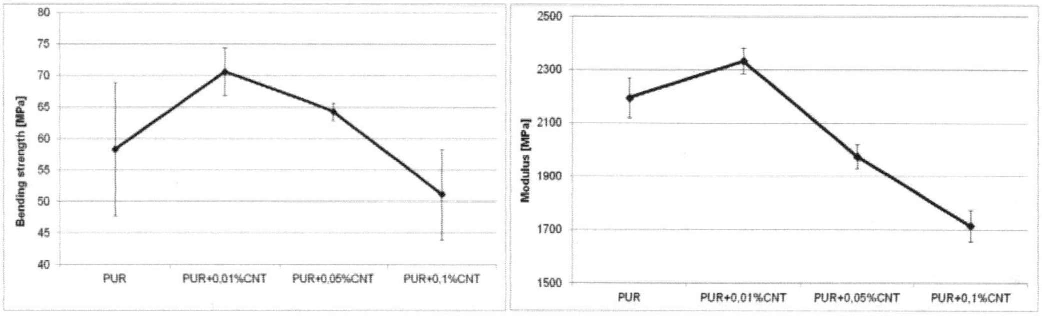

Fig. 3. Bending strength and modulus for polyurethane foam and its composites with carbon nanotubes

IOP Conf. Series: Materials Science and Engineering **64** (2014) 012007 doi:10.1088/1757-899X/64/1/012007

4. Acknowledgment

This research was financed by the National Centre for Research and Development within the project "New materials and technologies for lightweight generic components of electric low-emission concept vehicle".

References

[1] Peng-Cheng Ma, Naveed A. Siddiqui, Gad Marom, Jang-Kyo Kim, "Dispersion and functionalization of carbon nanotubes for polymer-based nanocomposites: A review", Composites: Part A 41 (2010) 1345–1367

[2] Iosif D. Rosca, Suong V. Hoa, "Highly conductive multiwall carbon nanotube and epoxy composites produced by three-roll milling", CARBON 47 (2009) 1958–1968

[3] Ewelina Ciecierska, Anna Boczkowska, Krzystof. J. Kurzydlowski, „Quantitative description of the spatial dispersion of carbon nanotubes in polymeric matrix", J Mater Sci (2010) 45:2305–2310

[4] Ning Hu, Zen Masuda, Go Yamamoto, Hisao Fukunaga, Toshiyuki Hashida, Jinghao Qiu, "Effect of fabrication process on electrical properties of polymer/multi-wall carbon nanotube nanocomposites", Composites: Part A 39 (2008) 893–903

[5] Marcelo Antunes, José Ignacio Velasco, "Multifunctional polymer foams with carbon nanoparticles", Progress in Polymer Science 2013

[6] Viktoriya Dolomanova, Jens Chr. M. Rauhe, Lars Rosgaard Jensen, Ryszard Pyrz, Ana Barros Timmons, "Mechanical properties and morphology of nano-reinforced rigid PU foam", Journal of Cellular Plastics 2011, 47: 81

[7] Dingxiang Yan, Ling Xu, Chen Chen, Jianhua Tang, Xu Jic, Zhongming Li, "Enhanced mechanical and thermal properties of rigid polyurethane foam composites containing graphene nanosheets and carbon nanotubes", Polym Int 2012; 61: 1107–1114

IOP Conf. Series: Materials Science and Engineering **64** (2014) 012008 doi:10.1088/1757-899X/64/1/012008

Effect of the number of layers of graphene on the electrical properties of TPU polymers.

B Galindo[1], S Gil Alcolea[2], J Gómez[2], A Navas[3],A Ortega Murguialday[1], M Pérez Fernandez[2] and R C Puelles[2]

[1] AIMPLAS, Instituto Tecnológico del Plástico, C/Gustave Eiffel 4,46980 , Paterna, Valencia, Spain .

[2] Avanzare Innovación Tecnológica S.L. Antonio de Nebrija 8, 26006, Logroño, Spain

[3] Instituto de Tecnologías Químicas de La Rioja (Inter-Química), San Francisco 11, 26370, Navarrete, Spain

E-mail: julio@avanzare.es

Abstract. In this paper graphene and few layers graphene were synthesized by a modified Hummers method using flake graphite powders as the starting material. The effect of the incorporation of graphene on the electrical properties of thermoplastic polyurethane (TPU) was investigated via two processing techniques: solution blending and melt compounding. When solution blending is used for the preparation of composites, the obtained electrical conductivity is higher, even at very low loads (0,25% w/w). Moreover, the single layer graphene shows 10.000 times higher electrical conductivity than few layers graphene.

1. Introduction

In the recent years graphene has attracted a great deal of attention due to its potential application in many different areas such as nanoelectronics, ultracapacitors, gas sensors, catalyst supports [1]. Moreover in polymer nanocomposites, graphene is probably the most promising nanofiller due to its high surface area, aspect ratio, tensile strength, thermal and electrical conductivity, EMI shielding ability, flexibility and transparency [2]. In particular, high values of electrical and thermal conductivity have been reported derived from ultra-fast transfer of charge carriers due to its high specific surface area and strong interaction between adjacent carbon atoms and monolayer structure [3].

Currently, a variety of techniques have been developed to prepare good quality graphene nanosheets (GNS), including the mechanical exfoliation, micromechanical cleavage, epitaxial growth (CVD), exfoliation of graphite in the liquid phase through direct sonication, and chemical wetting methods[4,5]. However some disadvantages of these techniques have been reported. Mechanical exfoliation even if high-uniform monolayer graphene flakes are obtained is not scalable and consequently not amendable for mass production of graphene. On the other hand epitaxial growth methods produce pure monolayer graphene sheets however it is a great challenge to transfer sample to insulating substrates for making useful devices.

Among the chemical methods, the Hummers method [6] has been the most popular approach for large scale production of graphene sheets, even though graphene sheets derived by this method contains a significant amount of defects due to the use of strong acids and oxidants (that largely disrupt the sp^2 bonded carbon network of graphene). The first step of Hummers method provides graphene oxide, which has several in-plane oxygen-containing groups. For reducing GO, two main approaches have been used in recent years: (1) chemical reduction and (2) thermal annealing. Since Boehm and co-workers [7] early reported a method to prepare graphene by chemical reduction, it has been focused on the synthesis and manipulation of single- and multilayered carbon species [8]. However, reduced graphene produced by the chemical reduction of exfoliated GO has poor electrical properties due to the presence of contaminants and structural defects. Thermal annealing is an alternative route for reduction of GO restoring graphene structure and increasing its conductivity.

Thermoplastic polyurethanes (TPUs) are linear block copolymers characterized by hard and soft segments. Thanks to their particular chemical structure, these materials are very versatile offering a wide range of service temperatures and harness options, excellent tear and abrasion resistance, good resistance to non-polar solvents, high compression and tensile strength [9]. These properties make TPUs suitable for several products such as automotive components, adhesives, medical devices, food processing equipments, and so on. In this regard, it is possible to increase and tailor some specific properties (such as mechanical or dielectric properties) of TPU resins through the inclusion of nanofillers such as graphene.

In this paper we have explored the effects of the number of graphene layers in the electrical properties of the resin through two processing techniques, solution and melt compounding.

2. Experimental

2.1. Materials

Flake Graphite powder Flake Graphite powder with a density of 0.640 g/cm^3 particle size of 500 µm obtained from Grafitos BARCO Company SL, Barcelona. For the preparation of graphene samples; $KMnO_4$ and concentrated H_2SO_4 sodium hydroxide and hydrochloric acid were obtained from COFARCAS (Burgos). Thermoplastic polyurethane (TPU) PEARLTHANE® D15 N70 (TPU; Merquinsa, Barcelona, Spain) with a density of 1,16 g/cm^3 (at 20ºC) was used as received. N,N-dimethylformamide (DMF; Scharlau, Barcelona S.L., Spain), methanol (MeOH; Scharlau, Barcelona, S.L., Spain), were used as received for the study.

2.2. Synthesis of graphene sheets

The graphene and few layers of graphene particles were synthesized by a modified Hummers method using flake graphite powders as the starting material. In this methodology, graphite flakes are suspended in a water solution of different strong oxidants (during 36 h to obtain few layers graphene and 48 h to prepare graphene monolayer); the resulting cake is washed and filtered to provide the graphite oxides. A subsequent step of controlled reduction by means of chemical reductants and thermal treatment provides the different qualities of graphene.

2.3. TPU/graphene nanocomposites preparation

Graphene and graphite TPU Nanocomposites (0.25% w/w) were prepared via 2 methodologies: Melt compounding and solution blending. Solution blending nanocomposites were prepared as follows: TPU was dissolved in DMF, on the other hand, the required quantity of graphene and graphite was dispersed in DMF and sonicated in an ultrasound bath. The TPU dissolution was added and the mixture was stirred in order to get a homogeneous solution. After that, it was poured slowly into methanol to precipitate the polymer-graphene nanocomposite. The material obtained was dried overnight at 70°C in a vacuum oven.

Melt compounding nanocomposites were prepared as follows: a Masterbatch (concentrated) compound was prepared through the solution methodology described before. The masterbatch obtained was diluted using an internal mixer (BRABENDER PL 2000,Germany) for 6 minutes at

190°C. The obtained nanocomposites via solution blending and melt compounding were compression moulded at 180°C to obtain a 500 micron film for electrical conductivity measurement

2.4. Characterization techniques

The morphology of graphene sample was obtained by scanning electron microscopy (SEM) images using a Hitachi S-2400. The morphology and thickness were characterized by atomic force microscopy (AFM) NT-MDT model NTEGRA PRIMA. The dimension and the number of layers were also evaluated using a transmission electron microscopy (TEM) experiments were performed on a JEOL model JEM-2010 electron microscope. For microscopy characterization, the samples were dispersed in isopropyl alcohol and sonicated with a Hielscher UP200S sonicator.

Raman spectra were recorded with a Raman Horiba Xplora using an excitation wavelength of 532 nm. Scans were recorded in the range from 400 to 3200 cm^{-1}. For lineal measurements the electrodes were at 1 cm of distance and 250 V polarization technique was used. TPU/nanocomposites were analysed with a scanning electron microscopy PHENOM World desktop SEM PRO-X.

3. Results and discussion

3.1. Graphene characterization

High-yielding and good-quality of few layers of graphene and single/bilayer of graphene are obtained by exfoliation of sheets of graphite using the modified Hummers method as shown by electron and atomic force microscopy (Figures 1,2,3 and 4). The coarse graphite powder is in fact composed of multiple layers of graphene; the exposure to strong oxidizing agents cause layers to spread obtaining graphitic oxides (GO). GO reduction via thermo-chemical treatment leads to a total exfoliation of graphite and in consequence graphene structures made of one or more layers are obtained maintaining XY plane while reducing the apparent density of the materials (Table 1).

Table 1. Properties of synthesized graphene products

Product	Reaction time. (*h*)	Apparent density. (*g/cm³*)	XY plane measured by SEM (*μm*)
Graphite	---	0.640	0-500
Few layer graphene	36	0.020	20x20
Mono/bilayer graphene	48	0.006	20x20

High resolution TEM is used to characterized the degree of exfoliation of graphite. Figure 1 shows TEM images of few layers graphene. The distinguishable red-white lines running almost parallel in some parts and in zigzag in other zones may be a reflection of the wrinkled nature of graphene. This wrinkle and folding nature is also seen in Figure 2 for the mono/bilayer graphene. SEM (Figures 3 and 4) images demonstrated the effectiveness of the modified Hummers method for the exfoliation of graphite flakes obtaining few and monolayer graphene products. The XY plane size has been reduced from 500 microns of the graphitic crystal to 20 x 20 μm size to graphene.

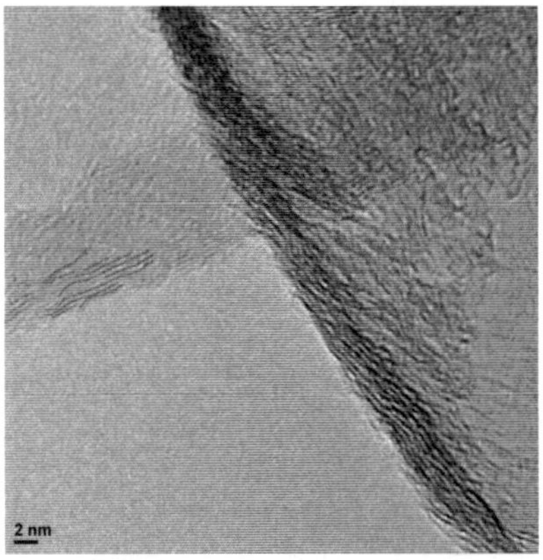

Figure 1. TEM image of few layers graphene

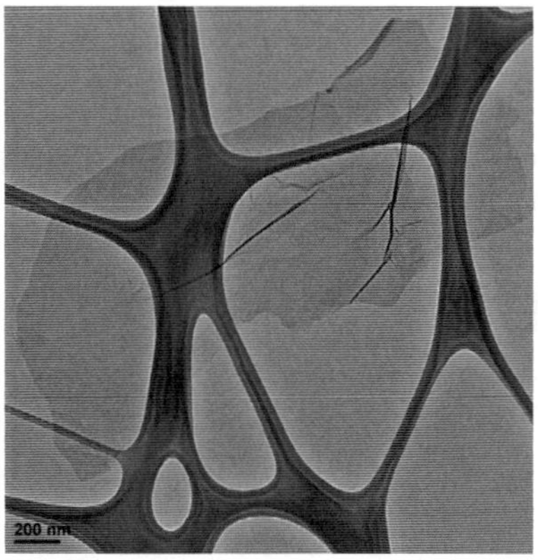

Figure 2. TEM image of mono/bilayer graphene.

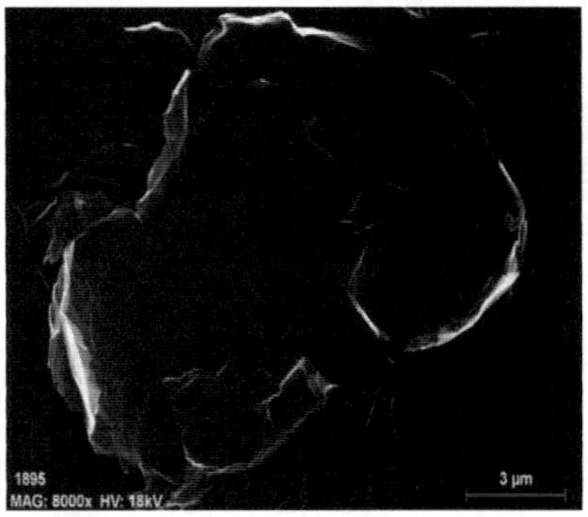

Figure 3. SEM image of few layers graphene

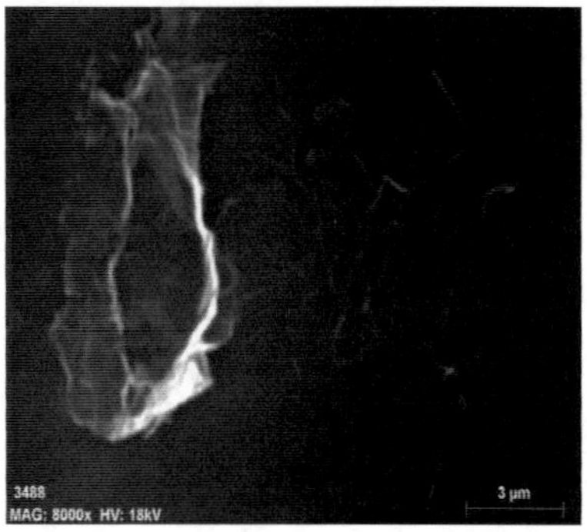

Figure 4. SEM image of mono/bilayer graphene..

The AFM is a powerful non-destructive technique for precisely identifying the number of graphene layers by measuring the step height profile. Figure 5 and 6 show a representative tapping mode AFM image of graphene and few layers of graphene sheets and the height profile in selected locations. The sizes and thicknesses vary, but as a whole, all the thicknesses of few layers graphene are roughly around 1 to 2 nm, and less than 1 nm for single layers graphene.

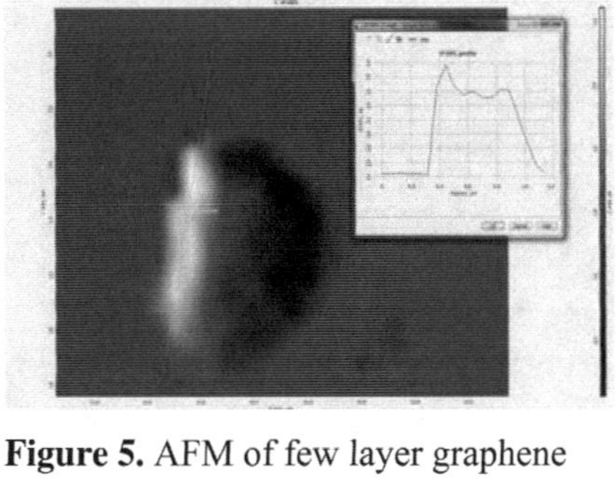

Figure 5. AFM of few layer graphene

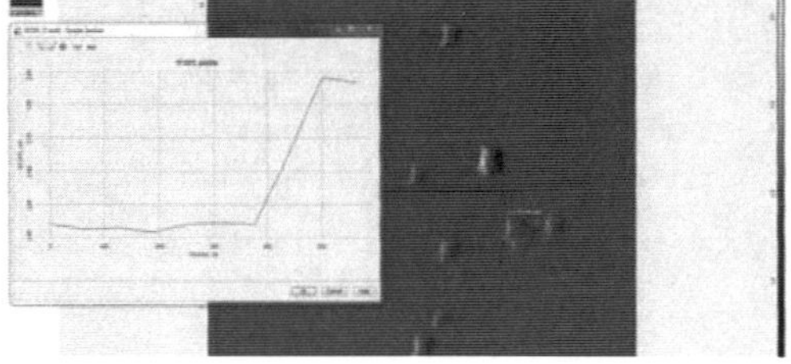

Figure 6. AFM of single/bilayer graphene..

Raman spectroscopy is a powerful tool for examining the quality of the graphene materials. The main features in the Raman spectra of carbonaceous materials are the so-called G and D peaks, which lie at around 1560 and 1360 cm^{-1} respectively. The Raman spectra of few layers and single layers graphene are displayed in Figure 7. There are three peaks at 1350 cm−1 (D peak), 1580 cm−1 (G peak) and 2714 cm−1 (2D peak). The D peak, is derived from the edge effects, disorder and lost of graphitic structure. In addition, the intensity ratio (ID/IG) is about 0.6 for the few layers and 0,4 for single layer one, which is much less than most chemical reduction reports [10].

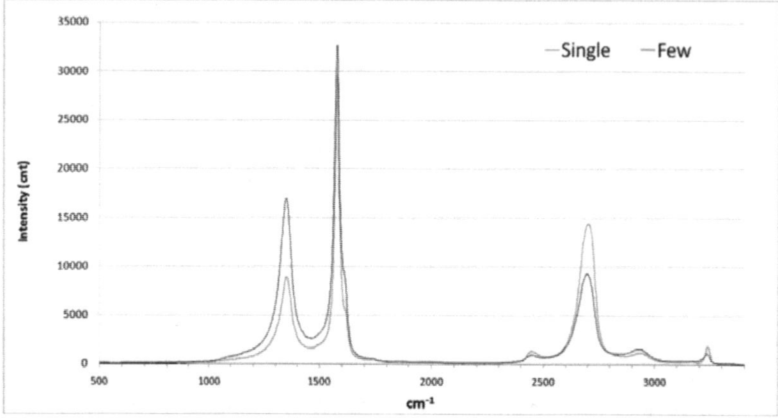

Figure 7. Raman spectra of Few layers and single layer graphene

3.2. TPU nanocomposites

The influence of two processing methodologies melt compounding and solution blending on the electrical performance has been studied on TPU graphene nanocomposites. Table 2 show the electrical properties of 0.25% w/w graphene TPU nanocomposites prepared by both methodologies.
The electrical conductivity is found to increase when the exfoliation of the graphitic structure is completed. Thus mono/bilayer TPU nanocomposites show 10.000 times higher conductivity than few layer graphene where bulk graphite composites were out of the measurement limit of the equipment. In addition solution blending methodology showed for the mono/bilayer nanocomposite 10 times higher electrical conductivity.

Table 2. Linear conductivity of nanocomposites

Methodology	Product	% filler.	linear conductivity ($Sxcm^2$)
Solution blending	Mono/bilayer graphene	0.25	$2,33.10^{-08}$
Melt compounding	Mono/bilayer graphene	0.25	$2,16.10^{-09}$
Solution blending	Few layer graphene	0.25	$1,25.10^{-12}$
Melt compounding	Few layer graphene	0.25	$1,26.10^{-12}$
Solution blending	Graphite	0.25	out-range
Melt compounding	Graphite	0.25	out-range

Figure 8 and 9 show the SEM microphotographs of cryofractured surface morphology of the TPU/mono-bilayer graphene nanocomposites in order to observe the dispersion and distribution of the nanofiller and the physical interaction with the TPU matrix. It can be observed that graphene has homogenously been dispersed among the TPU matrix.

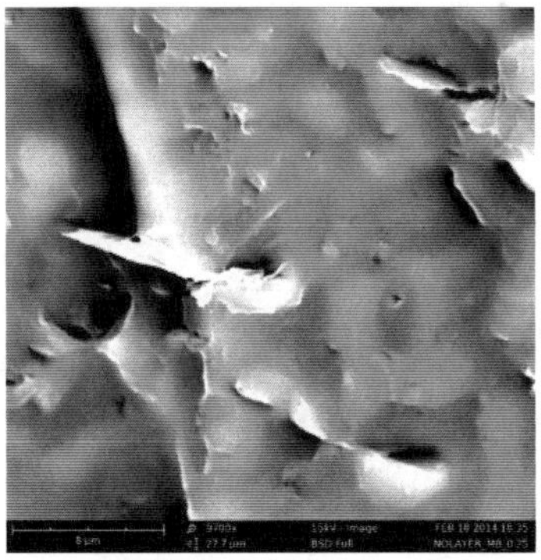

Figure 8. SEM image of mono/bilayer TPU produced by melt compounding

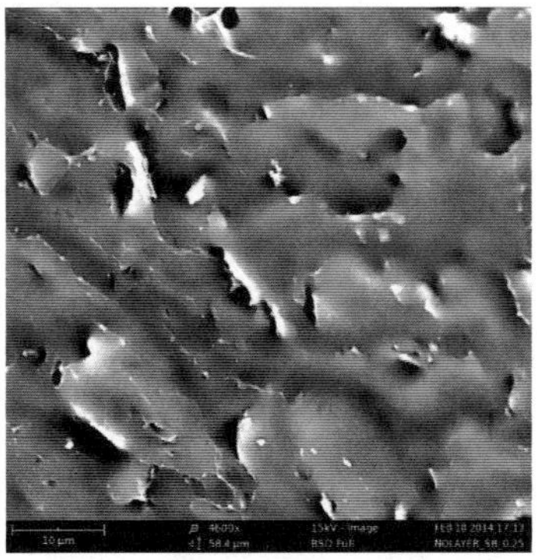

Figure9. SEM image of mono/bilayer TPU produced by solution blending

4. Conclusions

We have prepared graphene and few layers graphene using a modified Hummers method with highly restoration of the graphitic structure, as it can be seen by SEM, AFM and RAMAN measurements. Using solution blending and melt compounding, we have obtained TPU composites of graphene and few layers graphene. When solution blending is used for the preparation of composites, we observe higher electrical conductivity, even at very low concentration of graphene 0,25% w/w. We also observe that single layer graphene shows 10.000 higher electrical conductivity than few layers graphene.

Acknowledgments
This work was has been performed within the framework of the project NANOMASTER (from the European Community's Seventh Framework Programme FP7/2007-2013 under grant agreement n° 285718).

5. References

[1] Wang G X ,Shen X P, Yao J and Park J, 2009 Carbon **47** 2049
[2] Potts J R, Dreyer D R, Bielawski C W and Ruoff R.S 2011 *Polymer.* 5
[3] Geim A K, Novoselov K S 2007 *Nat. Mater* **6** 183
[4] Stankovich S, Dikin D A, Piner R D, Kohlhaas KA, Kleinhammes A, and Jia Y, et al. 2007, *Carbon* **45**:1558
[5] Dreyer R D , Murali S ,Zhu Y ,. Ruoff RS and. Bielawski C W. *Journal of Materials Chemistry* 2010
[6] Hummers WS and Offeman RE 1958 *J Am Chem Soc.* **80** 1339.
[7] Boehm H. P, Clauss A.; Fische G, Hofmann U *Proceedings of the Fifth Conference on Carbon*, 1962, 73.
[8] Park S and. Ruoff R. S 2009 *Nat. Nanotechnol* **4**,217
[9] Fernandez M, Landa M, Muñoz M E, Santamaría A 2010 *International Journal of Adhesion and Adhesives* **30** 609
[10] Fan Z J ,Kai W, Yan J, Wei T, Zhi L, Feng J, Ren Y, Song L , and Wei F 2011 *ACS nano* **5** 191

IOP Conf. Series: Materials Science and Engineering **64** (2014) 012009 doi:10.1088/1757-899X/64/1/012009

Compression-after-impact (CAI) performance of epoxy-carbon fibre-reinforced nanocomposites using nanosilica and rubber particle enhancement

B Nikfar [1,] and J Njuguna [2]
[1] Department of Automotive Engineering, Polytechnic of Turin, Turin, Italy
[2] IDEAS Research Institute, School of Engineering, Robert Gordon University, Aberdeen; Riverside East, Garthdee Road, Aberdeen AB10 7GJ, United Kingdom

Email: behzad.nikfar@studenti.polito.it; j.njuguna@rgu.ac.uk

Abstract. One of the problems in the design of automotive structures and body parts made by fibre reinforced composites is that these materials are susceptible to a small energy impact caused by for instance, accidental tool drop during maintenance or stone strike while in operation. This often lead to a barely visible impact damage which causes reduction in compressive strength of the composite part. To increase the impact tolerance of the composites, toughening agents like silica nanoparticles and rubber particles can be utilized to toughen the resin. To understand the effect of the particles enhancement, the impact tolerance was evaluated utilizing Compression After Impact (CAI) test after the impact induced by gas-gun impacting equipment. The results from CAI test after 20 J impact (high energy stone strike) shows about 30% improvement in residual compressive strength for the nanosilica enhanced composite compared to unmodified CFRP. Also C-scan results after 7 J impact shows about 50% smaller delamination area for the nano-enhanced composite.

1. Introduction

Polymer composites but composites can have very good energy absorption even better than metals if properly tailored. However, one of the problems in the design of structures and body parts made by carbon fibre reinforced composites (CFRC) is that these materials are susceptible to a small energy impact accidentally which may cause delaminations through the thickness. Dropped tool or stone strike can cause low-energy impact on race cars body and structures which results in significant reductions in residual compressive strength. The induced damage is called "barely visible impact damage" (BVID) and is a potential source of mechanical weakness of chassis and structures and reduces aerodynamic efficiency of the body parts (like the rear wing) since the efficiency of the different aerodynamic devices depends on boundary layers at the surface and the part's strength for inducing downforce [1].

To evaluate the composite's performance after a BVID caused by an accidental impact, the compression after impact (CAI) test is used. To improve the damage tolerance of structures, the constituents of the composite and their effects on impact tolerance should be studied. Most of the experiments to improve damage tolerance of composites were concentrated on either producing high strength fibres or toughening the resin matrix [1]. The effect of toughening agents is to increase the work of fracture within the composite. One of the methods to increase toughness of resin matrix like epoxy is to add nanoparticles like SiO2 nanoparticles with rubber particles to the resin matrix. This modifications of epoxy resins could cause some superior properties improvements like broadening of the glass transition temperatures, modest increases in the glassy modulus, low dielectric constant, and significant increases in key mechanical properties [2].

The experimental data presented here are from an investigation of the toughening effect of nanosilica particles on carbon fibre reinforced epoxy on damage caused by low and high velocity impacts. The delamination area was measured using C-scan ultrasonic techniques. Finally, the effects of the toughening agents in the matrix of the composite and the single and double step impacts on the CAI strength were investigated.

2. Experimental

2.1. Materials and Sample Manufacturing

Nanosilica (silica particles of the order 20nm diameter) enhanced epoxy-carbon fibre-reinforced samples and unmodified epoxy ones were manufactured. The two type resins were cured with a hardener and manufactured by vacuum injection moulding. The resins and hardener were blended together and the mixture was then injected into a mould using a vacuum at 70°C and cured in an oven for 2 hrs at 160°C and then rested at 40°C for 8 hrs to form the composites flat plates of 4 mm thickness. 8 unidirectional carbon fiber plies were used with stacking sequence of $[45^{\circ}/0^{\circ}/-45^{\circ}/90^{\circ}]_S$ as indicated in ASTM D7137 standard for CAI test. The volume fraction of the fibre was 55.7% (matrix volume fraction of 44.3%). The composite plates were then cut into 150x100 mm sections for impact testing as required in ISO 18352.

2.2. Impact Testing

Impact testing gives an indication of how a material will perform during a dynamic loading condition. The impact to be simulated is a stone strike vehicular composite structure. Experimental impact tests are carried out using a gas gun. Gas gun impact tester employs a 10 mm hemi-spherical 22 g projectile (simulating flying granite stone impact scenario) and has been conducted by adjusting the pressure in the gun to reach the desired impact velocity at 90 degrees angle to the impact surface. Single step impacted sample were impacted only at 20 J. Double step impacted samples were impacted in two steps: first step (7.5 J) and second step (20 J). For consistency, five samples were tested in each test. Ultrasonic C-scans were used to evaluate the damage.

2.3. Compression after Impact (CAI) Test

Residual compressive strength of polymer composites laminates damaged by an impact was evaluated using ASTM D7137 standard. This test is valid for multidirectional continuous fibre reinforced composite laminates which are symmetric and balanced with respect to the test direction [4]. A uniaxial compression test is performed using a balanced, symmetric laminated plate, which has been damaged and inspected prior to the application of compressive force. The damage state is imparted through out-of-plane loading caused by ballistic gas gun impact.

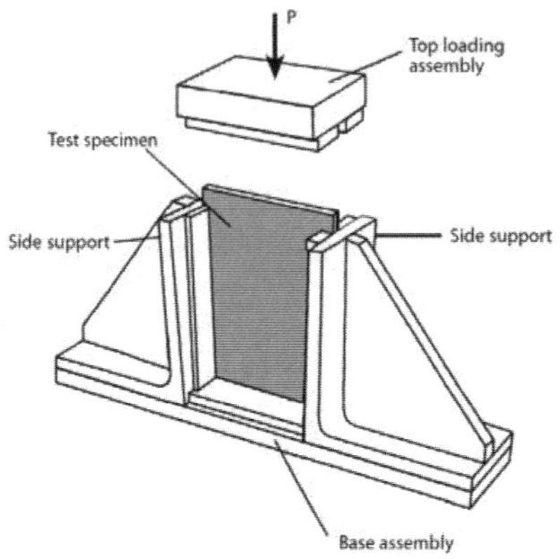

Figure 1. Schematic of compressive residual strength support fixture with specimen in place.

3. Results and Discussion

3.1. Impact results

Typical results from the C-scan results for epoxy/CFRC and SiO_2/rubber/epoxy CFRC after the first step impact simulating the low energy tool drop impact can be seen in Figure 3 and 4. The blue-white area indicates damage in that area which is the hit area of the projectile. Probably in the white area fibre breakage and delaminations exist in almost all interfaces through the thickness of the laminate and the blue area corresponds to splitting and delaminations of the back-face rather than internal damage. The areas with yellow colour indicate some possible additional damages. It can be obviously seen in the C-scan images that the delamination area diameter for the nanosilica enhanced samples is significantly smaller than that for the non-nano enhanced samples. Greater fracture energy is required for delamination initiation and growth for toughened epoxy composite.

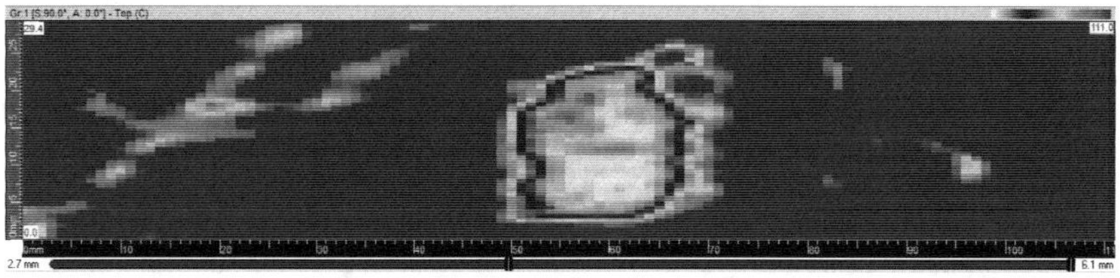

Figure 3. Ultrasonic C-scan for epoxy/CFRC typical sample.

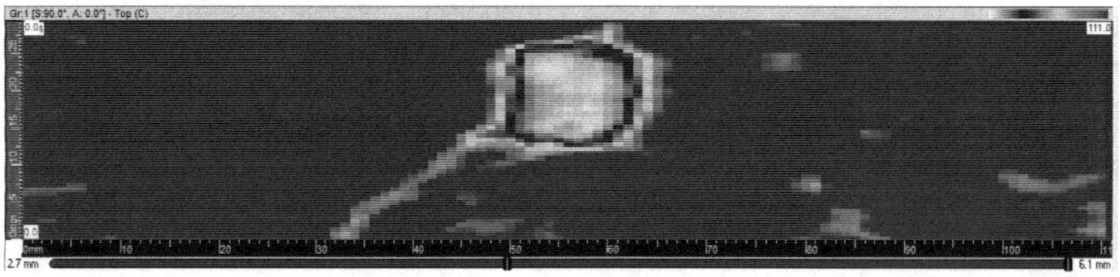

Figure 4. Ultrasonic C-scan images for SiO_2/rubber/epoxy CFRC typical sample.

Futher examination of the impacted specimens after second impact and although visible damages were observed after first impact to the naked eye inspection, both samples suffered from cracks damage at second impact. It is well known that the first step of the damage caused by an impact is matrix cracks which are generated by shear or tensile stresses mainly in the intermediate or back-wall layers, then delaminations grow from the crack tips between layers (Figure 5). This would mean that after the high energy second step impact, the damage area had increased and the cracks that were initiated in the back-wall layers had developed until the front-face layers due to significant delamination.

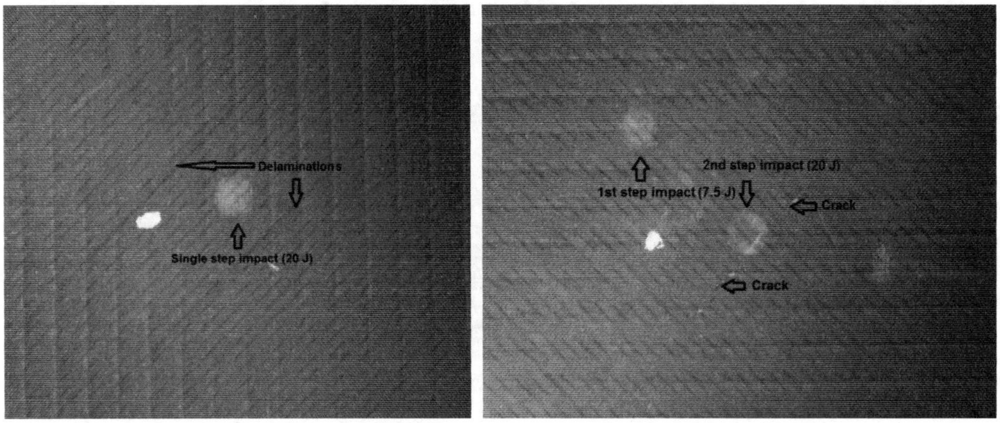

Figure 5. Typical images for epoxy/CFRC (left) and SiO_2/rubber/epoxy CFRC (right) samples demonstrating delamination visible by naked eye following impact tests.

3.2. Compression after Impact (CAI) test

The CAI test average results for single step impacted samples are listed in Table 1.

Table 1. Compression after impact results of the single step impacted samples .

Sample	F (kN)	CAI (MPa)	Impact Energy (J)
Epoxy/CFRC	53.348	133	20
SiO_2/rubber/epoxy CFRC	54.91	137	20

Comparing the nano enhanced samples and the unenhanced ones after the single step high velocity impact; no improvement in the residual compressive strength is observed which shows that the nanosilica enhancement and toughening is not useful in high energy impacts. That is because high velocity impact properties are fibre architecture dominant [5] and since the fibre lay-up for both types of samples are same; the residual compressive strength after the high velocity impact is almost equal. The CAI test average results for double step impacted samples after the second step impact are listed in Table 2. It can be understood from Table 2 that the nanosilica enhanced samples got more CAI strength than the SiO_2/rubber/epoxy CFRC samples after the second step high energy impact. As it was conclude from the C-scan results, using SiO_2 nanoparticles with rubber particles for matrix enhancement makes the epoxy/CRFC more damage tolerance and less susceptible to the low energy impact damage such as that caused by a tool drop on the structure.

Table 2. Compression after impact results of the double step impacted samples.

Sample	F (kN)	CAI (MPa)	1st Step Energy (J)	2nd Step Energy (J)
Epoxy/CFRC	43.833	109.25	7.5	20
SiO_2/rubber/epoxy CFRC	57.252	142.5	7.5	20

The present results show a greater influence of resin toughness over impact damage extent, than over growth of delaminations in compression. Comparing the average residual compressive strength of the nanosilica enhanced samples after the double step impact (142 MPa) with one for the standard samples (109 MPa), about 30% improvement in the strength can be observed. While comparing the average delamination area of the nanosilica enhanced samples ($\pi 14.2^2$ mm^2) with the standard ones ($\pi 20.6^2$), there is about 50% smaller delamination area for the nanosilica enhanced composite. These results agree to those of Cartie & Irving [6]. Considering the residual compressive strength of the SiO_2/rubber/epoxy CFRC samples after the single step impact and the double step one, it can be observed that SiO_2/rubber/epoxy CFRC samples after the single step impact have got less compressive strength than the double step impacted ones which is not expected [7,8]. Probably the damage caused by the low energy first step impact has a damping effect against the second step high energy impact for nanosilica and rubber enhanced composite. This effect is seen just in the enhanced samples and not in the unenhanced ones. Therefore this effect is emerged because of nano and rubber particle enhancement that increase toughness and impact resistance of the composite matrix, but need to be investigated further.

4. Conclusions

This study investigated impact tolerance of carbon fibre reinforced epoxy composite and the effect of silica nanoparticles and rubber particles enhancement on improvement of the impact resistance. The impact tolerance was evaluated utilizing compression after impact test after the impact by Gas-gun impacting equipment. Conclusions derived are listed below:

1. Ultrasonic C-scan of the specimens impacted at 7J shows that nanosilica and rubber enhanced specimens have about 50% smaller delamination area than the unenhanced specimens.
2. Compression after impact test results for single-step impacted specimens' show that nanosilica and rubber enhancement is not effective for impact resistance against high velocity impacts since high velocity impact properties are fibre architecture dominant and matrix properties has little effect.
3. Comparing the CAI test results for the double step impacted nano-enhanced and standard CFRP samples, about 30% improvement in residual compressive strength for the nanosilica and rubber particles enhanced composite is obtained for low velocity impacts. This impact resistant enhancement will help the rear wing to be more damage tolerant against low energy impact and keeps the aerodynamic efficiency of the part as high as possible.

References

[1] Cox P C (McLaren), McGrath G C (TWI) and Savage G M (McLaren) 1993 Designing with fibre reinforced composites to provide energy absorption mechanisms in damage tolerant Formula 1 cars, *ICCM*-**9**

[2] Njuguna J, Pielichowski K and Alcock J R 2007, Epoxy-Based Fibre Reinforced Nanocomposites, *Adv Eng Mater* **9**, No. 10, : 835-847.

[3] Shen Q, Omar M, Dongri S 2012, Ultrasonic NDE Techniques for Impact Damage Inspection on CFRP Laminates, *J Mater Sci Resear* **Vol. 1**, No. 1.p2.

[4] Standard Test Method for Compressive Residual Strength Properties of Damaged Polymer Matrix Composite Plates, *ASTM D 7137/D 7137M* – 07.

[5] Kinloch A J, Masania K, and Taylor A C 2009, The Fracture of Nanosilica and Rubber Toughened Epoxy Fibre Composites. *COMPOSITES & POLYCON 2009 American Composites Manufacturers Association*

[6] Cartie D D R, Irving P E 2002, Effect of resin and fibre properties on impact and compression after impact performance of CFRP, *Elsevier Composites: Part A* **33** 483-93.

[7] Njuguna, J (Ed.). *Structural Nanocomposites: Perspectives for Future Applications.* Springer Berlin Heidelberg. 2013.

[8] Njuguna J, Silva F, Sachse S. Nanocomposites for vehicle structural applications; In Nanofibers - Production, Properties and Functional Applications, Tong Lin (Ed.), 2011. ISBN: 978-953-307-420-7, DOI: 10.5772/23261.

IOP Conf. Series: Materials Science and Engineering **64** (2014) 012010 doi:10.1088/1757-899X/64/1/012010

Mechanical and moisture barrier properties of titanium dioxide nanoparticles and halloysite nanotubes reinforced polylactic acid (PLA)

J Alberton[1], **S M Martelli**[1], **F M Fakhouri,**[2,2] **and V Soldi**[1]
1 - Department of Chemistry, Federal University of Santa Catarina, Florianópolis, Brazil
2 - Faculty of Engineering, Federal University of Grande Dourados, Dourados, Brazil
3 - Faculty of Chemical Engineering, Campinas State University, Campinas, Brazil
E-mail: smmartelli@gmail.com; farayde@gmail.com

Abstract. Polylactic acid (PLA) has been larger used in biomedical field due to its low toxicity and biodegradability. The aim of this study was to produce PLLA nanocomposites, by melt extrusion, containing Halloysite nanotubes (HNT) and/or titanium dioxide (TiO_2) nanoparticles. Immediately after drying, PLLA was mechanically homogenized with the nanofillers and then melt blended using a single screw extruder (L/D = 30) at a speed of 110 rpm, with three heating zones in which the following temperatures were maintained: 150, 150 and 160ºC (AX Plásticos model AX14 LD30). The film samples were obtained by compression molding in a press with a temperature profile of 235 ± 5ºC for 2.5 min, after pressing, films were cooled up to room temperature. The mechanical tests were performed according to ASTM D882-09 and the water vapor permeability (WVP) was measured according to ASTM E-96, in triplicate. The tensile properties indicated that the modulus was improved with increased TiO_2 content up to 1g/100g PLLA. The Young's modulus (YM) of the PLA was increased from 3047 MPa to 3222 MPa with the addition of 1g TiO_2/100g PLLA. The tensile strength (TS) of films increases with the TiO_2 content. In both cases, the YM and TS are achieved at the 1% content of TiO_2 and is due to the reinforcing effect of nanoparticles. Pristine PLA showed a strain at break (SB) of 3.56%, while the SB of nanocomposites were significant lower, for instance the SB of composite containing 7.5 g HNT/100g PLLA was around 1.90 %. The WVP of samples was increased by increasing the nano filler content. It should be expected that an increase of nanofiller content would decrease the mass transfer of water molecules throughout the samples due to the increase in the way water molecules will have to cross to permeate the material. However, this was not observed. Therefore, this result can be explained considering the molecular structure of both fillers, which contain several hydroxyl groups in the surface, making the end material more hydrophilic.

1. Introduction

Polymer nanocomposites refer to multiphase materials where at least one of the constituent phases, com-monly the nanofiller, has at least one dimension in thenanoscale range (<100 nm) [1]. The advantages of nancomposites over traditional composites are numerous, including but not limited to enhanced mechanical properties, gas barrier, increased flame retardancy and thermal stability. This can be accomplished at low loading levels due to the large surface area and high aspect ratio of the nanoparticles. This effect greatly depends on the shape/aspect ratio of the particle, the dispersion of the nanoparticles in the polymer matrix, and the interaction between the filler and the bulk polymer .

Polylactide (PLA) is a biodegradable and bioresorbable polymer produced from corn. Because of its characteristics (mechanical properties, renewability, biodegradability and relatively low cost), it is successfully used in the packaging and textile industries as well as in medicine [2]. For many

[2] To whom any correspondence should be addressed.

applications of PLA, for example, as the bone internal fixation material, its mechanical performance should be improved in order to meet the requirement for strength and modulus [3]. Many types of nanoparticles, such as carbon nanotubes (CNT), hydroxyl-apatite (HA), nanoclay, silica, and zinc oxide, were incorporated into PLA to improve the physical and mechanical properties [4]

Halloysite (HNT) is a naturally occurring polymorph of kaolinite, $Al_2Si_2O_5(OH)_4.2H_2O$, with a dominant hollow tubular morphology [4, 5]. HNT has two different interlayer surfaces. The external surface is predominantly covered by siloxane (Si–O–Si) groups and some silanol (Si–OH) and aluminol (Al–OH) groups that are exposed at the edges of the tube, whereas aluminol groups are located in the internal surface [5]. HNTs have been widely used during recent years to reinforcing the polymer matrices such as epoxy resins [6] polypropylene, polyamide, polyvinyl alcohol and styrene rubber. In addition, HNTs is far less expensive than carbon nanotubes (CNTs) and can be mined from deposits as a raw material [6]. Titanium dioxide (TiO_2) belongs to the family of transition metal oxides [7]. TiO_2 is an inexpensive and chemically stable inorganic filler. When irradiated with ultraviolet (UV) light, TiO_2 has great potential as a photocatalyst. Besides the afore mentioned advantages, the presence of nanofillers can also control the biodegradation rate [8].

In this study, PLA filled HNT and/or TiO_2 nanocomposites were prepared through melt blending and characterized with the aim of to assess the mechanical and barrier properties of PLA films containing these nanoparticles.

2. Material and Methods

Poly(L-Lactide) (PLLA) was supplied by Cargill Dow SA (Minnetonka, MN, EUA) under commercial name NatureWorks® PLA *polymer* 2002D. The PLLA is composed by > 95% de L-lactide and < 5% de D-lactide. PLLA pellets have been dried for 2 hs at 90 °C before use. Titanium dioxide (TiO_2, anatase, 79.87 g/mol, nanopowder, <25 nm particle size, 99.7% trace metals basis) and Halloysite nanoclay ($Al_2Si_2O_5(OH)4.2H2O$, 294.19 g/mol, diam. x L 30-70 nm × 1-3 µm , surface area of 64 m^2/g, nanotube) were supplied by Sigma-Aldrich (Saint Louis, MO, EUA). Thirteen nanocomposites formulations were prepared according with Table 1.

PLLA nanocomposites were prepared according with the composition presented in Table1. Immediately after drying, PLLA was mechanically homogenized with the nanofillers and then melt blended using a single screw extruder (L/D = 30) at a speed of 110 rpm, with three heating zones in which the following temperatures were maintained: 150, 150 and 160ºC (AX Plásticos model AX14 LD30). After extrusion, PLLA nanocomposites powder were obtained by grinding the polymer using a mill (Marconi, model MA 048). The resulting material was compression molded (235±10ºC for 5 min) using a mold (Figure 1) in an thermoforming press (Ico Comercial SA (PHS Model 15t) with pressure control between 0-15 tones. Films were then cooled to room temperature and stored for further characterization. The mechanical tests were performed according to ASTM D882-09 and the water vapor permeability (WVP) was measured according to ASTM E-96, in triplicate.

Table 1. Formulations of nanocomposites based on PLLA

Sample	Nanoparticles (g/100g PLLA)	
	TiO_2	HNT
1	0	0
2	0.5	0
3	1.0	0
4	2.5	0
5	5.0	0
6	7.5	0
7	0	0.5
8	0	1.0
9	0	2.5
10	0	5.0
11	0	7.5
12	0.25	0.25
13	0.50	0.50

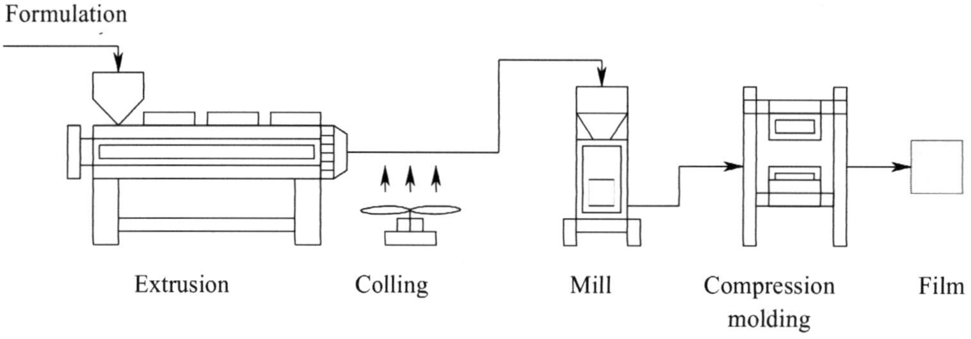

Figure 1. Film production

3. Results and Discussion

Mechanical and barrier properties of nanocomposites are listed in Table 2. As it can be seen, the Young's modulus (YM) of composites had a significant increase ($p \leq 0.05$) with the addition of TiO_2 up to 2.5g TiO_2/100 g PLLA. Regarding the TS, the better results were also achieved when adding 2.5g TiO_2/100 g PLLA (an increase of 13.5 % in relation to pristine PLLA). When adding more TiO_2 in the matrix the SB was drastically reduced (up to 39.3% for sample containing 7.5g TiO_2/100 g PLLA. The YM also presented an increase up to a 2.5g of filler /100g PLLA. The TS and YM increased up to 2.5g/100 g PLLA load and then decreased. Similar results were reported by Zhu et al.

(2011) where the TS of PLA composites increased up to a limit of 2 g TiO_2/100g of PLA, due to the reinforcing effect of nanoparticles and after this load the TS was reduced (5 and 10 g TiO_2/100g PLA), probably related to the aggregation of the nanoparticles in the polymer matrix [9].

Table 2. Mechanical Properties of PLA/TiO_2/HNT nanocomposites

Sample PLA/TiO_2/HNT	Tensile Strenght (MPa)	Strain at break (%)	Young's modulus (MPa)	WVP (g mm/m^{2}d kPa)
100/0/0	$53.66^f \pm 0.84$	$3.56^i \pm 0.08$	$3048^f \pm 43$	1.88
100/0.5/0	$57.69^{g,h} \pm 1.18$	$3.36^h \pm 0.09$	$3111^f \pm 35$	1.99
100/1.0/0	$58.28^h \pm 0.83$	$3.00^f \pm 0.08$	$3237^g \pm 31$	3.24
100/2.5/0	$57.10^{g,h} \pm 0.99$	$2.61^d \pm 0.13$	$3335^h \pm 39$	3.35
100/5.0/0	$56.49^g \pm 0.89$	$2.35^c \pm 0.10$	$3213^g \pm 32$	0.96
100/7.5/0	$52.74^f \pm 0.75$	$2.16^b \pm 0.10$	$2945^e \pm 33$	0.89
100/0/0.5	$43.06^e \pm 0.64$	$3.21^g \pm 0.09$	$2099^a \pm 48$	3.52
100/0/1.0	$40.06^d \pm 1.09$	$2.84^e \pm 0.07$	$2140^{a,b} \pm 49$	3.62
100/0/2.5	$38.48^c \pm 0.85$	$2.52^d \pm 0.10$	$2181^{b,c} \pm 50$	--
100/0/5.0	$36.54^b \pm 1.09$	$2.12^b \pm 0.11$	$2228^c \pm 40$	--
100/0/7.5	$32.77^a \pm 1.17$	$1.90^a \pm 0.08$	$2298^d \pm 48$	--
100/0.25/0.25	$41.40^d \pm 0.94$	$2.50^d \pm 0.09$	$2220^c \pm 40$	2.88
100/0.50/0.50	$38.55^c \pm 0.93$	$2.78^e \pm 0.10$	$2335^d \pm 53$	3.24

**Means and standard deviations of triplicates.
a-h Mean values in the same column with different letters are significantly different ($p \leq 0.05$).

The addition of HNT in the PLLA matrix did not provided any improvement in the mechanical properties of nanocomposites. The results showed a significant decrease in the TS, SB and YM values of samples. Studies of PLA composites prepares by injection molding showed an improvement of TS and YM properties when HNT was adding to the polymer [4, 10], which is different from the results obtained by the films prepared by hot pressing. As the shear rate in the hot pressing are extremely low compared to injection molding, the mechanical properties observed in the films can be related to the random orientation of the HNT into the PLLA matrix. In one of the references cited above [10], the results also showed a slight reduction in SB of composites when increasing the nano filler content. The addition of both TiO_2 and HNT did not presented a positive effect on the mechanical properties of composites.

The WVP of films loaded with TiO_2 increased up to a load of 2.5 g/100g PLLA and then decreased. The WVP of samples containing HNT increased by adding the nano filler. It should be expected that an increase of filler content would decrease the mass transfer of water molecules throughout the samples due to the increase in the way water molecules will have to cross to permeate the material. However, this was not observed. Therefore, this result can be explained considering the molecular structure of both fillers, which contain several hydroxyl groups in the surface, making the end material more hydrophilic.

4. Conclusions

In this study PLLA was filled with two different nano fillers, TiO_2 (nanoparticles) and HNT (nanotubes). The results showed significant differences in the mechanical and WVP properties of PLLA composites when loading different types and amounts of fillers. Comparing the TiO2 and HNT, we could observe a higher performance of TiO_2 as filler for PLLA, in relation to the mechanical properties of the films.

5. References

[1] J.-M. Raquez, Y. Habibi, M. Murariu, P. Dubois, Progress in Polymer Science, 2013, 38 1504-1542.
[2] R. Fryczkowski, B. Fryczkowska, W. Biniaś, J. Janicki, Composites Science and Technology, 2013, 89 186-193.
[3] K. Fukushima, D. Tabuani, G. Camino, Materials Science and Engineering: C, 2009, 29 1433-1441.
[4] M. Liu, Y. Zhang, C. Zhou, Applied Clay Science, 2013, 75–76 52-59.
[5] L.N. Carli, T.S. Daitx, G.V. Soares, J.S. Crespo, R.S. Mauler, Applied Clay Science, 2014, 87 311-319.
[6] P. Pasbakhsh, H. Ismail, M.N.A. Fauzi, A.A. Bakar, Applied Clay Science, 2010, 48 405-413.
[7] O. Carp, C.L. Huisman, A. Reller, Progress in Solid State Chemistry, 2004, 32 33-177.
[8] A. Buzarovska, A. Grozdanov, M. Avella, G. Gentile, M. Errico, J. Appl. Polym. Sci., 2009, 114 3118-3124.
[9] Y. Zhu, G.G. Buonocore, M. Lavorgna, L. Ambrosio, Polymer Composites, 2011, 32 519-528.
[10] K. Prashantha, B. Lecouvet, M. Sclavons, M.F. Lacrampe, P. Krawczak, J. Appl. Polym. Sci., 2013, 128 1895-1903.

IOP Conf. Series: Materials Science and Engineering **64** (2014) 012011 doi:10.1088/1757-899X/64/1/012011

Fabrication of nanofiber mats from electrospinning of functionalized polymers

Burcu Oktay[1], Nilhan Kayaman-Apohan[1], Serap Erdem-Kuruca[2]

1 Marmara University, Department of Chemistry, 34722 Göztepe-Istanbul, Turkey
2Istanbul University, Istanbul Medical Faculty, Department of Physiology, 34390 Çapa-Istanbul, Turkey
burcu_chemist@hotmail.com

Abstract. Electrospinning technique enabled us to prepare nanofibers from synthetic and natural polymers. In this study, it was aimed to fabricate electrospun poly(vinyl alcohol) (PVA) based nanofibers by reactive electrospinning process. To improve endurance of fiber toward to many solvents, PVA was functionalized with photo-crosslinkable groups before spinning. Afterward PVA was crosslinked by UV radiation during electrospinning process. The nanofiber mats were characterized by scanning electron microscopy (SEM). The results showed that homogenous, uniform and crosslinked PVA nanofibers in diameters of about 200 nm were obtained. Thermal stability of the nanofiber mat was investigated with thermal gravimetric analysis (TGA). Also the potential use of this nanofiber mats for tissue engineering was examined. Osteosarcoma (Saos) cells were cultured on the nanofiber mats.

1. Introduction

Electrospinning is a versatile and easy technique for preparation of nanofiber network structure. In the electrospinning setup consist of three main equipment; a syringe pump, a high voltage source and a ground collector. Basically, a polymer solution (or melt) is loaded into the syringe. The polymer solution pushes out of the syringe with constantly rate via the pump. Simultaneously, an electrical field is applied by the high voltage source [1]. The stable electrospinning jet composes of four region; the base, the jet, the splay and the collection. The shape of the base depends on the surface tension of liquid. The charging of the jet occur the base. The shape is called as the Taylor cone .With increasing electrical field, charging the pendant drop of the polymer liquid takes on a conical shape. Additionally, begin to evaporate and diameter of the jet decrease. In the next region, the jet forms to the splay into many small fibers [2]. Fiber diameters are affected by various parameters; flow rate, applied voltage, tip to collector ratio, molecular weight, viscosity, surface tension, solution conductivity and ambient parameters [3].

Nanofibers have exciting a new class of material which used for a wide range of applications from medical to consumer products such as, filtration, wound dressing, adsorbent, energy storage, protein separation, immobilization, drug delivery and composites. The main features of nanofibers are large surface area–to-volume ratio, good mechanical, strength, excellent flexibility, and high porosity. These properties make them suitable for many applications. [4,5]

Crosslinked nanofiber mats have been prepared with many processes. One of the processes is a reactive electrospinning. In this process, the photo-crosslinkable polymers are in-situ crosslinked with UV radiation during electrospinning process [6]. As it is known, poly(vinyl alcohol) nanofibers dissolve in many solvent due to its hydroxyl groups in structure. Our aim is to prepare insoluble PVA nanofiber mats. For this purpose, PVA was modified by methacrylate groups and then crosslinking of PVA was conducted by UV light attached electrospinning setup. The thermal properties, the morphological properties and the cell adhesive properties of the resultant nanofiber mats were characterized.

IOP Conf. Series: Materials Science and Engineering **64** (2014) 012011 doi:10.1088/1757-899X/64/1/012011

2. Experimental Details

2.1. Materials

Poly(vinyl alcohol) (Molecular weight 146000-186000, degree 99+% hydrolyzed), glycidyl methacrylate (97%) were purchased from Sigma Chem. Co. Freshly double distilled water was used throughout the experimental work.

2.2. Characterization

The surface morphology of the nanofibers was examined with scanning electron microscopy (SEM) on Philips XL30 ESEM-FEG/EDAX. Thermal degradation of the nanofiber mat was investigated using a Perkin–Elmer Thermogravimetric analyzer Pyris 1 TGA model.

2.3. Electrospinning process

PVA was functionalized by glycidyl methacrylate according to literature [7]. The methacrylated PVA solution was prepared by dissolving in distilled water at a concentration of % 10. The methacrylated PVA solution was electrostatically spun with in situ UV cured electrospinning. A nanofiber web was formed under following conditions: flow rate 3 mL/h, tip to collector distance 15 cm, voltage 25 kV. The reaction scheme was shown in Figure 1.

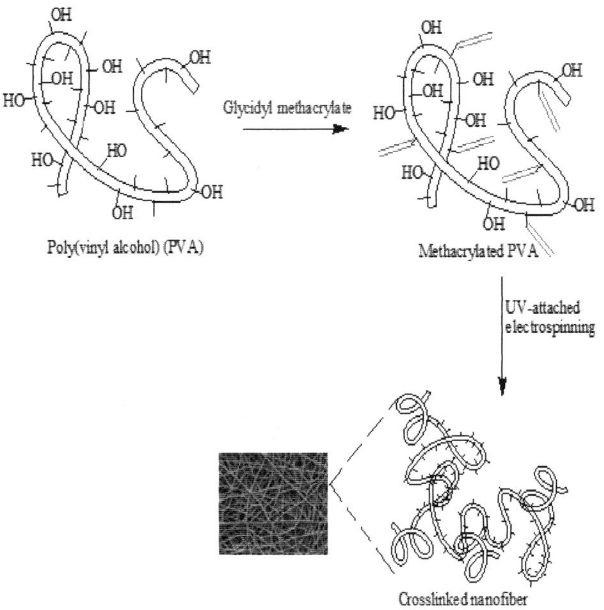

Figure 1: Preparation scheme of PVA based on nanofibers

3. Results and Discussions

The chemical structure of methacrylate modified PVA was confirmed with FT-IR. The new band observed at 1650 and 1715 cm^{-1} relating to methacrylate groups of PVA Also the characteristic band of PVA were occurred at 2090 and 1020 cm^{-1} due to C–H stretching and C–O stretching, respectively (figure not shown) [8].

IOP Conf. Series: Materials Science and Engineering **64** (2014) 012011 doi:10.1088/1757-899X/64/1/012011

The morphology of both the crosslinked and the cell seeded nanofiber mats were investigated by SEM. Morphological results show that, homogenous network of the crosslinked nanofiber with diameters in the range of 150- 200 nm was obtained (Figure 2). In figure 3 it can be seen, the linear PVA fibers were covalently crosslinked with each other.

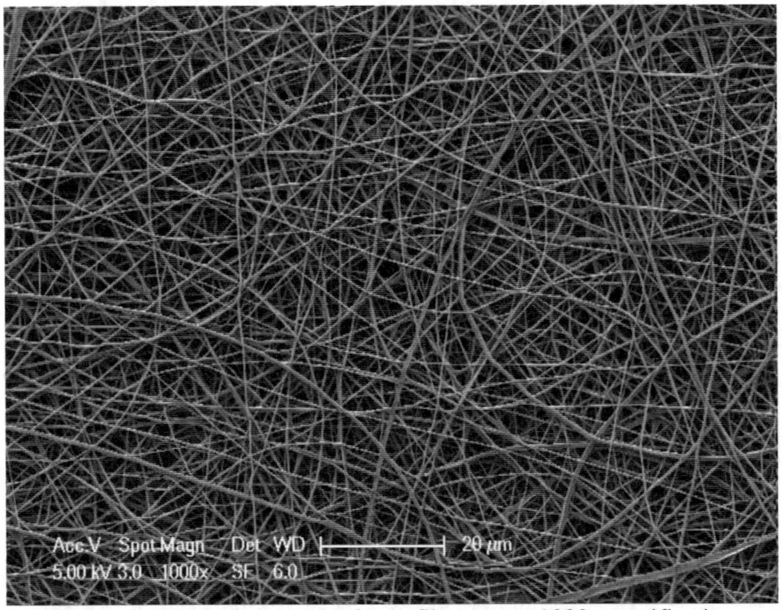

Figure 2: SEM micrograph of nanofiber mat at 1000 magnification

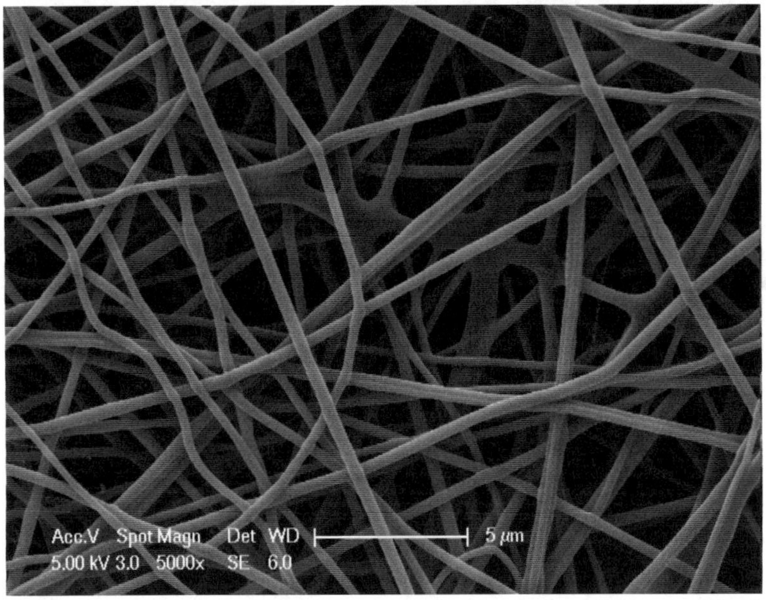

Figure 3: SEM micrograph of crosslinked nanofiber mat at 5000 magnification

The nanofibers were cultured with Saos (Osteosarcoma) cells during 24 hours in DMEM (Dulbecco's modified Eagle's medium). The cells adhered and spread on the nanofiber. The results showed that the nanofibers have the cellular compatibility for Saos cells (Figure 4).

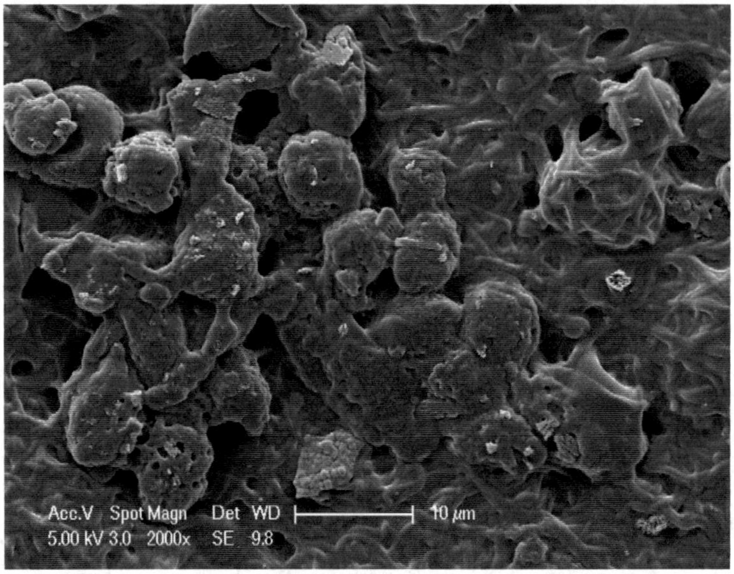

Figure 4: Saos cell culture of nanofiber mat examination

Thermal stability of crosslinked PVA nanofiber was investigated by TGA. As shown in Figure 5, TGA thermogram displayed two main decompositions. First weight loss was observed at around 280 °C due to decomposition of side chain of PVA. Second weight loss was monitored around 550 °C due to decomposition of main chain of PVA.

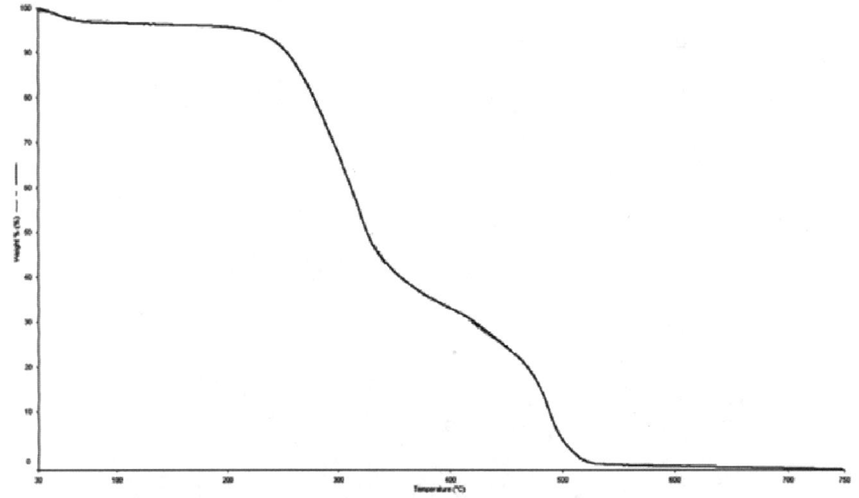

Figure 5: TGA curve of functionalized PVA at air atmosphere

Conclusion

In this study, we crosslinked nanofiber mats were prepared by electrospinning. The nanofibers became enduring toward to many solvents with UV radiation along electrospinning. Also, Osteosarcoma cells were cultured on the crosslinked nanofiber mats. The mats were characterized by SEM, TGA and FT-IR spectra. The results showed that, the crosslinked nanofiber mats could be used as a biomaterial a wide range.

Acknowledgements:

This work was supported by Marmara University, Commission of Scientific Research Project (M.U.BAPKO) under grant FEN-C-DRP-150513-0179

REFERENCES

[1] Baştürk E, Demir S, Danış Ö and Kahraman MV 2012 *J Apply Polym Sci* **127** 349.
[2] Pham QP, Sharma U and Mikos AG 2006 *Tissue Eng* **12** 1197.
[3] Jayesh Doshi J and Reneker DH *J of Electrostat* 1995 **35** 151.
[4] Oktay B,Baştürk E, Apohan NK, Kahraman MV *Polym Com* 2013 **34** 1321
[5] Wnek G, Marcus M, Simpson D and Bowlin G 2003 *NanoLeters*, **3** 213.
[6] Yuan J, Mo H, Wang M, Li L, Zhang J and Shen J 2012 *J Apply Polym Sci* **124** 1067
[7] Cavalieri F, Miano F, D'Antona P, and Paradossi G 2004 *Biomacromolecules* **5** 2439.
[8] Bastürk E, Kahraman MV 2012*Poly Composite* **33** 829.

IOP Conf. Series: Materials Science and Engineering **64** (2014) 012012 doi:10.1088/1757-899X/64/1/012012

New products made with lignocellulosic nanofibers from Brazilian amazon forest

L Bufalino[1], L M Mendes[1], G H D Tonoli[1], A Rodrigues[2], A Fonseca[1], P I Cunha[2], J M Marconcini[3]

[1]Department of Forestry Science, Federal University of Lavras, Campus Universitário, POB 3037, Lavras, MG, Brazil
[2]Department of Material Engineering, Federal University of São Carlos, Washington Luís road, Km 235, São Carlos, SP, Brazil
[3]Instrumentation Unit of Brazilian Agriculture Research Corporation – EMBRAPA, Quinze de Novembro st, 1452, São Carlos, SP, Brazil

E-mail: linabufalino@yahoo.com.br

Abstract. The biodiversity of the Amazon forest is undoubtedly rich; hence there is considerable variety of plant fibers regarding their morphological, chemical and structural properties. The legal exploration of the Brazilian Amazon is based on sustainable management techniques, but the generation of a relevant amount of plant wastes still can`t be avoided. The correct destination of such materials is a challenge that Brazilian companies have to face. In this context, the National Council of Science and Technology (CNPq) promoted the creation of investigation nets on sustainability of Brazilian agribusiness. The Brazilian Net on Lignocellulosic Composites and Nanocomposites was then created, with partnership between several national and international research institutions. Until the moment, the results showed that Amazon plant fibers that are discarded as residues have great potential to nanofiber production. Nanopapers with considerable high mechanical and physical strength, proper opacity and great crystalline index were produced by using a clean and simple mechanical method. Those materials are candidates to several uses such as packaging, substrates transparent conductive films, gas barrier films, solar cells and e-papers.

1. Introduction

Lately, there is a genuine interest and need to decrease society`s dependency on petroleum bio-based products. Inspired by this environmental awareness and new standards, the development of biodegradable materials with improved performance is crucial [1, 2].

This context, combined with new technology development in nanotechnology field, strongly favors the use of cellulose. The annual biomass production of cellulose in nature is about 1 trillion tons, making it a virtually inexhaustible source of raw material that may be obtained from non-wood plants, woods, bacteria and tunicate forms [3, 4].

Since the biodiversity of the Amazon forest is undoubtedly wide, this tropical environment certainly offers an enormous potential to provide cellulose sources. It should be mentioned that the legal exploration of the Brazilian Amazon is based on sustainable management techniques, but the

IOP Conf. Series: Materials Science and Engineering **64** (2014) 012012 doi:10.1088/1757-899X/64/1/012012

generation of a relevant amount of plant wastes still can`t be avoided. In this context, the National Council of Science and Technology (CNPq) promoted the creation of investigation nets on sustainability of Brazilian agribusiness. The Brazilian Net on Lignocellulosic Composites and Nanocomposites was then created, with partnership between several national and international research institutions. One of the goals of this partnership is to investigate the potential of wastes from wood processing for the production of nanopapers. Those materials are candidates to several uses such as packaging, substrates transparent conductive films, gas barrier films, solar cells and e-papers. Partial results from such efforts are given in this work.

2. Methodology

Veneers for saw wood of three certified Amazonian species were selected, namely *Cordia goeldiana, Brosimum parinarioides*, and *Parkia gigantocarpa*. Those species are legally used by a Brazilian panel company for tropical plywood production within the terms of sustainable management of the Amazon forest. Low-quality veneers for plywood production from the four species were reduced in a hammer mill and transformed into sawdust in a disintegrator.

2.1. Cellulose isolation

The sawdust was previously submitted to chemical modification through two stages: 1) Alkali treatment with NaOH at 5% in a digester, at 150°C for 30 min; 2) Bleaching with H_2O_2 at 24% and NaOH at 4% under mechanical stirring at 60°C for 2 h.

2.2. Nanopaper production

The bleached fibers were immersed in distilled water at a proportion of 1% (m:m). The fibers were then submitted to fibrillation in a grinder resulting in a suspension of structures known as microfibrillated celluloses (MFC`s). After 40 cycles the suspensions were poured on Petri dishes and the nanopapers were formed after about 5 days.

2.3. Nanopaper and MFC`s properties

The following analyses are presented:

- Scanning electron microscopy (SEM) investigation: The SEM micrographs of the *C. goeldiana* MFC's and raw sawdust were taken using a JEOL scanning electron microscope (SEM) model JMS 6510 with an acceleration voltage of 10 kV. The samples were previously sputter coated with gold before examination;

- Strength to water deterioration: Four 2 cm – diameter samples were obtained from the nanopapers of each species and previously dried at 100°C for 24 h. The samples were then weighted and immersed in water for 24 h under soft mechanical stirring. The samples were dried once again at the above mentioned conditions. After weighting, the percentage of mass loss was calculated;

- Tension strength: The analysis was carried out according to ASTM D882-12 (ASTM, 2012);

- X-ray diffraction: The diffractograms of the nanopapers were recorded on a Rigaku diffractometer model XRD 600 operating at 30 kV, 30 mA and CuKα radiation (λ = 1540 Å). The samples were placed on aluminum holders and scanned in 2θ/min ranges varying from 5 to 37.5° (2° min^{-1}). The extension of crystalline (CI) was estimated on the basis of areas under crystalline and amorphous peaks after appropriate baseline correction. The diffractograms were fit by placing Gaussian shaped peaks and the crystalline index (CI) was calculated by Eq. 3;

$$CI = (Ca/Ta)100$$

Where: Ca is the sum of the areas under the crystalline peaks; and Ta is the sum of the areas under the crystalline peaks and amorphous peak.

- Opacity change acquired by nanopapers is visually presented.

3. Results and Discussion

The difference between the raw wood sawdust and the obtained nanomaterial can be examined by SEM micrographs (figures 1 and 2). The raw sawdust presents bundles of fibers connected by amorphous bending materials which are intended to be removed by alkali and bleaching treatments. Afterwards, by only submitting the fibers to high mechanical shearing forces, disintegration of the fibers occurs, leading to the formation of MFC`s. Generally, MFC`s consists of nanofibril aggregates, whose lateral dimensions range between 10 and 30 nm [4]; hence those were successfully obtained previously to drying for nanopapers production.

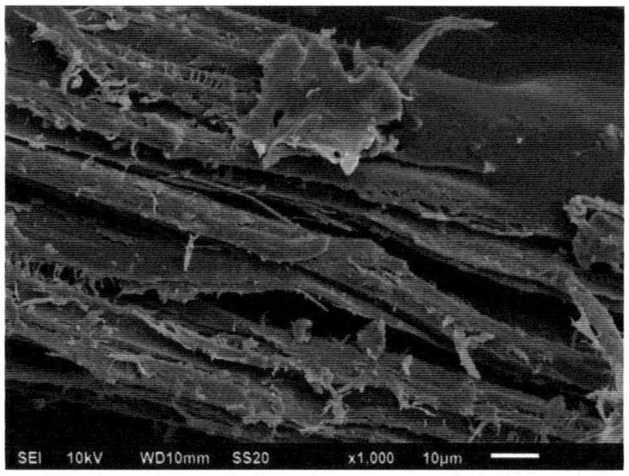

Figure 1. Micrograph of a fiber bundle from *C. goeldiana* raw sawdust (X 1000)

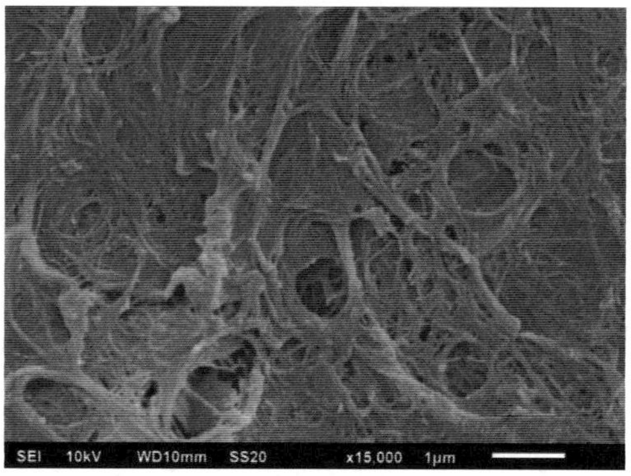

Figure 2. Micrograpn of microfibrilated fibers from *C. goeldiana* (X 15000)

Water deterioration analyses clearly showed high strength of the nanopapers produced from hardwood wastes. Among the four samples of each species, two from *B. parinaiodes* and *P. gigantocarpa* nanopapers showed no detectable mass loss, while the same happened for one sample obtained from *C. goeldiana* nanopapers. The remaining samples present mean deterioration ranging from 14 to 17% and the values were similar among nanopapers from different species (figure 3). Cellulose is a relatively stable polymer that does not readily dissolve in water [5]. As the nanopapers are basically comprised of MFC`s closely connect in a net, high water strength was achieved.

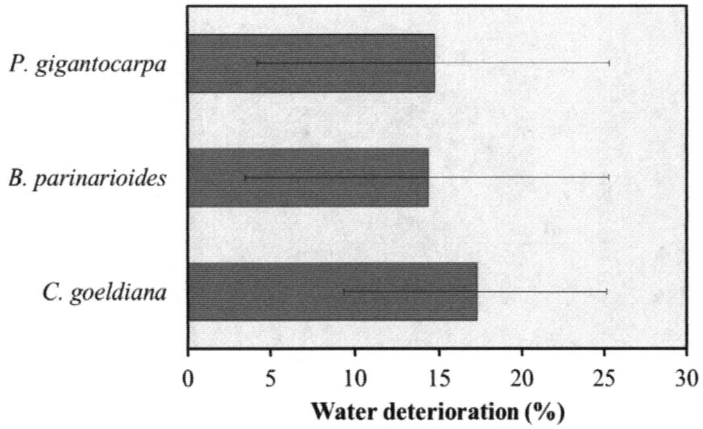

Figure 3. Water deterioration index of the nanopapers

Tension strength and crystalline indexes of the nanopapers are depicted in table 1. *C. goeldiana* nanopapers had the greatest mechanical strength. The nanopapers from this species were less bleached which suggests a higher amount of residual lignin which could act as a binder between the MFC's improving tension strength.

Table 1. Crystal index and mechanical strength of the nanopapers

Species	Tension Strength (MPa)	Crystalline index (%)
C. goeldiana	$50.7^{(7.6)}$	91
B. parinarioides	$37.0^{(6.6)}$	93
P. gigantocarpa	$42.0^{(6.6)}$	86

Crystalline indexes of the nanopapers ranged from 86 to 91%. The high and different values are related to previous removal of amorphous materials and the fibrillation process. The microfibrils consist of monocrystalline mesomorphous cellulose domains in a larger amount linked by amorphous domains that occur in smaller amount [1, 3, 6]. In cellulosic materials from several sources the length of crystallites ranges from 50-150 nm, while the disordered domains are 25-50 nm [3].

Changes in properties from micro to nano scale can be also clearly observed when a film made from cellulose bleached fibers is compared to a film made from cellulose nanofibers (figures 4, 5 and 6). It should be mentioned that the paper and nanopaper present in the pictures have exactly same chemical constitution and only a mechanical processing applied to achieve nanoscale was able to completely modify the opacity property.

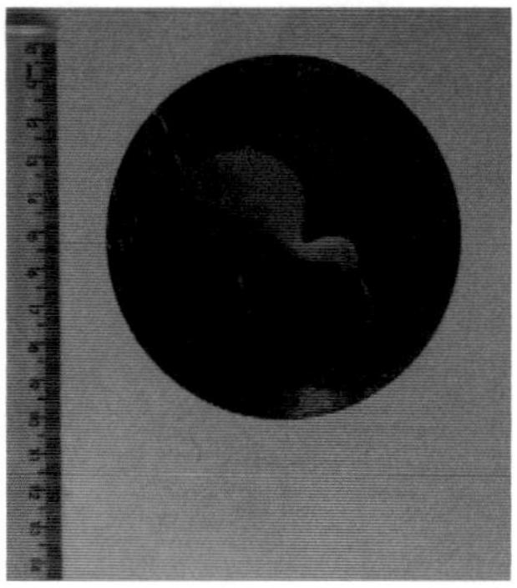

Figure 4. Uncovered picture

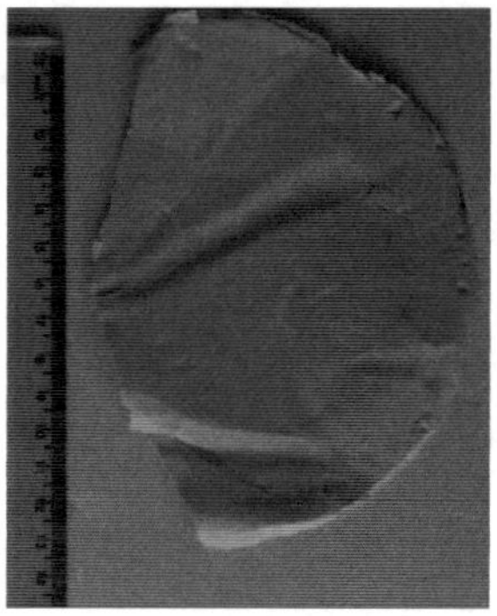

Figure 5. Picture covered by *B. parinarioides* paper

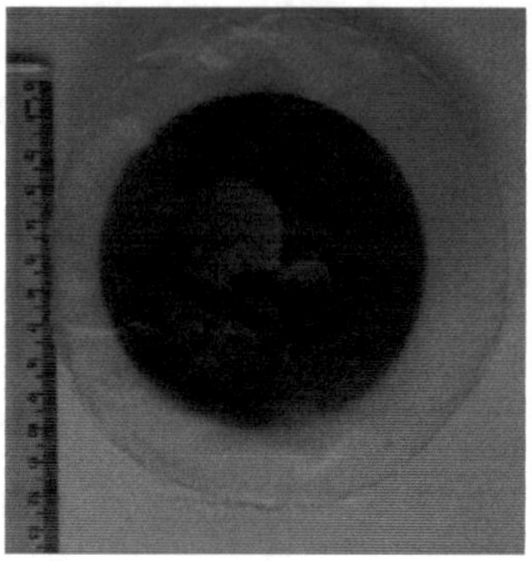

Figure 6. Picture covered by *B. parinarioides* nanopaper

4. Conclusion remarks

Nanopapers were produced from Amazon hardwood processing wastes in this work and showed promising results. High mechanical and physical strength were observed combined with high crystal index and opacity change. By using a simple and clean mechanical process it is possible to assure the utilization of wastes from Amazon wood processing avoiding them from injuring the environment.

Acknowledgments
The authors are grateful for the support of Coordination for the Improvement of Higher Level Personnel (CAPES), The Minas Gerais State Research Foundation (Fapemig), National Council for Scientific and Technological Development (CNPq), Brazilian Agriculture Research Corporation (EMBRAPA), and Brazilian Net on Lignocellulosic Composites and Nanocomposites (Religar).

References

[1] Oksman K Mathew A P Bondenson D Kvien I 2006 Manufacturing process of cellulose whiskers/polylactic acid nanocomposites. *Composites Science and Technology*. **66** 27762784

[2] Goetz L Mathew A Oksman K Gatenholm P Ragauskas A J 2009 A novel nanocomposite film prepared from crosslinked cellulosic whiskers. Carbohydrate Polymers. **75** 8589

[3] Ioelovich M 2008 Cellulose as nanostructured polymer: a short review. BioResources **3** 14031418

[4] Eichhorn S J Dufresne A Arangurem M Marcovich N E Capadona J R Rowan S J Weder C Thielemans W Roman M Renneckar S Gindl W Veigel S Keckes J Yano H Abe K Nogi M Nagakaito A N Mangalam A Simonsem J Benight A S Bismarck A Berglund L A Peijis T 2010 Review: current international research into cellulose nanofibres and nanocomposites. Journal of Materials Science **45** 133

[5] Lan W Liu C F Sun R G 2011 Fractionation of bagasse in cellulose, hemicelluloses and lignin with ionic liquid treatment followed by alkaline extraction. Journal of Agricultural and Food Chemistry **59** 8691

[6] Moran J I Alvarez V A Cyras V P Vazquez A 2008 Extraction of cellulose and preparation of nanocellulose from sisal fibers. Cellulose **15** 149159

IOP Conf. Series: Materials Science and Engineering **64** (2014) 012013 doi:10.1088/1757-899X/64/1/012013

Principle and equipment of polymer melt differential electrospinning preparing ultrafine fiber

Yang Weimin[1,2], **Li Haoyi**[1,2]

[1]College of Mechanical and Electrical Engineering, Beijing University of Chemical Technology, Beijing 100029, China
[2]State Key Laboratory of Organic–Inorganic Composites, Beijing University of Chemical Technology, Beijing 100029, China

E-mail: yangwm@mail.buct.edu.cn

Abstract: Two methods preparing polymer ultrafine fiber have been developed: solution electrospinning and melt electrospinning, among which, solution electrospinning is much simpler to realize in lab or industry. More than 100 institutions have made endeavors to research it and more than 30 thousand papers have been published. However, its industrialization was restricted in some extend because of existence of toxic solvent and low strength caused by small pores. Solventless melt electrospinning is environment friendly, but high melt viscosity, thick fiber diameter, low yield and complex equipment lead to less research on it. Aiming to solving the shortage of traditional needle nozzle equipment, we first proposed a melt differential electrospinning method preparing ultrafine fiber, through which fiber smaller than 1micrometer can be produced and a yield of 10-20g/h can be achieved by a needleless nozzle. Further more, principle and equipment of melt differential electrospinning are introduced.

Key words: melt electrospinning; differential equipment; needleless nozzle; fiber

1. Introduction

Reneker's[1] research work on electrospinning showed sound prospects for artificial nano-fiber preparation, which set the world on the research upsurge of polymer electrospinning in the past twenty years. As shown in figure1, on recent ten years the development of electrospinning research shows that number of published papers on solution electrospinning is nearly a hundred times that on melt electrospinning. The main reason lies behind this huge contrast is that the solution electrospinning experiment device is much simpler[2] while the melt electrospinning is relatively complex[3]. However, evaporation of solvent in solution electrospinning may cause small pores on the fibers[4] and environmental pollution[5], at the same time, polypropylene, polyethylene and some other materials have not found a suitable solvent at room temperature[6]. These facts limited the industrialized application of solution electrospinning technology. Now it is urgent to solve the bottleneck problem of melt electrospinning and realize green manufacturing technology of ultrafine fiber in industry.

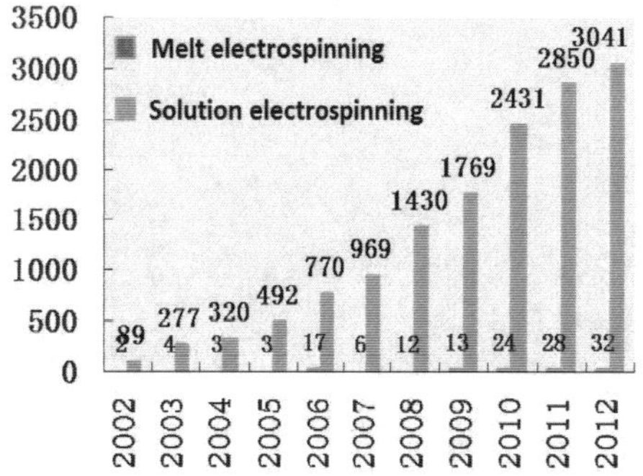

Figure 1. Comparison of number of published papers related electrospinning based on "SciFinder".

In 1936, Charles Norton's patent[7] showed the earliest description of melt electrospinning and solution electrospinning. However, half a century later, in 1981, the first scientific literature on melt electrospinning was just published[8]. After 20 years, the second article on melt electrospinning published by Reneker and Rangkupan [9] introduced the PP, PE, PET, PEN melt electrospinning in vacuum. In the last 10 years, more than 30 papers on melt electrospinning has been published. Representative results are as follows: Nobuo Ogata team of Fukui University [10] developed a laser heating melt electrospinning, they made the polymer bar heated and molten by laser, then electrospun fibers within 1μm; Jason Michael Lyons[11] and Cevat Erisken [12]tried melt electrospinning using single screw extruder and twin screw extruder respectively. Complex designing was involved in these devices to avoid electrical interference.

In order to elevate the electrospinning production, most researches focused on increasing the quantity of spinning needles. A multi-needle melt electrospinning equipment was produced by ITA Aachen of Germany as shown in figure 2 [13], however, the simple combination of fine needles was of low efficiency, high cost and complex hot runner. With the development of needleless technology in solution electrospinning, some needleless melt electrospinning device came to be published. Line-laser melt electrospinning proposed by the Ogata team[14] was an important evolution of laser melt electrospinning. They chose a 100×150mm sheet with thickness of 0.5 to 1 mm. When it was heated by a linear laser, a plurality of Taylor cone will formed at the edge of the sheets. It was 4cm to 6cm between the two Taylor cone (interjet distance), but sometimes Taylor cone may disappear because of un-uniformed spinning parameters. Finally the P(EVAL) and Nylon 6/12 fibers with diameter of 800 nm to 2.5 μm were produced at different parameters (voltage, flow velocity, and laser intensity). Komarek M and Martinova L[15] from Liberec University of science and technology in Czech, proposed a slit-type melt electrospinning device, but how to distribute melt uniformly in the slit remained a problem. Fang J et al[16] from Australia proposed a disc melt electrospinning device, but the device had high requirement of melt viscosity. In conclusion, breakthrough in melt electrospinning technology research may happen only when researches focused on the equipment innovation.

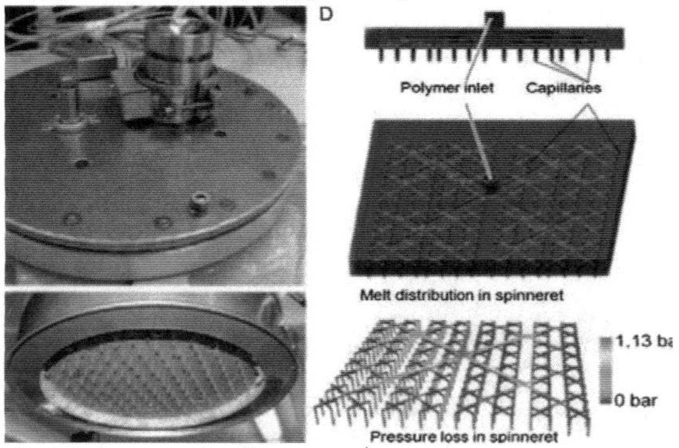

Figure 2. A multi-needle melt electro-spinning equipment produced by ITA Aachen of Germany.

2. The principle of polymer melt differential electrospinning

To develop polymer melt electrospinning technology, three problems should be considered: high voltage insulation and security issues of electrospinning system, fiber refinement, and productivity.

The high voltage insulation and security issues of electrospinning system brought many researchers various problems in device designing. Influenced by traditional solution electrospinning, majority of melt electrospinning experiments were proceeded with receiver plate grounded and the high voltage connected to capillary end. This may cause fire and since high insulated unites were involved. Actually the solution is easy: just to connect the positive high voltage electrostatic supplier to the spinning receiver plate directly, and ground the plasticizing system[17]. As Gaoqi L's comments[18], this Solution eliminated the obstacle of melt electrospinning to industry.

Melt electrospinning are nearly impossible to prepare nano-scale fiber in mass production using traditional needle-like nozzles, however, the author proposed a principle of melt differential electrospinning: first, the polymer melt was transformed from a cylindrical flow into a uniform ring-like flow, next, the ring-like flow was distributed to the umbellate nozzle, then a thin and uniform melt film formed on the umbellate circumferential surface, finally, when the applied high voltage surpassed a critical value, self-organized multiple jets around the rim of the umbellate nozzle were ejected to the receiver plate spontaneously(figure 3). Interjet distance (figure 4) depends on the electric field strength, material properties and melt viscosity. This process is similar to the so-called "Needleless electrospinning" in solution electrospinning[19], which abandons the traditional capillary tube or needle, applying a higher electric field to the free surface of the liquid to realize multijet self-organization.

As shown in figure 3, a typical melt differential electrospinning device consisted of five major components: melt inlet, melt distributor, umbellate nozzle, high voltage power supply, and receiver plate. Polymer melt was metered and supplied into melt inlet by a self-designed micro extruder, and the melt flow rate could be controlled in the range of 1 – 100 g/h. Melt distributor was designed to induce the single melt flow to the circumference of the umbellate nozzle uniformly. The nozzle was earthed and high positive voltage was loaded on the receiver plate using a high voltage power supply source at a value range of 0 – 80 kV. Receiver plate was a circular copper plate with diameter of 150 mm and thickness of 3 mm.

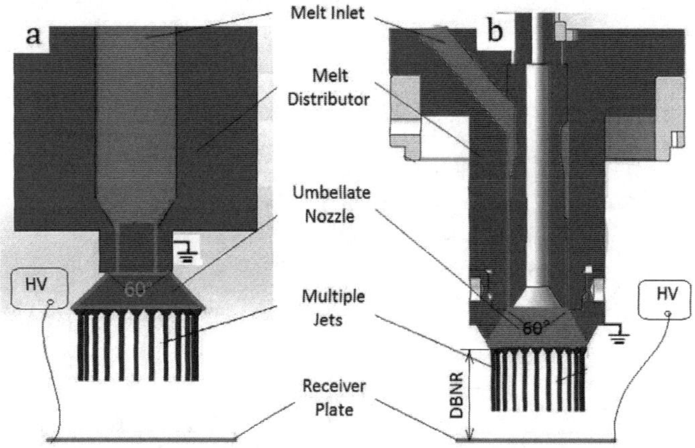

Figure 3. Structure diagram of the two melt differential nozzles.

Based on previous experimental study, it was found that in melt differential electrospinning, polymer melt feed rate had no evident influence on interjet distance. The main factors influencing the interjet distance were electric field strength and polymer melts viscosity. With the field strength increasing at a fixed melt feed rate, the self-organizing differential effect became stronger. It means that the constant volume of melt supply can be divided into tens of micro flows, thus more jets formed and then solidified. Without the consideration of thinning effect of electric field intensity on fiber diameter, smaller fiber diameter will be attained with the jets number increasing. Assuming melt flow rate is Q and the resultant single fiber diameter is D, after melt differential electrospinning process, the jet number increases to K, and fiber diameter will become $D/\sqrt{K}$.

Experimental results show that, for a polymer, the lower the melt viscosity or the higher the electric field strength, the shorter the interjet distance. Present experimental study shows that melt differential electrospinning can achieve the minimum interjet distance of 1.1mm (spinning parameters: applied voltage is 35kV, spinning distance is 70mm, nozzle temperature is 260 ℃, and spinning materials is Polypropylene), which is only less one third of that by needleless solution electrospinning from free surface (spinning parameters: applied voltage is 43kV, spinning distance is 80mm, and spinning material is the 4% polyvinyl alcohol in water). This result indicates a promising production.

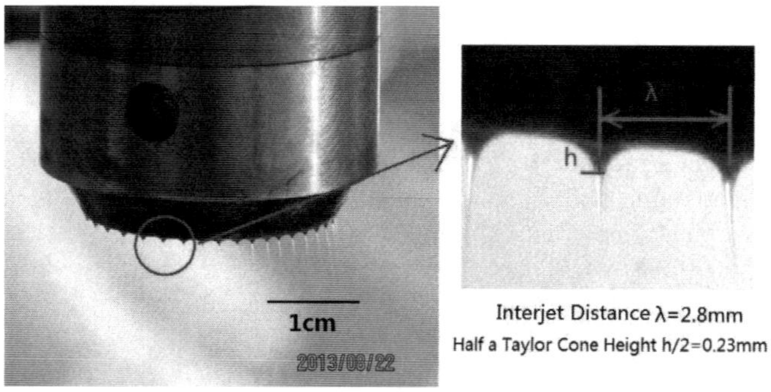

Figure 4. Picture of melt differential electrospinning process with an umbrella nozzle and the section enlargement of Taylor cones, where interjet distance λ was defined.

3. The pilot line of polymer melt differential electrospinning

Based on the principle of polymer melt differential electrospinning, a prototype equipment of pilot line was designed and operated. This machine can realize mass production of polymer nanofiber by melt electrospinning route. As shown in figure 5, the equipment includes nine components: a small twin screw extruder, a melt filter, a melt metering pump, a melt differential electrospinning box, air auxiliary equipment, suction device, a lapping machine, and rolling-coiling machine. The melt differential electrospinning(MD-ESP) process and resultant fiber diagram were compared with that of solution electrospinning(S-ESP) in figure 6.

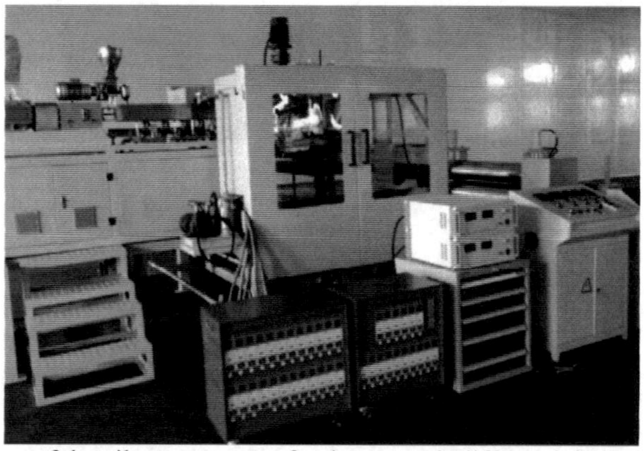

Figure 5. The photo of the pilot prototype of polymer melt differential electrospinning device.

The performance parameter of pilot line are as follows:

- Integration of the 32 differential nozzles;
- The jets number ejected from each nozzle (diameter of 26mm) can range from 50 to 100;
- The defined width is 0.8m, output is 300- 600g/h, and the average fiber diameter is 500-800nm;
- The thickness of non-woven fabrics is adjustable from 10μm to 1000μm;
- Continuous supply of polymer melt or polymer blend melt can be realize on line;
- Modularly extendable to the 6kg/h production line.

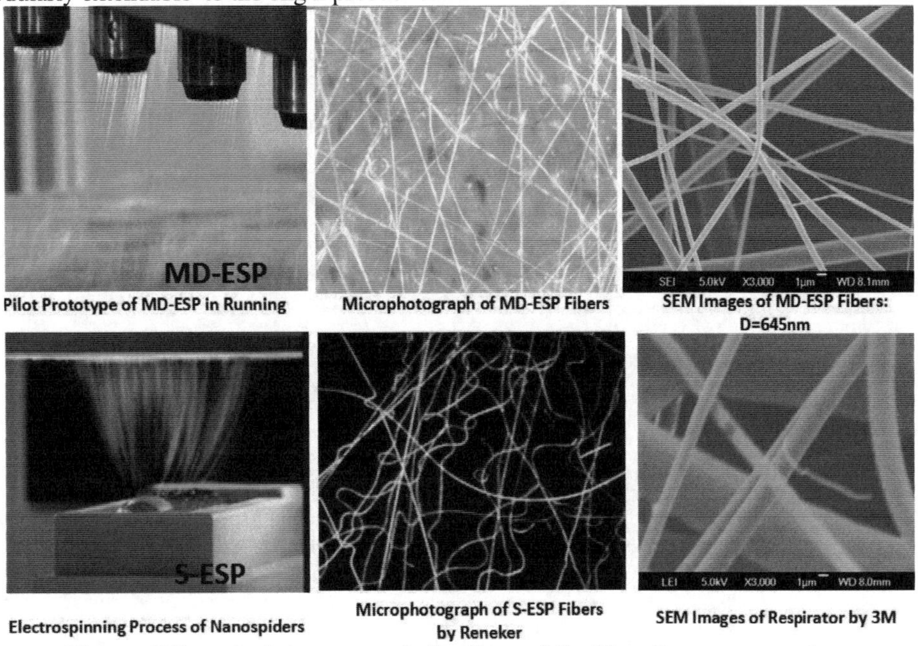

Figure 6. The spinning process of pilot line and the fiber diagram comparison.

[1] Darrell H R and Iksoo C 1996 *Nanotechnology* **7** 216
[2] Xing H. L, Chang L. S and Yi C. L 2007 *Langmuir* **23** 10920
[3] Eduard Z, Daehwan C and Yong L J 2010 *Polymer* **51** 4140
[4] Liwei H, Seetha S M and Jeffrey R M 2013 *J. Memb. Sci* **436** 213
[5] Seema A, Joachim H. W, and Andreas G 2009 *Adv. Mater* **21** 3343
[6] Deng R, Liu Y, Ding, Xie P, Luo L and Yang W 2009 *J. Appl. Polym. Sci* **114** 166
[7] Norton C L 1936 US patent No.2048651
[8] Larrondo L and St John Manley R 1981 *J. Polym. Sci. B. Polym. Phys* **19** 909
[9] Rangkupan R and Reneker D H 2003 *J. Min. Met. Mat. S* **12** 2
[10] Ogata N, Shimada N, Yamaguchi S, Nakane K and Ogihara 2007 *J. Appl. Polym. Sci* **105** 1127
[11] Lyons J and Ko F 2005 *Polym. News* **30** 170.
[12] Cevat E, Dilhan M K and Hongjun W 2008 *Nanotechnology* **19** 165302
[13] Hutmache D W and Dalton P D 2011 *Chem. Asian. J* **6** 44
[14] Shimada N, Yamaguchi S, Nakane K and Ogata N 2010 *J. Appl. Polym. Sci*, **116** 1282
[15] Komarek M and Martinova L 2010 "Design and Evaluation of Melt-Electrospinning Electrodes" Proceedings of 2nd NANOCON International Conference, Olomouc, 12-14 October 72
[16] Wang X, Niu H, Wang X and Lin T 2012 *J. Nanomater* **2012**
[17] Yang W, Deng R and Ding Y 2007 China Patent No. 2007101769158
[18] Gaoqi L2012 Japan processing research overview **24** 360
[19] Lukas D, Sarkar A and Pokorny P 2008 *J. Appl. Phys* **103** 084309

IOP Conf. Series: Materials Science and Engineering **64** (2014) 012014 doi:10.1088/1757-899X/64/1/012014

High-Aspect Ratio Bio-Metallic Nanocomposites for Cellular Interactions

Sneha Deodhar, Justin Huckaby, Miles Delahoussaye, and Mark A. DeCoster[1]
Biomedical Engineering and Institute of Micromanufacturing, Louisiana Tech University, Ruston, Louisiana, USA

E-mail: decoster@latech.edu

Abstract. We synthesized high aspect ratio composites with biological and metal components. Scanning electron microscopy (SEM) and Transmission Electron Microscopy (TEM) revealed linear morphology and smooth surface texture. SEM, TEM and light microscopy showed that composites have scalable dimensions from nano- to micro-, with diameters as low as 60 nm, lengths exceeding 150 μm, and average aspect ratio of 100. The structures are stable, remaining intact for over one year in dried form and in liquid, and did not aggregate, in contrast to metal nanoparticles such as iron and copper. Many metal nanoparticles are toxic to cells, limiting their use for biological applications. The bio-metallic composites characterized here showed lower toxicity compared to their precursor metal nanoparticles in brain tumor cell cultures. Due to these more biocompatible properties, we tested the ability of the composites to interact with cells. Zeta potential analysis indicated that composites carry a net negative charge (-24.3 ± 2.2 mV), while the starting metal nanoparticles measured (43.3 ± 2.4 mV). We labeled the composites with poly-l-lysine fluorescein isothiocyanate (PLL-FITC), which shifted the potential to 3.5 ± 2.9 mV. It was observed by fluorescence microscopy that composites smaller than cells were internalized by some cells and larger composites remained outside. Cells became fluorescent over time due to leakage of PLL-FITC from the composites which lost fluorescence over time. Higher biocompatibility, low aggregation, and ability to control size distribution of the linear composites may make them ideal vehicles to deliver drugs or other materials to cells, and may be used as a scaffolding material for cells.

Introduction

As the synthesis and characterization of nanomaterials has expanded, a greater diversity of different size and shape nanomaterials have become available for interaction with cells. These now include, for example, the most typical nanoparticles of a range of compositions and sizes [1], but also, high aspect ratio nanomaterials such as nanowires and nanorods [2]. Compared to nanoparticles, higher-aspect ratio nanomaterials and micromaterials provide biophysical benefits such as reversal of the coffee-ring effect [3] and may also serve as connectors for cellular scaffolding or device construction [4]. New biological applications are developing from nanotechnology as a result of the inherent chemical, optical, magnetic, and electronic properties of high aspect ratio nanoparticles [5].

With nanomaterials of many different compositions and sizes, biocompatibility and cytotoxicity can be an issue, since the surface area to mass ratio is greatly increased in nanomaterials, thus making surfaces potentially more reactive and thus potentially more toxic to cells. Furthermore, a benefit but also potential concern of nanomaterials is that due to their small size, they may actually enter cells and

[1] To whom any correspondence should be addressed.

thus have short-term and long-term effects beyond the short evaluation period that may be initially carried out. Thus, we need to understand short-term and long-term effects of nanomaterials when biological interactions are envisioned.

One approach to understanding impacts of nanomaterials on cells is to characterize nanomaterial biocomposites [6]. Biocomposites have some biological component to the nanomaterial, and thus the material has been potentially processed in some way by cells or biological molecules that might make the biocomposite more biocompatible than the starting material(s). In the work here, we characterized composites that are scalable to have both nano- and micro-dimensions, while maintaining a high aspect ratio of approximately 100, with diameters of as small as 60 nm and lengths exceeding 150 microns.

Natural and synthetic hybrid organic-inorganic structures have been prominent in research and industry for decades. Biology and nanotechnology advancements have been attempting to create more biocompatible nanostructures for use in medicine, drug delivery, biomaterials, and tissue engineering. The hybrid structures are developed from the combination of organic and inorganic compounds to utilize the properties of each component in a single nanocomposite. Proper selection of organic and inorganic starting materials can generate bio-metallic nanocomposites for cellular interactions. Organic compounds contribute their easy shaping and processing properties while inorganic compounds provide mechanical and thermal stability to the final hybrid structure [7]. Current examples of hybrid organic-inorganic nanocomposites for cellular interactions include filling composites in dental applications [7] and biomaterials in bone tissue engineering applications such as the repair of osseous and chondral defects [8]. Hybrid nanocomposites for dental fillings retain tooth-like properties of hardness, elasticity, and thermal expansion behavior as well as minimum shrinkage and non-toxic interaction. Mechanical properties of chitosan were improved by introducing an inorganic compound into the hybrid structure scaffold [8]. Hybrid organic-inorganic structures can be synthesized by a variety of strategies including conventional sol-gel chemistry, the use of specific bridged and polyfunctional precursors, hydrothermal synthesis, assembling or dispersion of well-defined nanobuilding blocks, self-assembling procedures, and integrative synthesis [7]. Here we characterize a novel metallic/organic biocomposite which has a high aspect-ratio and favorable biocompatibility compared to the starting material (copper).

Materials and Methods

1.1. Composite Structures: Organo-Metallic nano and microstructures were obtained from Nanogaia, LLC (www.nanogaia.com). Composites were synthesized by a proprietary process and were certified to contain copper nanoparticles as starting materials and amino acids as the biological component by elemental analysis using EDAX on a scanning electron microscope (SEM). The purpose of the current study was to characterize the properties of these high aspect ratio structures and their interaction with cells. Future studies will include descriptions of synthesis of the structures.

1.2. Cell Cultures: The rat brain tumor cell line (CRL 2303) from ATCC was used in this study. According to the ATCC protocol, these cells were maintained in Dulbecco's Modified Eagle Medium containing 10% fetal bovine serum, 0.5% penicillin-streptomycin and 1% amino acid solution (Sigma-Aldrich) at 37°C, 70% humidity and at a constant pressure of 5% carbon-dioxide. During regular passaging, cell viability was checked by trypan blue exclusion and cells were plated at a density of 25,000 cells per well in a 24 well plate in 0.5ml of media.

1.3. Nanomaterials: Samples were prepared by adding specified amount of material to the appropriate volume of Locke's solution to make stock solutions of concentration of 1 mg/ml. Twenty five microliters of stock solution was added to different wells containing 475 μl of media to obtain effective concentrations of 50 μg/ml and 25 μg/ml respectively (or as indicated). After loading the sample, the plate was gently agitated and was incubated for 18 hours.

1.4. Zeta Potential measurements: Zeta potential measurements were carried out using Zeta Plus Zeta Potential Analyzer (Brookhaven Instruments Corporation, Holtsville,NY,USA). For this, 50μl of the aqueous dispersion of sample (0.1mg/ml concentration) was diluted up to 2ml with DI water and the analysis was carried out at 25°C.

1.5. Fluorescent labelling of composites: Layer-by-layer polyelectrolyte coating procedures were performed to label composites fluorescently. The fluorescent polyelectrolyte used was poly-l-lysine fluorescein isothiocyanate (PLL-FITC, Sigma Aldrich) at a final concentration of 0.5 mg/ml. Composite structures were combined with 0.5 mg/ml PLL-FITC in a mechanical shaker for 60 minutes at 25°C. The solution was centrifuged at 13,500 rpm for 6 minutes to remove supernatant. Fluorescently labelled composites were washed with sterile DI water to remove excess polyelectrolyte before final suspension and addition to cells.

1.6. Cell viability measurements: The MTT assay was utilized as we have previously described [9]. After exposure, culture medium was removed and 400 µl of MTT solution of 1.25mg/ml concentration in Locke's was added to each well. The plates were further incubated at 37°C for 60 minutes. The MTT solution was then discarded from the wells and the formazan crystals were dissolved in 250 µl of the 91% isopropyl alcohol. From each well, 200µl of formazan solution was transferred to a 96 well plate, and then read on a Multiskan Spectrum plate reader (Thermo Scientific) at 570 nm. The mean percentage viability was determined from the data of three representative experiments.

1.7. Microscopy: Scanning Electron Microscopy (SEM). Hitachi S-4800 Field Emission SEM was used to record images at 15 kV. The sample was prepared by placing a 20 µl drop of structure suspension on a carbon coated metallic stage. The samples were placed at ambient temperature and allowed to dry before viewing under the electron microscope.

Results

2.1 Microscopic Characterization of Composites:

Composite structures were characterized by electron and white light microscopy. As shown for composites examined by scanning electron microscopy (SEM, Figure 1) and transmission electron microscopy (TEM, Figure 2) structures demonstrated smooth edges with elongated, high-aspect ratio features, scaling down to nanometers in the diameter axis and many microns in length. These composites when examined by light microscopy showed scaling of many microns in length with a few microns in the diameter axis (See Fig. 5).

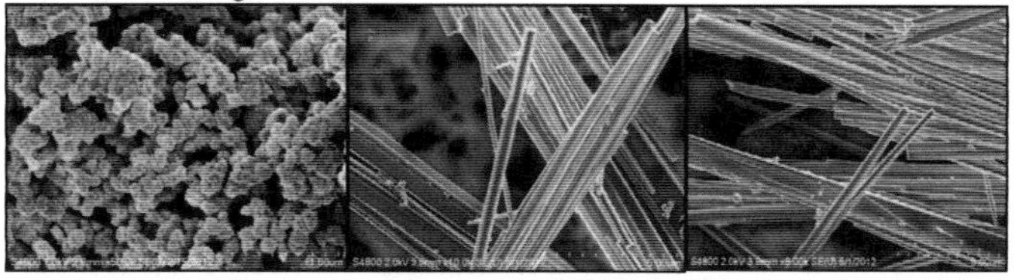

Figure 1 *SEM images of starting material copper nanoparticles (left), and metallic biocomposite structures shown at the micro-scale (middle and right panels). Scale bars= 1 micron (left panel) and 5 microns (middle and right panels).*

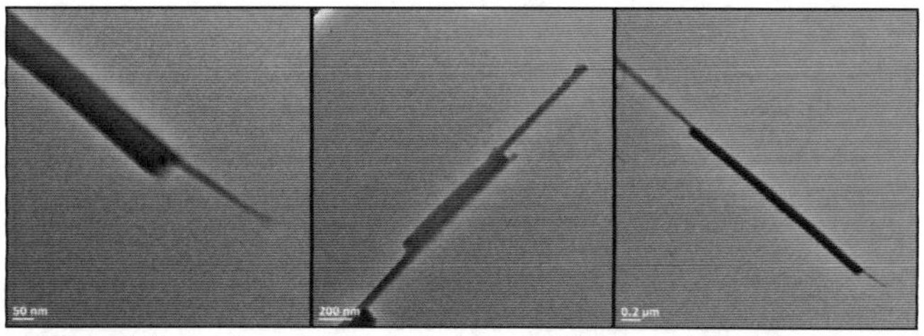

Figure 2 *TEM images of the biometallic composites shown at the nanoscale.*

2.2 Charge characterization of starting metal and composites:

Copper nanoparticles unaltered from the commercial source (Sigma Aldrich), as expected, demonstrated a positive Zeta Potential (Figure 3). This charge distribution was driven to much more negative values upon biocomposite formation of the high aspect ratio structures (squares, fig. 3). Since these negative values might not be amenable to interaction with cells, due to the negative charge of cell membranes,

IOP Conf. Series: Materials Science and Engineering **64** (2014) 012014 doi:10.1088/1757-899X/64/1/012014

we then coated the biocomposites with FITC-labeled poly-l-lysine (triangles, fig. 3), which re-distributed the charge distribution to values near 0.

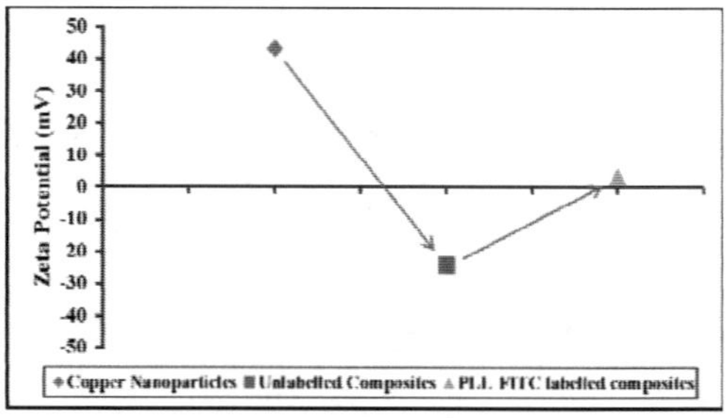

Figure 3. Charge characterization as measured by Zeta potential. Materials were prepared and Zeta potential measured as described in methods. The starting material was copper nanoparticles (diamond), which was then formed into a composite (squares). This composite was finally labeled with FITC-PLL (triangles). Zeta potential was measured at each step, demonstrating changes in charge distribution.

2.3 Labeling of composite structures and interaction with cells:

We tested for the ability of FITC-labeled PLL-coated biocomposites to interact with cells by adding the fluorescent high-aspect ratio structures to CRL-2303 glioma cells and capturing epifluorescence microscopy images over 4 days (Figure 4).

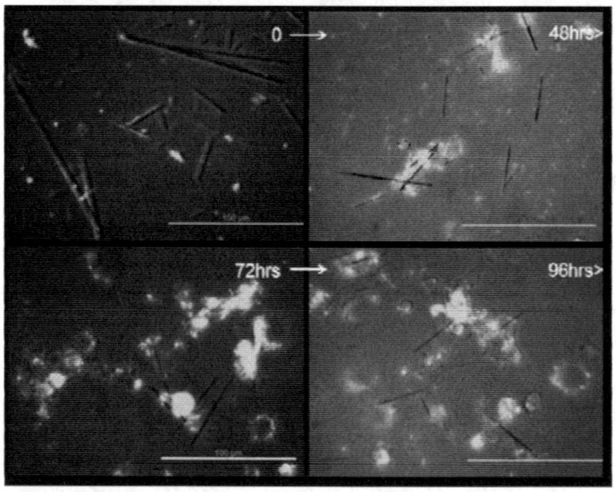

Figure 4. Interaction of fluorescently labeled biocomposite structures with cells. Images were obtained with both white light and fluorescence illumination. Metal components of the composites (dark linear structures) as well as cells and fluorescent areas could all be seen simultaneously. Scale bar indicates 100 microns.

During this period, we observed that most of the long linear structures disappeared or were degraded, and cells, which continued to grow and proliferate, became more fluorescent with time (compare 0 hours to 96 hours in Figure 4). It also appeared that smaller particulates and/or fluorescent label from the biocomposites entered the cells over time.

2.4 Comparative Metallic Nanoparticle processing by cells:

We compared the ability of cells in vitro to process metallic nanoparticles of two different compositions: iron and copper, to test for biocompatibility but also for addressing long-term changes in the fate of nanomaterials when interacting with biological environments. As shown in Figure 5, CRL-2303 brain tumor cells in control conditions grew aggressively in the cell culture after 3days in vitro (DIV), as indicated by more cell coverage than after 1 DIV (left panel top vs left panel bottom).

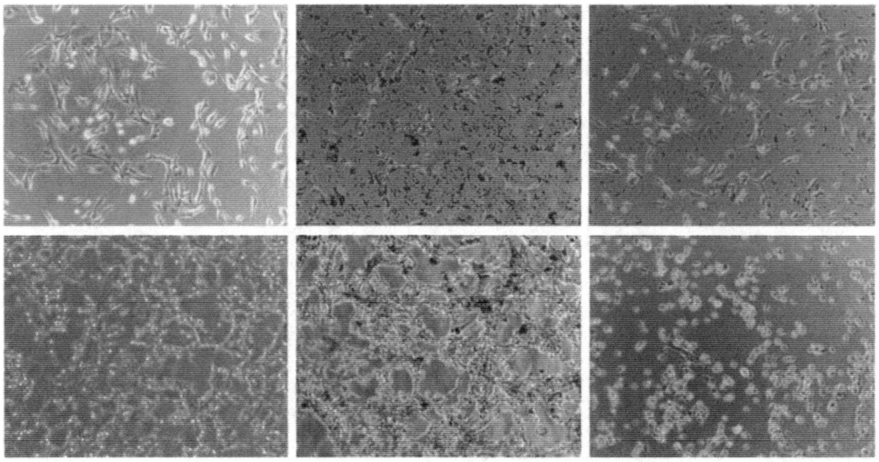

Figure 5*. Nanoparticle processing by cells in vitro. From left to right, control cells (left, media only), iron oxide nanoparticles (middle), and copper nanoparticles (right panels). Top row images= cells at 1DIV and bottom row= at 3 DIV. All images=100x magnification using 10 microscope objective.*

Addition of 50 ug/ml of iron oxide nanoparticles did not appear to greatly affect cell growth (middle column), while addition of 50 ug/ml of copper nanoparticles was very toxic to cells (right panels).

2.5 Comparative Biocompatibility testing of biocomposites and starting material:

As a final measure of biocompatibility of different starting materials and biocomposite structures, we directly compared the toxicity or lack of toxicity of copper nanoparticle starting material, and their derived biocomposite structures, at two different concentrations, compared to cells treated with growth media only, using the metabolic MTT assay (Figure 6).

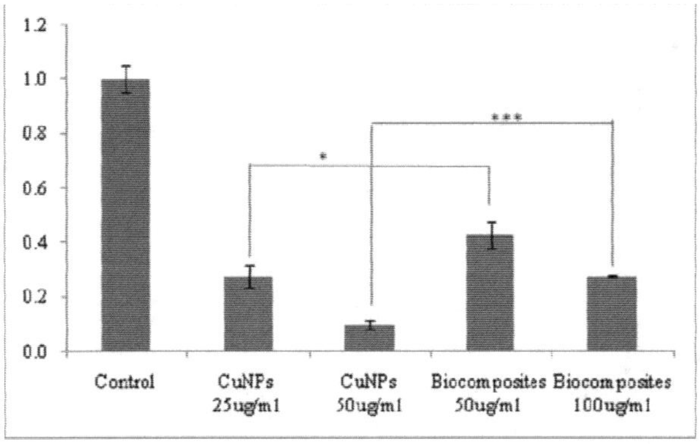

Figure 6. Toxicity testing of starting materials and biocomposites: *The metabolic activity of rat glioma cells 18 hours after treatment with composite structures and copper nanoparticles (CuNPs) is shown. Metabolic activity was quantified by MTT assay. Values were normalized with controls. (*$P<0.05$ and ***$P<0.001$; two tailed paired t-test).*

Discussion

Copper nanoparticles transformed into a biologically-derived nanocomposite with very high aspect ratios were characterized. Zeta potential analysis (Fig. 3) indicated that as we transformed the size, shape, and charge distribution from copper nanoparticles to a linear biocomposite, the zeta potential went from positive to negative values. With the goal of developing nanocomposites that would better bind to cells, we functionalized the biocomposites with FITC-labelled PLL, which as expected, moved the charge distribution to a less negative value, in fact, generating structure with slightly positive zeta potentials (Fig. 3). This strategy was followed to build on the long-standing observation of binding of poly lysine to negatively charged cells [10]. These labelled structures had utility in interacting with cells, demonstrating association and possibly entry into the cells and/or degradation over time as indicated by epifluorescence microscopy tracking over a 4 day period (Figure 4).

The very small size and positive charge of copper nanoparticle starting material (Figure 3) may explain their toxicity to cells, which we have reported previously [11]. When added to growing glioma cells in culture, it was interesting to note that dark, metallic particulates apparent at the start of the experiment (1 DIV), were processed by the cells, and disappeared. This was a toxic interaction, as cells failed to grow (Fig. 5). In contrast, at the same concentration (50 ug/ml), iron oxide nanoparticles did not prevent cell growth, and remained relatively inert, as shown by phase microscopy after 3DIV (Fig. 5). In these fast-growing cancer cell lines, due to the presence of the pH indicator phenol red, the color of the culture media could easily be used as a marker for cell growth or cell death. Thus, control cells shifted from red (basal) to yellow (acid), indicating growth, as did the iron oxide treated cultures, but copper treated cells did not shift pH (Fig. 5). This could provide additional utility for assessing toxicity in cell assays. It was encouraging to note that biocomposites were less toxic than the starting nanoparticle material (Fig. 6). Future studies will measure the biocompatibility of PLL-labelled composites, which appeared to interact with cells (Fig. 4). It is anticipated that size, composition, and charge distribution are all likely components for determining interaction and compatibility with cells.

References

[1] Jain PK, Lee KS, El-Sayed IH, El-Sayed MA. Calculated absorption and scattering properties of gold nanoparticles of different size, shape, and composition: applications in biological imaging and biomedicine. The journal of physical chemistry B 2006;110:7238-48.

[2] Li Z, Xiong Y, Xie Y. Selected-control synthesis of ZnO nanowires and nanorods via a PEG-assisted route. Inorg Chem 2003;42:8105-9.

[3] Yunker PJ, Still T, Lohr MA, Yodh AG. Suppression of the coffee-ring effect by shape-dependent capillary interactions. Nature 2011;476:308-11.

[4] Patolsky F, Timko BP, Yu G, Fang Y, Greytak AB, Zheng G, et al. Detection, stimulation, and inhibition of neuronal signals with high-density nanowire transistor arrays. Science 2006;313:1100-4.

[5] Bauer LA, Birenbaum NS, Meyer GJ. Biological applications of high aspect ratio nanoparticles. Journal of Materials Chemistry 2004;14:517-26.

[6] Sapsford KE, Tyner KM, Dair BJ, Deschamps JR, Medintz IL. Analyzing nanomaterial bioconjugates: a review of current and emerging purification and characterization techniques. Anal Chem 2011;83:4453-88.

[7] Sanchez C, Julián B, Belleville P, Popall M. Applications of hybrid organic-inorganic nanocomposites. Journal of Materials Chemistry 2005;15:3559-92.

[8] Depan D, Surya PK, Girase B, Misra RD. Organic/inorganic hybrid network structure nanocomposite scaffolds based on grafted chitosan for tissue engineering. Acta biomaterialia 2011;7:2163-75.

[9] Joshi CK, B.; Newman, J.J.; DeCoster, M.A. Cell morphological changes combined with biochemical assays for assessment of apoptosis and apoptosis reversal. In: Mendez-Vilas A, editor. Current Microscopy Contributions to Advances in Sciences and Technology. Spain: Formatex; 2012. p. 756-62.

[10] Yavin E, Yavin Z. Attachment and culture of dissociated cells from rat embryo cerebral hemispheres on polylysine-coated surface. The Journal of cell biology 1974;62:540-6.

[11] Parekh G, Pattekari P, Joshi C, Shutava T, Decoster M, Levchenko T, et al. Layer-by-layer nanoencapsulation of camptothecin with improved activity. International journal of pharmaceutics 2014.

Acknowledgements: This work was funded in part with support from Louisiana Board of Regents PKSFI Contracts No. LEQSF (2007-12)-ENH-PKSFI-PRS-04 and LEQSF-EPS (2013)-OPT-IN-27, and the James E. Wyche III endowed professorship (to M. DeCoster). We thank Dr. Weilie Zhou of the University of New Orleans for assistance with Transmission Electron Microscopy.

IOP Conf. Series: Materials Science and Engineering **64** (2014) 012015 doi:10.1088/1757-899X/64/1/012015

Low Temperature Synthesis of Composition Poros Materials from Mortar Sands

O.V. Kazmina[1] , S. N. Volland[2], M. A. Dushkina[1]
[1]National Research Tomsk Polytechnic University, Tomsk, Russian Federation
[2] Technical University, Darmstadt, Germany
E-mail: kazmina@tpu.ru

Abstract. It is established that the eliminations of construction sand with the content of SiO2 about 70 wt. % and particle size less than 60 μm are suitable for the production of a foamglass-crystal material on the basis of the low-temperature frit, which was synthesized at the temperature 900 °C. The obtained foamglass-crystal material exceeds foamglass (by 3.0 times) and clayite (by 1.5 times) by strength and is characterized by low value of water absorption (0.1 %).

Introduction

The production of lightweight and efficient building materials, in particular of lightweight granulated materials, is a constant and modern subject of scientific investigations. One of the high-efficient heat-insulating materials that meet the requirements of environmental safety is foam glass. Raw material for production of foamed glass can serve various types of waste. One of the wastes which contaminates territory as collecting sludge ponds are fine dispersive sand screenings which are formed as result of mining and processing of quartz sands [1].

The aim of work is to investigate the possibility of preparation of foam material similar to foam glass on the basis of frit synthesized from quartz sands screenings. Industrial technology of foam glass is based on the use of glass cullet or frit with composition of sheet and container glasses. High dispersiveness of quartz sand wastes allows to suppose the possibility of low temperature synthesis of frit. Therefore one of assigned tasks of this research was development of blend compositions suitable to obtain frits at temperatures not exceeding 900 °C, excluding traditional glass melting process.

When one selects chemical composition of low temperature frit, it is necessary to meet the following requirements. The first condition which defines the component composition of a blend is provision sufficient quantity of glass former (60 – 75 mas %) and the alkali metal oxides (13 – 22 mas %). At that it is necessary to take into account that foaming occurs at viscosity of 10^5-10^7 Pa×s, therefore it is necessary to use the glasses which reach such viscosity at temperatures of 750 – 900 °C. It is possible to estimate preliminary viscosity properties of the glass by its composition using viscosity modulus, which value can be in range 1.6 – 1.8 [2]. The second selection condition is formation of not less than 70 % of melting at temperature not exceeding 900 °C, what has been stated by results of previous experiments [3]. The third condition is hydrolytic stability of glass (not below the 3rd class) and the presence of active oxidizing component in sufficient quantity to carry out foaming reactions, for example, presence of SO_3 in quantity not less than 0.2 %. One of the main requirements to all blend components are their dispersiveness; particle size must not exceed 100 μm.

[1] To whom any correspondence should be addressed.

Glass composition corresponding to the above mentioned requirements, selected by the state diagram of Na_2O–CaO–SiO_2, has the following content of components: Na_2O-14, CaO-13, SiO2-73 mass %. The composition of sand under investigation according to results of chemical analysis is presented by relatively low quantity of main glass forming oxide (SiO2) and sufficiently high quantity of the alkali metal oxides, aluminas, and iron oxides (Table. 1). According to data of X-ray phase analysis sand contains together with quartz and illite, tiff, plagioclase, and feldspar in about equal proportions. High dispersiveness of sand is confirmed by results of screen analysis according which 50 % of sand is presented by fraction of particles less than 60 μm that meets necessary requirements.

Table 1. Characteristics of the surveyed Sands

<table>
<tr><td colspan="10">Chemical composition, content of oxides, mas %</td></tr>
<tr><td>SiO_2</td><td>TiO_2</td><td>Al_2O_3</td><td>Fe_2O_3</td><td>MnO</td><td>CaO+MgO</td><td>Na_2O+K_2O</td><td>P_2O_5</td><td>SO_3</td><td>орг.</td></tr>
<tr><td>69,67</td><td>0,47</td><td>7,12</td><td>3,93</td><td>0,09</td><td>7,75</td><td>2,66</td><td>0,15</td><td>0,03</td><td>0,16</td></tr>
<tr><td colspan="10">Mineralogical structure, mass %</td></tr>
<tr><td>Quartz</td><td>Illite</td><td>Plagioclase</td><td colspan="2">Potassium feldspar</td><td>Calcite</td><td>Rutile (anatase)</td><td colspan="2">Ferric hydroxide</td><td>Organic</td></tr>
<tr><td>50</td><td>15</td><td>5 – 10</td><td colspan="2">10</td><td>10 – 15</td><td>1</td><td colspan="2">5</td><td>< 1</td></tr>
<tr><td colspan="10">Granulometric structure, mas %</td></tr>
<tr><td colspan="4">< 2 μm</td><td colspan="3">2 – 63 μm</td><td colspan="3">>63 μm</td></tr>
<tr><td colspan="4">6,8</td><td colspan="3">49,7</td><td colspan="3">43,5</td></tr>
</table>

To obtain frit of selected composition one prepared a blend containing sand under investigation and fluxing addition in form of calcined soda in quantity of 80 and 20 mas %, accordingly. Calculated composition of glass has following oxide content, mass. %: SiO_2 – 66.3; Na_2O – 13.4; CaO – 6.5; Al_2O_3 – 6.8; Fe_2O_3 – 3.7; K_2O – 1.8; MgO – 0,8; MnO – 0.1; TiO_2 – 0.5. Value of viscosity module, calculated according to the formula, is 1.8, that falls into the recommended interval.

$$M_в=(M_{SiO2}+2M_{Al2O3})/(2M_{Fe2O3}+M_{CaO}+M_{MgO}+2M_{K2O}+2M_{Na2O}), \qquad (1)$$

where $M_в$ is viscosity module; $M_{Rm}O_n$ is quantity of the corresponding oxides, mass %.

Preparation of foam crystalline material has been carried out by two stage technology proposed in the paper [4]. The first stage includes synthesis of frit consisting of glass (not less than 70 %) and residual crystalline phase (not more than 30 %), that is provided by chemical composition of a blend. The second stage includes preparation of foam forming blend from frit powder with obtaining finished foam material.

Results of DTA showed that at heating of the blend of selected composition up to 1000 °C one observed endo-effects corresponding to removal of hygroscopic and crystallization water of the blend (82 and 245 °C), polymorphic transitions of quartz (571 °C), melting of forming double salts and eutectics (733 and 797 °C). The main mass losses occur at temperature range of 500 – 800 °C, corresponding to silication reactions. At temperature of 800 °C thermogravimetric curve comes out on horizontal that shows total binding of sodium carbonate and completion of silication reaction according to equation (2).

$$n\,SiO_2 + Na_2CO_3 \rightarrow Na_2O \bullet nSiO_2 + CO_2\uparrow \qquad (2)$$

It is stated by data of thermal analysis that thermal treatment of the blend under investigation at temperatures of 800 – 900 °C secures total completion of silication processes. To confirm this supposition one carried out X-ray phase analysis of frit prepared at temperatures of 800 – 900 °C (with step of 25 °C). Results of quantitative X-ray phase analysis show that frit consists of glass phase in range from 66 to 72 %, remaining crystalline phase presented by residual quartz (d=3.342 nm; 2Θ = 26,7o), feldspar and silicate $Na_2O \bullet CaO_3 \bullet SiO_2$.

Differential – thermal analysis of frit showed no exothermic effects both at heating up to 1000 °C, and at cooling to 200 °C, that testifies absence of crystallization processes. DTA of foam forming mixtures prepared on the basis of frits prepared at various temperatures (from 800 to 900 °C) with addition of soot (0.5 %) showed presence of exo- and endo-effects corresponding to processes of oxidation of gas forming agent and melting of glassy phase. Increase of frit preparation temperature biases exo-effect temperature in higher field, so for admixture from frit (800 °C) effect occurs at 463.7 °C, while for frit (900 °C) value is 498.6 °C, that is conditioned by differences of phase composition of frits. According to quantitative XFA results when sintering temperature of frit increased from 800, 825, 850 to 900 °C, content of crystalline phase decreased from 34, 32, 29 to 27 %.

Bias of oxidation temperature in higher field is favorable for foaming process since probability of early burning out of gas forming agent decreases. Total mass losses for frit (800 °C) is 5 times more in comparison with foam forming admixture prepared from frit (900 °C). In addition, thermograms of admixtures of low temperature frits (800 and 825 °C) show some endo-effects, thermograms of frits with temperatures 850 and 900 °C show one endo-effect that points on large inhomogenity of low temperature frits and the presence of some phases differing with melting temperature.

Foaming of pellets (d=10 mm), prepared from foam forming admixtures (specific surface 6000 cm^2/g), has been performed in temperature range 900 – 975 °C (with step 25 °C) with exposure at maximal temperature in 20 minutes. It is stated that foam material with relatively uniform fine porous structure is obtained at foaming temperature of 950 °C. While macrostructure of the samples obtained at 900 – 925 °C has dense crust with thickness more than 3 mm, and the samples obtained at 975 °C are deformed in consequence of vitrification and show random structure with the presence of large pores with size more than 5 mm. It is stated by XFA data that if foaming temperature increases from 900 to 975 °C quantity of crystalline phase in finished foam material decreases from 23 % to 8 %.

One traces connection of foam material macrostructure with temperature of preparation of frit, from which foam forming admixture is obtained. The foam material obtained from frit at 900 °C is the most optimal from point of view of pore size and uniformity of their distribution.

Values of the main physical – mechanical properties of samples prepared from frit synthesized at various temperatures (Fig. 1) allow to make a note of following laws in changes of properties. If temperature of frit synthesis increases from 800 to 900 °C, density of pellets decreased on an average in two times, strength – in four times, and water absorption – in 13 times.

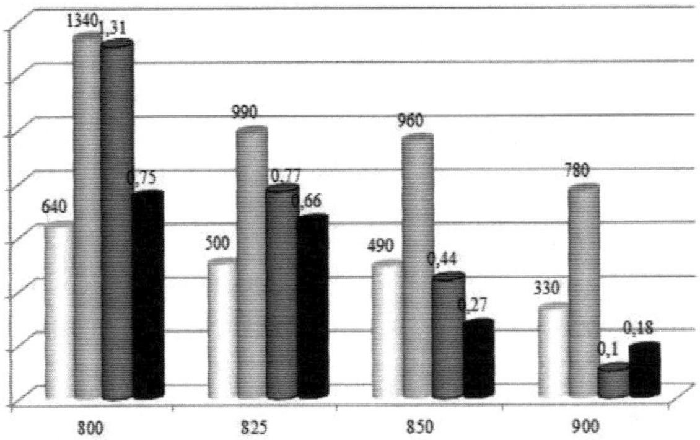

Fig. 1 – Properties of the foam material received from frit, synthesized at: a – 800; b – 825; c – 850; d – 900 °C

Comparative characteristic of the samples prepared on the basis of mortar sand screenings with other heat insulating materials shows that the material occupies intermediate position between foam glass and expanded clay (Table 2). The value of strength coefficient of the foam crystalline material, which is ratio of strength to density, exceeds one of foam glass and expanded clay. Material differs with high strength, low water absorption and relatively low heat conduction coefficient that allows to recommend its as heat – insulating and construction material.

Table 2. The comparative characteristic properties of foams

The granulated material	Bulk density, kg/m^3	Strength at compression in the cylinder, MPa	Heat conduction coefficient, W/m.K	Water absorption volume, mass %	Strength coefficient
Foamglass-crystal	300 – 450	4,5 – 5,5	0,09 – 0,10	0,5 – 0,7	1,3
Foamglass from a cullet	100 – 250	1 – 1,5	0,06 – 0,08	5	0,7
Clayite	300 – 800	0,6 – 4,5	0,10 –0,16	8 – 20	0,5

Conclusions.

1. Screenings of mortar sands with content of SiO_2 about 70 mas % and size of particles less than 60 μm are suitable for preparation of foam crystalline material from low temperature frit synthesized at temperature of 900 °C. Glass phase of frit is not crystallized in range of foaming temperature, has sufficient viscosity and temperature of its pyroplastic state coincides with temperature of oxidation of gas forming agent.
2. If the synthesis temperature increases from 800 to 900 °C quantity of glass phase in frit increases from 66 to 73 %. The crystalline phase of frit is presented by residual quartz and feldspar as well as appearance of new phase in form of silicate $Na_2O{\bullet}CaO{\bullet}3SiO_2$. Content of crystalline phase in finished foam material decreases from 23 to 8 % in dependence on increase of foaming temperature from 900 to 975 °C.
3. Frit synthesized at 900 °C and foaming temperature 950 °C are optimal for preparation of foam material with improved properties. Prepared foam crystalline material has higher strength up to 5.5 MPa at bulk density 450 kg/m^3 and low water absorption not exceeding 1 mas %.

Researches has been carried out in the frame of joint Russian – German project no. 01DJ12087 of the Ministry of education and science of Germany BMBF on equipment of the Nanocentre of National Research Tomsk polytechnic university.

References

[1] Schmitz M., Röhling S., Dohrmann R., Waschschlamm (2012). Ein vernachlässigtes heimisches Rohstoffpotenzial. *Gesteins Perspektiven*, No. 8, pp. 16-18.
[2] Kazmina O.V., Vereshchagin V.I. , Abiyaka A.N. , and Popletneva Yu.V. 2009 Estimation of glass viscosity and glass ceramic compositions in temperature interval of their foaming. *Glass and ceramics*, No.7, pp. 6-9.
[3] Kazmina O.V., Vereshchagin V.I., Semukhin B.S., Abiyaka A.N. 2009 Low-temperature synthesis of quenched cullet from mixtures based on siliceous components for foamed material production *Glass and Ceramics*. No. 10. pp. 5 – 8.

IOP Conf. Series: Materials Science and Engineering **64** (2014) 012016 doi:10.1088/1757-899X/64/1/012016

Thermal and Dielectric Properties of High Performance Polymer/ZnO Nanocomposites

Divij Vaishnav[1] and R. K. Goyal[2]

[1] Research Fellow, Department of Metallurgy and Materials Science, College of Engineering, Pune – 411005, INDIA
[2] Associate Professor, Department of Metallurgy and Materials Science, College of Engineering, Pune – 411005, INDIA

Email: rkgoyal72@yahoo.co.in

Abstract. Zinc oxide (ZnO) filled high performance poly(aryletherketon) (PAEK) matrix nanocomposites were studied for the application in electronic applications. The nanocomposites were prepared using planetary ball milling process followed by hot pressing. Experimental density of the nanocomposites was close to those of theoretical density indicating porosity free samples. Scanning electron microscopy showed excellent dispersion of nano sized (< 100 nm) ZnO particles into the PAEK matrix. X-ray diffraction (XRD) confirmed that the size of ZnO crystallites is about 58 nm. Thermogravimetry analyzer (TGA) showed significant increase in thermal stability and char yield of the nanocomposites with increasing ZnO content in the matrix. The dielectric constants of the nanocomposites increased significantly compared to those of pure PAEK.

1. Introduction

Polymer matrix composites with discontinuous fillers (or dispersants) are widely used for the applications like die attachment, electrically and thermally conductive, adhesives encapsulations and thermal interface materials. It is well know that polymers are easily processable and need low processing temperatures However, compared to ceramics they have lower electrical and thermal conductivity, lower stiffness, and higher coefficient of thermal expansion (CTE). The field of nanotechnology is one of the most popular areas for current research and development. At present, easily availability of nanomaterials offers the promise of developing polymer matrix nanocomposites (PNCs) with tailored thermal, mechanical and electrical properties for a particular application. PNCs are polymer matrix composites (PMCs) containing fillers with at least one dimension in the range of 1 nm to 100 nm. There are several ceramic particles like Al_2O_3, AlN, SiO_2, BN, ZnO etc [1-5]. which have been added into the polymer matrices such as ultra high molecular weight polyethelyene (UHMWPE) [4], Polyvinyl alcohol [5], emeraldine base polymer [6], Low density polyethylene [7], Polycaprolacton [8], Polyester imide [9], Polyamide–imide [10], Silicon rubber [11], Polymethylmethacrylate [12-13], Nylon-6 [14], Polypropylene [15], Polystyrene [16], Polyimide [17]. It was reported that the addition of ZnO into the polymer matrix increases the thermal stability, dielectric properties, and wear resistance of the composites.

Polyaryletherketone (PAEK) is a high temperature semi-crystalline thermoplastics polymer which has high glass transition and melting point In addition, it shows very good impact strength at low temperatures, high mechanical fatigue strength, and a very low tendency to creep as well as good sliding and wear properties. The chemical resistance is also very good [18]. Despite its good mechanical, tribological and chemical properties, hardly any work on the PAEK based composites is reported in the literature. In view of above, the aim of this study is to study the effect of addition of ZnO nanopowder on the dielectric constant and thermal stability of the PAEK matrix.

2. Experimental procedure

2.1 Materials

PAEK powder was donated by Gharda Chemicals Ltd., Thane (Mumbai), India. The ZnO nanopowder was purchased from M/s Sigma Aldrich Chemicals Ltd., USA. According to manufacturer, its particle size is < 100 nm.

2.2 Preparation of PAEK/ZnO nanocomposites

The different weight percentages (0 to 30 wt. %) of ZnO were added into the PAEK matrix using planetary ball milling with 5 h mixing time. The ball to powder ratio was about 10:1. Mixed powder was taken out from the mill after 5 h and dried in a vacuum oven at 180 °C for 4 h. The weighted powder was filled into the tool steel die for hot pressing. The samples were hot pressed at 350 °C under 45 MPa pressure. The thickness and diameter of the samples were 1.5 mm and 13 mm, respectively. Since, properties of nanocomposites vary with volume fraction of reinforcing particles, volume fraction of ZnO was determined using equation 1.

$$V_f = \frac{W_f}{[W_f + W_m(\rho_f/\rho_m)]} \tag{1}$$

where, V_f is the volume fraction, W_f is the weight fraction, and ρ_f is the density of ZnO particles. W_m and ρ_m are the weight fraction and density of PAEK matrix, respectively. The wt. % and corresponding vol. % of the ZnO into the nanocomposites is shown in Table 1.

Table 1 The wt% and vol% of ZnO added to the PAEK matrix

Sample code	% ZnO in the PAEK matrix	
	By wt.	***By vol.***
NC-0*	0	0
NC-5	5	1.19
NC-10	10	2.49
NC-20	20	5.44
NC-30	30	8.97

** NC-X, where X represents wt% ZnO added into the PAEK matrix.*

2.3 Characterization

2.3.1 Density Measurement

The theoretical density (ρ_{th}) of the nanocomposites was calculated using rule of mixture as shown in equation (2). The densities of ZnO and PAEK were considered as 5.61 g/cc and 1.29 g/cc, respectively.

$$\rho_{th} = \rho_f V_f + \rho_m V_m \tag{2}$$

Where, V_m is the volume fraction of matrix. The experimental density (ρ_{ex}) was determined by Archimedes Principle using equation (3).

$$\rho_{ex} = \left[\frac{W_{air}}{(W_{air} - W_{liq})} \times \rho_{liq}\right] \tag{3}$$

where, W_{air} and W_{liq} are the weight of samples in air and liquid medium (ethanol), respectively. The density (ρ_{liq}) of ethanol is (0.79 g/cc).

2.3.2 X-Ray Diffraction (XRD)

XRD pattern of the pure ZnO nano-sized powder was recorded on Philips X'Pert PANalytical PW 3040/60. Ni filtered Cu kα radiation (λ = 1.54) generated at 40 kV and 30 mA was used for diffraction angle (2θ) ranging from 10° to 90°. Analysis was carried out at room temperature.

2.3.3 Scanning Electron Microscopy (SEM)

SEM (JOEL: JSM-6360A) was carried out to examine the dispersion of ZnO nanoparticles in the PAEK matrix. Before SEM, 30 wt% ZnO nanocomposites sample was fractured in liquid nitrogen for about 15 minutes. Sample surface was coated with a thin layer of platinum using sputter coater (JEC 60) to avoid its charging.

2.3.4 Dielectric properties

The dielectric constant was obtained from the measurement of capacitance (C) using Precision Impedance Analyzer (Wayne Kerr Electronics 6515B, UK) at frequency ranges from 1 kHz to 15 MHz at room temperature. The dielectric constant (ε) was evaluated using equation (4),

$$\varepsilon = \frac{(c \times t)}{(\varepsilon_o \times S)} \tag{4}$$

where, S is the surface area, C is the capacitance and t is the thickness of sample. The ε_o is the permittivity of free space (8.854×10^{-12} F/m) i.e. real part. The dielectric loss tangent/dissipation factor which is the ratio of dielectric loss to dielectric constant was obtained directly from the instrument. Before measurements silver paste was applied on both sides of the samples.

2.3.5 Thermal properties

Thermogravimetry analysis (TGA) was done to find out thermal stability and char yield of the nanocomposites using TGA (Perkin Elmer TGA 4000). The samples were heated from 50 °C to 1000 °C at a heating rate 10 °C/min in N_2 atmosphere. The temperatures of 10 wt% and 30 wt% loss were represented as the degradation temperature of T_{10} and T_{30}, respectively. The maximum decomposition temperature (T_{max}) was taken as the temperature corresponding to maximum of the peak obtained from the first order derivative curve of TGA (i.e. DTG). The % char yield was determined at temperature of 1000 °C.

3. Results and discussion

3.1 Density of nanocomposites

Figure 1 shows the experimental and theoretical density of nanocomposites as a function of ZnO content. It is observed that the density of the samples increased from 1.29 g/cc for pure PAEK to 1.68 g/cc for 30 wt% (9.0 vol.%) ZnO nanocomposite. It is due to higher density (i.e. 5.61 g/cc) of pure ZnO. The experimental density of the nanocomposite containing 5 wt% (~ 1.19 vol.%) ZnO is slightly higher than that of theoretical density. This may be due to the nucleating effect of ZnO nanoparticles which probably increases the crystallinity of the matrix. Thereafter, the experimental density of the nanocomposites is close to those of theoretical density of the samples.

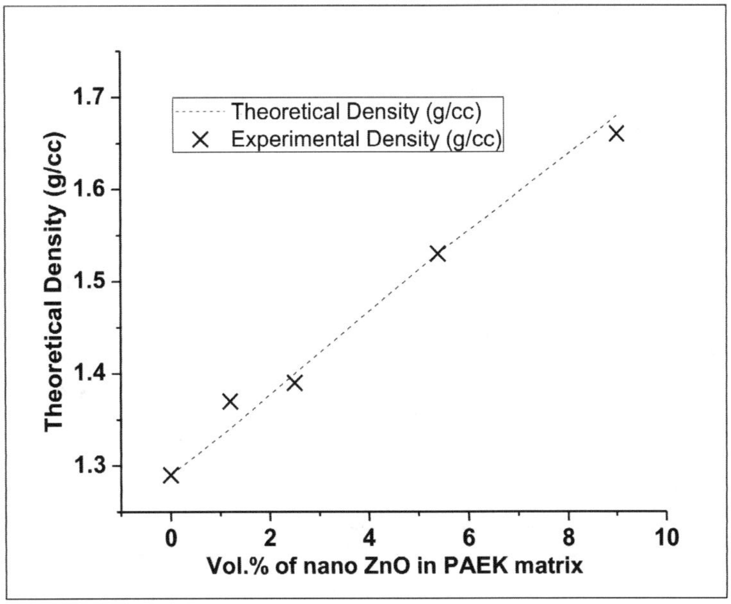

Figure 1 Density of PAEK nanocomposites as a function of ZnO.

3.2 X-Ray Diffraction (XRD)

Figure 2 shows diffraction intensity versus diffraction angular position (2θ) in the range 30-70° for the major crystallographic reflection for the ZnO nano powder. Pure ZnO powder shows sharp Bragg peaks at 31.837°, 34.502°, 36.334°, 47.650°, 56.726°, 63.012° and 68.114° corresponding to diffraction planes of (010), (002), (011), (012), (110), (013) and (112) indicating that it has an hexagonally wurtzite crystal structure [13]. The crystallite size of ZnO nanopowder was calculated using Scherrer's equation (Equation. 5),

$$\text{Crystallite size} = \frac{k \times \lambda}{\beta \times Cos\, \theta} \tag{5}$$

where, k is the shape factor (for cubic materials, it is 0.9), λ is the Cu kα wavelength of X-ray, β is the full width at half maxima and θ is the Bragg's angle. The powder size was about 58 nm.

IOP Conf. Series: Materials Science and Engineering **64** (2014) 012016 doi:10.1088/1757-899X/64/1/012016

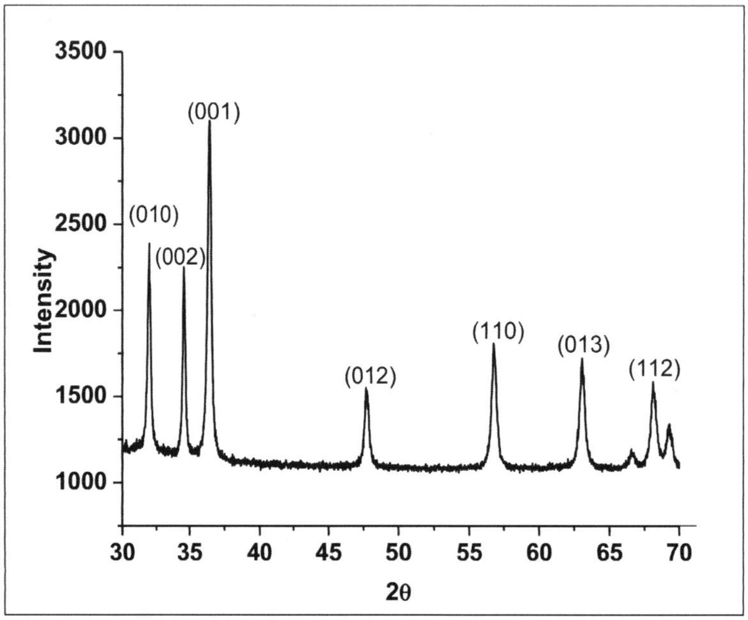

Figure 2 XRD pattern of pure ZnO nano sized powder.

3.3 Scanning electron microscopy (SEM)

Figure 3(a) shows the SEM image of pure ZnO nanopowder. It can be seen that the size of ZnO particle is smaller than 100 nm and some aggregates are also seen. Figure 3(b) shows SEM image 30 wt% ZnO filled PAEK nanocomposite (NC-30) at a magnification of ×10,000. It can be clearly seen that the ZnO nanoparticles are almost uniformly dispersed in the matrix. However, ZnO aggregates were also observed at some location due to its hydrophilic nature [9] and higher specific surface area.

3.4 Dielectric Properties

Figure 4 shows the dielectric constant of PAEK/ZnO nanocomposites. For a given frequency, dielectric constant of nanocomposites increases with increasing ZnO content in the matrix. It is well known that ZnO is a polar ceramic. When a high frequency AC field is applied, the ZnO particle is in a quick and strong polarization and hence it results in higher dielectric constant for the nanocomposites [7]. Figure 5 shows the dissipation factor of the nanocomposites. It decreases for the nanocomposites containing up to 10 wt% ZnO, thereafter it increases with increasing ZnO content in the matrix.

3.5 Thermogravimetry analysis (TGA)

Figure 6(a) shows the residual weight of the nanocomposites as a function of temperature. It is observed that pure PAEK has a degradation temperature (T_{10}) of 567 °C. It was observed that the T_{10} of the nanocomposites increases significantly with increasing ZnO content and also the char yield at 1000 °C increased significantly from 53 % for pure PAEK (NC-0) to 67.5 % for NC-30 nanocomposites. Therefore, the incorporation of ZnO in polymer matrix improved the thermal stability of the nanocomposites. The increase in thermal stability could be due to strong interaction and/or interfacial bonding between the matrix and ZnO nanoparticles, which hindered the segmented movement of the PAEK [9]. Figure 6(b) shows the derivative curve of the nanocomposites as a function of temperature.

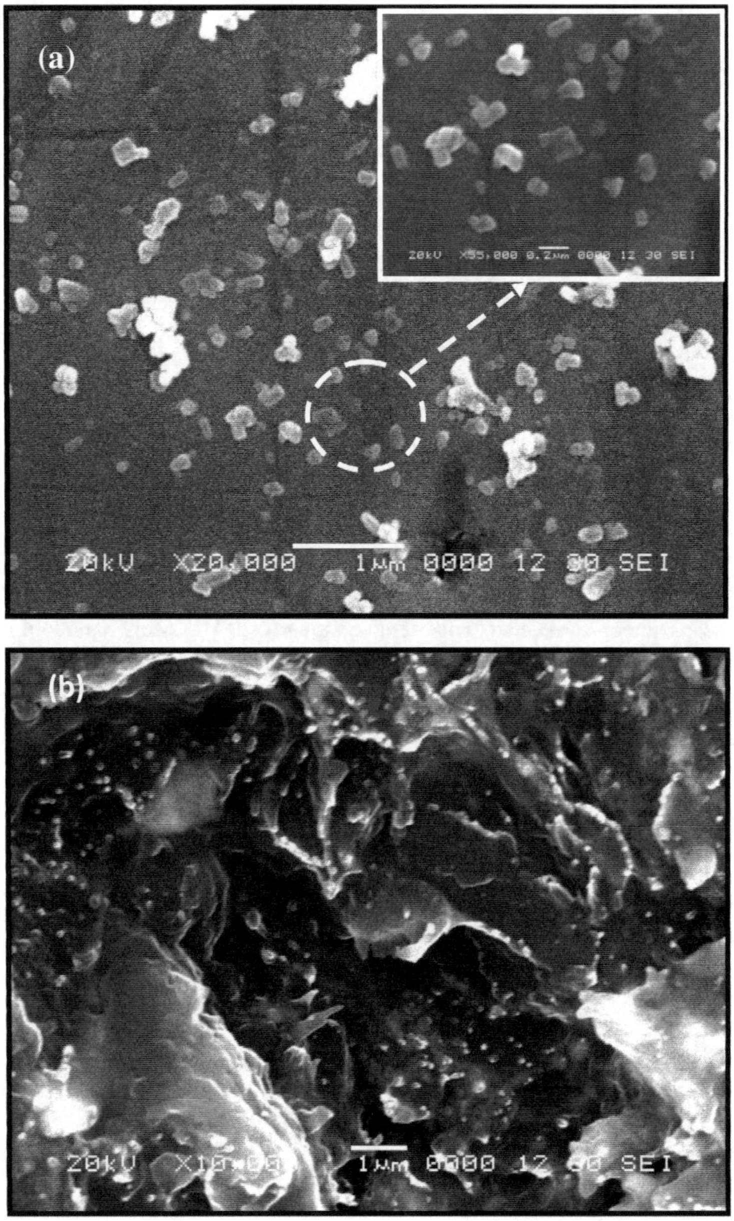

Figure 3 SEM images of (a) pure ZnO nanopowder, inset shows high magnification image (b) 30 wt. % ZnO (NC-30) filled PAEK nanocomposites.

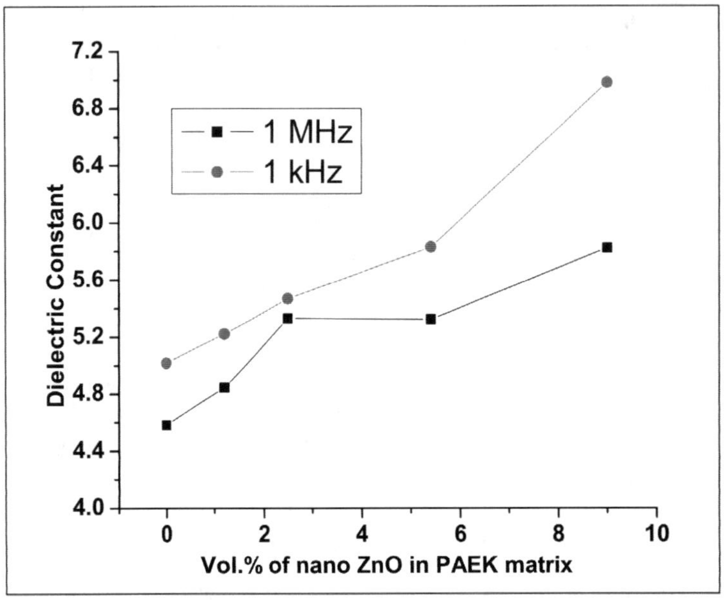

Figure 4 Dielectric constant versus vol% ZnO content for PAEK/ZnO nanocomposites

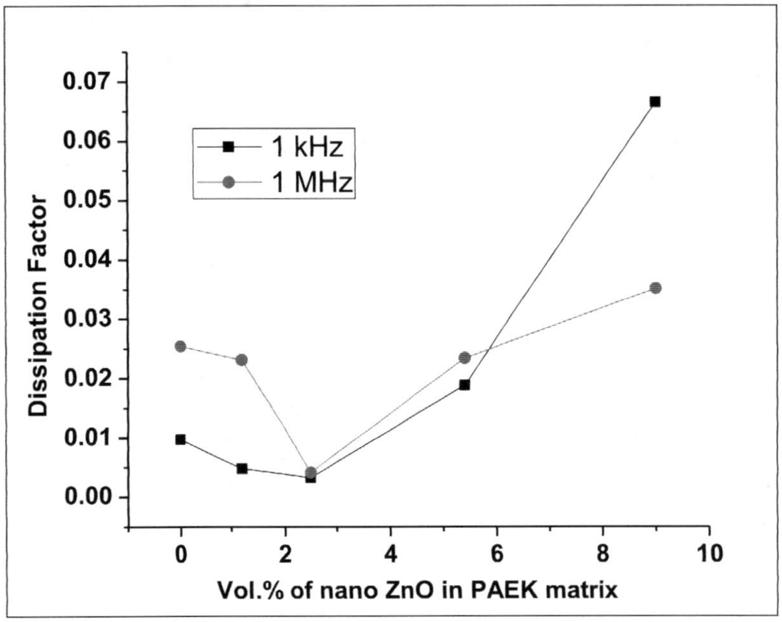

Figure 5 Dissipation factor versus vol.% ZnO content for PAEK/ZnO nanocomposites

When major weight loss begins, there is a sharp increase in the derivative which then reaches a maximum rate of weight loss. After the maximum, the derivative curves tend to level off somewhat. It shows that the maximum decomposition of the nanocomposites increases from 579 °C for the pure matrix to 592 °C for NC-30.

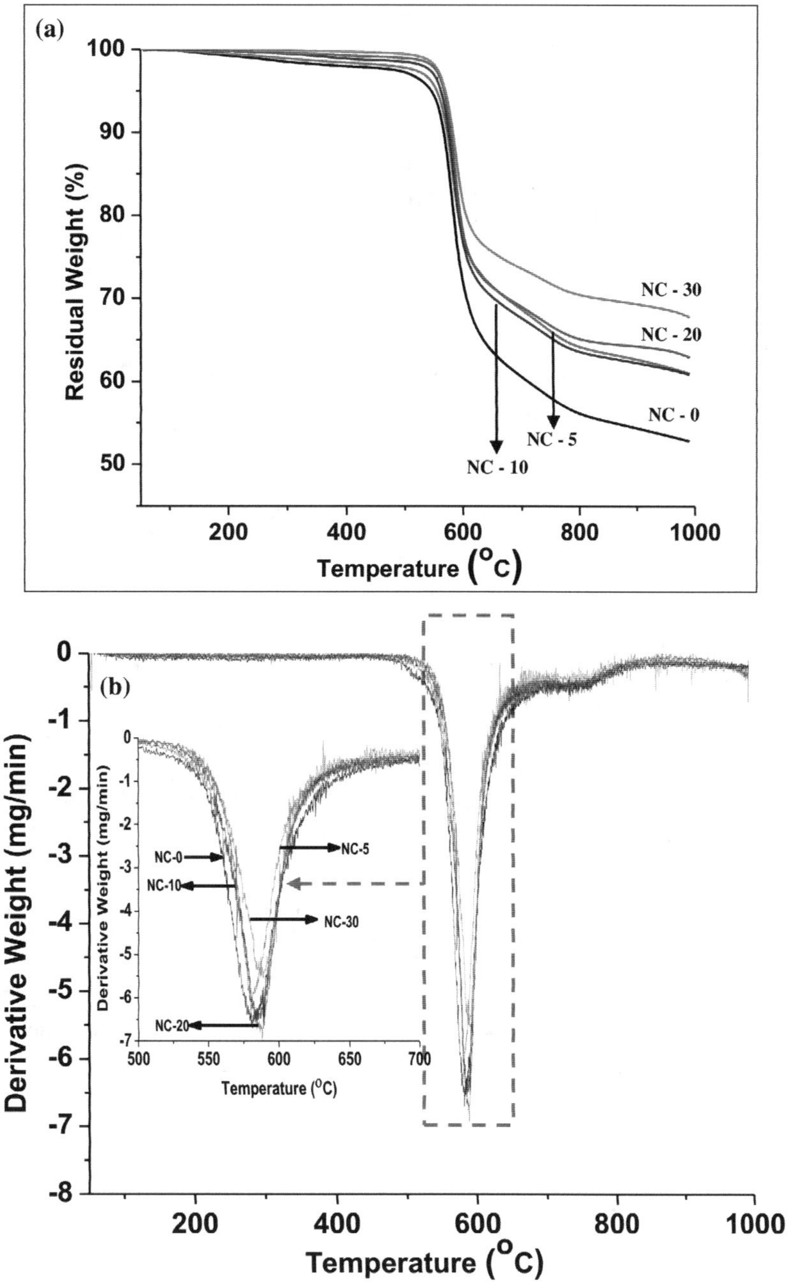

Figure 6 (a) TGA curves and (b) DTG curves for the PAEK/ZnO nanocomposites under nitrogen atmosphere

4. Conclusion

High performance PAEK/ZnO nanocomposites were successfully prepared using planetary ball mill followed by hot pressing. XRD shows that ZnO powder has hexagonal crystal structure and its crystallite size is about 58 nm. SEM shows good dispersion of ZnO nanoparticles into the matrix. Dielectric constant and the thermal stability of the nanocomposites increased significantly with increasing content of ZnO in the matrix.

5. Acknowledgement

We greatly acknowledge to Dr. Mathew Abraham, Polymer Division, M/s Gharda Chemicals Ltd., Thane (Mumbai), India for donating commercial grade PAEK powder for this research. Divij is thankful to Professor S.T. Vagge, Head of the department of Metallurgy and Materials Science, College of Engineering, Pune for his kind moral support. Authors are also thankful to Dr. A. D. Sahasrabudhe, Director, College of Engineering, Pune for his moral support to faculty and students.

6. References

1. Goyal RK, Tiwari AN, Mulik UP, Negi YS. "Thermal, mechanical and dielectric properties of high performance PEEK/AlN nanocomposites", *J Nanosci Nanotechnol 2009,* **9** 6902–09.
2. R. Goyal, A.Kapadia, "Study on phenyltrimethoxysilane treated nano-silica filled high performance poly(ehteretherketone) nanocomposites", *Composites: Part B 2013*, **50** 135-43
3. R. Goyal, Y. Negi, A. Tiwari, "Preparation of high performance composites based on aluminium nitride/poly (etheretherketone) and their properties", *European Polymer Journal 2005,* **41** 2034-44
4. B. C. Chang, H.M. Akil, R. Nasir, "Comparative study of micro and nano-ZnO reinforced UHMWPE composites under dry sliding wear", *Wear 2013*, **297** 1120-27.
5. A. Roy, S. Gupta, S. Sindhu, A. Parveen, P. Ramamurthy, "Dielectric property of novel PVA/ZnO hybrid nanocomposites films", *Composites 2013*, **47** 314-19
6. M. Singla, R. Sehrawat, R. Nidhi, K. Singh, "Dielectric behaviour of emeraldine base polymer-ZnO nanocomposites film in the low to medium frequency", *J. Nanopart Res. 2011*, **13** 2109-16
7. Z. Dang, L. Fan, S. Zhao, C.Nan, "Preparation of nanosized ZnO and dielectric properties of composites filled with nanosized ZnO", *Materials science and engineering 2003*, **99** 386-89
8. M. Ahmad, A. Fatehi, A. Zakaria, S. Mahmud, S. Mohammadi, "Fabrication of an electrically-resistive varistor-polymer composite", *Int. J. Mol. Sci. 2012*, **13** 15640-652
9. A. Abdolmaleki, S. Mallakpour, S. Borandeh, "Preparation, characterization and surface morphology of novel optically active poly(ester-imide)/functionalized ZnO bionanocomposites via ultrasonication assisted Process", *Applied surface science 2011*, **257** 6733-35
10. S. Mallakpour, M. Madani, "Use of silane coupling agent for surface modification of zinc oxide as inorganic filler and preparation of Poly(amide-imide)/ZnO nanocomposite containing phenylalanine moieties", *Bull. Mater. Sci. 2012*, **3** 333-39
11. L.Sim, S.Ramanan, H. Ismail, K.Seetharamu, T. Goh, "Thermal characterization of Al_2O_3 and ZnO reinforced silicon rubber as thermal pads for heat dissipation purposes", *Thermochimica Acta 2005*, **43** 155-65
12. W. Viratyaporn, R. Lehman, "Effect of nanoparticles on the thermal stability of PMMA nanocomposites prepared by in situ bulk polymerization", *J. Therm Anal Calorim 2011*, **103** 267-73
13. R. Hong, J.Qian, J.Cao, "Synthesis and characterization of PMMA grafted ZnO nanoparticles", *Powder Technology 2006*, **163** 160-8

14. J. Zheng, R. Siegel, C. Toney, “Polymer crystalline structure and morphology changes in Nylon-6/ZnO nanocomposites”, *Polymer Science 2003*, **41** 1033-50
15. J. Tang, Y. Wang, H.Liu, L. Belfiore, “Effect of organic nucleating agents and zinc oxide nanoparticle on isotactic polypropylene crystallization”, *Polymer 2004*, **45** 2081-91
16. E. Tang, S. Dong, “Preparation of styrene polymer/ZnO nanocomposites latex via miniemulsion polymerization and its antibacterial Property”, *Colloid Polym Sci 2009*, **287** 1025-32
17. S. Hsu, W. Whang, C. Hung, P. Chiang, Y. Hsiao, “Effect of the polyimide structure and ZnO concentration on the morphology and characterization of polyimide/Zno Nanohybrid Films”, *Macromol. Chem. Phys. 2005*, **206** 291-98
18. Rosato, V.Dominick, V.Donald, “Plastic product material and process selection handbook”, *Elsevier (2004)*

IOP Conf. Series: Materials Science and Engineering **64** (2014) 012017 doi:10.1088/1757-899X/64/1/012017

Core-shell composite metal catalysts incased into natural ceramic nanotubes

V Vinokurov[1,3], A Berberov[1], D Afonin[1], H Borzaev[1], E Ivanov[1], P Gushchin[1], Y Lvov[2]

[1]Gubkin Russian State University of Oil and Gas, 65 Leninsky prospect, Moscow, Russia

[2]Institute for Micromanufacturing, Louisiana Tech University, 911 Hergot Ave, Ruston, LA 71272, USA

[3]Corresponding author

E-mail: vladimir@vinokurov.me

Abstract. The bimetallic halloysite nanotubes were prepared by the injection of halloysite-containing aerosols into the microwave plasma reactor. Nanotubes contain metal nanoparticles formed from the metal salt solution in the lumen of nanotubes and the iron oxide nanoparticles at the outer surface of nanotubes. Such halloysite composites may be sputtered onto the surface of the porous carrier forming the nanostructured catalyst, as was shown by the pure halloysite sputtering onto the model porous ceramic surface.

1. Introduction

Nanotubes of different nature intercalated with metal and metal oxides are very efficient as mesoporous media for advanced catalysis. Natural halloysite clay nanotubes available in large quantities at low price are very promising for this purpose. Halloysite tubes have diameter of around 50 nm, inner lumen of 15-20 nm and length of 1-2 μm [1]. Traditional intercalation method usually include wetting nanotubes with aqueous metal salts following with chemical treatment for reduction [2,3]. The reaction volume being restricted within nanoscale, one can produce metal nanoparticles fitting the tube lumen exploiting such tubule nanoreactor. Previously, we have applied this approach for the synthesis of 5-10 nm diameter particles of Ag, Au and Fe_3O_4 incased into such core-shell structures [4].

2. Experimental

In the present work, even better results on core-shell nanocatalytic systems were obtained by introducing the aerosols of aqueous nitrate salts of iron, nickel or cobalt mixed with the clay nanotubes into the low temperature argon or carbon dioxide plasma. Treatment was performed during 2 sec preventing the halloysite decomposition though allowing for metal nanoparticle formation on and into the tubes.

The microwave plasma reactor was assembled from coaxial waveguide with the central nozzle electrode and the magnetron (P=1 kW, ν=2.45 GHz). The high electromagnetic field density near the nozzle electrode opening ensures the ionization of the wide range of gases (air, argon, nitrogen, methane, carbon dioxide, etc.). The energy conversion efficiency by the reactor is up to 60%, while the plasma temperature is ca. 2500 K.

3. Results and discussions

The iron oxide nanoparticles formed in the tube lumen have diameter of ca. 3 nm and nanoparticles formed on the tube outer surface have larger diameters of ca. 10 nm (see figure 1). The reason of this diameter difference may be related with steric hindrance, different component concentrations inside and outside the nanotube, as well as with the different chemistry and the possible temperature gradient in the tube interior and outer surface.

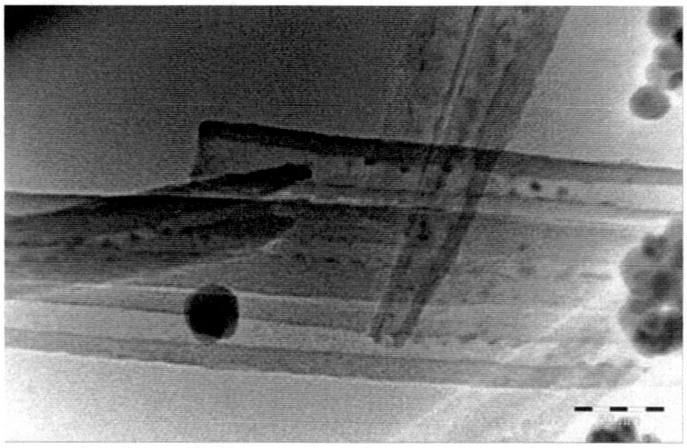

Figure 1. TEM micrograph of the CoO nanoparticles in the interior and Fe_2O_3 at the outer surface of halloysite tube.

The elaborated method allowed the deposition of different metals at the inner and outer surfaces of ceramic nanotubes. As the outer layer of metal nanoparticles is formed from the eroded iron nozzle electrode material, spraying of the aerosols of cobalt nitrate-loaded halloysite into the microwave plasma produces bimetallic core-shell nanotubes.

In the present work, there was made an attempt to sputter pure halloysite on a porous ceramic surface by direct aerosol injection of its aqueous suspension in plasma beam. An even nanoparticles layer thicker than 10 μm was formed on a ceramic surface as a result. SEM micrographs show that halloysite particles do not lose their shape and do form an even close-packed layer on a surface of the porous catalyst support in microwave plasma conditions, as one can see in figure 2.

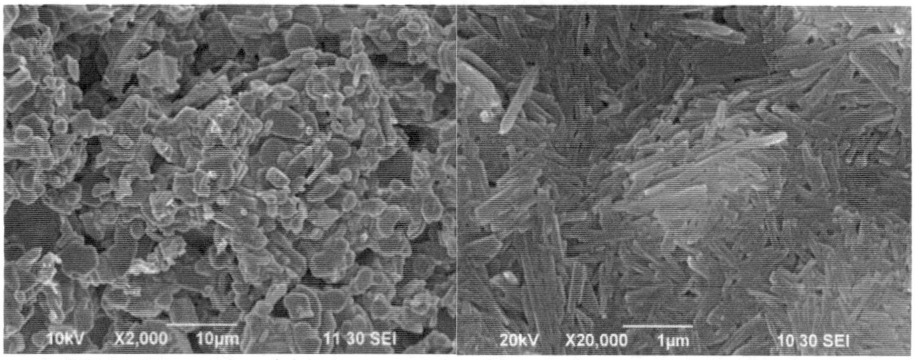

Figure 2. SEM micrographs of the same ceramic surface before (at the left) and after (at the right) the plasma sputtering with halloysite.

The proposed method of halloysite sputtering on a solid porous surface may be implemented for the advanced nanostructured catalysts preparation, given the possibility of preliminary preparation of the bimetallic core-shell halloysite nanotubes. This approach is quite perspective for the preparation of new high-efficiency catalysts for the organic and petrochemical syntheses.

4. Acknowledgements

The work was supported by the President of Russia (project 14.124.13.7577-MK).

References

[1] E. Abdullayev, Y. Lvov, Halloysite clay nanotubes for controlled release of protective agents. J. Nanosci. Nanotech., 2011, 11 10007-10026.

[2] Varade, D. & Haraguchi, K. Synthesis of highly active and thermally stable nanostructured Pt/clay materials by clay-mediated in situ reduction. Langmuir, 2013, 29, 1977–1984.

[3] Zhen-Yu Cai, Huan dai, Yi Liu, Jian-Xin Mao, Xin-Zhi Chen, Chao-Hong He. Halloysite nanotubes supported gold catalyst for cyclohexane oxidation with molecular oxygen. Advances in Chemical Engineering and Science, 2011, 1, 15-19.

[4] Kirill A. Arapov and Yuri M. Lvov. Iron Oxide (III) Nanoparticle Synthesis in Halloysite Clay Nanotube Lumen. Polymeric Materials: Science & Engineering, 2011, 104, 338.

Silver nanoparticles on Zinc Oxide thin film: An insight in fabrication and characterization

M K Hossain[1,$], Q A Drmosh[2,3,$], Z H Yamani[2,3] and N Tabet[3,4]
[1]Center of Research Excellence in Renewable Energy, King Fahd University of Petroleum and Minerals, Dhahran, 31261, Kingdom of Saudi Arabia.
[2]Center of Research Excellence in Nanotechnology, King Fahd University of Petroleum and Minerals, Dhahran, 31261, Kingdom of Saudi Arabia.
[3]Department of Physics, King Fahd University of Petroleum and Minerals, Dhahran, 31261, Kingdom of Saudi Arabia.
[4]Qatar Environment and Energy Research Institute, Qatar Foundation, Qatar.

[$]E-mail: drmosh@kfupm.edu.sa, kamalhossain@kfupm.edu.sa

Abstract. In this work, a simple two-steps process has been explained to fabricate silver (Ag) nanoparticles on Zinc Oxide (ZnO) thin film followed by their characterizations. The underneath layer ZnO thin film, as an example, was also investigated how the properties change during the course of nanoparticles fabrication. ZnO thin film was sputtered on standard glass substrate followed by further sputtering of an ultrathin Ag layer. Subsequently the specimen was treated at high temperature in inert environment. A periodic observation at specific temperature intervals confirmed the formation of Ag nanoparticles on ZnO thin film. Field-emission scanning electron microscopic (FESEM) observations revealed the size distribution of as-fabricated Ag nanoparticles in the range of 50-250 nm. Elemental analysis was also confirmed by SEM-aided energy dispersion spectroscopy. The underneath layer ZnO thin film was found to go through recrystallization, stress relaxation, and grain growth during the annealing process. Further treatment to ZnO only film showed a variation in surface topology with reference to those with Ag nanoparticles on ZnO. Such a system was also analysed with finite different time domain (FDTD) analysis. A typical model was considered and FDTD simulation was carried out to understand the trend of absorption depth profile within the absorbing layer involved in plasmonics solar cell.

1. Introduction

Metal nanoparticles exhibit unprecedented and exciting properties strongly dependent on their size, shape and inherent electrons distribution with reference to those of macro-scaled counterpart [1]. Amongst them, silver nanoparticles (Ag-NPs) have attracted huge interest because of their unique physical, chemical and biological properties. Enhanced electrical and thermal conductivity, surface-enhanced Raman scattering, chemical stability, catalytic activity, enhanced photocurrent induction, etc. are some of distinguished physio-chemical properties of Ag-NPs responsible behind their potential value in microelectronics, sensing, optoelectronics, corrosion and biomedical [2–9]. Apart from these, Ag-NPs are extremely popular in a wide range of consumer products such as plastics, soaps, pastes, food, textiles, etc. because of their excellent anti-bacterial and anti-fungal activity [10–13].

Many methods are reported so far using chemical, physical and photochemical routes. Every route carries some pros and cons with common issues such as costs, uniformity, controllability, purity, etc. Amongst the existing methods, the wet chemical synthesis is known to be the best [14–18]. The formation of colloids from the reduction of silver salts involves two stages *viz.* nucleation and subsequent nanoparticles growth thereof. Therefore, purity, cost and other issues remain same. The photo-induced synthetic strategies are generally consisted of photo-physical and photochemical processes. In photo-physical, the NPs are synthesized through break-down of bulk metals into nanoscale particles, whereas photochemical route gets NPs into ionic precursors [19–20]. In such techniques, purity, scalability and uniformity become a real challenge. NPs can be synthesized by evaporation–condensation, thermal-decomposition, arc discharge, etc. Narrow size distribution and almost zero contamination probability are the two key advantages in this approach with reference to other routes. The physical synthesis process of Ag-NPs usually utilizes the external energies to produce Ag-NPs with nearly narrow size distribution [21]. The physical approach can permit producing large quantities of Ag-NPs samples in a single process. This is also the most useful method to produce Ag-NPs with different size and shape. However, primary costs for investment of equipment are very high.

In this work, a simple two-steps process was explained to fabricate Ag-NPs on Zinc Oxide (ZnO) thin film. ZnO thin film was sputtered on standard glass substrate followed by further sputtering of an ultrathin Ag layer. Thereafter the specimen was treated at high temperature in inert environment. Size distribution of as-fabricated Ag-NPs in the range of 50-250 nm was confirmed by field emission scanning electron microscopic (FESEM) observations followed by SEM-aided energy dispersion spectroscopy (EDS) for elemental analysis. The underneath layer ZnO thin film was found to go through recrystallization, stress relaxation, and grain growth during the annealing process. A typical simulation for Ag-NPs dispersed on substrate was carried out using Lumerical Solutions to understand the trend of absorption depth profile induced inside the substrate. Such a system in fact shows the contribution of Ag-NPs to the absorbing layer of plasmonics solar cell.

2. Experimental

Thin layers of metallic silver were obtained by DC sputtering technique using an automatic sputter coater (model NSC-4000). The deposition of Ag-ZnO layers was carried out in two steps. First, ZnO thin films were deposited on glass substrate using reactive plasma of pure oxygen (O_2) and pure Zinc target (99.999 %). The deposition time was maintained ~25 minutes at 100 watt plasma power. In the second step, Ag was sputtered under pure Argon (Ar) using a pure silver target (99.99%) received from Semiconductor wafer, Inc. with deposition time of ~10s. Standard procedure was maintained to keep the targets neat and clean. The targets were ultrasonicated for 15 min in acetone followed by 15 min ultrasonic cleaning in methanol. The targets were plasma etched for 1 min before starting each deposition process. The

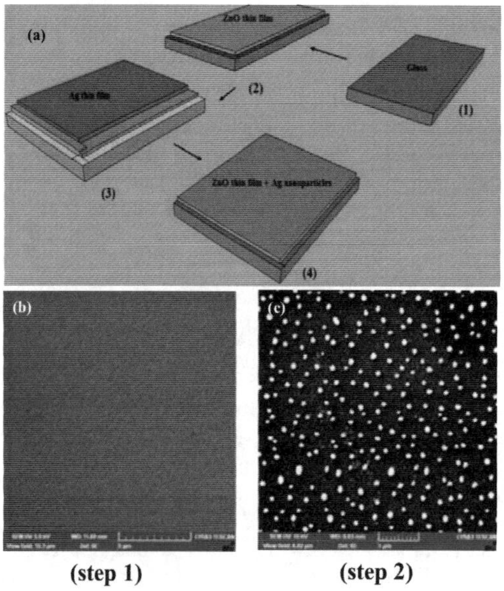

Figure 1. (a) A free-hand schematic of two-steps process adapted herewith to fabricated Ag nanoparticles on ZnO thin film and (b)-(c) SEM micrographs of each individual steps. Individual scale bar show the image size.

target substrate distance was maintained fixed at 10 cm. The base pressure in the chamber was less than 2×10^{-6} torr and the working pressure was set to 7×10^{-3} torr by adjusting the gas flow at 80 sccm (pure O_2 for ZnO deposition and Ar for silver deposition). The samples were annealed in a tube furnace (MTI OTF-1200X) at different temperatures ranging from 200°C to 700°C under a constant flow of Ar atmosphere. FESEM (FEI Nova-Nano SEM-600) was applied for morphological studies followed by SEM-aided EDS. A typical system of Ag-NPs on Si was modelled and absorption depth profile was simulated by FDTD solutions (Lumerical Solutions Inc.).

3. Results and discussion

Ag-NPs on ZnO were fabricated in simple two steps process, *viz.* step1: sputtering of ultrathin Ag layer on ZnO thin film and step2: subsequent heat treatment under a constant flow of Ar. A free hand schematic was shown in Fig.1a wherein the process started with glass substrate to ZnO and Ag sputtering to heat treatment. Sputtering of Ag layer on ZnO thin film showed a relatively smooth surface topology as shown in SEM image of Fig.1b with reference to a clear and visible formation of Ag-NPs after the treatment as shown in Fig. 1c.

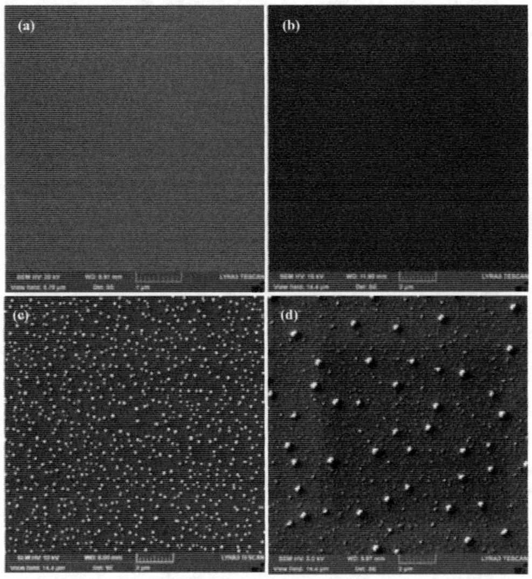

Figure 2. (a)-(d) FESEM micrographs of Ag-NPs on ZnO after heat treatment for 4 hrs in Ar atmosphere at T = 200°C, 400°C, 600°C and 700°C respectively. Individual scale bar show the image size.

This is very important to understand the evolution of Ag-NPs from ultrathin Ag layer on ZnO. A systematic approach was taken to observe the sequences of Ag-NPs formation on ZnO thin film. The ultrathin layer of Ag on ZnO was treated at specific temperature for 4 hours in Ar atmosphere and the morphology was investigated by FESEM. It was revealed that at 200°C the Ag film remained as it was as shown in Fig. 1b and Fig. 2a. The nucleation was found to start at annealing temperature of 400 °C and the average nanoparticles size increased with increasing annealing temperatures as shown in Fig. 2b-d. To the extend, the density of nanoparticles was observed to decrease from *ca.* 7.5×10^{8} particles/cm^2 (at 600 °C) to *ca.* 3.125×10^{7} particles/cm^2 (at 700 °C) in addition to broad size distribution.

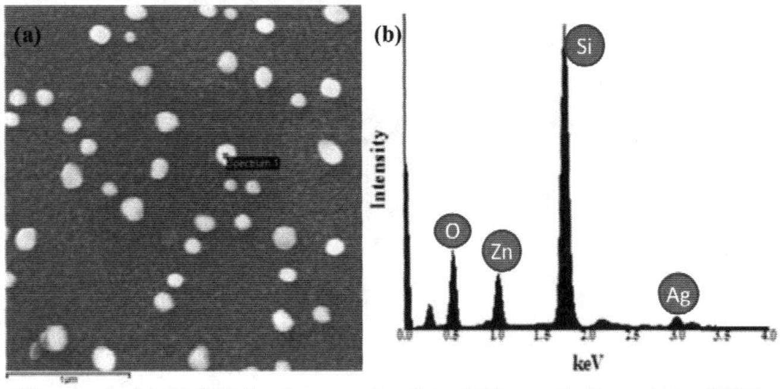

Figure 3. (a) FESEM micrograph of Ag-NPs on ZnO used for SEM-EDS analysis and (b) EDS spectrum showing the presence of Ag, Zn, O and Si peaks. Scale bar show the image size.

The elemental analysis of as-fabricated Ag-NPs on ZnO thin film was carried out by FESEM-aided EDS. As shown in Fig. 3c, the peaks for Zn, O and Si were obvious in addition to Ag Lα and Ag

Lβ peaks at 2.98 keV and 3.15 keV respectively. The positions of the Ag peaks coincide with the FESEM database as well as the values reported previously [22].

Since ZnO thin film was always there under the ultrathin Ag layer, it was apparent that underneath layer would also go through the same treatment during the course of fabricating Ag-NPs. That's why, an addition investigation was carried out to realize the characteristics of ZnO thin film with and without ultrathin Ag layer. FESEM micrographs of ZnO thin film without Ag ultrathin layer confirmed that the surface topography became rough after the heat treatment in addition to some small pores thereof. Fig. 4a and Fig. 4b show the SEM images of ZnO thin film without ultrathin Ag layer before and after the treatment respectively. X-ray diffraction of such specimen also confirmed the [0001] directional growth of ZnO and grain growth after the treatment. ZnO thin film with Ag ultrathin layer showed that the optical band gap decreases from 3.31 eV to 3.24 eV with increasing annealing temperature. The detailed measurements and interpretation is out of the context and explained somewhere else [23].

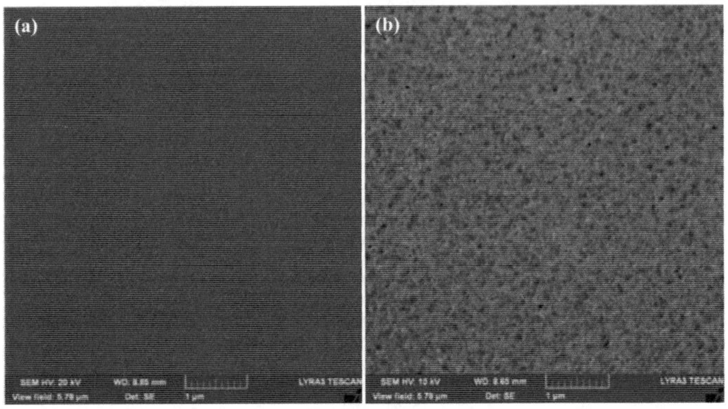

Figure 4. SEM micrograph of ZnO thin film without ultrathin Ag layer (a) before treatment and (b) after treatment. Individual scale bar show the image size.

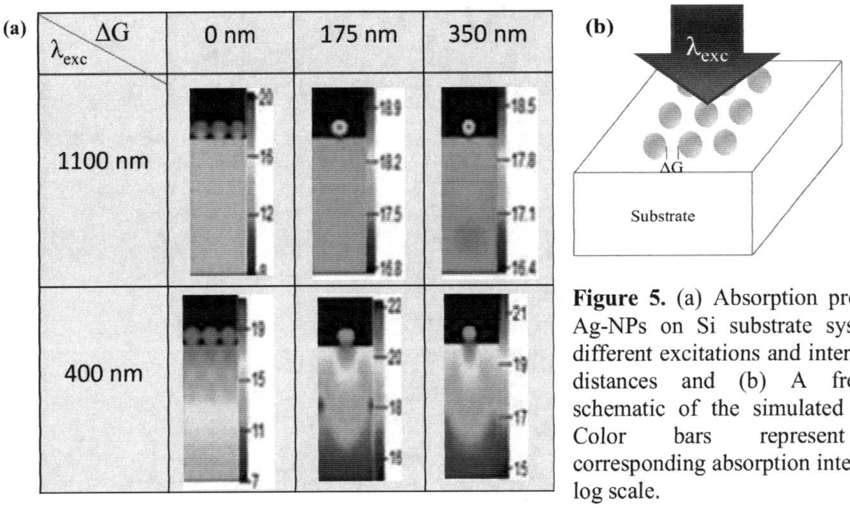

Figure 5. (a) Absorption profile of Ag-NPs on Si substrate system at different excitations and interparticle distances and (b) A free-hand schematic of the simulated model. Color bars represent the corresponding absorption intensity in log scale.

Strong optical scattering associated with metal nanoparticles due to plasmonics can be a prime factor to engineer photon propagation into and within a solid-state semiconductor device, such as plasmonic solar cell. Until recently studies have focused on the increase of absorption in Si based solar cells through the surface plasmon effect [24-27]. It possesses at least three-ways contribution to the solar cell, *viz.* i) subwavelength scattering element to trap free plane waves into absorption layers, ii) surface plasmon polaritons modes to confine and guide the light in absorption layers and (d) nanoantenna to store incident energy efficiently in absorbing layers. FDTD is a convenient technique to analyse and predict inherent characteristics of any system. Considering the benefit of as-fabricated Ag-NPs, a typical model was used to understand the trend of absorption depth profile as can be seen in Fig. 5. The model consisted of Ag-NPs of 100 nm diameter dispersed on silicon substrate of more than a micron depth. The interparticle distance was varied from 100 to 275 to 450 nm. A free-hand schematic of this model is shown in Fig. 5b. Since the absorption peak of Ag-NPs are known to be around 400 nm and that of silicon around 1100 nm, the absorption depth profiles were kept limited to these excitations only. A tabulated representation of absorption depth profiles for such conditions are shown in Fig. 5a. It is noteworthy that the absorption is significantly enhanced and broadened in silicon layer due to the presence of Ag-NP only when the system is excited at 400 nm (i.e. near the absorption band of Ag-NPs).

4. Conclusion

Ag-NPs on ZnO thin film was fabricated in a simple two-steps process followed by their topographic characterizations. The nanoparticles were observed to be visible at annealing temperature of 400 °C and the average nanoparticles size increased with increasing annealing temperatures. The density of nanoparticles showed a trend to decrease with increasing annealing temperatures. The peak positions of Ag, Zn, O and Si were found consistence with the reported values in SEM-EDS measurements. FESEM micrographs of ZnO thin film without Ag ultrathin layer confirmed that the surface topography became rough after the heat treatment in addition to some small pores thereof. Keeping in mind the benefit of as-fabricated Ag-NPs, a typical model, Ag-NPs on silicon, was considered and FDTD simulation was carried out to understand the trend of absorption depth profile within the absorbing layer involved in plasmonics solar cell. It was observed that at higher energy but near the absorption band of Ag-NPs, the absorption intensity of silicon layer gets enhanced and broadened significantly due to the presence of Ag-NP.

5. Acknowledgement

Authors thank to Center of Research Excellence in Renewable Energy (CORERE) and Center of Research Excellence in Nanotechnology, King Fahd University of Petroleum and Minerals (KFUPM) for experimental support. Dr. Fahhad AlHarbi (QEERI, QF) and Ayman Wajeeh Mukhaimer (CORERE, KFUPM) are also acknowledged for their support. QAD is thankful to the government of Yemen for financial support to his PhD work.

6. References

[1] Hornyak G L, Tibbals H F, Dutta J and Moore J J 2008 *Introduction to Nanoscience and Nanotechnology* (CRC Press of Taylor and Francis Group LLC)
[2] Dutta J and Hofmann H 2004 *Encyclopedia of Nanosc.& Nanotech.* **9** 617
[3] De M, Ghosh P S and Rotello V M 2008 *Adv. Mater.* **20** 4225
[4] Lu A-H, Salabas E L and Ferdi S 2007 *Angew. Chem. Int. Ed. Engl.* **46** 1222
[5] Ghosh C R and Paria S 2012 *Chem. Rev.* **112** 2373
[6] Alshehri A H, Jakubowska M, Mlozniak A, Horaczek M, Rudka D, Free C and Carey J D 2012 *Appl. Mater. Interfaces* **4** 7007
[7] Chou C-W, Hsu S-H, Chang H, Tseng S-M and Lin H-R 2006 *Poly. Deg. and Stab.* **91** 1017
[8] Hossain M K 2013 *Mat. Sci. Forum* **754** 143
[9] Hossain M K, Willmott G R, Etchegoin P G, Blaikie R J and Tallon J L 2013 *NanoScale* **5** 8945
[10] Ahamed M, Alsalhi M S and Siddiqui M K 2010 *Clin. Chim. Acta* **411** 1841
[11] Garcıa-Barrasa J, . Lopez-de-luzuriaga J M and Monge M 2011*Cent. Eur. J. Chem.* **9** 17
[12] Fabrega J, Luoma S N, Tyler C R, Galloway T S and Lead J R 2011 *Environ. Inter.* **37** 517
[13] Dallas P, Sharma V K and Zboril R 2011 *Adv. Colloid Interface Sci.* **166** 119
[14] Turkevich J, Stevenson P C and Hillier J 1951 *Discuss. Faraday. Soc.* **11** 55

[15] Kimling J, Maier M, Okenve B, Kotaidis V, Ballot H and Plech A *2006 J. Phys. Chem. B* **110** 15700
[16] Brust M, Walker M, Bethell D, Schiffrin D J and Whyman R 1994 *Chem. Commun.* **7** 801
[17] Manna A, Chen P, Akiyama H, Wei T, Tamada K and Knoll W 2003 *Chem. Mater.* **15** 20
[18] Perrault S D and Chan W C W 2009 *J. Am. Chem. Soc.* **131** 17042
[19] Christy A J and Umadevi M 2012 *Adv. Nat. Sci.: Nanosci. Nanotechnol.* **3** 035013
[20] Ghosh S K, Kundu S, Mandal M, Nath S and Pal T, *J. Nanopart. Res.* **5** 577
[21] Siegel J, Kvıtek O, Ulbrich P, Kolska Z, Slepicka P and Svorcık V 2012 *Mater. Lett.* **89** 47
[22] Dengler S, Kübel C, Schwenke A, Ritt G and Eberle B 2012 *J. Opt.* **14** 075203
[23] Manuscript submitted in J. Vac. Sc. Tech.
[24] Nagel J R and Scarpulla M A 2010 *Opt. Express* **18** A139.
[25] Lim SH, Mar W, Matheu P, Derkacs D and Yu E T 2007 *J. of App. Phy.* **101** 104309
[26] Atwater H A and Polman A 2010 *Nat. Mat.* **9** 205
[27] Hossain M K, Kitahama Y, Huang G G, Kaneko T and Ozaki Y 2008 *App. Phys. B* **93** 165

IOP Conf. Series: Materials Science and Engineering **64** (2014) 012019 doi:10.1088/1757-899X/64/1/012019

Dialkyldiselenophosphinato-metal complexes – a new class of single source precursors for deposition of metal selenide thin films and nanoparticles

Sajid N Malik [1,3], Masood Akhtar [2], Neerish Revaprasadu [2], Abdul Qadeer Malik[1] and Mohammad Azad Malik[3*]

[1] School of Chemical and Materials Engineering (SCME), National University of Sciences and Technology (NUST), Islamabad 44000 Pakistan

[2] Department of Chemistry, University of Zululand, KwaDlangezwa South Africa

[3] Schools of Materials, The University of Manchester, Manchester

M13 9PL United Kingdom

*Corresponding Author: Azad.Malik@manchester.ac.uk

Abstract. We report here a new synthetic approach for convenient and high yield synthesis of dialkyldiselenophosphinato-metal complexes. A number of diphenyldiselenophosphinato-metal as well as diisopropyldiselenophosphinato-metal complexes have been synthesized and used as precursors for deposition of semiconductor thin films and nanoparticles. Cubic $Cu_{2-x}Se$ and tetragonal $CuInSe_2$ thin films have been deposited by AACVD at 400, 450 and 500 °C whereas cubic PbSe and tetragonal CZTSe thin films have been deposited through doctor blade method followed by annealing. SEM investigations revealed significant differences in morphology of the films deposited at different temperatures. Preparation of $Cu_{2-x}Se$ and In_2Se_3 nanoparticles using diisopropyldiselenophosphinato-metal precursors has been carried out by colloidal method in HDA/TOP system. $Cu_{2-x}Se$ nanoparticles (grown at 250 °C) and In_2Se_3 nanoparticles (grown at 270 °C) have a mean diameter of 5.0 ± 1.2 nm and 13 ± 2.5 nm, respectively.

1. Introduction

Semiconducting materials like Cu_2Se, PbSe, CdSe, $CuInSe_2$ (CIS), $CuGaSe_2$ (CGS), $CuIn_{(1-x)}GaxSe_2$ (CIGS), Cu_2ZnSnS_4 (CZTS), $Cu_2ZnSnSe_4$ (CZTSe) and $Cu_2ZnSn(S/Se)_4$ (CZTSSe) are the leading absorber materials for solar photovoltaic applications. Solar cells based on potassium doped CIGS thin films have already demonstrated 20.8 % power conversion efficiency (PCE) [1]. Similarly, CZTS/Se materials are being explored as earth abundant solar absorber materials. Several approaches have been employed for the deposition of photo-absorbing semiconductor materials. CIGS based thin films showing maximum efficiency have been deposited via co-evaporation process, whereas hydrazine based solution processing technique has given best results for CZTSSe materials [2]. However, these processes either require high vacuum resulting in higher processing cost and significant loss of materials or involve use of toxic chemicals like hydrazine. Large scale applicability of these techniques is also limited by the difficulty in achieving precise control over the structure, morphology and stoichiometric composition of the deposited material over larger substrate area. Consequently, up-scaling of these technologies is presently a significant research challenge.

Chemical vapour deposition (CVD) is a process which employs thermal degradation of vaporized chemical reagents (i.e., precursors) to form thin solid films of target materials onto suitable substrates. Presently, it is regarded as the vital technology for commercial production of solar cells, light-emitting diodes (LEDs), photodetectors, injection lasers, quantum-well lasers, hetero-junction bipolar transistors and various other electronic and optoelectronic devices [3]. The major hindrances to the application of CVD include the unfavourable volatility of the precursors and difficulty in precisely controlling stoichiometry of the deposits. These limitations can be circumvented by aerosol assisted chemical vapour deposition (AACVD) method as aerosols in a stream of inert carrier gases, are used for delivery of the precursors onto substrate. There is no stringent requirement of volatility for AACVD precursors. They should however, be soluble in some solvent in order to generate their aerosols by employing any suitable technique like ultrasonic modulation.

Our previous research efforts involve the synthesis of metal-organic complexes belonging to xanthates, dithiocarbamates, diselenocarbamates, thiourea, selenourea, thiobiurets *etc* as single source precursors for deposition of high quality semiconductor thin films and nanoparticles. Use of single source precursors in AACVD process further simplifies these issues as it ensures the availability of all requisite elements of semiconductor material in judiciously designed precursor molecule. We have previously synthesized a number of selenophosphinate ligands and reported their use for deposition of semiconductor thin films and nanoparticles [4]. This report deals with an efficient and reproducible synthesis of diphenyldiselenophosphinato-metal precursors and their use for deposition of semiconductor photo-absorbing thin films and nanoparticles for solar photovoltaics applications.

2. Experimental

Diphenylphosphine, Se poweder ~ 100 mesh, indium (III) chloride, gallium (III) chloride, copper (II) chloride, copper (I) chloride, tin (II) chloride, zinc (II) chloride, toluene and dicloromethane were purchased from Sigma-Aldrich and used as received. Ethanol and methanol used as solvents in these reactions were also used without carrying out further drying steps. All synthetic manipulations were carried out under an inert atmosphere of nitrogen using standard Schlenk line techniques. ^{1}H NMR spectra were recorded on Bruker AC400 FT-NMR spectrometer and a Kratos Concept 1S instrument was used to record mass spectra. Elemental analyses were performed by University of Manchester microanalysis facility. Melting points were recorded using a Stuart melting point apparatus and are uncorrected.

2.1. Deposition of thin films

Glass slides (1 x 3 cm) were used as substrates for the deposition of thin films. Any possible contamination on the substrates was removed by thoroughly cleaning the substrates and sonicating them in acetone for half an hour. An improvised AACVD kit was used for the deposition of thin films. Typical deposition experiment involved dissolving 150 mg of the precursor (or a suitable combination of precursors while depositing the ternary and the quaternary thin films) in 15 mL toluene taken in a two-necked 100 mL round-bottom flask. The flask was connected with a gas inlet which allowed the argon carrier gas to flow through the solution and carry the aerosols generated by a PIFCO ultrasonic humidifier. A piece of reinforced tubing connected the flask to the reactor tube and argon flow rate was controlled at 150 mL/min through a Platon flow gauge. The reactor tube containing 6-8 glass substrates was positioned in a Carbolite furnace. Carrier gas transferred aerosols generated in the flask to the hot zone of the reactor. Thermolysis of both the solvent and the precursor at the hot substrate surface due to thermally induced reactions lead to deposition onto substrates. Deposition experiments were performed at different temperatures i.e., 400, 450 and 500 °C at a constant argon flow rate on substrates for 1.5 h. Some of the thin films were deposited through doctor blade method followed by annealing at different temperature under argon gas flow. p-XRD patterns were recorded on a Bruker D8 AXE diffractometer (Cu-Kα) from 20 to 80 degrees with a step size of 0.05. Morphology and microstructure of thin films were investigated by using a Philips XL 30 FEGSEM and film composition was studied by EDX analysis using a DX4 instrument.

2.2 Growth of nanoparticles

Nanoparticles were grown by thermolysis of molecular precursors, respectively in HDA/ODE and TOP. Colloidal method, previously reported for growth of $CuIn_{(1-x)}Ga_xSe_2$ nanoparticles, was employed with slight modification [5]. UV-Vis-NIR absorption spectra of as grown nanoparticles were recorded in dry toluene using HELIOS λ spectrophotometer. Bruker D8 AXE diffractometer (Cu-Kα) was used to obtain p-XRD patterns of the films of nanoparticles obtained through drop cast method. Phase identification was performed by comparing the p-XRD patterns with the standard ICDD patterns using DIFFRACplus EVA software package. Size and morphology of as synthesized nanoparticles was determined by high resolution transmission electron microscopy (HRTEM). TEM samples were prepared by evaporating a dilute toluene solution of the nanoparticles on carbon coated copper grids (S166-3, Agar Scientific) and TEM images of the nanoparticles were recorded on Philips CM 200 and Tecnai F30 TEM, operated under accelerating voltage of 200 kV and 300 kV, respectively. Stoichiometric composition of the materials was determined by EDX analysis.

3. Results and discussion

3.1. Synthesis of precursor complexes

Synthesis of dialkyldiselenophosphinates has been carried out by many research groups involving diverse synthetic strategies. Lindner *et al* synthesized $[Ph_2PSi(Me)_3]$ through lithiation of diphenylphosphine chloride and this intermediate was subsequently reacted with trimethylsilylchloride at -78°C [6]. Insertion of metallic Se into this species yielded $[Ph_2PSe]_2Se$. As similar intermediates (R_2PSiCl_3) can also be synthesized via Benkeser reaction [7], we had previously synthesized a number of diisopropylselenophosphinates and used these compounds as ligands for synthesis metal-organic complexes. However, low reaction yields and lower solubility of these ligands in MeOH, the reaction medium for complexation reaction, remained as issues. In order to overcome these problems, our research efforts were diverted towards synthesis of $[R_2PSe_2]$- anion which was in-situ stabilized by using excess triethylamine (NEt_3) to yield $[R_2PSe_2]^-[HNEt_3]^+$ (where R= Ph, ^{i}Pr, ^{t}Bu). These ligand salts were isolated in good yield and undergo metathesis reaction with metal halides to give corresponding dialkyldiselenophosphinatometal complexes. We have used this approach for synthesis dialkyldiselenophosphinato-matal complexes of a number of metals including Cu, Zn, Pb, In, Ga, Cd and Ni and these complexes have been used as single source precursors for deposition of binary metal selenide thin films and nanoparticles [8]. This method still required longer reaction periods (~ 30 h) under drastic reflux and sometimes suffered with low-yield.

Our quest to develop even simpler and reproducible synthetic strategy for preparation of dialkyldiselenophosphinato-metal complexes led us to a newer method based on the synthesis of potassium diselenophosphinate by reaction of diphenylphosphine, elemental selenium and potassium hydroxide as reported by Gusarova *et al* [9]. This diselenophosphinic ester salt was then reacted with metal halides to give corresponding diphenyldiselenophosphinato-metal complexes. Diphenyldiseleno-phosphinato- metal complexes with general formula $M_x[Ph_2PSe_2]_y$ (where M = Cu, In, Ga, Pb, Zn, and Sn) were thus synthesized and characterized accordingly.

3.2. Deposition of thin films

In a recent report, we have described the deposition of $Cu_{2-x}Se$ and In_2Se_3 thin films onto glass substrates by AACVD using the diisopropyldiselenophosphinato-metal complex $Cu_4[^iPr_2PSe_2]_4$ and $In[^iPr_2PSe_2]_3$ as single source precursors [10]. Ternary $CuInSe_2$ and $CuGaSe_2$ thin films were also deposited by using 1:4 molar ratios of $Cu_4[^iPr_2PSe_2]_4$ and $In[^iPr_2PSe_2]_3$, $Cu_4[^iPr_2PSe_2]_4$ and $Ga[^iPr_2PSe_2]_3$ complexes, respectively. Furthermore, the same complexes were also used for deposition of CIGSe ($CuIn_{0.7}Ga_{0.3}Se_2$ thin films by using 1:2:2 ratio of the copper, indium and gallium complexes, respectively. Present work deals with the deposition of $Cu_{2-x}Se$ and $CuInSe_2$ thin films by AACVD process. Furthermore, PbSe and CZTSe thin films have been deposited by doctor blading the precursor complexes onto glass substrates and subsequent annealing of the as deposited thin films in an argon atmosphere at various temperatures.

Figure 1A shows the p-XRD patterns of the $Cu_{2-x}Se$ thin films deposited at 400 °C, 450 °C and 500 °C by AACVD process. XRD analyses revealed that the berzelianite phase of cubic $Cu_{2-x}Se$ (standard ICDD pattern 06-0680) has been deposited at all the temperatures. All the films showed a preferred orientation along (111) plane while other major peaks were observed at (220), (311) planes at 44.6 and 52.9 on 2-theta axis. Sharp and narrow peaks demonstrated good crystallinity of the deposited material. Microstructural studies through SEM images showed growth of randomly shaped grains having distinct grain boundaries for films deposited at 400 °C (Fig. 1B) and 450 °C (Fig. 1C). larger grains having triangular faceting were observed in the thin film deposited at 500 °C (Fig. 1D). EDX analyses showed that the films had a Cu deficient morphology and this deficiency was even more pronounced for thin film deposited at 500 °C.

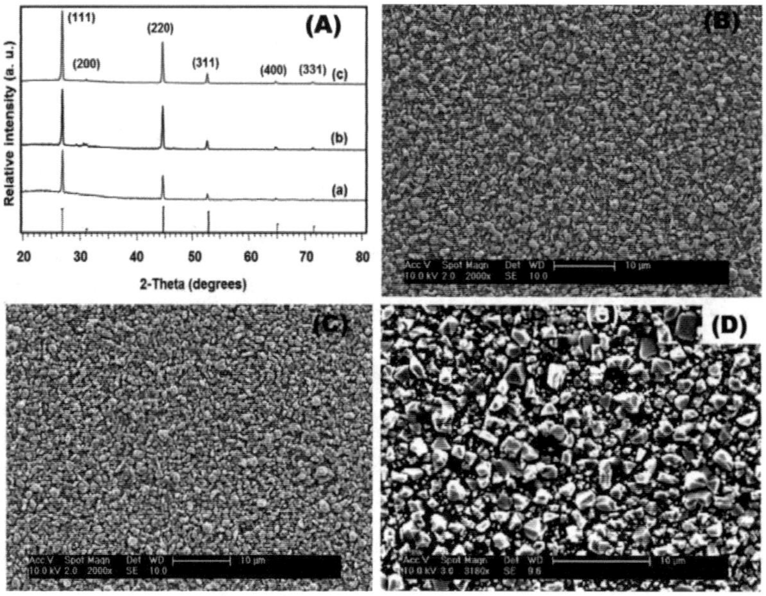

Figure 1. (A) p-XRD patterns of $Cu_{2-x}Se$ thin films deposited at (a) 400 °C (b) 450 °C and (c) 500 °C by AACVD using $Cu_4[Ph_2PSe_2]_4$ complex as single source precursor. Vertical lines below show standard ICDD pattern 06-0680 for berzelianite phase of $Cu_{2-x}Se$. (B) SEM image of as deposited thin film at 400 °C (C) at 450 °C and (D) at 500 °C.

Deposition of $CuInSe_2$ has been carried out by AACVD using 1:4 molar ratio of $Cu_4[Ph_2PSe_2]_4$ and $In[Ph_2PSe_2]_3$, respectively at 450 °C and 500 °C. p-XRD patterns of as deposited thin films are shown in Figure 2A which revealed deposition of tetragonal phase of $CuInSe_2$ (standard ICDD pattern 040-1487). Preferred orientation of deposited material along (112) plane was observed which is typical for copper chalcogenide materials deposited through CVD. SEM imaging used to study microstructure of thin films revealed deposition of rice like grains deposited at 450 °C (Fig. 2B) whereas films deposited at 500 °C exhibited somewhat spherical grains with distinct grain boundaries. 1:1:2 atomic ratio for Cu, In and Se was demonstrated by EDX analyses in both the films.

Lead selenide (PbSe) thin films have been deposited by doctor blading the complex $Pb[Ph_2PSe_2]_2$ onto glass substrate and then annealing the films in argon atmosphere at different temperatures. Weakly adhered dark brown films were obtained at all deposition temperatures. Figure 3A shows the p-XRD patterns for as deposited thin films at 400 °C, 450 °C and 500 °C. Phase analyses demonstrates that these p-XRD patterns can be indexed to clausthalite phase of cubic PbSe (standard ICDD pattern 06-0356). Thin films deposited at all temperatures had preferred orientation along (200) plane. SEM image of the thin film deposited at 400 °C (Fig. 3B) showed deposition of small sized cubic PbSe crystallites with poor coverage of the film. However, densely populated films were obtained at 450 °C (Fig. 3C) and the film deposited at 500 °C (Fig. 3D) showed linearly arranged streaks of PbSe cubes fully covering the glass substrate. EDX analyses demonstrated 1:1 ratio of the lead and selenium atoms within the error limits of EDX technique. No individual differences were observed in the stoichiometric composition of individual grains.

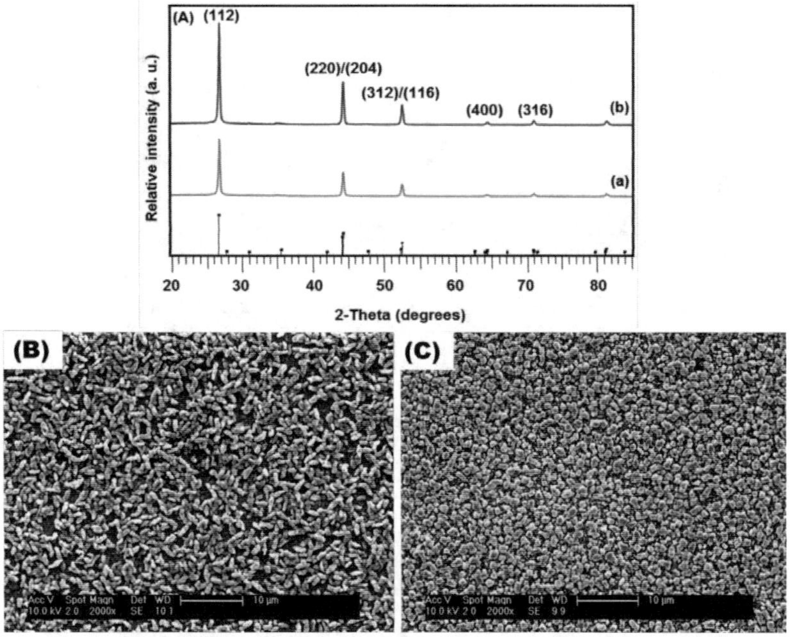

Figure 2. (A) p-XRD patterns of $CuInSe_2$ thin films deposited by AACVD using 1:4 molar ratio $Cu_4[Ph_2PSe_2]_4$ and $In[Ph_2PSe_2]_3$ complexes at (a) 450 °C and (b) 500 °C indexed with standard ICDD pattern 040-1487 for tetragonal $CuInSe_2$ (B) SEM image of as deposited PbSe thin film at 450 °C and (C) at 500 °C.

Thin films of the quaternary material CZTSe were deposited by the doctor blade method using 0.5:1:1 molar ratio of $Cu_4[Ph_2PSe_2]_4$, $Zn[Ph_2PSe_2]_2$ and $Sn[Ph_2PSe_2]_2$ precursors, respectively. The films were later annealed at 500 °C under argon atmosphere for 2 hours. Figure 4A shows p-XRD pattern of CZTSe of the as deposited film which fairly matches with the standard ICDD pattern 052-0868 for tetragonal phase of $Cu_2ZnSnSe_4$. Relatively broader peaks suggest deposition of small sized crystallites and absence of any peaks from binary or ternary phases indicates deposition of monophasic CZTSe onto thin films. Morphological studies of the films by SEM imaging (Fig. 4B & 4C) revealed deposition of nano-sized globular clusters onto thin films. Stoichiometric ratio of Cu, Zn, Sn and Se was found to be $Cu_{1.91}Zn_{1.13}Sn_{1.00}Se_{4.26}$ as determined by EDX analyses. These values are fairly close to the expected $Cu_2ZnSnSe_4$ stoichiometry. Selected area EDX elemental mapping (Fig. 4D) confirmed uniform distribution of Cu, Zn, Sn and Se throughout the film.

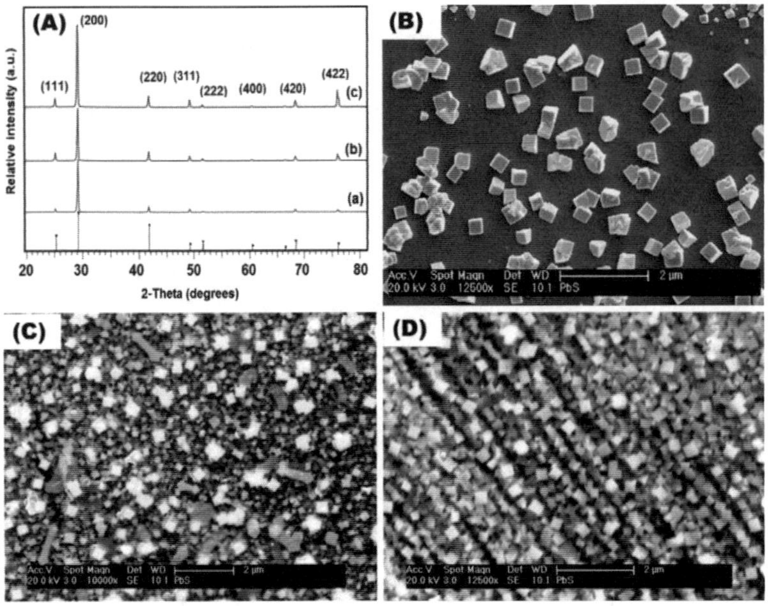

Figure 3. (A) p-XRD patterns of PbSe thin films deposited by doctor blading the suspension of $Pb[Ph_2PSe_2]_2$ complex and annealing at (a) 400 °C (b) 450 °C and (c) 500 °C indexed with standard ICDD pattern 06-0356 for cubic phase of PbSe (B) SEM image of as deposited PbSe thin film at 400 °C (C) at 450 °C and (D) at 500 °C.

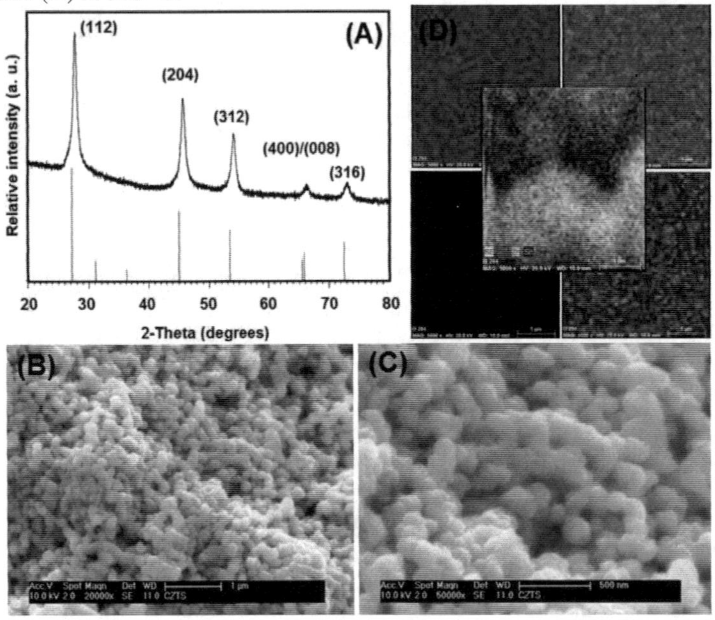

Figure 4. (A) p-XRD patterns of CZTSe thin film deposited by doctor blading the suspension of $Cu_4[Ph_2PSe_2]_4$, $Zn[Ph_2PSe_2]_2$ and $Sn[Ph_2PSe_2]_2$ complexes onto glass substrates and annealing at 500 °C. Vertical lines below show standard ICDD pattern 052-0868 for tetragonal CZTSe (B) & (C) SEM images of as deposited CZTSe thin film (D) EDX elemental mapping showing uniform distribution of Cu, Zn, Sn and Se in the film.

3.3. Growth of nanoparticles

Colloidal method was chosen for growth of semiconductor nanoparticles from dialkyldiselenophosphinato-metal precursors as this method provides better control over crystallite size and stoichiometric composition of the nanoparticles. In a previous report, we have described colloidal preparation of $CuInSe_2$, $CuGaSe_2$ and $CuIn_{1-x}Ga_xSe_2$ nanoparticles from diisopropyldiselenophosphinato-metal precursors [5]. In present work, we report the preparation of $Cu_{2-x}Se$ and In_2Se_3 nanoparticles from diisopropyldiselenophosphinato-metal complexes as single source precursors.

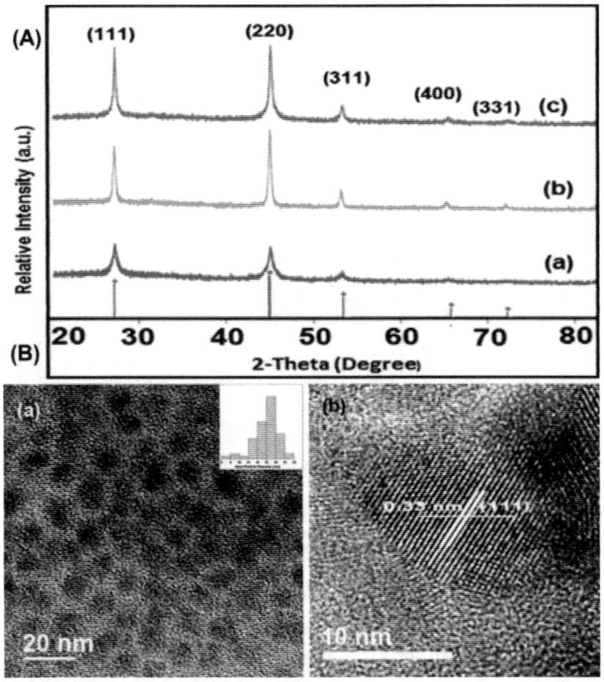

Figure 5. (A) p-XRD patterns of cubic $Cu_{2-x}Se$ nanoparticles synthesized from (a) 0.15 g (0.11 mmol) precursor at 250 °C (b) 0.3 g (0.22 mmol) precursor at 250 °C and (c) 0.15 g (0.11 mmol) precursor at 300 °C indexed with the standard ICDD pattern 006-680 for cubic $Cu_{2-x}Se$ (B) (a) TEM and (b) HR-TEM images of HDA capped $Cu_{2-x}Se$ nanoparticles grown from 0.3 g (0.22 mmol) of $[Cu_4({}^iPr_2PSe_2)_4]$ precursor at 250 °C.

Thermolysis of 0.3 g (0.22 mmol) of $Cu_4[{}^iPr_2PSe_2]_4$ in HDA/TOP system at 250 °C yielded brown $Cu_{2-x}Se$ nanoparticles. UV-Vis absorption spectrum of as grown nanoparticles showed band edge at 860 nm. XRD patterns of the HDA capped nanoparticles prepared at different temperatures as compared to standard ICDD pattern for cubic $Cu_{2-x}Se$ phase (ICDD 00-006-0680) are shown in Figure 5A. Mean diameter of the nanoparticles was calculated using Scherrer equation and it was found that the mean diameter of nanoparticles synthesized from 0.22 mmol precursor at 250 °C was 14.4 nm, which is in good agreement with the results obtained from TEM images. It was observed that higher precursor concentration and reaction temperature leads to increase in mean diameter of the nanoparticles, as evident from the stronger XRD peaks. Figure 5B (a) shows TEM images of the HDA capped $Cu_{2-x}Se$ nanoparticles obtained from thermolysis of 0.22 mmol of $[Cu_4({}^iPr_2PSe_2)_4]$ precursor at 250 °C. Well defined $Cu_{2-x}Se$ nanocrystals were observed in TEM images while HR-TEM image (Figure 5B (b) clearly shows lattice fringes. Interplaner distances were found to be 0.33 nm which are consistent with the known value for d_{111} interplaner distance of cubic $Cu_{2-x}Se$ nanoparticles.

Indium selenide nanoparticles were grown by thermolysis of the [In(^{i}Pr$_2$PSe$_2$)$_3$] precursor at 270 °C in HDA/TOP system. The TEM image of nanoparticles grown at 270 °C for 2 hour showed well formed equi-axed crystallites with an average size of 5.0 ± 1.2 nm (Figure 6B a & b). Crystallographic phase of the nanoparticles was studied by powder X-ray diffraction. XRD pattern of the HDA capped nanoparticles (Figure 6A) corresponds to the hexagonal In_2Se_3 phase (ICDD pattern 00-012-0117). Preferred orientation along (102) lattice plane was observed in the material and broader diffraction peaks indicated formation of very small nanoparticles. This was also supported by TEM images of the nanoparticles. Previously we have synthesized InSe nanoparticles with an average size of 6.4 nm by thermolyzing single source precursor [In(Se$_2$CNEt$_2$)$_3$] in TOPO at 250 °C [11]. However, in this study, phase present in the nanoparticles could not be defined due to poor crystallinity of the material. In a later report, hexagonal 'wafer-like' nanocrystals with average diameter of 141 nm (± 24 nm) were synthesized by reaction of In(OAc)$_3$ and Se in HDA at 300 °C [12]. In this report, longer reaction durations (18 h) were employed which might be the cause of the formation of larger nanocrystals.

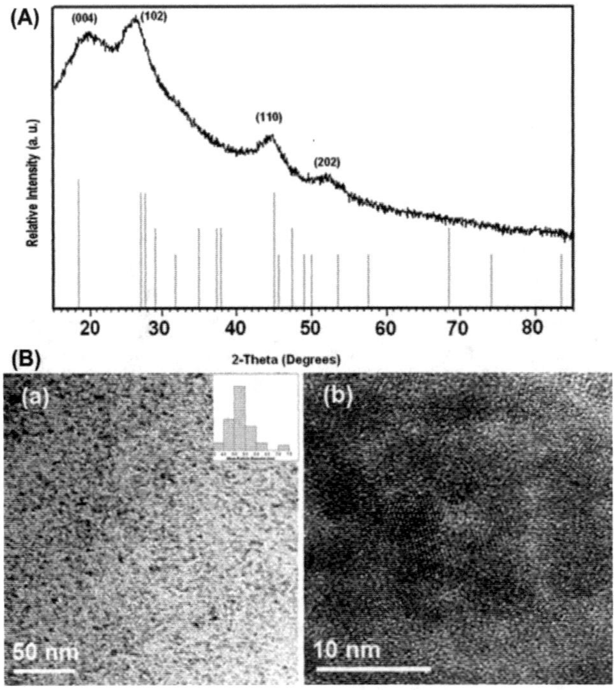

Figure 6. (A) p-XRD pattern of HDA capped In_2Se_3 nanoparticles grown at 270 °C from [In(^{i}Pr$_2$PSe$_2$)$_3$] precursor. Vertical lines below show standard ICDD pattern 00-012-0117 for hexagonal In_2Se_3 (B) (a) TEM and (b) HR-TEM images of HDA capped In_2Se_3 nanoparticles grown at 270 °C.

4. Conclusions

We have devised a new strategy for facile and reproducible synthesis of dialkyldiselenophosphinato-metal complexes. These complexes have been successfully utilized as single source precursors for deposition of binary metal selenide thin films and nanoparticles. Deposition of ternary and quaternary monophasic thin films has also been carried out using these precursors. These dialkyldiselenophosphinatometal complexes are a useful addition to the family of selenide molecular precursors as they exhibit fair air and moisture stability. Furthermore, they undergo clean decomposition in AACVD experiments as well as in HDA/TOP system yielding phase pure metal

IOP Conf. Series: Materials Science and Engineering **64** (2014) 012019 doi:10.1088/1757-899X/64/1/012019

selenide thin films and nanoparticles. Future work will involve deposition of thin films and nanoparticles of multinary semiconductor materials like $Cu_2ZnSn(S/Se)_4$.

References

[1] Jackson P, Hariskos D, Wuerz R, Wischmann W and Powalla M 2014, *Phys. Status Solidi* **8(3)** 219-22.
[2] Winkler MT, Wang W, Gunawan O, Hovel HJ, Todorov TK, Mitzi DB 2014 *Energy Environ. Sci.* **7** 1029-36.
[3] Dupuis, R D *J.* 2000 *Selected Topics in Quantum Electronics* **6(6),** 1040.
[4] Nguyen CQ, Adeogun A, Afzaal M,. Malik MA and O'Brien P 2006 *Chem Commun*, 2179-2181.
[5] Malik SN, Mahboob S, Haider N, Malik MA and O'Brien P 2011 *Nanoscale*, **3** 5132-39.
[6] Horn HG and Lindner HJ 1988 *Chemiker-Zeitung*, **112** 195.
[7] Benkeser RA, 1971 *Acc. Chem. Res.* **4** 94.
[8] Nguyen CQ, Afzaal M, Malik MA, Helliwell M, Raftery J and O'Brien P 2007 *J. Organomet. Chem.,* **692** 2669.
[9] Trofimov BA, Artem'Ev AV, Gusarova NK, Malysheva SF, Fedorov SV, Kazheva ON, Alexandrov GG and Dyachenko OA 2009 *Synthesis*, 3332-3338.
[10] Mahboob S, Malik SN, Nguyen CQ, Haider N, Malik MA and O'Brien P 2014 *J. Crystal Growth*, **394**, 39-48.
[11] Revaprasadu N, Malik MA, Carstens J and O'Brien P 1999 *J. Mater Chem.*, 9 2885-2888.
[12] Tabernor J, Christian P and O'Brien P 2006 *J Mater Chem.***16** 2082-2087.

IOP Conf. Series: Materials Science and Engineering **64** (2014) 012020 doi:10.1088/1757-899X/64/1/012020

Effect of silica nanoparticles on polyurethane foaming process and foam properties

A B Francés and M V Navarro Bañón

Wood and Furniture Technological Centre, CETEM, Yecla, Murcia, Spain

E-mail: ab.frances@cetem.es

Abstract. Flexible polyurethane foams (FPUF) are commonly used as cushioning material in upholstered products made on several industrial sectors: furniture, automotive seating, bedding, etc. Polyurethane is a high molecular weight polymer based on the reaction between a hydroxyl group (polyol) and isocyanate. The density, flowability, compressive, tensile or shearing strength, the thermal and dimensional stability, combustibility, and other properties can be adjusted by the addition of several additives. Nanomaterials offer a wide range of possibilities to obtain nanocomposites with specific properties. The combination of FPUF with silica nanoparticles could develop nanocomposite materials with unique properties: improved mechanical and thermal properties, gas permeability, and fire retardancy. However, as silica particles are at least partially surface-terminated with Si-OH groups, it was suspected that the silica could interfere in the reaction of poyurethane formation.The objective of this study was to investigate the enhancement of thermal and mechanical properties of FPUF by the incorporation of different types of silica and determining the influence thereof during the foaming process. Flexible polyurethane foams with different loading mass fraction of silica nanoparticles (0-1% wt) and different types of silica (non treated and modified silica) were synthesized. PU/SiO_2 nanocomposites were characterized by FTIR spectroscopy, TGA, and measurements of apparent density, resilience and determination of compression set. Addition of silica nanoparticles influences negatively in the density and compression set of the foams. However, resilience and thermal stability of the foams are improved. Silica nanoparticles do not affect to the chemical structure of the foams although they interfere in the blowing reaction.

1. Introduction

Polyurethanes are polymeric materials that have a wide range of applications in industry and households because of their numerous valuable properties and the ability to control many of their parameters during preparation. Flexible polyurethane foams (FPUF) are widely used in many industries, including the furniture, construction and automotive fields. These materials are also commonly used in mattresses and upholstery foams. FPUF have many advantages, such as a wide range of flexibility and hardness, good cushioning, high durability during use, low price and low density [1-4].

The main reaction in polyurethane foaming process is the polycondensation of poly-hydroxy compounds (polyols) with diisocyanates. Foaming phenomenon occurs when a small amount of blowing agent (water) is added during polymerization process. Water reacts with isocyanate groups giving carbamic acid, which spontaneously lose CO_2, generating the foam bubbles [1].

FPUF are formed as a result of the chemical reactions between components. The urethane linkage is formed as a result of the exothermic reaction between the isocyanate groups with the hydroxyl groups from polyol [3].

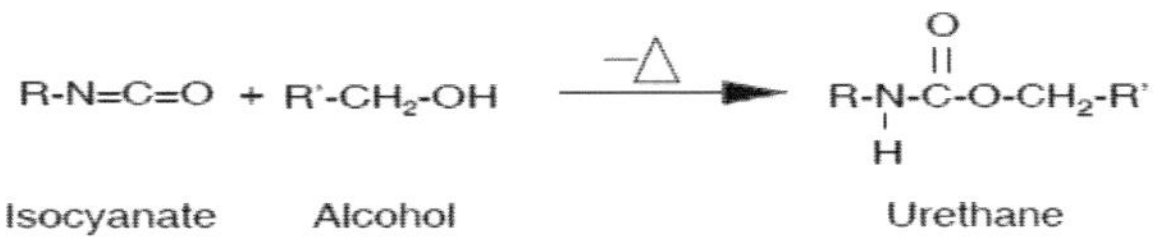

The reaction between water and the isocyanate leads to the production of carbon dioxide, which causes foaming of the mixture. The scheme of the foaming reaction is as follows:

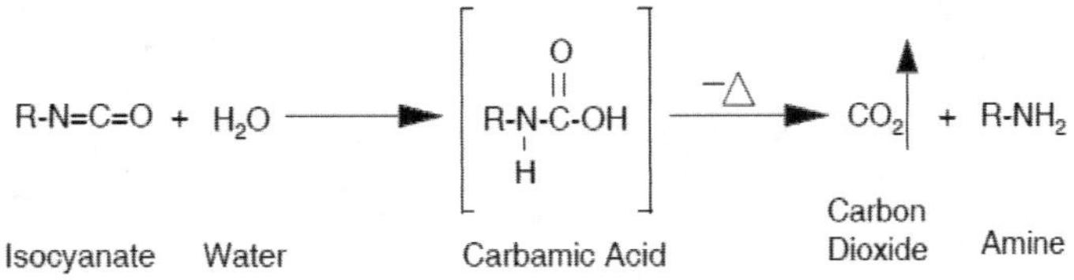

When the isocyanate component is mixed with the polyol component the chemical reaction starts. The reaction is exothermic meaning it produces heat. As the chemical reaction progresses the viscosity of the mixture will also increase. This increase continues during the foaming process until the reacting liquid turns into the polyurethane solid at the end of the reaction.

This foaming process was controlled by measuring different times, usually measured in seconds, with the time beginning as soon as the isocyanate and polyol components are first mixed together. Concepst such as cream time, gel time, rise time and tack-free time were taken into account.

- **Cream time**- this is the first event measured, and is the time interval between the mixing of all the foam ingredients and the time at which the clear mixture turns creamy and starts to expand.
- **Rise time**- this is when the reacted foam has reached its largest volume, or maximun height. It is the time between the start of the finale mixing and the time of complete expansion of the foaming mass.

In recent years, polymer nanocomposites have attracted much attention because of their high potential to achieve great improvement in properties including dimensional stability, gas permeability, flame retardancy, and mechanical and physical properties by adding a small amount of nanoparticles in the polymer matrices [1,5-6]. Furthermore, it has been shown that incorporation of nanoparticles with their large surface area and fine dimensions as well as very close contact between particles and the polymer matrix could alter foam morphology and properties.

2. Experimental

2.1. Materials

FPUF were prepared by the one-step method. The polyol used to produce the FPUF was Alcupol F-4811 from Repsol. Toluene diisocyanate (TDI T80) from Perstorp was used as an isocyanate. Water was used as a blowing agent. Four types of FPUF were tested in this study. The first foam was pure flexible polyurethane foam (FPUF), and the others were made with the addition of silica nanoparticles and modified silica nanoparticles as fillers. Silica nanoparticles of 12 nm size were obtained from Sigma Aldrich, whereas modified silica nanoparticles were obtained from Wacker Silicones (HDK H2000).

2.2. Foam preparation

Polyurethane foam were prepared by one-step and free-rising method. The experiments were as following: weight the polyol, catalysts, surfactant and blowing agent and mix them with a mechanical stirrer at 2200 rpm for 60 seconds. Then, rapidly add TDI into the mixture and continue to stirring for another 7-10 seconds at the same speed. Pour the mixture into a mold and allow the foam to rise and set at room temperature for 24 hours. Sonication and mechanical stirring were used to aid SiO_2 nanoparticles thoroughly mixed with the polyol before adding the other chemicals.

2.3. Formulation of foams

For this work, varying concentration of silica and modified silica nanoparticles ranging from 0 to 1 wt% respect polyol weight were introduced into the formulation as shown in Table 1:

Table 1. Concentration of the nanomaterials into the polyurethane formulation

Samples	SiO_2 nanoparticles dispersed in polyol (%wt)	SiO_2 nanoparticles dispersed in isocyanate (%wt)	Treated SiO_2 nanoparticles dispersed in polyol (%wt)
0	0	0	0
0.1S	0.1	0	0
0.5S	0.5	0	0
1S	1	0	0
0.1Si	0	0.1	0
0.5Si	0	0.5	0
1Si	0	1	0
0.1SH	0	0	0.1
0.5SH	0	0	0.5
1SH	0	0	1

2.4. Characterization

2.4.1. *Physical and mechanical properties measurements.* The apparent density tests were carried out according to ISO 845:2010. The samples were measured and weighted. The value of the apparent density was calculated as a mass/volume ratio. Set compression tests were conducted according to ISO 1856:2001. Three samples of 100x100x50 mm of each foam were compressed to 75% of its original height for 22 hours at 70°C, and the compression set is the permanent deformation remaining when the force is removed. Resiliencie was determined according to ISO 8307:2008 by ball rebound test.

2.4.2. *Thermogravimetric analysis (TGA)* . TGA was used to analyse the thermal stability of the prepared foams. The TGA were performed using a TA Instruments Q500. Samples of ~5 mg were heated to 800 °C at a heating rate of 20 °C/min in a nitrogen atmosphere.

2.4.3. Fourier transform infrared spectrocopy (FT-IR). FT-IR was performed using a Bruker Tensor 27 equiped with an ATR accessory with diamond crystal. The samples were scanned 100 times with a resolution of 4 cm^{-1} over the wavenumbers range from 4000-400 cm^{-1}.

3. Results and discussion

3.1. Reaction times

Figure 1 illustrates the cream time and the rise time variation for the flexible polyurethane foam containing silica nanoparticles obtained.

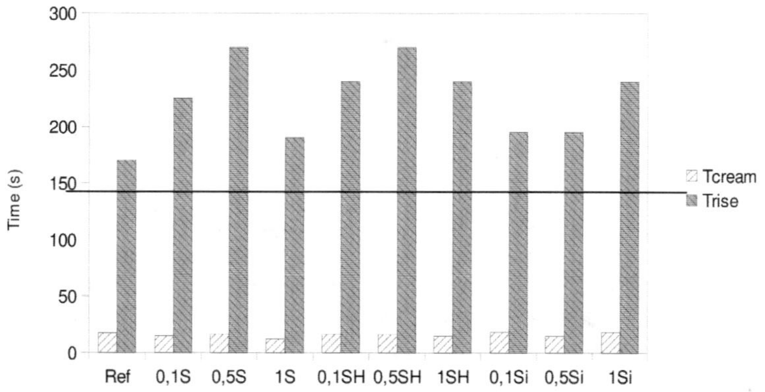

Figure 1. Cream and rise time variation

Variation of cream and rise time shows a uniform increment in rise time while silica nanoparticles concentration increases untill 0.5% wt. This implies that the foam rises slower when the filler composition increases indicanting that the blowing reaction between TDI and water occurs slower. It could therefore be due to the reaction between silanol groups from the silica nanoparticles and the NCO groups from the isocyanate. This effect is less pronounced in the case of the addition of modified silica nanoparticles in which silanol groups have been eliminated. When the concentration of silica nanoparticles added is over 0.5% wt, agglomerates are formed so they no react with the NCO groups.

3.2. Physical and mechanical properties

Samples from foam produced were cut into various sizes suitable for their respective tests.

Figure 2 presents the physical and mechanical properties of the studied foams. In general, the addition of fillers into a flexible polyurethane foam caused an increase in its apparent density. The resilience increased in all foams containing non treated silica nanoparticles. However, the addition of modified silica nanoparticles did not present any effect in the resilience of the foams. The addition of only 0.1% wt silica and modified silica nanoparticles to the flexible polyurethane foam raised the compression set value higher than that of foam without any filler.

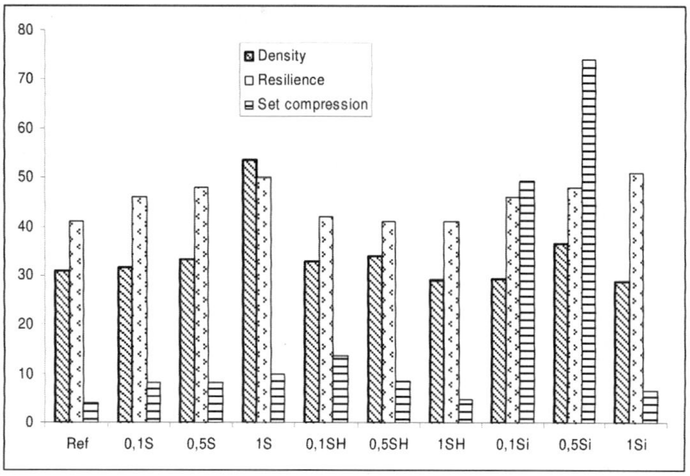

Figure 2. Physical and mechanical properties of synthesized foams

Summarizing, addition of fillers to the flexible polyurethane foam, led to deterioration in the physical and mechanical properties. It indicates that the presence of fillers in foams causes some inappropiate changes in the structure of flexible polyurethane foam. This may be related to the changes in foaming process, occurring as a result of the presence of fillers.

3.3. Thermogravimetry

TGA was performed to assess the effect of the addition of silica nanoparticles and hydrophobic silica nanoparticles on the thermal stability of the flexible polyurethane foams.

The decomposition of polyurethanes occurs in two well-defined stages. The temperature corresponding to 5% mass loss is considered to be the initial temperature of the sample's decomposition process. The temperature of 50 % mass loss corresponds to the temperature range of the decomposition of soft segments. T_{max1} is defined as the temperature of the maximum rate of decomposition of rigid segments in polyurethane. T_{max2} corresponds to the temperature of the maximum rate of decomposition of soft segments of the polyurethane [3]. Table 2 summarizes the results of the selected parameters obtained from the TGA thermograms.

Table 2. Results of the TGA

Sample	5% mass loss (°C)	50% mass loss (°C)	T_{max1} (°C)	T_{max2} (°C)
Ref	241.9	351.8	286.8	370.3
0.1S	244.6	356.4	287.1	374.5
0.5S	246.3	360.3	288.1	375.4
1S	247.1	364.4	289.0	380.4
0.1Si	242.9	351.9	286.4	371.1
0.5Si	246.0	359.1	288.2	374.5
1Si	249.0	360.6	289.4	379.0
0.1SH	235.3	351.0	283.1	373.6
0.5SH	238.8	359.4	283.5	375.1
1SH	237.0	356.5	283.8	371.5

Temperature at which a 5% wt mass loss is produced increased in the foams with the addition of silica nanoparticles in comparison to the value for FPUF without fillers when they are dispersed both in polyol and isocyanate. Higher values of this temperature were observed in foams with the addition of silica nanoparticles. However, this value decreased with the addition of modified silica nanoparticles.

The temperature of 50% mass loss increased with the presence of each investigated nanoparticle. The largest increase was observed in the foam with silica dispersed in polyol in concentration of 1%.

The addition of polyol and isocyanate dispersed silica nanoparticles increased the value of T_{max1}, but the addition of hydrophobic silica caused this value to reduce lower than the value of FPUF without fillers.

The addition of any filler led to an increase in the value of T_{max2}.

The results obtained from TGA indicate that the thermal stability of the foams modified with the addition of researched fillers is higher than that of foams without modification. The addition of silica nanoparticles dispersed into polyol produced the best results for the thermal stability. This result indicates that this type of modification could be an effective way to improve the thermal stability of FPUF. Furthermore, it should be noted that these changes are not large, and it is possible that higher contents of fillers should be examined to asses the influence of filler content on the properties of FPUF, and, if it is necessary, adjust the formulation of the foam to improve its physical and mechanical properties.

3.4. ATR-FTIR spectroscopy

FT-IR is a research technique that allows the chemical structure of a tested material to be analysed. In this study, the influence of the added fillers on the chemical structure of the tested foams was investigated. Typical peaks observed in polyurethanes are discussed below. The peak at 3294 cm^{-1} is assigned to the N-H stretching vibrations. The peaks at 2973 and 2869 cm^{-1} are from the asymmetric and symmetric stretching vibrations of CH_2 respectively. The peak at 2276 cm^{-1} is assigned to the N=C=O group. The peak at 1717 cm^{-1} is assigned to the stretching vibrations of hydrogen-bonded C=O groups. The peak at ~1530 cm^{-1} is from amide II (urethane N-H bending + C-N stretching). The peak at 1080 cm^{-1} is assigned to the C-O-C stretching vibrations. Figure 3 presents the FT-IR spectra of the foams containing silica nanoparticles dispersed in the polyol studied. The addition of this type of nanomaterials to FPUF does not cause changes in their chemical structure.

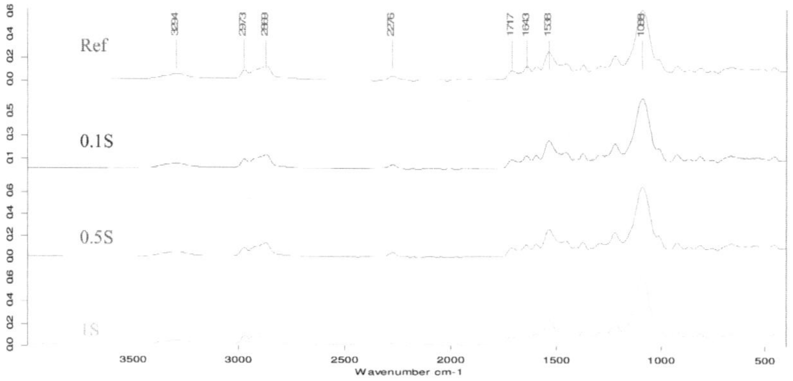

Figure 3. FT-IR spectra of the foams containing silica nanoparticles

4. Conclusions

This study has shown that silica nanoparticles addition in foam formulation decrease the rate of the blowing reaction between TDI and water with increase rise time. However the density and set compression properties of foam are worsened. On the contrary, resilience of foam which contain silica nanoparticles is improved. Moreover, silica do not affect to the chemical structure of the foams.

Silica nanoparticles dispersed in polyol caused the best improvement in the thermal stability of the foams. Improvement in the thermal properties is sometimes associated with a better fire behaviour and perhaps the silica could act as a good flame retardant in flexible polyurethane foam formulations.

References

[1] Jahanmardi R, Kangarlou B and Dibazar A 2013 Effects of organically modified nanoclay on cellular morphology, tensile properties, and dimensional stability of flexible polyurethane foams *J Nanostruct Chem* **3** 82

[2] Liang K and Shi S Q 2010 Soy-based polyurethane foam reinforced with carbon nanotubes *Key Engineering Materials* **419-420** 477

[3] Ubowska A 2010 Montomorillonite as a polyurethane foams flame retardant *Arch Combust* **30** 4

[4] Chao C and Wang J H 2001 Comparision of the thermal decomposition behavior of a non-fire retarded and a fire retarded flexible polyurethane foam with phosphorus and brominated additives *J Fire Sci* **19** 137

[5] Wolska A, Gozkzikiewiez M and Ryszkowska J 2012 Thermal and mechanical behaviour of flexible polyurethane foams modified with graphite and phosphorous fillers *J Mater Sci* **47** 5627

[6] Zammarano M, Krämer R H *et al* 2008 Flammability reduction of flexible polyurethane foams via carbon nanofiber network formation *Polym Adv Technol* **19** 588-595

Treatment of natural wood veneers with nano-oxides to improve their fire behaviour.

A.B. Francés Bueno[1], M.V. Navarro Bañón[1], L. Martínez de Morentín[2], J. Moratalla García[3]

[1] Furniture and Wood Technological Center (CETEM).
[2] L'Urredera Technological Center.
[3] Technology Institute Furniture Wood and Packaging (AIDIMA).
E-mail: v.navarro@cetem.es

Abstract. Conventional flame retardants used to improve fire behaviour of wood based materials are commonly based on halogenated and/or nitrogenated chemicals. These chemicals are toxic and can harm the environment and human health. Some works describe the incorporation of nanomaterials to the polymeric systems to improve their fire behaviour. The aim of this work was to analyze the effect of several treatments based on the use of nanomaterials on the properties of natural wood veneers and mainly on their fire behaviour. Firstly, several modes for treating pine veneers (immersion, spraying and impregnation) were evaluated using a commercial flame retardant to select the most effective treatment. The treatment selected as the most effective was immersion in a bath of flame retarding agent for 30 minutes at standard conditions. Afterward, pine veneers were treated by immersion in aqueous dispersions which contained 3wt% of the following nanoparticles: SiO_2, TiO_2 and ZrO_2, respectively. The effect of each treatment on the properties of veneers was analyzed. The results obtained showed that the treatment based on the use of 3wt% SiO_2 aqueous dispersion was the most effective to improve the fire behaviour of pine veneers.

1. Introduction

Wood veneers are commonly used for manufacturing veneered and plywood boards which are greatly applied in the furniture and construction sectors. Wood veneers are joined themselves or onto the surface of wood based panels using an adhesive, usually urea melamine formaldehyde adhesive.

The use of veneered and plywood boards has some limitations from the fire behaviour point of view. On the one hand, it is common that these veneered boards do not achieve the fire reaction classification required for determined applications. For example, in Spanish public building projects is common the requirement of an M-1 classification for wood based panels according to the UNE 23721:90. Flame retardants which can be found in the market are effective for improving the fire reaction of veneers but those are not commonly used. Wood veneers are usually joined without any treatment.

On the other hand, commercial flame retardants which can be used to improve the fire behaviour of wood veneers are based on halogenated and/or nitrogenated compounds and these chemicals have a considerable toxicity and can harm the environment and human health. The treatment of wood based materials with these conventional flame retardants produces materials which pass standardization fire tests according to the normative in force; however the use of these chemicals as flame retardants involves several limitations: (i) a harmful exposure to these substances that are toxic for human health and the environment during their handling, (ii) the release of toxic gaseous substances when these flame retardants burn which are identified as the main cause of death during a fire [2] and (iii) the required high concentration of these chemicals produces a negative effect on physical and mechanical properties of wood based materials.

The use of nanomaterials to enhance properties or impart innovative properties of conventional materials has been extended during the last decades. The small size and great surface area of nanomaterials make them more effective at low concentrations than other conventional additives. Some works describe the incorporation of nanomaterials to the polymeric systems, mainly into thermoplastic materials, to improve their fire behaviour [3-6]. However, the use of flame retardant nanomaterials has not been exhaustively studied for wood based materials; works detected related with this research field are limited [7-9].

2. Experimental

2.1. Materials

To analyze the effect of each nanoparticle on the fire behaviour of wood veneers, pine veneers with 0.59 ±0.02 mm of thickness, 488 ± 28 kg/m^3 of density and 3.2 ± 0.2 percentage of water were used.

Nanoparticles of silica (SiO_2), titanium dioxide (TiO_2) and zirconium dioxide (ZrO_2) were used for preparing aqueous dispersions which were used for treating pine veneers to improve their fire behaviour (Table 1). These nanoparticles and aqueous dispersions of them were prepared by L'Urederra Technological Center. These aqueous dispersions were prepared with a 1 wt% and 3wt% of each nanoparticle. Besides, a commercial flame retarding agent based on nitrogen compounds was used to optimize the best mode for carrying out the treatment of veneers.

Table 1. Properties of nanoparticles

Nanoparticles	SiO_2	TiO_2	ZrO_2
BET surface area (m^2/g)	216	120	87
Diameter (nm)	13	12	12

2.2. Experimental Procedure

A commercial flame retarding agent was used to optimize the methodology for treating pine veneers. Three methods were carried out for treated veneers at environmental conditions:

(i) Impregnation. A rod commonly used for applying conventional varnish was used for treating veneers by impregnation.
(ii) Spraying. A HVLP spray gun with a nozzle of 1 mm and working at 4 bars was used for treating veneers by spraying.
(iii) Immersion. Veneers were treated in a bath for 30 minutes.

To determine the optimal methodology for treating pine veneers and, therefore, the most effective treatment, chemical composition of two sides of treated pine veneers was analyzed by Attenuated Total Reflectance Infrared Spectroscopy (ATR-IR). An infrared spectrophotometer Tensor 27 (Bruker) provided with an ATR accessory with diamond crystal was used. 100 scans at 4 cm^{-1} resolution were collected for each sample.

Optical Microscopy (Moritex, MSX-500Di) were used to analyze esthetical modification originated on the veneers surface by each treatment.

Thermogravimetric Analysis using a Q500 thermobalance (TA Instruments) with a constant heating rate of 10 °C/min until 1000 °C and under nitrogen atmosphere was used to determine the effect on veneers thermal properties of each treatment.

To evaluate the fire behaviour of veneers (non-treated and treated), fire test method according to the UNE 23721:90 standard were carried out.

3. Results and discussion

Surface chemical composition of pine veners is constituted by typical chemical groups of cellulose and lignine: O-H groups (st O-H at 3335 cm^{-1}, δ O-H at 1645 and 1024 cm^{-1} and C-O at 1263 cm^{-1}), C=O groups (st C=O at 1745 cm^{-1}) and CH_2 and CH_3 groups (st C-H at 2922 and 2855 cm^{-1}), Figure 1(b). Typical infrared absorption bands of commercial flame retardant are showed in Figure 1(b). This commercial flame retardant is mainly based on ammonium salts.

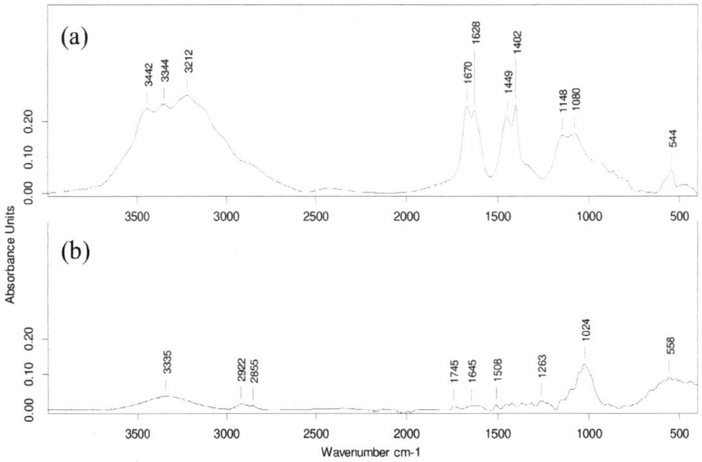

Figure 1. ATR infrared spectra of (a) commercial flame retarding agent and (b) pine veneer.

The treatment of pine veneers by immersion in a bath containing flame retarding agent for 30 seconds is more effective than other treatments (Figure 2(a)). The penetration of flame retardant is limited for impregnating and spraying treatments according to the chemical composition obtained by infrared analysis on each side of veneer (Figure 2(b) and 2(c)).

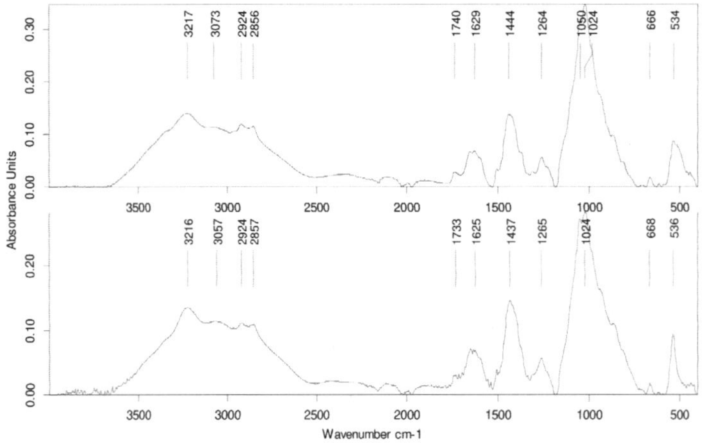

(a)

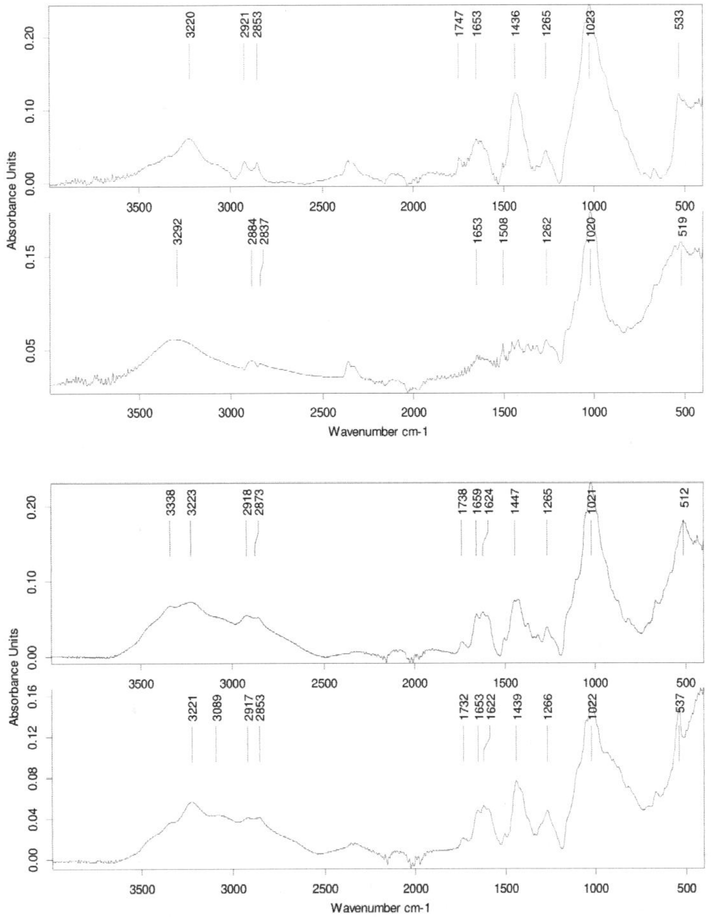

(b)

(c)

Figure 2. Infrared spectra of each side of treated veneers by (a) impregnation, (b) spraying and (c) immersion in commercial flame retarding agent.

All treatments created chemical modification in both side of pine veneers as show obtained infrared spectra (Figure 2). Fire tests, according to the UNE 23721:90 standard, were carried out to obtain more information about the more effective treatment. The results obtained showed that immersion treatment was the more effective than others to improve fire behaviour of pine veneers. Non treated pine veneer was classified as M4 while pine veneer treated with commercial flame retardant obtained a M1 classification according to the UNE 23721 fire tests.

The best way for carrying out the treatment is by immersion of pine veneers in the flame retarding agent bath. Therefore, treatments of pine veneers with nano-oxides (SiO_2, TiO_2 and ZrO_2) dispersion were carried out by immersion.

The treatment of pine veneers with aqueous nanoparticle dispersion modifies the chemical composition on surface pine veneers as it can be observed in Figure 3. Besides of infrared absorption bands characteristic of cellulose and lignin, typical absorption bands of Si-O links are detected on 3wt% SiO_2 treated pine veneer infrared spectrum (Figure 3a). Infrared spectra of pine veneers treated with 3wt% TiO_2 and 3wt% ZrO_2 aqueous dispersion also show infrared absorption bands characteristics of these compounds. The modification introduced on the surface of pine veneer after treating them with nanoparticle aqueous dispersion is mainly due to the deposition of solid nanoparticles on the surfaces (Figure 4).

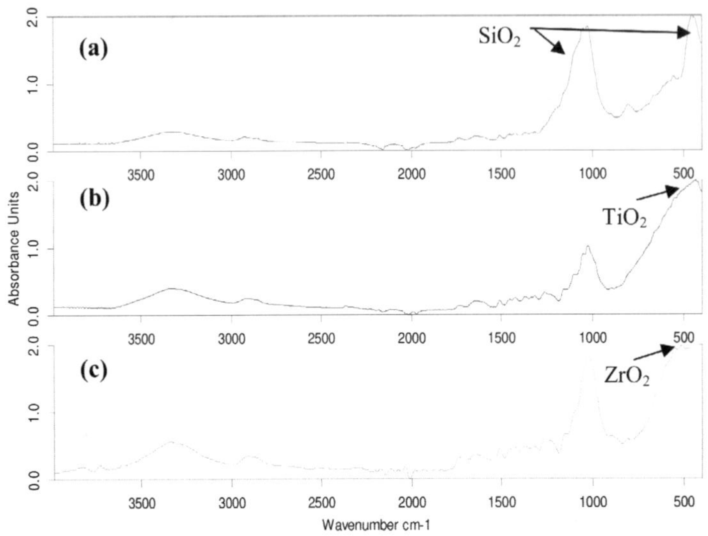

Figure 3. ATR infrared spectra of: (a) 3% SiO_2, (b) 3wt% TiO_2 and (c) 3wt% ZrO_2 aqueous dispersion treated pine veneers.

The treatment of veneers with aqueous nanoparticles dispersion produces a deposition of aggregates and/or agglomerates of nanoparticles which can be observed by microscopic analysis and as consequence a colour change of the veneer is produced by the treatment being more marked on the sample treated with 3wt% TiO_2 dispersion.

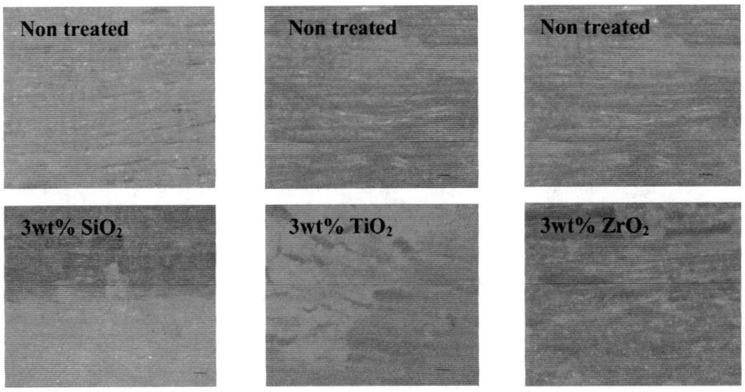

Figure 4. Images taken with optical microscope at 100x magnification of samples.

The treatment of pine veneer with aqueous nanoparticle dispersions produces a noticeable modification on their thermal properties which are sometimes related with the fire behaviour. The most effective treatment for modifying thermal properties of veneers is that based on the use of 3wt% SiO_2 aqueous dispersion. The results obtained by thermogravimetric analysis of 3wt% SiO_2 treated veneer show a weight loss on the second thermal decomposition lower than that for non-treated or treated with the other dispersions. Besides, the amount of solid residue obtained is higher than that obtained for non treated veneer or treated with aqueous nanoparticle dispersions based on 3wt% TiO_2 or 3wt% ZrO_2 (Table 2).

Table 2. Results obtained by thermogravimetrical analysis of non-treated and treated pine veneers (*Td: thermal decomposition temperature, WL: weight loss*).

Sample	Td1 (°C)	WL1 (wt %)	Td2 (°C)	WL2 (wt %)	Td3 (°C)	WL3 (wt %)	Residue (wt %)
Non treated	83.1	8.6	319.0	56.4	468.3	34.5	0.5
3% SiO_2	83.1	6.4	320.3	51.3	467.8	27.6	14.7
3% TiO_2	78.1	7.2	316.5	55.2	464.1/490.5	34.9	2.7
3% ZrO_2	86.9	7.8	319.0	56.0	460.3/494.3	33.2	3.0

After testing the samples according to the UNE 23721:90 standard, the treatment based on the use of 3wt% SiO_2 aqueous dispersion was the more effective to improve the fire behaviour of pine veneers such as shows the classification obtained for each sample (Table 3).

Table 3. Results of fire tests

Treatment of pine veneer	Classification according to UNE 23721:90 standard
Non-treated	M4
Treated with 3wt% SiO_2 dispersion	M1
Treated with 3wt% TiO_2 dispersion	M3
Treated with 3wt% ZrO_2 dispersion	M3

4. Conclusions

The results obtained showed that immersion of pine veneers was the most effective treatment to achieve a good penetration of flame retarding agent.

The treatment of pine veneers with a 3wt% SiO_2 aqueous dispersion is the most effective to modify thermal properties of pine veneers and also to improve their fire behaviour.

As future works it would be interesting to analyze the effect of adding nanooxides in the formulation of wood based panels on their fire behaviour.

References

[1] Peraza Sánchez F. 2001 Protección preventiva de la Madera. Ed. Asociación de Investigación Técnica de la Madera y Corcho, 211-233.

[2] Takashi K., Fangming, D. et al. 2005 Nanoparticles networks reduce the flammability of polymer nanocomposites *Nature Materials* **4** 928-933.

[3] F. Laoutid, L. Bonnaud et al. 2009 New prospects in flame retardant polymer materials: From fundamentals to nanocomposites *Materials Science and Engineering* **63** 100-125.

[4] Laoutid, F., Bonnaud, L., Alexandre, M., Lopez-Cuesta, J. M., Dubois, Ph. 2009 New prospects in flame retardant polymer materials: From fundamentals to nanocomposites *Materials Science and Engineering* **3** 100-125.

[5] Goodarzi, V., Ali, S., Torabi, M., Motahari, S. 2008 Improvement of Thermal and Fire properties of Polypropylene *Journal of Applied Polymer Science* **110** 5 2971-2979.

[6] Wang, Z., Han, E., Ke, W. 2006 Effect of nanoparticles on the improvement in fire-resistant and anti-ageing properties of flame-retardant coating *Surface and Coatings Technology* **200** 20 5706-5716.

[7] Guo G., Park C.B., Lee Y.H., Kim Y.S., Sain M. Flame retarding effects of nanoclay on wood fiber composites 2007 *Polymer Engineering and Science* **47** 3 320-336.

[8] Giudice, C. A., Pereyra, A. M. 2009 Silica nanoparticles in high silica/álcali molar ratio solutions as fire-retardant impregnants for woods *Fire and Materials* **34** 4 177-187

[9] Miyafuji, H., Saka, S. 1997 Fire-resisting properties in several TiO_2 wood-inorganic composites and their topochemistry *Wood Science and Technology* **31** 6 449-455

IOP Conf. Series: Materials Science and Engineering **64** (2014) 012022 doi:10.1088/1757-899X/64/1/012022

Selective enrichment and sensitive detection of candidate disease biomarker using a novel surfactant-coated magnetic nanoparticles

RY Capangpangan[1,2†], MAC dela Rosa[3,4†], CH Chang[5], WC Wang[6], J. Peng[6], SJ Shih[7], MH Chiang[5], DL Tzou[5], CCLin[1] and YJ Chen[3,5]*

[1]Department of Chemistry, National Tsing Hua University, Hsinchu Taiwan
[2]Molecular Science and Technology, Taiwan International Graduate Program
[3]Department of Chemistry, National Taiwan University, Taipei Taiwan
[4]Nanoscience and Technology Program, Taiwan International Graduate Program
[5]Institute of Chemistry, Academia Sinica, Taipei Taiwan
[6]En Hou Polymer Chemical Ind. Co., Ltd., Taiwan
[7]Department of Material Science and Engineering, National Taiwan University of Science and Technology, Taipei Taiwan

E-mail: yujuchen@gate.sinica.edu.tw, reycaps80@gate.sinica.edu.tw

[†]*contributed equally to this work*
[*]*To whom any correspondence should be addressed*

Abstract. In this study, novel surfactant-coated magnetic nanoparticles were synthesized and evaluated for enrichment performance towards the sensitive detection of disease biomarkers. Surfactants with phosphate ester groups (RD35A and RD66) were used as a coating to reduce aggregation and to enhance the nanoparticle dispersion. Importantly, sensitive enrichment of the target proteins using the antibody-functionalized magnetic nanoparticles (Ab@MNP) was obtained, with a five-fold increase in recovery compared to uncoated magnetic nanoparticles. Similarly, phosphopeptide enrichment using the NTA@MNP in standard samples showed that the nanoparticles could selectively enrich phosphorylated peptides.

1. Introduction

Within the past few decades, synthesis and application of various nanomaterials is actively explored, as these materials are implicated in several fields which include electronics, bio (chemical) sensing and disease biomarker screening and detection [1-5]. Among all of the diverse classes of nanomaterials, magnetic nanoparticles received considerable attention due to its remarkable properties such as magnetic property, biocompatibility and excellent surface functionality which attracted a growing significance of these materials in various applications especially in biomedicine and bioengineering [1]. In spite of its unique and immeasurable applicability, there is still a great deal of interest to investigate new surface protection strategies to enhance the nanoparticle dispersion and to understand the behaviour of these nanomaterials, particularly the aggregation behaviour which influences its performance towards bioanalytical applications.

Aggregation phenomenon of magnetic nanoparticles is one of the crucial factors which can affect the performance for their desired application [6]; hence, alters its analytical performance and detection sensitivity. Commonly, attractive magnetic forces between particles draw the particles to come close in proximity to each other and result in aggregation [7]. Thus, protection of surface of the nanoparticle is of great importance in order to minimize the attractive forces between particles. A range of protection strategies were reported, including metal inorganic coating [8], organic/functional group [9] and polymer/surfactant coating [10-11]. Among all of these, surface protection using surfactants were extensively investigated because of its suitability and ease in providing active surface functionality towards different target ligands.

This paper describes the use of new surfactants with phosphate ester groups as a protective layer in the surface of the nanoparticles. To the best of our knowledge, this is the first report which utilizes this unique class of surfactants which minimizes magnetic attractive forces by enhancing electrostatic repulsion with optimized amphiphilicity of the resulting particles. Preliminary results showed that a significant improvement in the nanoparticle dispersion, though an in-depth investigation on the effect of these surfactants to the aggregation behaviour of the nanoparticles is still underway. Likewise, initial application performance of these surfactant-stabilized nanoparticles was demonstrated by fabricating an efficient nanoprobe for the target disease biomarker enrichment and detection. Similarly, Fe-IMAC (Immobilized metal ion affinity chromatography) nanoprobe was constructed using the Nitrilotriacetic acid (NTA)-functionalized MNP for the selective phosphopeptide enrichment.

2. Experimental

All chemicals and reagents were used as received without further purification unless otherwise stated. $FeCl_2.4H_2O$ ($\geq$ 99%), HCl ($\geq$37%), NaOH ($\geq$ 98%), and NH_4OH (25%) were obtained from Sigma-Aldrich; $FeCl_3$ and Suberic acid bis(N-hydroxy-succinimide ester (DSS, $\geq$ 95%) were obtained from Sigma; Aminopropyltrimethoxysilane (APS, $\geq$ 97%), ethanolamine ($\geq$ 99.5%) and N_α-N_α-bis(carboxymethyl)-L-lysine (NTA-Lysine, $\geq$ 97%) from Aldrich; 1-propanol ($\geq$ 99%) and tetraethyl orthosilicate (TEOS, $\geq$ 98%) from Acros Organics; dimethyl sulfoxide (DMSO) from Thermo Scientific, and Tris (99.8 – 100%) from PlusOne. Surfactants (RD35A and RD66) were supplied by En Hou Polymer Chemical Ind. Co., Ltd. (Taiwan) and amine methoxyethylene glycol (MEG) was provided by Chun-Cheng Lin of National Tsing Hua University. Standard alpha-fetoprotein (AFP) was obtained from Abnova and Anti-AFP antibody was obtained from Thermo Scientific. α-casein and β-casein were purchased from Sigma.

2.1 Synthesis of magnetic nanoparticles (MNPs) and fabrication of nanoprobes

Core magnetic nanoparticles (MNP) and amine-functionalized MNP (N009) were synthesized as adopted from [12]. Similarly, Ab@MNP and NTA@MNP were fabricated adopted from our previous works [5] and [13], respectively with slight modifications. For the surfactant coating, to a solution of 1-propanol (80%), the required amount of surfactants (RD35A and RD66) were added and sonicated for 30 minutes. In a separate flask, the previously prepared core magnetic nanoparticles (50 mg) were dissolved in 25% 1-propanol solution (10 mL) and were sonicated for 30 minutes. The two mixtures were then mixed and stirred vigorously for 4 hours at room temperature. The mixture was transferred to a pre-heated silicon bath (~55 ^{o}C) and NH_4OH (20.8 mmol) and TEOS (1 mmol) was added sequentially and continuously stirred for another 4 hours. APS (1 mmol) was then added and incubated for overnight. The particles were then isolated by centrifugation, washed with 1-propanol three times, then with water three times, and the residue dried in vacuum and stored for further use. Two different batches of the surfactant-stabilized NH_2@MNP (with different amount of surfactants) were synthesized, 11-NH_2@MNP and 5-NH_2@MNP for Batch 1 and Batch 2, respectively.

2.2 Protein and phosphopeptide Enrichment

Antibody@MNPs (Ab@MNP, 2 μL, 10 mg/mL) were added to a standard protein solution in phosphate buffered saline (PBS, pH ~ 7.4, 1-5 ng/mL). The solution was incubated at room temperature for 1 hour in a rotary mixer. After extraction, the MNPs were separated using a magnet, washed twice with Tween-TBS (TTBS, 200 μL) and then twice with deionized water (200 μL). For LOD determinations, AFP was detected by liquid chromatography-mass analysis using multiple reaction monitoring (MRM) mode. For % recovery experiments, SDS-PAGE followed by silver staining was carried out. Similarly, phosphopeptide enrichment was carried out using the Fe-IMAC method and Ni-NTA beads as described in [13]. Different loading buffers were prepared and 0.5 μg α, β-casein was used to demonstrate the applicability of the NTA@MNP for the enrichment of phosphopeptides by MALDI-TOF analysis.

3. Results and Discussion

3.1 Physical properties and surface morphology of core and functionalized nanoparticles

The size and morphology of the core and functionalized magnetic nanoparticles (MNPs) were investigated by TEM (Figure 1). TEM images revealed that the synthesized surfactant-coated MNPs had a core-shell structure with an average core diameter of 21 nm. Although an aggregate of iron oxide nanoparticles per shell was noticeable, production of a stable core-shell morphology was proven. Moreover, the images also show that an empty shell, probably an empty silica particle formed after silica coating, was observed. The production of empty silica nanoparticles is reportedly difficult to prevent completely especially during silica coating process [14]. Furthermore, preliminary results showed that the rate of sedimentation or the decay time (data not shown) of the core and the uncoated-NH_2@MNPs (N009) were shorter compared to the surfactant-stabilized magnetic nanoparticles. This suggests that the new surfactants can help improve the dispersity of the nanoparticles in an aqueous environment.

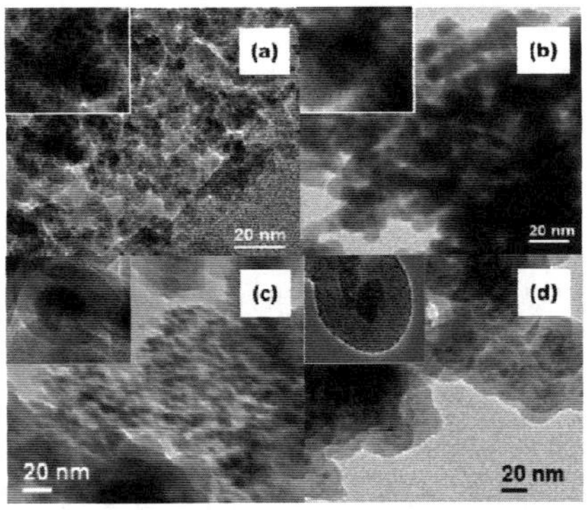

Figure 1. TEM images of magnetic nanoparticles (a) core MNPs and (b) N009 NH_2@MNP (c) RD66 and (d) RD35A

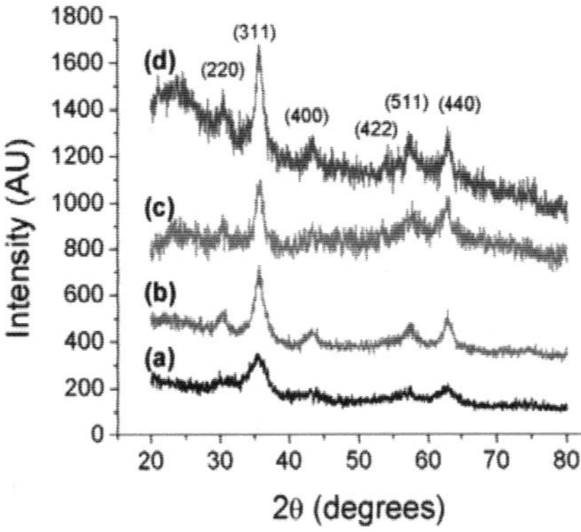

Figure 2. XRD pattern of (a) core MNPs and (b) N009 NH_2@MNP (c) RD66 and (d) RD35A

The crystal structure of the nanoparticles was monitored using X-ray diffraction (XRD) analysis as shown in Figure 2. XRD data revealed that even after surface modification of the nanoparticles, its crystal structure was not altered. Indeed, the obtained XRD peaks were still consistent with that of the standard Fe_3O_4 diffraction pattern. Importantly, the amorphous nature of the nanoparticle after surfactant and silica coating can be inferred based on the XRD pattern as shown in Figure 2c and d and is also consistent with the TEM result which confirms the presence of silica shell structure. On the other hand, FT-IR and EDX spectra of the core and functionalized magnetic nanoparticles are shown in Figure 3 and 4, respectively. Characteristic peaks for the Si-O and NH band were observed in the functionalized nanoparticles which were not observed in the core MNP's. Moreover, elemental analysis using EDX showed that a higher Si peak was noted in all of the functionalized nanoparticles which are consistent with the IR results.

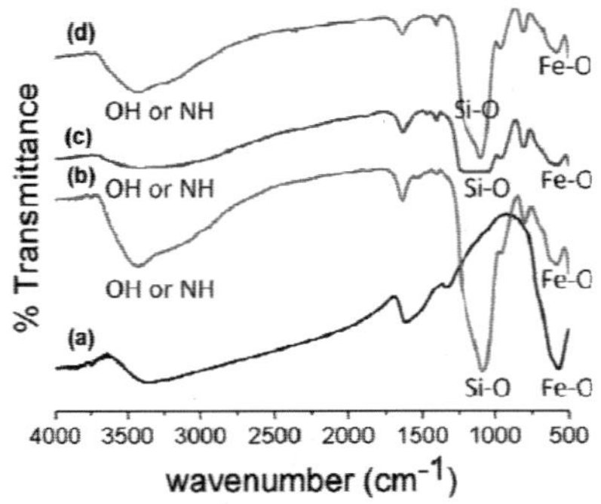

Figure 3. IR spectra of (a) core MNPs and NH_2@MNP (b) N009 (c) RD66 and (d) RD35A

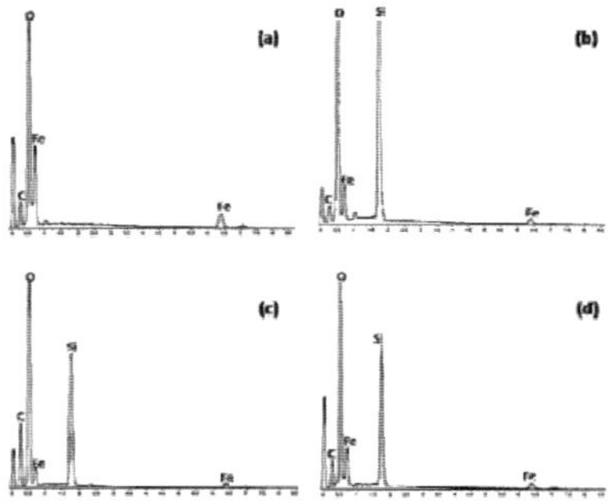

Figure 4. EDX spectra of (a) core MNPs and NH_2@MNP (b) N009 (c) RD66 and (d) RD35A

3.2 Nanoprobe-based enrichment and detection application
(a) Alpha-fetoprotein (AFP) enrichment and detection

To demonstrate the potential application of the newly synthesized surfactant-stabilized nanoparticles towards biomarker enrichment and detection, an antibody-conjugated nanoprobe was fabricated. Alpha-fetoprotein (AFP) was chosen as a model protein since this is the widely known biomarker for hepatocellular carcinoma. Three different batches of AFP-based nanoprobes were designed which includes the Ab@RD66, Ab@RD35A and Ab@N009. To evaluate the specificity of the fabricated nanoprobes, enrichment was conducted in a mixture of proteins (AFP, Lactoferrin and myoglobin). The results showed (Figure 5) that the percentage recovery of the nanoprobes based on the surfactant-stabilized nanoparticles were exceptionally higher compared to the nanoprobe without surfactant. As shown, about 5.5-fold enhancement and 3-fold enhancement for Batch 2 and Batch 1, respectively, was noted for the surfactant-stabilized nanoprobes compared to that of uncoated nanoparticles (N009). Further, it is noticeable that the said enrichment enhancement is not correlated to the antibody density (shown in Figure 6); instead other factors may contribute for the observed enhancement. Plausibly, the dispersity of the nanoparticles may play an important role in the observed performance of the surfactant-stabilized nanoprobes towards the target protein enrichment. Hence, an experiment which focuses on understanding the aggregation nature of these nanoprobes and relating it in its enrichment performance is underway.

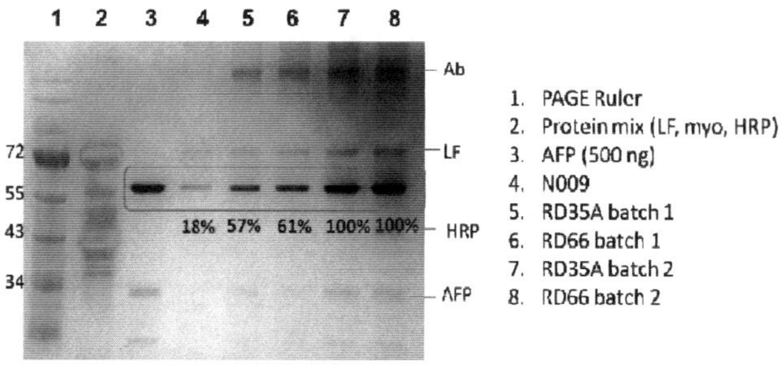

Figure 5. SDS-PAGE of the target protein after nanoprobe enrichment in a protein mix sample (*LF- Lactoferrin; myo – myoglobin; HRP – horseradish peroxidise*)

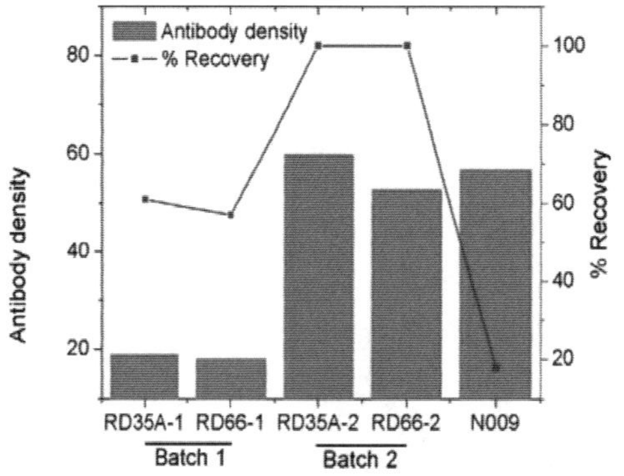

Figure 6. Correlation of antibody density to percent recovery

Furthermore, multiple reaction monitoring (MRM) analysis was conducted to evaluate the limit of detection (LOD) of these nanoprobes. All data were processed in MultiQuant software (V 2.1, AB Sciex). The most intense transition was used for peak area integration while the other 5 transitions were used to confirm the protein identity. Results showed that the surfactant-stabilized nanoprobes, especially the one coated with RD66 surfactant, gave the best performance among all other nanoprobes fabricated. The obtained LOD for the RD66-coated nanoprobe was able to detect to as low as 0.5 ng/mL (7.3 pM) with excellent S/N ratio (128), which is far lower than the AFP concentration of healthy controls (20 ng/mL) [15]. Further condition optimization and validation is currently being conducted to completely demonstrate the general applicability of these nanoprobes towards screening and detection of different disease biomarkers.

(b) Phosphopeptide enrichment

As a proof-of-concept, NTA-functionalized magnetic nanoparticles were fabricated specifically for phosphopeptide enrichment. Standard α- and β-casein (0.5 μg) sample and 12 μg of the NTA nanoprobe were used in the enrichment process following the sequential elution with the loading buffers and Fe^{3+}-metal activation. It is noteworthy, moreover, that the newly-fabricated nanoprobes were unblocked nanoprobes, which means that no blocking agent was added as compared to some of the reported literatures [13]. Normally, conjugation of blocking agent to the nanoprobe was made to get rid of the non-specific binding to other analytes in the mixture. However, the result shows that selective enrichment of the target phosphopeptides were still detected even the use of the unblocked NTA nanoprobe (as shown in Figure 8). However, in order to improve the selectivity and enrichment efficiency, further condition optimization, particularly of the loading buffer, nanoprobe amount and enrichment protocol should be further investigated. Lastly, fabrication of NTA-nanoprobe with the optimized conjugation of the desired blocking agent is in progress.

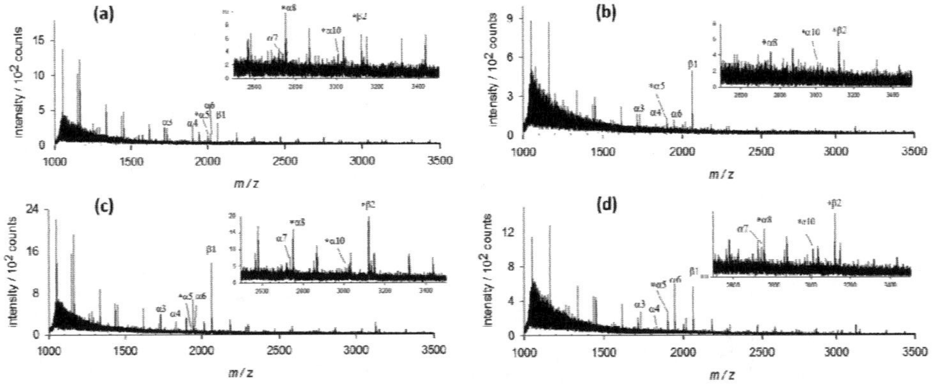

Figure 7. Representative MALDI spectra of mixture of α- and β-casein purified from Fe-NTA nanoprobe (a) RD35A-2 (b) RD66-2 (c) RD35A-1 and (d) RD66-1 to enrich phosphopeptide (*Note: * means multiply-charged phosphopeptides*)

4. Conclusion

It has been demonstrated that RD35A and RD66 surfactants successfully protects the surface of the iron oxide magnetic nanoparticles. Characterization revealed that surfactant coating did not alter the surface morphology and properties of the nanoparticles. Most importantly, it has been shown the potential application of the nanomaterials as efficient nanoprobes for both target protein disease biomarker and phosphopeptide enrichment and detection. Although demonstration of the potential application of these nanoprobes has been illustrated successfully, detailed and in-depth study for validation in the enrichment and detection aspect is necessary and is currently being investigated. Understanding the effect of these surfactants to the nature of aggregation of magnetic nanoparticles is likewise in progress.

5. References

[1] Deng Y H, Wang C C, Hu J H, Yang W L and Fu S K 2005 Investigation of formation of silica-coated magnetite nanoparticles via sol-gel approach *Colloids & Surf.* A*: Physicochem. Eng. Aspects* **262** 87-93

[2] Wanunu M, Dadosh T, Ray V, Jin J, McReynolds L and Drndić M 2010 Rapid electronic detection of probe-specific microRNAs using thin nanopore sensors *Nat. Nanotechnology* **5** 807-814

[3] Fu A H, Hu W, Xu L, Wilson R J, Yu H, Osterfeld, S J, Gambhir, S S and Wang S X 2009 Protein-functionalized synthetic antiferromagnetic nanoparticles for biomolecule detection and magnetic manipulation *Angew. Chem. Int. Ed.* **48** 1620-24

[4] Wang J, Munir A, Zhu Z and Zhou H S 2010 Magnetic nanoparticle enhanced surface plasmon resonance sensing and its application for the ultrasensitive detection of magnetic nanoparticle-enriched small molecules *Anal. Chem.* **82** 6782-89

[5] Lin P C, Chen S H, Wang K Y, Chen M L, Adak, A K, Hwu J R, Chen Y J and Lin C C 2009 Fabrication of oriented antibody-conjugated magnetic nanoprobes and their immunoaffinity application *Anal. Chem* **81** 8774-82

[6] Eberbeck D, Wiekhorst F, Steinhoff U and Trahms L 2006 Aggregation behaviour of magnetic nanoparticle suspensions investigated by magnetorelaxometry *J. Phys.: Condens. Matter* **18** S2829-46

[7] Rosicka D and Sembera J 2013 Changes in the nanoparticle aggregation rate due to the additional effect of electrostatic and magnetic forces on mass transport coefficients *Nanoscale Res. Lett.* **8**:20

[8] Wang L, Park H Y, Lim S I, Schadt M J, Mott D, Luo J, Wang X and Zhong C J 2008 Core@shell nanomaterials: gold-coated magnetic oxide nanoparticles *J. Mater. Chem.* **18** 2629-35

[9] Abu-Reziq R, Alper H, Wang D and Post M L, 2006 Metal supported on dendronized magnetic nanoparticles: highly selective hydroformylation catalysts *J. Am. Chem. Soc.* **128** 5279-82

[10] Kainz Q M and Reiser O (in press) Polymer- and dendrimer-coated magnetic nanoparticles as versatile supports for catalysts, scavengers and reagents *Acct. Chem. Res.*

[11] Qin H, Xu D and Yang S 2011 Dextran-coated Fe_3O_4 magnetic nanoparticles as a contrast agent in thermoacoustic tomography for hepatocellular carcinoma detection *J. Phys.: Conf. Ser.* **277**, 012028

[12] Lin P C, Chou P H, Chen S H, Liao H K, Wang K Y, Chen Y J and Lin C C 2006 Ehtylene glycol-protected magnetic nanoparticles for a multiplexed immunoassay in human plasma *Small* **2** 485-89

[13] Wu H T, Hsu C C, Tsai C F, Lin P C, Lin C C and Chen Y J 2011 Nanoprobe-based immobilized metal affinity chromatography for sensitive and complementary enrichment of multiply phosphorylated peptides *Proteomics* **11** 2639-53

[14] Narita A, Naka K and Chujo Y 2009 Facile control of silica shell layer thickness on hydrophilic iron oxide nanoparticles via reverse micelle method *Colloids & Surf.* A: *Physicochem. Eng. Aspects* **336** 46-56

[15] Polanski M and Anderson N L 2006 A list of candidate cancer biomarkers for targeted proteomics *Biomarker Insights* 1-48

6. Acknowledgements

Special thanks to En Hou Polymer Chemical Ind. Co., Ltd. (Taiwan) for providing the RD66 and RD35A surfactants

IOP Conf. Series: Materials Science and Engineering **64** (2014) 012023 doi:10.1088/1757-899X/64/1/012023

Waste to Want: Polymer nanocomposites using nanoclays extracted from Oil based drilling mud waste

Urenna V. Adegbotolu[1], James Njuguna, Pat Pollard and Kyari Yates
Robert Gordon University, Aberdeen; School of Engineering, IDEAS Research Institute, Riverside East, Garthdee Road, Aberdeen AB10 7GJ, United Kingdom

E-mail: u.v.ekeh@rgu.ac.uk

Abstract. Due to the European Union (EU) waste frame work directive (WFD), legislations have been endorsed in EU member states such as United Kingdom for the Recycling of wastes with a vision to prevent and reduce landfilling of waste. Spent oil based drilling mud (drilling fluid) is a waste from the Oil and Gas industry with great potentials for recycling after appropriate clean-up and treatment processes. This research is the novel application of nanoclays extracted from spent oil based drilling mud (drilling fluid) clean-up as nanofiller in the manufacture of nanocomposite materials. Research and initial experiments have been undertaken which investigate the suitability of Polyamide 6 (PA6) as potential polymer of interest. SEM and EDAX were used to ascertain morphological and elemental characteristics of the nanofiller. ICPOES has been used to ascertain the metal concentration of the untreated nanofiller to be treated (by oil and heavy metal extraction) before the production of nanocomposite materials. The challenges faced and future works are also discussed.

1. Introduction

Oil based drilling mud also called oil based drilling fluid and non aqueous drilling fluid, is a mixture of base oil (petroleum), bentonite clay or lime, barite and other chemical additives used for oil and gas exploration [1]. Drilling fluids are used to remove and bring to the surface, cuttings generated by the drilling bit as the well is drilled. However, drilling fluids (see figure 1) have become a potential source of pollution thereby increasing the environmental footprint.

The main concern about oil based drilling fluids emanate from the base oil used to produce them and the reservoir hydrocarbons they may interact with during drilling (also for water based fluids). The oils (Diesel or mineral oil) contain polycyclic aromatic hydrocarbons (PAH) [2]. The PAHs are cytotoxins. PAH is a colourless, sweet smelling and highly flammable hydrocarbon. Under European Union directive of dangerous substances (Directive 79/831/EEC); amended from Directive 67/548/EEC, EU regulation 1907/2006 - REACH (Registration, evaluation, authorisation and restriction of chemicals) information on substances notified under Directive 67/548/EEC and DSEAR – Dangerous substance and explosives atmosphere regulation 2006, petroleum oil is described as a carcinogenic, mutagenic and toxic due to PAH presence [3].

[1] To whom any correspondence should be addressed.

During Exploration and production (E&P) activities, the drilling fluid and cuttings generated are brought back to the surface (the top-side of the platform). At this location the drill cuttings are separated from the drilling fluid using the solids control equipment before it is tested for rheology and reused for further drilling activities. However, when the rheology of the drilling fluid is lost, the use of the drilling fluid is discontinued. Figure 2 illustrates the fate of the used drilling fluid during and/or after the drilling process.

Figure 1. Spent (used) Drilling Mud [2]

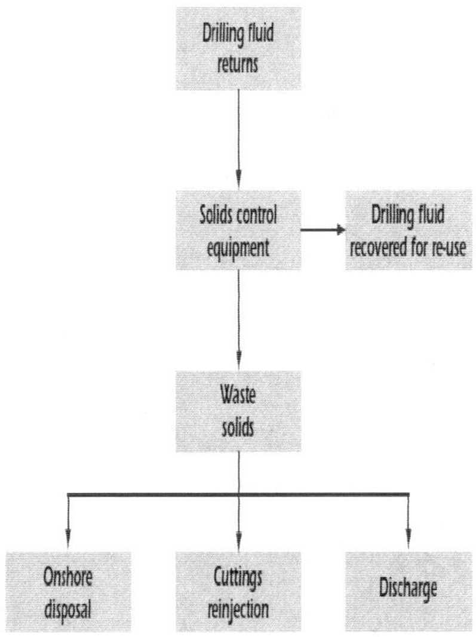

Figure 2. Fate of the Used Drilling Fluid [2]

Different methods have been proffered as solutions in the management of hazardous wastes generated from oil and gas E&P activities. However, some solutions are highly energy and power consuming, generating loads of greenhouse gases and other forms of environmental hazards. Some of these methods include: landfilling, road spreading, phytoremediation, bioremediation, subsurface drilling waste re-injection, ocean dumping, mechanical, chemical, thermal, stabilisation and solidification. Drilling fluid wastes have been used to make different products such cement base products, levees, tar road bases [6]. In our pervious research, we had developed an eco-sustainable (environmentally friendly), biodegradable non-toxic chemical treatment for the extraction of oil and water from the spent drilling mud through phase separation to obtain oil pure water and solid for reuse and recycling.

1.1 Environmental legislation supporting Recycling

According to the European Parliament, under Article 4 of the EU Waste Framework Directive (WFD) (2008/98/EC) the waste management hierarchy for waste prevention and management involve the following: prevention, preparing for re-use, recycling, recovery, e.g. energy recovery, and disposal.

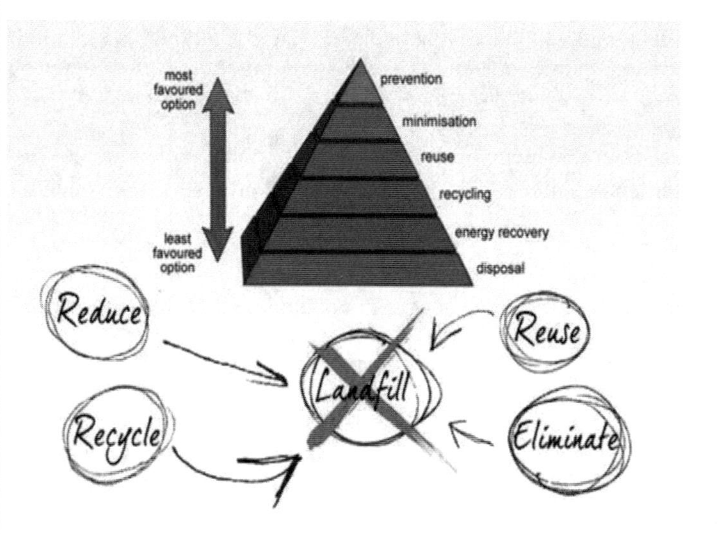

Figure 3. Waste minimisation triangle

Recycling is a method of waste management in the waste minimisation process of the WFD (waste framework directive) waste management heirarchy as shown by the waste triangle in figure 3. Recycling is one of the key ways the UK oil and gas industry can reduce its carbon footprint. Recycling of drilling fluids is a way of minimising waste to the environment via landfill sites to groundwater system, land, flora and fauna [6]. It is carried out to ensure that waste drilling mud can be converted into other products for further use. Technically, there are two kinds of recycling: Closed loop and Open loop recycling [5].

Recycling of drilling fluid ensures the safe use of drilling fluid as product from 'cradle to cradle' rather than 'cradle to grave' which discourages and minimises the use of the landfill disposal option. As common practise in waste management, the waste have to be sorted, separated, and treated for impurity removal before recycling. The raw waste material, spent (used) drilling fluid is sorted to remove most drill cutting, then, the waste is separated into oil, clean water and solid by phase separation treatment and could be modification before it can be reutilised. This work aims at recycling this oil and gas waste to an essential commodity, a want – a nanofiller for the Nanocomposite material industry.

1.2 Nanocomposite

Nanocomposite material is a unique of material made up of a synergy of different materials such as polymers (biological or synthetic), nanofillers, and metals. Nanocomposites are vital in today's world of sustainable innovations. Thus, the wide variety of engineered nanocomposite materials available for use today from metallic to plastic based nanocomposite materials. They offer high performance, cheaply fabricated and sustainable materials for different applications for construction, tool, household materials and more [12].

1.2.1 Nanofillers: These are materials added to polymers and distributed within the matrix of the polymer to produce a nanocomposite. They are used in the composite industry to enhance and modify

IOP Conf. Series: Materials Science and Engineering **64** (2014) 012023 doi:10.1088/1757-899X/64/1/012023

the state of different polymeric materials and thus, improve performance, functionality, value, efficacy, benefits, reduce cost and effectiveness of the polymer product [11]. Some of the properties that can be improved or controlled include: mechanical properties, rheology, transparency, flame retardant, electro conductivity or magnetism. Nanofiller may be spherical, flaky, rod-shaped particles. They have a nano dimension of less than 100nm.

In selecting appropriate nanofiller for use, a good understanding and application of the particle size, geometry, chemical modifications or functionality is very vital. Fillers have extensively been used in the production of paints, coatings, sealants, adhesives, concrete, rubbers, and plastics products. Some examples of carbon nanotubes, clay (layered silicates), calcium carbonate, magnesium hydroxide, silicon oxide, glass nano fibers, metallic nanoparticles, grapheme, barium sulphate [13].

1.2.2 Polymer

The polymer of interest in this work is Polyamide 6. Polyamide 6 (PA6) is also called Nylon 6. Gorrasi et al, 2002 used a combination of PA6 and montmorillonite (clay nanofiller) to produce nanocoposites and study nanofiller-polymer interaction. The choice of PA6 is for two reasons: its high temperatures of melting/decomposition and its thermoplastic nature which encourages further recycling. It has a density of 1.15g/cm^3, melting point / glass transition of >220°C and thermal degradation temperature of >300°C. It is tough, elastic and lustre. It is wrinkle free, resistant to abrasion, chemicals sua10^{13}/10^{10}Ohm h as dilute acids and alkalis. It could have a high tensile strength of 3000/1000MPa, elongation at break of 70/200%, Stress at yield of 80/45MPa. It could have a dielectric constant at 1MHz and 50Hz at 3.5/70 and 4.0/12/0 respectively. The dissipation factor at 50 Hz and 1MHz is 100/300 1 E-4 and 310/3000 1E-4.Dielectric strength is 15kV/mm. the thickness of its electrical strength is 1.0mm while Volume and surface resistivity are 10^{13}/10^{10}Ohm.m and 1013/1010 Ohm. Thermal: conductance is 0.23W/Km; while linear thermal expansion along cross to the direction of flow is 70/10 (10-6/K) [14]. PA6 finds application in the automotive and building industries.

2. Materials and methods

Polyamide 6 was obtained for the plastic shop. Spent oil based drilling fluid was donated by oil and gas company, Aberdeen, UK. SEM/EDXA experiment was carried out using Zeiss EVO LS10 variable pressure scanning electron microscope/energy dispersive x-ray. Perkin Elmer ICPOES was used to carry out the metal analyses. Smear test using plain white paper was carried out to ascertain oil free nature of the treated solid (drilling mud). Particle size analysis was carried using the Malvern particle sizer.

3. Results and Discussion

3.1 Influence of nanofiller geometry

Bentonite clay is in the montmorillonite which are known as layered sillicates. Their nature enables them to disperse into separate layers as shown in the SEM image in figure 4. These layers have gaps between them called interlayer spaces or gallery. These spaces allow for ion exchange with organic and inorganic cations which can alter the spaces. The spaces also allow for the satisfactory interphase adhesion between the polymer with filler which could occur in three forms: (i) Intercalated nanocomposites; (ii) intercalated, flocculated nanocomposites and (iii) exfoliated nanocoposite)as shown in figure 5. Clay fillers could contain sodium, potassium, aluminium and magnesium ions which accounts for their hydrophilic nature. Thus, best interaction with this kind of fillers will be the hydrophilic polymers [15]. For better interaction with hydrophobic polymers the fillers are treated with organophilic or hydrophobic tailed surfactants with cation surfactant treatment to reduce surface energy of silicate, increase layer spaces and raise the fillers wettability by the polymer. Studies by [4] showed that PA6 combination with clay could result in exfoliated nanocomposite. Furthermore studies by Cho and Paul, 2001, and Fornes et al 2002 signifies that surfactant treatment could enhance dispersion of clay matrix of the filler and enhance clay polymer interaction [7,8].

Figure 4. SEM image of Treated (Oil-free) spent oil based drilling fluid (2μm, Mag = 13.18K X)

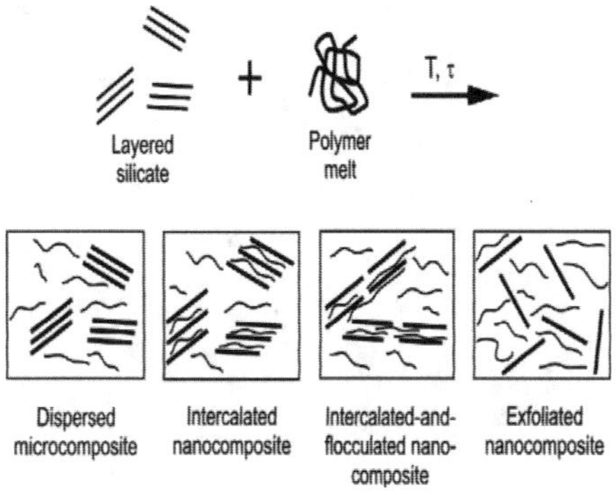

Figure 5. Types of nanofiller – polymer interaction.[15]

3.2 Effect of filler

Information on the MSDS of the formulated oil based drilling fluid from the manufacturer indicates that it contains bentonite clay and barite as weighting agents, and the main solid content. Some other drilling fluid may contain other weighting agents and solids such as Calcium carbonate, Hematite. During the drilling process, as the drill bits crushes the formation the drilling fluid carries the drill cutting of various sizes to the surface to be removed from the drilling mud. However, some of the cuttings are crushed into very fine particles which cannot be removed by the shale shaker. Thus these very fine cuttings (about 0.02 um) are collected as part of the spent oil based drilling mud (drilling fluid) for disposal. Depending on the drilled formation, the fine size cuttings may be from sand stones or lime stone, hence, increasing the concentration of silicate or calcium/carbonate respectively in the spent drilling mud. Nanoclay introduction through reinforcement with 5 wt% Claytone in polymer nanocomposite manufacture improved the optical transparency, mechanical and thermal properties [9].

3.3 EDXA elemental composition of proposed nanofiller

Energy dispersive x-ray analysis (EDXA) was carried out to ascertain the elemental composition of the treated oil based drilling fluid. The result shown in figure 6 indicates that the nanofiller contained Na,Ca, Al, Mg and Mn were attributed to have come from the ion exchange surface of the clay. Si was inferred to arise from the silicate matrix of the bentonite clay used in the formulation of the drilling mud (fluid). Fe was suggested to have come from oil and gas piping, Ba was inferred to be arising

IOP Conf. Series: Materials Science and Engineering **64** (2014) 012023 doi:10.1088/1757-899X/64/1/012023

from the Barium sulphate weighting agent added to the drilling fluid during formulation. C, carbon and O, Oxygen were expected to have arisen from the organic, eco-sustainable biodegradable cleaning agent used to extract or remove the oil (petroleum hydrocarbon) from the drilling mud solid. Cl was attributed to have been contributed from the sea water NaCl used as wash water or electrolyte additive in cleaning agent.

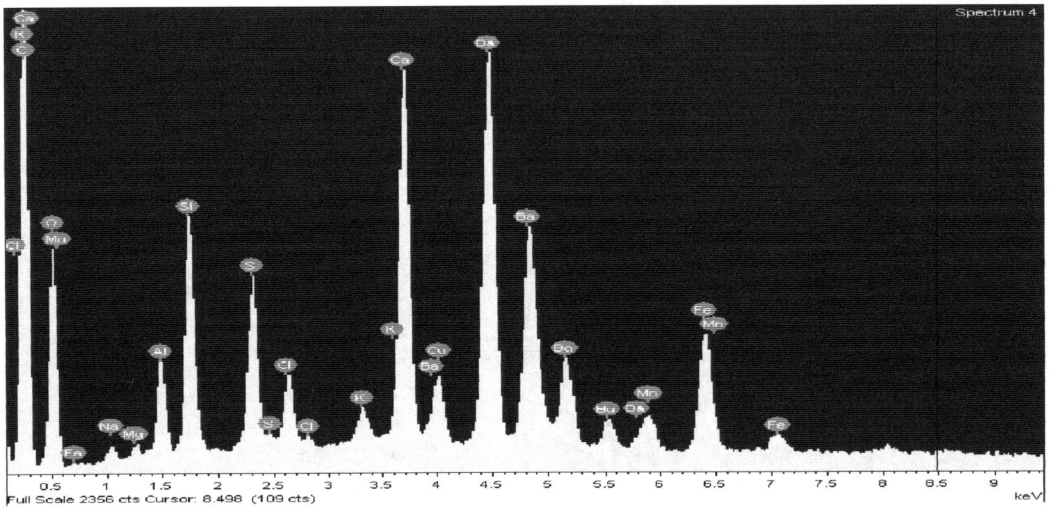

Figure 6. EDXA graph of spent oil based drilling mud

3.4 Particle size of the spent drilling fluid solid for nanocomposite manufacture

Particle size analysis report showed that about 0.33-0.40% of the sample were under 100nm. This was attributed to improperly dispersed solids after oil extraction and drying. Thus, for use of the treated oil based drilling mud waste solids as nanofiller, they would have to be pre-granulated before use as nanofiller in order to get a fine dispersion of the filler within the polymer matrix, produce a fine textured nanocomposite material and most importantly for the nanofiller particle size to qualify as a nanoparticle. A second alternative would be to treat and dry the treated drilling mud in such a way as to prevent the formation of solid lumps.

3.5 Heavy metal content of untreated spent (used) oil based drilling fluids

Due to the health concerns regarding the heavy metals likely to be present in the untreated drilling fluid, it was important to carry out a proper quanitative metal analysis to find out the metal concentration in the untreated nanofiller. The information would enable researchers justify the possibilities of using the nanofiller without treatment without possibilities of environmental harm arising, or the importance for the cleaning treatment agent to wash off, extract or remove the heavy metals present to as low as practicably possible. Currently, the heavy metal concentration of treated drilling mud is not reported. Table 1 gives the concentration of the heavy metal in the untreated oil based drilling mud sample. The information in Table 1 would enable treatment formulator and synthesisers to establish the content of treatment (oil/heavy metal removal agent) required for appropriate metal extraction, and the number of washes required to reduce heavy metal concentration to as low as practically possible, the kind of polymer to used in the production of nanocomposite material and finally, specific application of the nanocomposite material. Other metals and elements analysed by the ICPOES include: Ba (20448 mg/kg), Ca (2122mg/kg), Na (437.6mg/kg), Mg (238.43mg/kg), Mn (94.38mg/kg), Al (877.12mg/kg), Fe (1561.82mg/kg), k(136.74mg/kg), Sr (26.75mg/kg) Si (14.86mg/kg) and S (625.23mg/kg)

Table 1. Heavy metal concentration (mg/kg) of the untreated oil based drilling mud.

Heavy Metals	Concentration (mg/kg)
As	0
Cd	0
Cr	2.87
Cu	5.08
Ni	1.62
Zn	20.18
Hg	0
Pb	30.44

3.6 Possible applications

The high mechanical strength of PA6 (nylon 6) has ensured its utility for making different durable materials such as strings for musical instruments for instance guitar and violin. It is also used in making brushes and adhesives. It is used in packaging and clothing. According to [16] Toyota Japan has used it for the manufacture of auto parts such as the body. Other auto parts include dashboard, oil and water tanks. It is used in car parts as a result of its high melting point. The aim of this work is to recycle the treated oil-free drilling mud into a nanocomposite material for use in the oil and gas and automotive industries. By adopting this process, it is expected that unremoved heavy metals can be entrapped/ engulfed inside the naocomposite material. This process could be seen as a solidification and stabilisation method for the remaining (unextracted) heavy metals. The research findings and analytical results in this work suggests that the nanocomposite material of PA6 and treated oil based mud could be used to make pipelines, oil and gas pipeline (pipe-in-pine application), corrosion inhibition pipe liners, seals, rig floor and wall structure and, unexposed automotive parts such as oil tanks. These applications were chosen so as to prevent the reach of vulnerable people.

4. Conclusion

There is a real potential for recovered and treated solids from spent oil based drilling fluids to be recycled as a nanoclay fillers and used in the production of engineering nanocomposites materials. This application would certainly promote the European waste framework directive by diverting the recovered drilled solid from landfill sites to manufacturing plants for the manufacture and production of nanocomposite materials. It could be used to save money spent on the purchase of resins and other expensive materials used as nanofillers for engineered nanocomposite materials. Our future work entails analyses on the performance characteristics of nanocomposite materials produced from PA6 and the treated waste drilling mud nanofillers as well as working out a policy and advocacy framework for the full adoption of spent drilling mud waste as nanofillers.

Acknowledgement
The authors wish to thank Petroleum Trust Development Fund (PTDF), Nigeria for funding this project.

References

[1] Caenn R Darley H C H and Gray G R 2011 *Composition and Properties of drilling and completion fluids* (Sixth Edition) Pages 39-89
[2] OGP – International Association of Oil and Gas Producers 2003 *Environmental Aspects of the Use and Disposal of Non-Aqueous Drilling Fluids Associated With Offshore Oil & Gas Operations.* Report No. 342.
[3] Eur-lex 2013 *Dangerous substance directive.* Available: http://www.legislation.gov.uk/uksi/2002/2776/pdfs/uksi_20022776_en.pdf. Accessed: 7th January 2013.
[4] VanderHart D L Asano A and Gilman J W 2001 *Solid-state NMR investigation of paramagnetic nylon-6 clay nanocomposites. 1. Crystallinity, morphology, and the direct influence of Fe3+ on nuclear spins.* Chemistry of Materials, 13, pp. 3781–3795
[5]CIWM, 2012
[6] Chandler G 2009 *Reduce, Reuse, Recycle: What to do with drilling waste: Far North Oil & Gas.* Available online at http://gchandler.ehclients.com/index.php/graham/farnorth2_energy/reduce_reuse_recycle_what_to_do_with_drilling_waste/ Assessed 23.02.12
[7] Cho JW and Paul D R 2001 *Nylon 6 nanocomposites by melt compounding.* Polymer, 42, pp. 1083–1094
[8] Fornes T D Yoon P J Hunter D L Keskkula H and Paul D R (2002) *Effect of organoclay structure on nylon-6 nanocomposite morphology and properties.* Polymer, 43, pp. 5915–5933
[9] Hema R Ng P N and Amirul A A 2012 *Green nanobiocomposite: reinforcement effect of montmorillonite clays on physical and biological advancement of various polyhydroxyalkanoates.* Polymer Bulletin Springer-Verlag
[10] Bruzaud S and Bourmaud A 2007 *Thermal degradation and (nano)mechanical behavior of layered silicate reinforced poly(3-hydroxybutyrate-co-3-hydroxyvalerate) nanocomposites.* Polym Test 26:652–659
[11] Gorrasi G Tortora M Vittoria V Galli G Chiellini E 2002 *Transport and mechanical properties of blends of poly(ε-caprolactone) and a modified montmorillonite-poly(ε-caprolactone) nanocomposite.* J Polym Sci Polym Phys 40:1118–1124
[12] Camargo P H C Satyanarayana K G Wypych F 2009 *Nanocomposites: synthesis, structure, properties and new application opportunities.* J Mater Res 12:1–39
[13] EU ObservatoryNANO 2011 *Nanofillers improving performance and reducing cost.* available online at http://www.nanopinion.eu/sites/default/files/observatorynano_briefing_no.21_nanofillers_-_improving_performance_reducing_cost.pdf ObservatoryNANO Briefing 21 assessed on the 11.06.14
[14] PA6 2014 *Polyamide PA6* availabe onlinehttp://www.rolec.de/download/Polyamid_PA6_en.pdf
[15] Pesetskii S S Bogdanovich S P Myshkin N K 2013 *Chapter 5 - Tribological behavior of polymer nanocomposites produced by dispersion of nanofillers in molten thermoplastics.* Tribology of Polymeric Nanocomposites (Second Edition) Friction and Wear of Bulk Materials and Coatings. Pages 119–162, Butterworth-Heinemann, Oxford
[16] Yebra-Rodriguez A Alvarez-Lloret P Yebra A Cardell C and Rodriguez-Navarro A B 2011

Influence of processing conditions on the optical and crystallographic properties of injection molded polyamide-6 and polyamide-6/montmorillonite nanocomposites. Applied Clay Science 51, 414–418
[17] Ekeh-Adegbotolu U V, Ekeh O M and Wegwu M O 2012 *Cleanup of Crude Oil Polluted Sites Using Arachis hypogaea L. (Groundnut) and Biostimulants.* SPE-162968-MS

IOP Conf. Series: Materials Science and Engineering **64** (2014) 012024 doi:10.1088/1757-899X/64/1/012024

Effect of a gelatin-based edible coating containing cellulose nanocrystals (CNC) on the quality and nutrient retention of fresh strawberries during storage

F. M. Fakhouri [1,4], A. C. A Casari [11]M. Mariano[2], F. Yamashita [3] L. H. Innocnentini Mei [4], V. Soldi[2] and S. M. Martelli [1]

[1-] Faculty of Engineering, Federal University of Grande Dourados, BR.

[2-] Department of Chemistry, Federal University of Santa Catarina, BR.

[3-] Department of Food Science and Technology, Londrina State University, BR.

[4-] Faculty of Chemical Engineering, Campinas State University BR.

E-mail: farayde@gmail.com ; smmartelli@gmail.com

Abstract. Strawberry is a non-climacteric fruit with a very short postharvest shelf-life. Loss of quality in this fruit is mostly due to its relatively high metabolic activity and sensitivity to fungal decay, meanly grey mold (*Botrytis cinerea*). In this study, the ability of gelatin coatings containing cellulose nanocrystals (CNC) to extend the shelf-life of strawberry fruit (*Fragaria ananassa*) over 8 days were studied. The filmogenic solution was obtained by the hydration of 5 g of gelatin (GEL) in 100 mL of distillated water containing different amounts of CNC dispersion (10 mg CNC/g of GEL or 50 mg of CNC/g of GEL) for 1 hour at room temperature. After this period, the solution was heated to 70 ºC and maintained at this temperature for 10 minutes. The plasticizer (glycerol) (10g/100g of the GEL) was then added with constant, gentle stirring in order to avoid forming air bubbles and also to avoid gelatin denaturation until complete homogenization. Strawberries (purchased at the local market) were immersed in the filmogenic solution for 1 minute and after coated were dried at 15 ºC by 24 hours. The strawberries were then kept under refrigeration and characterized in terms of their properties (weight loss, ascorbic acid content, titratable acidity, water content). The results have shown that samples covered with GEL/CNC had a significant improvement in its shelf-life. For instance, for the control sample (without coating) the weight loss after 8 days of storage was around 65%, while covered samples loss in the range of 31-36%. Edible coating was also effective in the retention of ascorbic acid (AA) in the strawberries, while control sample presented a fast decay in the AA content, covered samples showed a slow decay in the AA concentration. Moreover, the use of GEL/CNC edible coating had an antimicrobial effect in the fruits.

1. Introduction

In the last two decades, many studies were presented in scientific literature, using edible films, and coatings to extend shelf life and enhance the quality of fresh, frozen and processed food. A variety of polysaccharides, proteins and lipids were used either singularly or combined to produce compound films (KESTER, FENNEMA, [1]).

Edible coating can be defined as a layer of edible material formed around food or placed between its components (KROCHTA, MULDER-JOHNSTON, [2]). According to Krochta *et al.* (1994), edible films are different from edible coatings for being formed before its application on the product, whereas the coatings are formed during their application [3].

This technology is used to inhibit the migration of moisture, oxygen, carbon dioxide, aroma, lipids and can also introduce additives to food, such as antioxidants, antimicrobial agents and aroma, enhancing the mechanical integrity and/or the handling characteristics of a food product (KROCHTA, JOHNSTON, [2]). Many deteriorating chemical and enzymatic reactions such as lipid oxidation, browning due to Maillard reactions and enzymatic browning, are directly influenced by humidity and water activity on food, which cause changes in color, odor, texture and nutritional value of food.

[1] farayde@gmail.com

Consequently, in order to extend the storage period of many different kinds of food, it is necessary to slow the migration of moisture (KESTER, FENNEMA,1989a [4]).

The permeability to gases like oxygen and carbonic gas can also influence, in a significant way, food stability during the storage period. A primary way of deterioration due to oxygen permeability, which occurs in many food products involves the oxidation of lipids, vitamins, composite of flavor and pigments (KESTER, FENNEMA b [5]).

The ripening of the fruit is accompanied by a series of physical and biochemical processes, which result in synthesis and degradation of pigments, conversion of starch into sugar, loss of firmness, production of volatiles and increase of respiration of climacteric fruits (Andrews and Li [6]). Physico-chemical parameters and sensorial alterations are usually employed to accompany the biochemistry of ripening or to evaluate the processes of preservation.

Recent studies are exploring the potential of surface edible coatings to maintain and extend the quality and the shelf-life of fresh products, as well as to reduce the quantity of non-degradable disposable packaging. Among the studied fruits are: bananas (Banks [7], citric fruits (Hagenmaie and Baker [8]), apples (Banks et al. [9]), mangoes (Dhalla and Hanson [10]), pears (Amarante et al. [11]), avocados (Jonhston and Banks [12]), strawberries (Garcia et al. [13] [14], Tanada-Palmu [15] and guavas (Fakhouri and Grosso, [16], Oliveira and Cereda [17], Tanada-Palmu et al. [18]).

The use of surface coatings can exert effects similar to the storage under controlled atmosphere. Usually, the effect of reducing the quantity of O2 and elevating the content of CO2 over the respiration rate, as well as the synthesis of C2H4, are pointed out as the primary causes for the beneficial effects of coatings on the storage of fruits and vegetables (Kader [19]). The coatings exert effect on the permeability of the peel to gases by blocking the pores on the surface, in greater or smaller proportions.

The modification of the internal atmosphere of the fruits and leafy vegetables, obtained from the coatings, depends on the permeability of the film, its thickness and the covering on the surface of the fruit, (Cisneros – Zevallos and Krochta [20]). Besides, the permeability throughout the coating and the film is influenced by several factors, including morphology, density, chemical structure and polymeric orientation. The kind of solvent, the plastifier and the drying rate also influence the coefficient of permeability (Mchugh et al. [21]).

In the present work, were developed biofilms formed from filmogenic solutions of gelatin and nanocrystals. The filmogenic solutions were later applied as coatings in strawberries, which is a fragile and of ripening fruit and also a highly nutritious food (Poling [22]). In addition, this fruit was chosen due to its (i) easiness of handling and storage (due to its reduced volume); (ii) wide surface of transpiration in relation to its weight, (iii) frailty; and (iv) the strawberry does not have a protective membrane, so the use of coatings would make difficult the loss of water, noting that it possesses 85% of water in its composition.

2. Material and Methods

2.1 Materials

Gelatin, the biopolymer used to obtain the coatings, was supplied by Leiner Davis Gelatin, Cellulose nano crystals (CNC) were produced from eucalyptus fiber and glycerol (Synth) was used as plastifier. The strawberries used were supplied by a small-scale producer in the borough of Dourados-MS, Brazil.

2.2 Methods

2.2.1. *Production of the filmogenic solution:*

The filmogenic solution of gelatin was elaborated through the hydration of 10g of gelatin in 100 mL of distilled water with 0, 1 and 5g of CNC/ 100 g GEL. The mixture was heated at 70°C for 10 minutes. Afterwards, the glycerol was added under smooth mechanical agitation (10g/100 g).

Production of CNC from eucalyptus fibers: The material has been provided in milled form.
i) Step 1: 10 grams of fibers have been submerged in 200 mL of NaOH solution (10% w/w) for a period of 4 hours at 60°C. After this period, the solution was filtered and washed until a pH of 7.0 was achieved. Then the material was dried at 40°C.
ii) Step 2: 300 mL of a combined solution of NaOH (5% w/w) + $CaO(Cl)_2$ (2,5% w/w) was added to the dried fiber obtained in step 1. The material was submitted to a rest of 4 hours at 45°C. After this period, the solution was filtered and washed until a pH of 7.0. Then the material was dried at 40°C.
ii) Step 3: The extraction of the CNC was done using the treated fiber material. For that, the fiber obtained from step 2 was treated with 250 ml of sulphuric acid 60% for 30 minutes at 55°C under vigorous stirring. After 30 minutes, hydrolysis reaction was stopped and the sample was centrifuged until the CNC were in suspension (Figure 1).
iv) Step 4: The suspended solution went through dialysis for 72 hours.

2.2.2. *Strawberry coating:*

For the application of edible coatings, strawberries were picked ripe and immediately transported to the department of Food Technology Department of FEA/UNICAMP. The strawberries were then selected as to for a homogeneous lot. Afterwards, were hygienized with sodium hypochlorite solution and then submitted to the recoating with filmogenic solutions. Then four lots were formed: (i) LC – control lot (without coating), (ii) AC1 – stawberries coated with filmogenic solution containing 1% of nanocrystals, (iii) AC2 –strawberries coated with 5% of nanocrystals. All the dried strawberries were dried for 12 hours at 25°C.

2.2.3. *Physical Chemical characteristics*

2.2.3.1 *Mass Loss*

Strawberries were weighed in semi-analytical weighing machines

2.2.3.2. *Soluble solids, pH, Total Titrable Acidity, Humidity and Total Solids*

Determined by the utilization of AOAC Methods, 1997

2.4. *Microorganism grown.*

The dilution plating was used to detect yeasts and molds in the strawberries. This method provides maximum exposure of the cells to atmospheric oxygen and allows greater uniformity in the development of the colonies increasing accuracy of the results and facilitating the subsequent isolation of pure cultures.

2.5. *Transmission electron microscopy.*

The shape and dimensions of the CNC were assessed through transmission electron microscopy (TEM). To perform this, a drop of the nanocrystal suspension to be analyzed was diluted in 2 ml of iso-propyl alcohol and placed in an ultrasonic bath (UltraCleaner 1600 A) for 10 minutes. Then, the alcoholic dispersion was deposited on a special grade for 24 hs. After, each film was covered with a 2% solution of uranyl acetate (for coloring) and dried again at room temperature. The images were obtained using a machine model JEM-1011 TEM.

2.6. *Statistical Analysis*

Made using Statistica® 5.5 (Stasoft, USA).

3. Results and Discussion

All the filmogenic solutions presented themselves transparent and homogeneous. The filmogenic solution obtained presented nanocrystals in solution (Figure 1).

IOP Conf. Series: Materials Science and Engineering **64** (2014) 012024 doi:10.1088/1757-899X/64/1/012024

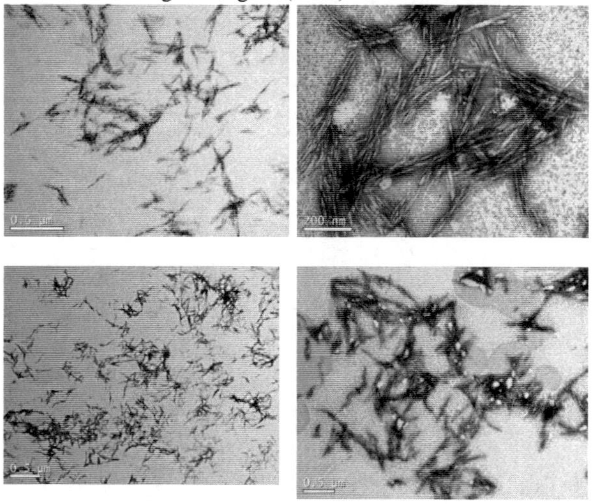

Figure 1. TEM of CNC in suspension

Visually the coatings adhered uniformly to the fruits, presenting a natural bright, without rupture zones. Moreover, the use of coating kept the natural flavor of fruits. The appearance of fruits can be seen in Table 1.

Table 1. Images of strawberries (coated and control) during storage

Storage time (days)	Control	1% CNC	5% CNC
1			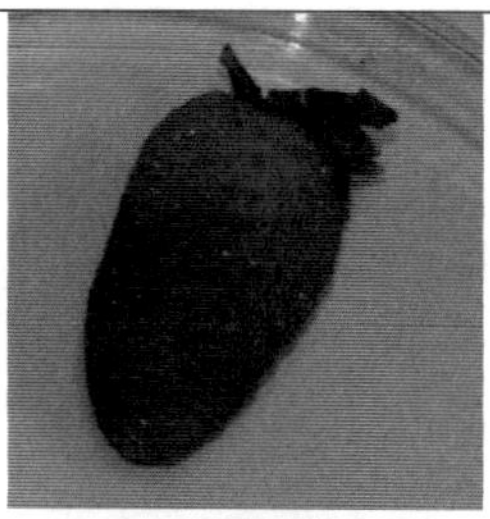
8	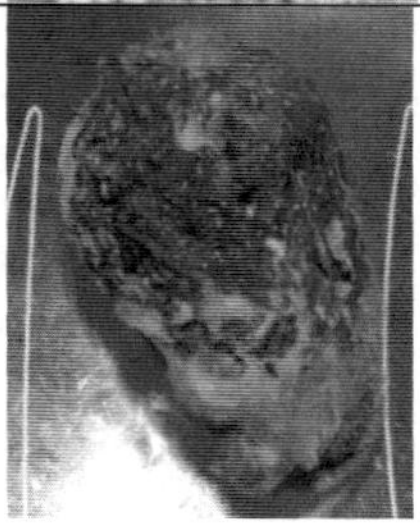		

In all coated fruits, the weight loss was lower than in control fruits (Figure 2). While control fruits lost 65.8 % of total weight, the maximum loss in coated fruits was 35%, observed in trial 3, coated with films based on gelatin and CNC. On the control fruit, the loss of 35% was observed on the 5th day of storage .

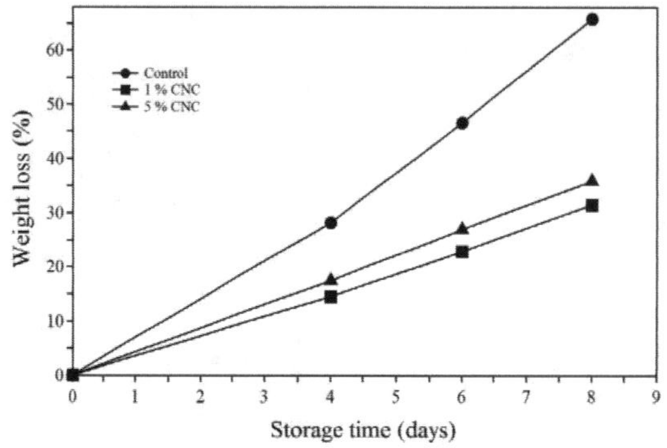

Figure 2. Weight loss (%) of strawberries during storage

4. Conclusion

The ability of gelatin containing CNC coatings to extend the shelf-life of strawberry fruit (Fragaria ananassa) were studied, mainly for industrial applications. The coating was effective in protect fresh strawberries during the storage period, the aspect of coated fruits was uniform aspect. This type of treatment could be used to extend the conservation period of strawberries to support long periods of time between the fruit harvest and consumer.

References

[1] Kester, J. J., And Fennema, O. Resistance Of Lipid Films To Water Vapor Transmission. *J. Am. Oil Chem. Soc.* **66**, pp. 1139-1146, 1989.

[2] Krochta, J. M., De Mulder-Johnson, C. M. Edible And Biodegradable Polymer Films: Challenges And Opportunities. *Food Technology*, **51**, pp. 61-74, 1997.

[3] Krochta, J. M., Baldwin, E.A., Nisperos-Carriero, M. O. **Edible Coatings And Films To Improve Food Quality.** A Technomic Publishing Co., 1994

[4] Kester, J. J., Fennema, O. R. Edible Films And Coatings: A Review. *Food Technology*, **40**: pp.47-59, 1989, a.

[5] Kester, J. J., Fennema, O. R. Edible Films And Coatings: A Review. *Food Technology*, **40**: pp.47-59, 1989, b.

[6] Andrews, P.K., Li, S. Parcial Purification And Characterization Of β - D – Galactosidase From Sweet Cherry, A Nonclimateric Fruit. *Journal Agricultural And Food Chemistry*, **42**, pp.2177-2182, 1994

[7]. Banks, N. H. Studies Of The Banana Fruit Surface In Relation To The Effect Of Tal Pro Long Coating Onf Gaseous Exchange. *Scientia Hort.*, 24, pp.279-286, 1984.

[8] Hagenmaier, R. D., Baker, R. A. Reduction In Gas Exchange Of Citrus Fruit By Wax Coatings. *J. Agr. Food Chem.*, **41**, pp.283-287, 1993.

[9] Banks, N. H., Eliatem, S. M. and Hammat, M.T. The Oxygen Affinity Of Ethylene Production By Slices Of Apple Fruit Tissue. Acta Hort., **157**, pp.257-260, 1985.

[10] Dhalla, R., Hanson, S. W. Effect Of Permeable Coatings On The Storage Life Of Fruits. Ii. Pro-Long Treatment Of Mangoes (Mangifera Indica L. Cv. Julie). *Int. J. Food Sci. Technol.*, **23**, pp.107-112, 1988.

[11] Amarante, C., Banks, N. H., Silva, G. Gas exchange and ripening behaviour of coated pears. p 193-202. In: Proc. **Australian Postharvest Horticulture Conference**, Univ. Western Sidney. Hawkesbury, NSW, Australia, 1997.

[12] Johnston, J. W., Banks, N. H. Selection Of A Surface Coating And Optimization Of Its Concentration For Use On "Hass" Avocado (Persea Amarica Milli) Fruit. *New Zealand J. Crop. Hort. Scie*, **26**, pp. 143-151, 1988.

[13] Garcia M. A., Martino M. N., Zaritzky N. E. Starch-Based Coatings: Effect On Refrigerated Strawberry (Fragaria Ananassa) *Quality. J Sci Food Agric* **76**, pp. 411-420, 1998.

[14] Fakhouri F. M , Grosso C.R.F. Efeito De Coberturas Comestíveis Na Vida Útil De Goiabas In Natura (Psidium Guajava L.) Mantidas Sob Refrigeração Brazilian Journal Of *Food Technology*, **l6**, 2, pp. 203-211, Jul/Dez 2003

[15] Tanada-Palmu, P.S. ; Fakhouri F. M.; Grosso C. R. F. ,Filmes Biodegradáveis: Extensão Da Vida Útil De Frutas Tropicais. In : **Biotecnologia, Ciência E Desenvolvimento**, pp. 7-14, Jun2.002.

[16] Andrews, P.K., Li, S. Parcial Purification And Characterization Of β - D – Galactosidase From Sweet Cherry, A Nonclimateric Fruit. *Journal Agricultural And Food Chemistry*, **42**, pp.2177-2182, 1994.

[17] Kader, A. A. Regulation of fruit physiology by controlled modified atmospheres. Acta *Horticulturae*, **398**, pp.59-67, 1995.

[18] Cisneros-Zevallos, L., Krochta, J. M. Internal Modified Atmospheres Of Coated Fresh Fruits And Vegetables: Understanding Relative Humidity Effects. *Journal Of Food Science*, **67**, pp. 1990-1995, 2002.

[19] Mchugh T. H, Krochta, J. M. Permeability Properties Of Edible Films. In: Krochta, J. M., Baldwin, E. A., Nisperos-Carriedo, M. (Eds). **Edible Coatings And Films To Improve Food Quality.** Technomic Publishing Co, Inc.Lancaster Usa, 1994.

[20] Poling, E. B. Blackberries. Journal of Small Fruit and Viticulture **14 (1-2)**, pp38-69, 1996.

[21] Macfie, H. J., Bratchell, N. Designs To Balance The Effect Of Order Of Presentation And First-Order Carry-Over Effect In Hall Tests. *Journal Of Sensory Studies*, **4,** pp 129-148, 1989.

[22] Briston, J. H. Plastic films , Third Edition. New York, NY: John Wiley & Sons, Inc, 1988.

IOP Conf. Series: Materials Science and Engineering **64** (2014) 012025 doi:10.1088/1757-899X/64/1/012025

Nanocharacterization of titanium nitride thin films obtained by reactive magnetron sputtering

V V MERIE[1], M S PUSTAN[1], C BÎRLEANU[1] and G NEGREA[2]
[1] Technical University of Cluj-Napoca, Faculty of Machines Building, Department of Mechanical Systems Engineering, 103-105 Muncii Avenue, Cluj-Napoca, Romania
[2] Technical University of Cluj-Napoca, Faculty of Materials and Environmental Engineering, Department of Materials Science and Engineering, 103-105 Muncii Avenue, Cluj-Napoca, Romania

E-mail: Violeta.Merie@stm.utcluj.ro; vio1919@yahoo.com

Abstract. Titanium nitride thin films are used in applications such as tribological layers for cutting tools, coating of some medical devices (scalpel blades, prosthesis, implants etc.), sensors, electrodes for bioelectronics, microelectronics, diffusion barrier, bio-micro-electromechanical systems (Bio-MEMS) and so on. This work is a comparative study concerning the influence of substrate temperature on some mechanical and tribological characteristics of titanium nitride thin films. The researched thin films were obtained by reactive magnetron sputtering method. The experiments employed two kinds of substrates: a steel substrate and a silicon one. The elaboration of titanium nitride thin films was done at two temperatures. First, the obtaining was realized when the substrates were at room temperature, and second, the obtaining was realized when the substrates were previously heated at 250 °C. The elaborated samples were then investigated by atomic force microscopy in order to establish their mechanical and tribological properties. The nanohardness, roughness, friction force are some of the determined characteristics. The results marked out that the substrate which was previously heated at 250 °C led to the obtaining of more adherent titanium nitride thin films than the substrate used at room temperature.

1. Introduction

The development of new materials and new technologies is needful in order to meet the growing necessities of the customers. A special attention must be paid to the link between the structure, the processing and the properties of the materials studied for a specific application. Some researchers aimed at studying the titanium nitride due to its wide range of properties. A low friction coefficient, good resistance to corrosion, high hardness, good adhesive wear, high electrical conductivity, high evaporation temperature, biocompatibility and chemical and metallurgical stability are some of the most important characteristics of titanium nitride [1-9].

The properties mentioned above make titanium nitride suitable for applications within the biomedical, space, microelectronics, cosmetics industries and so on [10-16]. One of the applications is the fabrication of miniaturised devices such as microelectromechanical systems (MEMS). The high melting point and the covalent bonding represent two difficulties that hinder the elaboration of such materials by forming and sintering [11]. Chung-Fon Hsieh and his co-workers reported the use of titanium nitride for manufacturing a micro-gap discharge device. In this case, titanium nitride acts as a robust electrode in the device previously mentioned [12].

A widely used application of titanium nitride is that of coating material such as diffusion barrier in the microelectronics industry, hard and protective coatings for the mechanical tools or decorative coatings [13]. Machunze and Janssen studied the use of titanium nitride as diffusion barrier layers in integrated circuit (IC) technology or as wear protective coatings [14]. A cosmetic gold-coloured purposes and wave length selective transparent optical films are other applications of titanium nitride coatings [15]. According to Lawand and his co-workers, titanium nitride is also used in the biomedical industry as coatings of some medical devices such as scalpels [16].

The researchers tried to elaborate titanium nitride thin films through different methods. Plasma-assisted CVD (Chemical Vapour Deposition), magnetron sputtering, RF reactive sputtering, reactive evaporation, laser ablation, electron cyclotron resonance (ECR), PVD (Physical Vapour Deposition), ion-beam deposition, arc deposition are some techniques that were employed for the deposition of titanium nitride thin films on different substrates [10, 15].

The present research is a study concerning the influence of two different parameters on the tribological and mechanical properties of some titanium nitride thin films obtained by reactive magnetron sputtering. The two parameters whose influence was attended are: the material of the substrate and the temperature of the substrate (the temperature at which the deposition of the thin films was done).

2. Materials and experimental procedure

2.1. Materials

Two different materials were employed for the substrates on which the deposition of titanium nitride thin films was realized. First, titanium nitride coatings were deposited on silicon Si (100) substrate. After that a X30WCrV9-3 tool steel was used as material for the substrate. The chemical composition of the tool steel used is given in table 1.

Table 1. The chemical composition of the used tool steel.

Steel	Chemical composition (%)					
	Carbon	Silicon	Manganese	Chromium	Vanadium	Tungsten
X30WCrV9-3	0.25÷0.35	0.10÷0.40	0.15÷0.45	2.50÷3.50	0.30÷0.50	8.50÷9.50

The obtaining of titanium nitride thin films was done using a titanium target with purity of 99.95 %. The process took place in an atmosphere formed by a mixture of argon and nitrogen.

2.2. Experimental procedure

The deposition of titanium nitride thin films was obtained by reactive magnetron sputtering method. The experiments started with the preparation of the substrates. For that purpose, the tool steel substrates were first heat treated (hardened) at 1150 °C in oil and then they were annealed at 600 °C for an hour.

After that the substrates (both the silicon and the tool steel surfaces) were cleaned and washed with acetone and isopropyl alcohol. Then they were introduced in an ultrasonic bath for removing any possible remaining impurities.

The reactive magnetron sputtering process of titanium nitride coatings was realized providing the following conditions and parameters:

- target material: titanium with a purity of 99.95 %;
- atmosphere: mixture of argon and nitrogen;
- atmosphere pressure: 1 mtorr;
- discharge current: 350 mA;

- substrate bias: - 40 V;
- nitrogen flow rate: 1.2 cm^3/min;
- deposition time: 17 minutes.

The deposition of titanium nitride thin films was carried out at two different temperatures in order to study the influence of deposition temperature on the obtained layers. For that purpose, some layers were deposited on substrate at room temperature, while the rest were obtained on substrates that were previously heated at 250 °C.

The following notation was established for the obtained samples in order to facilitate the interpretation of the results: "Substrate material _ Deposition temperature" (table 2). "RT" stands for room temperature.

Table 2. The notation of the investigated samples.

Sample	Substrate material	Deposition temperature (°C)
Steel_RT	X30WCrV9-3	Room temperature
Steel_250	X30WCrV9-3	250
Silicon_RT	Silicon (100)	Room temperature
Silicon_250	Silicon (100)	250

The as-obtained samples were characterized from the mechanical and tribological point of view at nanoscale. The nanocharacterization of the researched materials was realized using two kinds of microscopes: a NT 206 and a XE 70 atomic force microscope respectively. The tests carried out on the NT 206 atomic force microscope with a n-silicon CSC38 cantilever were done when relative humidity was 19 %, temperature was 25 °C and scanning frequency was 0.7 Hz. The tribological characteristics of studied samples were determined with this kind of microscope. The tests performed with the XE 70 atomic force microscope for determining the mechanical properties employed a silicon nitride cantilever. Its characteristics are: cantilever stiffness of 144 N/m, tip radius of 25 nm, tip height of 109 μm, tip thickness of 24 μm and cantilever length of 782 μm respectively. The XEI Image Processing Tool for SPM data used the Oliver and Pharr method for the interpretation of the obtained data.

3. Theoretical formula

The determination of tribological properties implied among other things the determination of friction force. This parameter can be calculated based on the data provided by the atomic force microscope as follows [17, 18]:

$$F_f = \frac{d_z \cdot r \cdot G \cdot h^3 \cdot b}{l^2 \cdot s}$$

where d_z represents the tips deflection, r is a constant (r = 0.33), G is shear modulus (for silicon: G = $53.9 \cdot 10^{-3}$ $N/\mu m^2$), h, b and l are the dimensions of the cantilever, while s is tips height (s = 18 μm). The values of tips deflection are taken from the data offered by the program used for interpreting the images taken by the Surface Scan program. Data processing was performed using the Surface Explorer program.

According to the data given by the producer, the dimensions of the cantilever used for investigations are:

- thickness: h = 1 μm;
- width: b = 32.5 μm;
- length: l = 350 μm.

4. Results and discussions

The topography of the obtained titanium nitride thin films on both steel and silicon substrates was studied by atomic force microscopy. The 3D images obtained for the samples deposited on steel and silicon surfaces at 250 °C are presented in figure 1. The samples obtained on silicon substrates shows much smoother surfaces as the samples obtained on steel substrates.

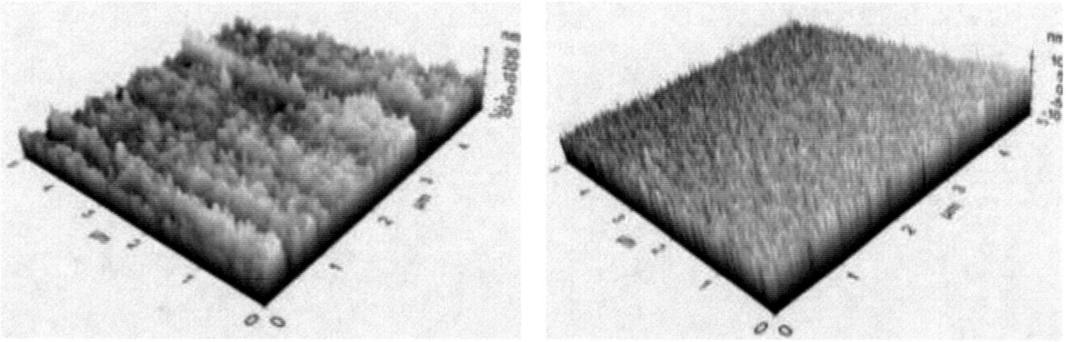

(a) (b)

Figure 1. 3D images of thin films for (a) Steel_250; (b) Silicon_250 samples respectively.

The roughness of the studied samples was determined based on the data given by the NT 206 atomic force microscope in order to characterize the surfaces of the elaborated thin films. Its fluctuation for all four kinds of samples is graphically given in figure 2. When the samples were elaborated on the same substrate (either steel or silicon), the preheating the substrate led to the decrease of thin films roughness. Taking into account the average value determined for this parameter, the decrease of this parameter was about 18 % for the samples deposited on steel substrates and 22 % for the samples deposited on silicon substrates respectively. Regarding the influence of deposition temperature, the average values of roughness for the samples elaborated at room temperature were relatively equal, while the average values of roughness for the samples elaborated at 250 °C varied in the sense that the samples deposited on steel surfaces were characterized by a roughness higher with almost 6 % than the samples elaborated on silicon surfaces. The higher values for surface parameter were marked out for the samples deposited on steel substrate at room temperature, while the smallest values were determined for the samples deposited on silicon substrate at 250 °C.

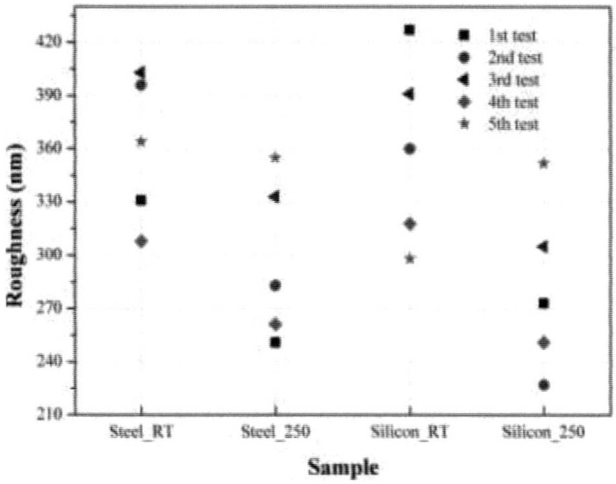

Figure 2. Roughness of titanium nitride thin films obtained by reactive magnetron sputtering.

Concerning the friction force, the fluctuation of this parameter for the researched samples is presented in figure 3. When the samples were deposited on the same surface, the use of substrate preheating determined a significant increase of friction force for both steel and silicon substrates. Thus this increase for the samples elaborated on steel substrate was about 30 % while the increase for the samples elaborated on silicon substrate was about 24 %. As regards substrate influence on the friction parameter of studied titanium nitride thin films, the samples elaborated on silicon surface were determined to be characterized by a higher value of friction force both for the samples deposited at room temperature and 250 °C respectively. When the deposition was realized at room temperature, the friction force was about 15 % higher for the samples elaborated on silicon surfaces. The same effect was highlighted when deposition temperature was 250 °C, the increase of friction force being about 10 %. The higher values for the friction parameter were determined for the samples deposited on silicon substrates at 250 °C while the smallest values were marked out for the samples deposited on steel substrates at room temperature.

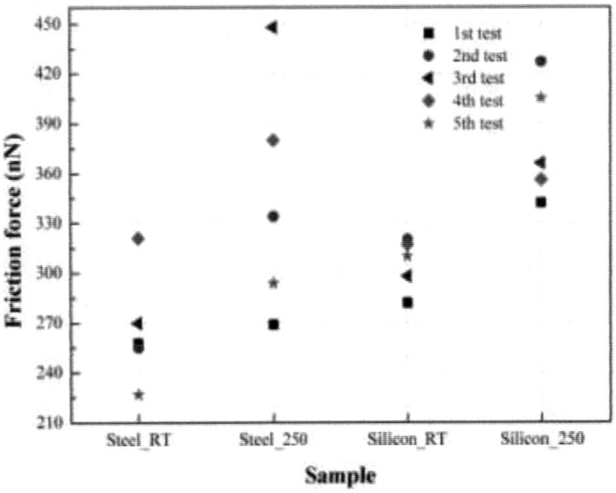

Figure 3. Friction force of titanium nitride thin films obtained by reactive magnetron sputtering.

The determination of the nanohardness for the studied titanium nitride thin films was another aim of the present research. The tests carried out for that purpose highlighted a significant influence of substrate and its deposition temperature on the nanohardness of researched coatings. Figure 4 presents the fluctuation of nanohardness for all kinds of researched samples. The preheating of the substrate at 250 °C led to the increase of this parameter for the samples deposited on steel and silicon surfaces with 41 % and 19 % respectively. The values of nanohardness for the samples deposited on silicon substrate were determined to be higher with 62 and 37 % respectively when the deposition was realized at room temperature and 250 °C respectively. The smallest and the biggest values of this parameter were established for the samples elaborated on steel substrate at room temperature and the samples elaborated on silicon substrate at 250 °C respectively.

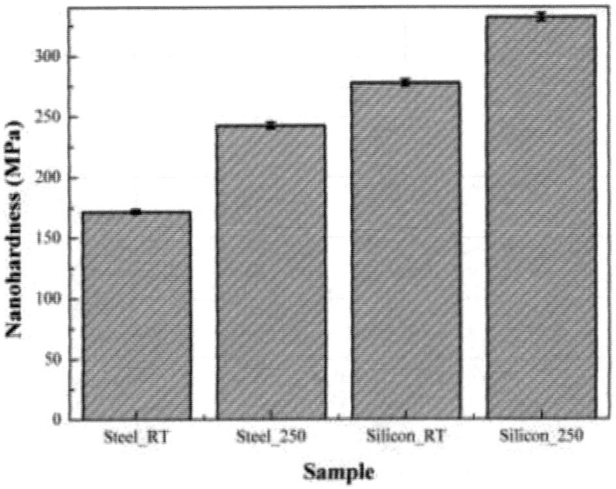

Figure 4. Nanohardness of titanium nitride thin films obtained by reactive magnetron sputtering.

5. Conclusions

Concerning the nanocharacterization of titanium nitride thin films obtained by reactive magnetron sputtering on silicon and steel substrates at two different temperatures, we can conclude that:

- the substrate and the its deposition temperature have an important effect on the properties of the elaborated thin films;
- the preheating of substrate at 250 °C leads to the decrease of thin films roughness both for steel and silicon substrate; the smallest and the biggest values of roughness were marked out for the titanium nitride thin films deposited on silicon surfaces at 250 °C and the thin films deposited on steel surfaces at room temperature respectively;
- the titanium nitride coatings obtained on the substrates previously heated at 250 °C are characterized by a better adherence in regard to the thin films obtained at room temperature;
- the preheating of the substrates at 250 °C leads to the increase of friction force for the samples obtained on both steel and silicon surfaces; the biggest values of friction parameter were determined for the thin films deposited on silicon surfaces at 250 °C;
- the nanohardness of samples obtained either on steel or silicon surfaces increases when the substrate is preheated at 250 °C; when the deposition temperature is kept constant, the samples deposited on silicon surfaces are characterized by higher values of the nanohardness than the samples deposited on steel surfaces;
- the temperature of 250 °C was established to be the optimal temperature for obtaining titanium nitride thin films by reactive magnetron sputtering.

6. References

[1] Yazdani A, Soltanieh M, Aghajani H and Rastegari S 2011 *Vacuum* **86** 131
[2] Karagkiozaki V, Logothetidis S, Kalfagiannis N, Lousinian S and Giannoglou G 2009 *Nanomed. Nanotechnol. Biol. Med.* **5** 64
[3] Vasu K, Ghanashyam Krishna M and Padmanabhan K A 2011 *Thin Solid Films* **519** 7702
[4] Serban V A, Rosu R A, Bucur A I and Pascu D R 2013 *Appl. Surf. Sci.* **265** 245
[5] Subramanian B, Ashok and Jayachandran M 2008 *Appl. Surf. Sci.* **255** 2133
[6] Po-Jui Su and Haochih Liu B 2013 *Thin Solid Films* **529** 317
[7] Cui X, Yu Z, Ma M and Chu P K 2009 *Surface & Coatings Technology* **204** 418
[8] Borah S M, Pal A R, Bailung H and Chutia J 2008 *Appl. Surf. Sci.* **254** 5760
[9] Jeyachandran Y L, Narayandass Sa K, Mangalaraj D, Areva S and Mielczarski J A 2007 *Mat. Sci. Eng., A* **445–446** 223
[10] Hussain T, Ahmad R, Khan I A, Siddiqui J, Khalid N, Bhatti A S and Naseem S 2009 *Nucl. Instrum. Methods Phys. Res., Sect. B* **267** 768
[11] Sanchez O, Hernandez-Velez M, Navas D, Auger M A, Baldonedo J L, Sanz R, Pirota K R and Vazquez M 2006 *Thin Solid Films* **495** 149
[12] Hsieh C F and Jou S 2006 *Microelectron. J.* **37** 867
[13] Bavadi R and Valedbagi S 2012 *Materials Physics and Mechanics* **15** 167
[14] Machunze R and Janssen G C A M 2009 *Thin Solid Films* **517** 5888
[15] Kim S H, Park H, Lee K H, Jee S H, Kim D J, Yoon Y S and Chae H B 2009 *Journal of Ceramic Processing Research* **10(1)** 49
[16] Lawand N S, French P J, Briaire J J and Frijns J. H. M. 2012 *Procedia Eng.* **47** 726
[17] Merie V, Pustan M and Birleanu C 2013 ACTA *Technica Napocensis. Series: Applied Mathematics and Mechanics* **56(IV)** 709
[18] Pustan M, Rochus V, Wu L, Noels L and Golinval J C 2010 *ECNF First European Conference on Nanofilm* (Liege, Belgium)

Acknowledgments

This work was supported by the Space Technology and Advanced Research Grant STAR 2012, project number 30036 (REDEMS).

Biomaterial Co-Cr-Mo Alloys Nano Coating Calcium Phosphate Orthopedic Treatment

N.Palaniappan[1]., Gajendra Kumar Inwati[2] ., Man Singh *
Central University of Gujarat

palaniappancecri@rediffmail.com

Abstract

The modern study a thermal martensitic transformation of biomedical Co–Cr–Mo alloys and ultimately offers large elongation to failure while maintaining high strength. In the future study, structural evolution and dislocation slip as an elementary process in the martensitic transformation in Co–Cr–Mo alloys were investigated to reveal the origin of their enhanced phase stability due to nitrogen addition and coating of calcium phosphate specimens with and without nitrogen addition were prepared. The N-doped alloys had a single-phase matrix, whereas the N-free alloys had a duplex microstructure. Irrespective of the nitrogen content, dislocations frequently dissociated into Shockley partial dislocations with stacking faults. The Nano range coating of calcium phosphate function as obstacles to the glide of partial dislocations and consequently significantly affect the kinetics of the martensitic transformation. As a result, the formation of marten site plays a crucial role in plastic deformation and wear behavior, the developed nanostructures modification associated with nitrogen addition must be a promising strategy for highly durable orthopedic implants.
Key words, implant, Cr, Co, Mo alloys, Ca-phosphate coating, SEM, corrosion, nanostructures.

Introduction

The Co-Cr-Mo alloys with exceptional corrosion and wear resistances have been used in orthopedic implants. The artificial hip and knee joints although as cast method Co -Cr -Mo alloys are practically use for these applications their strength and elongation to failure are sometimes low due to a coarse microstructure solidification defects precipitations. The improve the mechanical reliability of biomedical of biomedical Co-Cr –Mo alloys to obtain orthopedic .the hard precipitation phase with size from 10-20 nm to several 10 of micrometers The nucleation and growth of calcium phosphates on chromium alloys like Cr,Co Mo investigated because of their relevance to orthopaedic application . It is well known that Ca-P coatings have led to better clinical success rates in long term than uncoated chromium alloys Cr, Co Mo for implants. these advantages are due to superior initial rates of Osseo integration the apatite coating deposited on different biomaterials (metals and their alloys polymers) can reduce fibrous encapsulation promote bone growth enhance direct bone contact and has also been shown to promote differentiation of bone marrow stromal cell along oestrogenic lineage (1) the design of novel coating Ca-P coatings involves also addition of small amount of metal or metals alloys the nano particles exhibiting antibacterial activity . Such nano particle incorporated to Ca-P coatings may impart antibacterial activity, which makes them promising to be applied in hard tissue replacement against postoperative infections. However it is worth to mention that the size and amount of such nanoparticles should be properly chosen in order to prevent their toxic effect to living cell. The coating method is two step procedures, based on chemical and electrochemical methods aimed at activating chromium alloys surface for subsequent deposition of calcium phosphate coating is presented. The combined effects of surface topography and chemistry of chromium alloys subtracts on calcium phosphate formation are discussed. The calcium phosphate coating produced by both biomimetic as well as electrochemical methods are compared with respect to their physicochemical characteristics and biological evaluation. The biomimicry a peptide motif bases on the amino acids sequence of pp was designed. This peptide is composed of three repeats of DSS amino acids. The control Ca-P deposition to achieve a biomimetic coating .such a coating would be expected to have high compatibility in a mineralizing tissue site and thereby further improve the cellular compatibility of alloys. Corrosion study (CH instruments Inc. 900D) with a three electrode cell configuration was used. In this study the temperature was maintained at 37^c platinum was recorded. A potentiodynamic polarization cure was recorded with a scanning rate of 1mv s-1 the potential window was set from 1.5.0V to 2.0V Cr,Co, Mo alloys as well as 3 DSS peptide – Cr-Co-Mo alloys were used as a working electrode , with one surface connected to an electrical wire the other surface exposed to the media (surface area 1 cm^2)the corrosion potential .

.

Materials and Method

The nitrogen – doped alloy Co-29, Cr- 6, Mo-0.2 (wt. %) that conforms with the ASTM F75 standard was prepared by high frequency induction melting in an Ar atmosphere, Cr 2N Powder was used as the nitrogen source, the Co-29, Cr-6 Mo alloys without N doping and with 0.24wt% N were also prepared by casting for comparison. The chemical compositions of the alloys were determined by conventional chemical and gas analyses. The metallic elements were determined using an inductively coupled plasma optical emission spectroscopy .carbon was determined by using a combustion infrared absorption method. N determined was by He carrier fusion – thermal conductivity, these alloys are respectively referred to as 0.20N, N-free, 0.24N alloys .the homogenizing heat treatment at 1498 K for 43.2 Ks (12h) and hot forging are 1473K followed by water quenching, the cast ingots N free 0.24 alloys heat treated at 1473K by hot caliber rolling followed by water quenching to modify atomic structure to Nano range the 0.20 N alloys.

Coating method;

The Cr, Co, Mo alloys have cut 4cm width 5 cm length, cleaned by different grate carbon emery sheet 600, 1200 and washed with acetone under sonication. the calcium phosphate prepared w/v method, the first sodium bicarbonate and phosphate solution with peptide and ascorbic acids solution - calcium phosphate coating both solution together mixing after apply by using galvanization method 1.5 m.v to 2 mv continuously apply 3 hour coating formed after it had been taken out placing in oven 300^c continuously 24 hour in side of the oven .the calcium phosphate coating alloy cooling in room temperature . A thin layer of micro-porous calcium phosphate (CaP) coating was processed by electrolytic deposition (5) onto the surface of titanium as the drug carrier (5).the titanium plate substrate was finally polished with anAl_20_3 paste (o.3mm) , while the surface of the titanium cylinder .(6) The calcium phosphate coatings(7) for metallic medical implants that are used to replace the bone tissues synthesized by various methods are considered in the structural and morphological aspects. The methods of deposition of such coatings on metal substrates. The osteoconductive property of HA coating is able to enhance bone (8) ingrowth and to reduce early migration of both hib and knee prostheses.

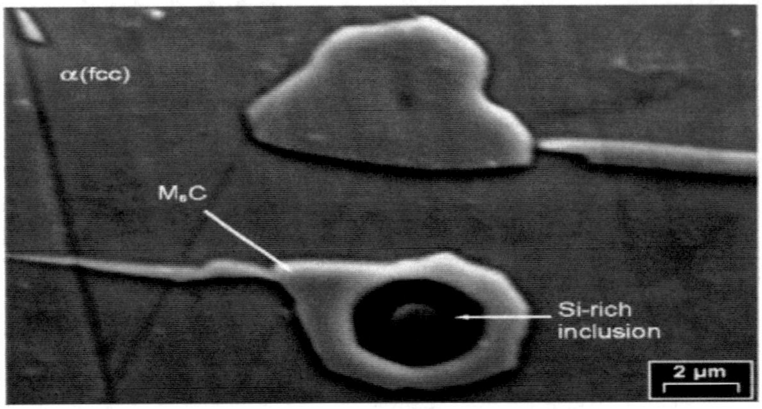

Figure 1. SEM image of Cr, Co, Mo

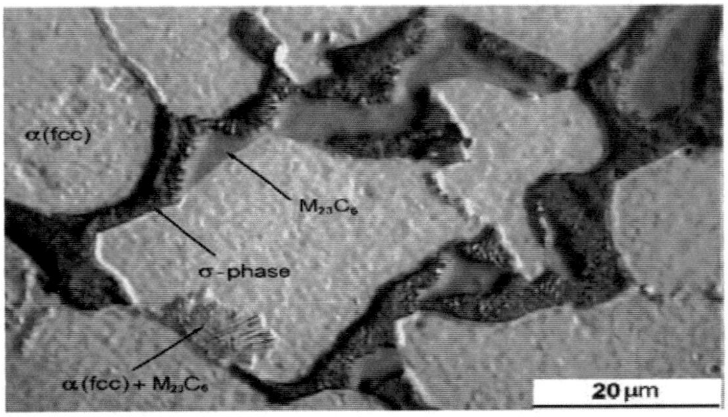

Figure 2. Optical micrograph of the main phase detected on C 1 image by color metallography

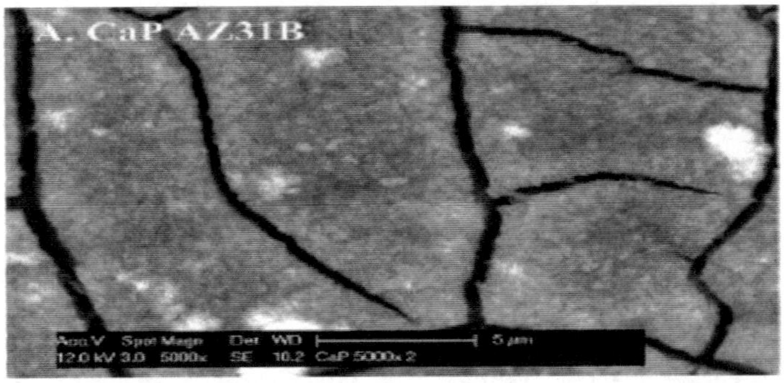

Figure 3. Calcium phosphate coated Cr-Co-Mo alloys SEM

Table 1. Corrosion parameter obtained by the mixed potential theory.

a	*Alloys*	*E couple (V) Vs Ag/AgCl 3M*	*I couple* (μAm^{cm-2})
b	*Cr-Co-Mo*	*-0.230*	*0.550*

Table 2. Chemical composition of the alloys used in the present study (wt %).

a	*Co- Alloy*	*Cr*	*Mo*	*Ni*	*Mn*	*Si*	*C*	*N*
b	*N free Bal*	*28.7*	*6.3*	*<0.01*	*<0.01*	*<0.1*	*0.002*	*0.001*
c	*0.20N Bal*	*28.5*	*5.9*	*<0.01*	*0.22*	*0.16*	*0.001*	*0.20*

Applications of Calcium phosphate nano coating Cr-Co-Mo alloys all the biological fields

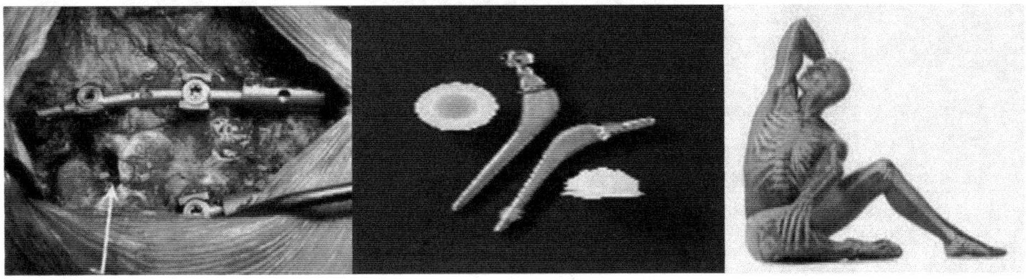

(a) **(b)** **(c)**

Figure 4. Potential Applications of the Cr-Co-Mo Alloys in Domains of the biological Fields **a-thigh fracture implant Cr-Co- Mo, b-chest replacement bone, c-humane bone part**

Conclusion

The nitrogen addition is a promising strategy for improving the mechanical properties of biomedical properties of biomedical Co-Cr-Mo alloys. The novel strategy for developing nano architecture Co-Cr-Mo orthopedic implant alloys with optimized martensitic transformation behavior. The metal alloys Co, Cr, Mo also have a significant corrosion resistance in biological media. Due to its comparable strength and cost effectiveness Co-Cr-Mo alloy may also be a good alternative as an implant materials. The bio active DSS and peptide with calcium phosphate nano range coating on the Co-Cr- Mo implants were obtained these material are unusual corrosion resistance and anti-bacterial activity friendly to body function.

Acknowledgement;

I would like thanks to Dean of chemical science central university Gujarat.

Reference;

[1] Gristina AG, Costerton JW 1985 Bacterial adhesion to biomaterials and tissue. J. Bone

Joint Surgery, **67**A 264-273

[2]Michaela G recu, Gabriela Novac, Daniela ionita, 2011 Camelia Ungureanu, *Rev. Chem.* **62**, No. 3

[3] Park Jiu H, Lee Y.-K, Kim K.-M, Kim K.-N, 2005 bioactive calcium phosphate prepared on H_2O_2 treated titanium substrate by electro deposition. Surf. Coat. technol. **195**. 252-257

[4] Ricardo Luiz Perez Teixeira., Geralda Cristina Durães de Godoy, 2004, *Mat. Res.* vol.**7** no June

[5] Kazemzadeh-Narbat M, Kindrachuk J, Duan K, Jenssen H, Hancock RE, Wang R. 2010 Biomaterials.**36**. 9519-26.

[6] Kyosuke Ueda1. Takayuki Narushima. Takashi Goto. Tomoyuki Katsube. Hironobu Nakagawa. Hiroshi Kawamura. and Masayuki Taira. 2007 Materials Transactions, 48,307 -312

[7] Valentin M Ievlev. 2013 *Russ. Chem. Rev.* **82** 131

[8] Soren Overgaard. 200 *Acta Orthop Scand* .**71**.1-74

IOP Conf. Series: Materials Science and Engineering **64** (2014) 012027 doi:10.1088/1757-899X/64/1/012027

Highly transparent and luminescent nanostructured Eu_2O_3 doped ZnO films

R Sreeja Sreedharan[1], V Ganasan[2], C Sudarsanakumar[3], Radhakrishna Prabhu[4] and V P Mahadevan Pillai[1*]
[1]Department of Optoelectronics, University of Kerala, Kariavattom, Thiruvananthapuram, Kerala 695581, India.
[2]UGC-DAE Consortium for Scientific Research, Khandwa Road, Indore-452017, Madhyapradesh, India.
[3]School of Pure and Applied Science, Mahatma Gandhi University, Kottayam, Kerala 686 560, India.
[4]School of Engineering, Robert Gordon University, Aberdeen, U.K
IOP Publishing, Dirac House, Temple Back, Bristol BS1 6BE, UK

[*]E-mail: vpmpillai9@gmail.com

Abstract. Zinc oxide is a wide, direct band gap II-VI oxide semiconductor. Pure and Eu-doped ZnO films are prepared by RF Magnetron sputtering at different doping concentrations (0.5, 1, 3 and 5 wt %). The films are annealed at 500 ^{0}C in air for two hours. The structural, morphological and optical properties of the films are characterized using XRD, micro-Raman, AFM, UV-Visible and photoluminescence spectroscopy. The thickness of the films is measured using stylus profilometer. XRD analysis shows that all the films are highly c-axis oriented exhibiting a single peak corresponding to (002) lattice reflection plane of hexagonal wurtzite crystal phase of ZnO. The micro-Raman spectra analysis reveals the presence of E_2 high mode in all the samples which is the intrinsic characteristic of hexagonal wurtzite structure of ZnO. The appearance of LO modes indicates the formation of defects such as oxygen vacancies in the films. AFM micrographs show uniform distribution of densely packed grains of size with well defined grain boundaries. All the films exhibit very high transmittance (above 80%) in the visible region with a sharp fundamental absorption edge around 380 nm corresponding to the intrinsic band edge of ZnO. All the films show PL emission in the UV and visible region.

1.Introduction

Zinc Oxide (ZnO) is a wide, direct band gap (E_g = 3.3 eV at room temperature) II-VI oxide semiconductor with a large excitonic binding energy of 60 meV at room temperature, making it suitable for exciton-related applications.[1-2] It is highly transparent in the visible region with a sharp cut-off around 380 nm. The band structure and optical properties of ZnO are similar to GaN, which is commonly used for the fabrication of photonic devices in the UV range. Owing to these characteristics, ZnO has attracted much attention for a variety of applications in the field of short wavelength optoelectronic devices such as blue and ultraviolet light emitters, laser diodes, phosphorescent display and photo detectors.[3] It is widely used as transparent conducting electrodes, window material, flat panel displays, gas sensors etc.[4] Several deposition techniques such as pulsed laser deposition (PLD)[5], molecular beam epitaxy (MBE)[6], spin coating[7], chemical vapor deposition

(CVD)[8], spray pyrolysis[9], sputtering[10] etc are employed for the preparation of high quality ZnO thin films. This paper reports the preparation of Eu-doped ZnO films by RF magnetron sputtering and their structural, morphological and optical characterisation using various techniques such as XRD, micro-Raman, AFM, UV-Visible and photoluminescence spectroscopy.

2.Experiment

Europium oxide doped ZnO films are deposited on cleaned quartz substrate at room temperature (300 K) using RF (13.56 MHz) magnetron sputtering technique. Commercially available ZnO powder (Aldrich - purity 99.99%) is mixed with europium oxide powder (Aldrich - purity 99.99%) at different doping concentrations viz.,0, 0.5, 1, 3 and 5 wt % and the pressed pellets are used as the target for sputtering. The sputter chamber is initially evacuated to a pressure of 5×10^{6} mbar. Then argon is admitted into the chamber and argon pressure is maintained at 0.02 mbar. The target is powered through a magnetron power supply (Advanced Energy, MDX 500).The sputtering is carried out under constant RF power of 150 watts for 30 minutes on quartz substrate kept at a distance of 5 cm from the target. The as-deposited films are annealed in air at 500 ^{0}C for two hours. The films thus prepared with different Eu_2O_3 doping concentrations viz.,0, 0.5, 1, 3 and 5 wt % are designated as E0, E0.5, E1, E3 and E5 respectively

The crystalline quality and crystallographic orientation of the films are investigated using X-ray diffraction analysis (Brucker D8 Advance X-ray diffractometer) using Cu Kα1 radiation of wavelength 1.5406 A° in the 2Θ range 20-70°. The phonon vibrational modes of the films are recorded using micro-Raman spectrometer (Labram HR-800) with 514 nm argon ion laser as an excitation source at a spectral resolution of 1cm^{-1}. The surface morphology of the samples is investigated using AFM (Digital Instruments Nanoscope III) analysis. The transmittance, reflectance and absorption spectra of the films in the spectral range 200-900 nm.are measured using JASCO v-550, UV-visible double beam spectrometer. The thicknesses of the samples are measured using Dektak stylus profilometer. Photoluminescence spectra of the samples are recorded using Horiba Jobin Yvon Fluralog III modular spectro flourometer.

3 Results and Discussion

3.1. XRD Analysis

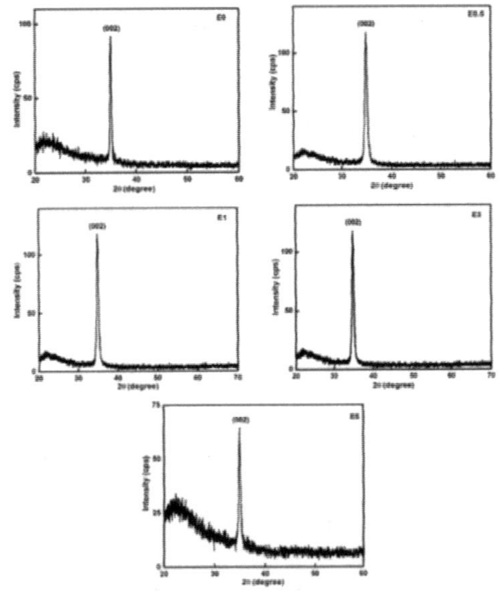

Figure 1: X-ray diffraction patterns of Eu_2O_3 doped ZnO films for different Eu_2O_3 doping concentrations

IOP Conf. Series: Materials Science and Engineering **64** (2014) 012027 doi:10.1088/1757-899X/64/1/012027

The X-ray diffraction analysis shows that all the films present a single sharp intense peak corresponding to (002) lattice reflection plane of hexagonal wurtzite structure of ZnO (JCPDS card No-75-0576)[11] presenting a single crystalline like structure with c-axis orientation due to lowest surface free energy of (002) lattice reflection plane.[12] There is a slight enhancement in intensity of (002) peak with moderate Eu_2O_3 doping concentrations (up to 3 wt%). But in 5wt% Eu_2O_3 doped film, there is a reduction in intensity of (002) peak. The FWHM of the Eu_2O_3 doped films are found to be greater than that of undoped film. The stress introduced due to the mismatch of ionic radii of Zn (Zn^{2+} = 0.74A°) and Eu (Eu^{3+} = 1.07A°) can be the reason for enhanced FWHM of the (002) peak in Eu_2O_3 doped ZnO films compared to the undoped ZnO film.[13-14]
The c-values of all the films are found to be greater than the JCPDS value indicating that the crystallites are in a state of tensile stress. The increase in the out of plane c lattice constant indicates the expansion of the lattice due to the substitution of Eu^{3+} at Zn^{2+} sites. The introduction of dopant and lattice mismatch between film and substrate introduces strain in the film which can affect the structure and properties of the film to some extent. The positive value of stress indicates that the crystallites are in a state of tensile stress and is found to be decreasing from 5.54 - 4.93GPa with Eu doping. The average size of the crystallites in the films are estimated using the following Debye-Scherrer' formula and is found to be in the range 16-27 nm showing the nanostructured nature of the films.

3.2. Micro-Raman analysis

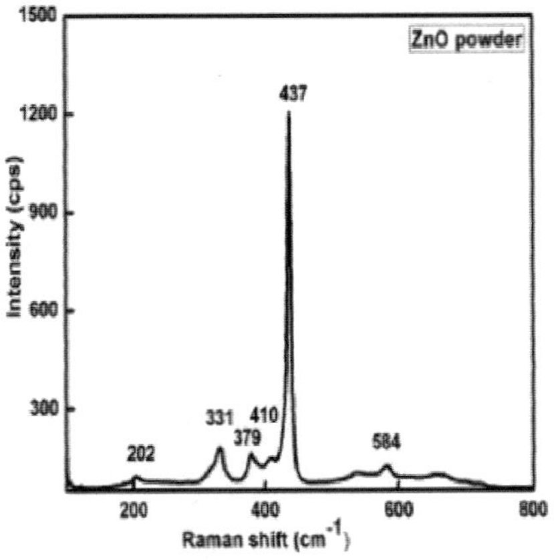

Figure 2: Micro Raman spectra of ZnO powder

Figure 2 shows the micro-Raman spectrum of bulk ZnO powder. The Raman spectrum of ZnO powder presents a very intense band at 437 cm^{-1} corresponding to E_2 high mode which is the intrinsic characteristic of hexagonal wurtzite structure of ZnO. This mode is related to the vibration of oxygen atoms and is considered as the Raman fingerprint of ZnO.[15-17] The Raman spectrum also shows medium intense band at 331 cm^{-1} and weak bands at 202, 379, 410 and 584 cm^{-1}. The band at 584 cm^{-1} can be due to E_1 (LO) mode which is associated with the formation of various crystal defects such as oxygen vacancy, zinc interstitials or other defect states.[18] The bands at 331 cm^{-1} and 379 cm^{-1} can be due to A_1 (TO) mode and E_1 (TO) mode respectively. The micro-Raman spectra of undoped ZnO and Eu_2O_3 doped ZnO films (figure not included) present an intense Raman band around 437 cm^{-1} which supports the presence of wurtzite ZnO phase in the films. The presence of E_1 (LO) mode at 581 cm^{-1} with appreciable intensity in all the films is an indication of the presence of various crystal defects such as oxygen vacancy, zinc interstitials or other defect states in them. Compared to the Raman spectrum of ZnO powder (bulk), the Raman spectra of films present a broad spectral feature. Here the ZnO is deposited on amorphous quartz substrate and the films can have residual stress. The broad nature of Raman spectra of the films can be due to this residual stress. The measured thickness of the films is in the range 83-94 nm and hence the spectral contribution from the substrate can also be expected in the Raman spectra of the films. The medium intense band around 488 cm^{-1} can be due to the vibration from the quartz substrate. Compared to the Raman spectra of other films, the Raman spectrum of 5% Eu_2O_3 doped film presents broader spectral feature. This can be attributed to the decline in crystallinity due to the higher doping concentration in this film as observed in the XRD analysis. Also the medium intense band observed in the Raman spectrum of E5 can be due to intrinsic host lattice defects such as oxygen vacancies and zinc interstitials.[19]

3.3.AFM analysis

The AFM (3D) images of undoped and Eu_2O_3 doped ZnO films are shown in Figure 3. AFM image of undoped ZnO film presents uniform distribution of densely packed grains of almost same size (around 45 nm) with well defined grain boundaries. The surface morphology of the Eu_2O_3 doped films presents the distribution of grains of varying sizes throughout the film. The films with Eu_2O_3 doping concentration 0.5 and 1 wt% show higher values of rms surface roughness compared to that of the other films. The variation of the grain size with Eu_2O_3 doping concentration obtained from the AFM analysis keeps the same trend as obtained from XRD analysis.

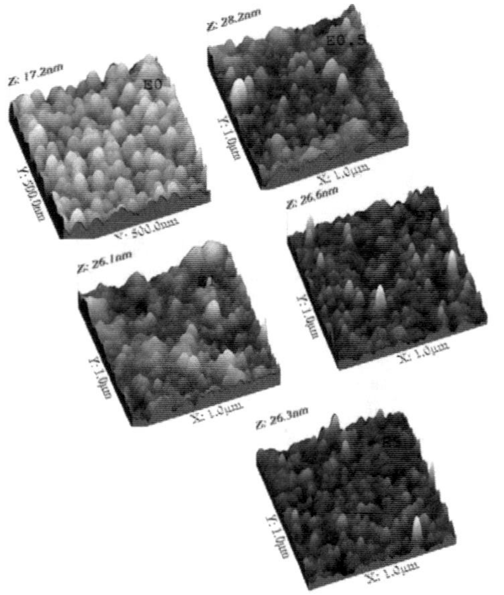

Figure 3. AFM (3D) pictures of ZnO films with different Eu_2O_3 doping concentrations

3 . 4.Optical properties

Figure 4 shows the transmittance spectra of undoped and Eu_2O_3 doped ZnO films recorded in the wavelength range 200 – 900 nm. Average transmittance of the films in the wavelength range 400 – 900 nm are calculated and found that all films show very high transmittance above 85%. For all the films the transmittance exhibits a sharp reduction at around 380 nm corresponding to the fundamental absorption of ZnO. The sharp absorption onset in the UV- region and high transmittance in the visible region indicates the good crystalline and optical quality and direct band gap nature of the films. The oscillations observed in the transmission and reflection spectra (figure not given) of the films can be due to the interference of light arising from the difference in refractive indices of the film and the substrate. This interference of multiple reflections arising from the film and substrate surface shows the thickness uniformity of the films. These oscillations in the spectra indicate that smooth films are formed on quartz substrate as evident from the AFM analysis.[20-21] This reduces the scattering loss at the surface. [22]

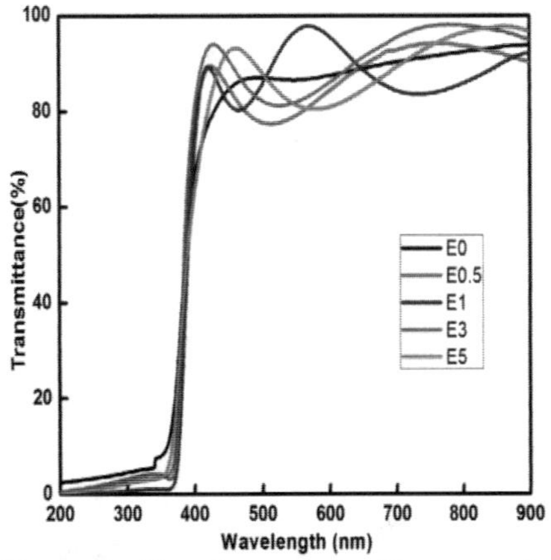

Figure 4: Optical transmittance spectra of undoped and Eu_2O_3 doped ZnO films

The optical band gap of the films are calculated using the Tauc plots. Undoped film shows a band gap of 3.24 eV and Eu_2O_3 doped films show higher values of band gap energy. There are several reasons for the shift of band gap in the films such as improvement or reduction in crystallinity, modification in barrier height due to the change in crystallite dimension, quantum size effect, change in the density of impurities, tensile or compressive strain in the films etc.[23] It can be expected that the contribution from Eu^{3+}ions on substitutional sites of Zn^{2+}ions and Eu-interstitial atoms determine the widening of the band gap caused by increase in carrier concentration. Typically the blue shift of the absorption edge of the Eu_2O_3 doped ZnO films is associated with an increase in the carrier concentration blocking the lowest states in the conduction band, the well known Burstein-Moss effect.[24]

3.5.Photoluminescence Spectra

The PL spectra of undoped and Eu_2O_3 doped ZnO films recorded using excitation wavelength of 325nm. The undoped film shows PL emission in the UV and visible regions. The intense UV emission in ZnO film arises from the near band edge emission due to the free excitonic emission of ZnO. The weak visible emission at 450, 540 and 570 nm can be due to the defects such as oxygen vacancy, Zn interstitials, antisite oxygen etc in the films.[25-26]In the doped films the intensity of the UV emission decreases and the emissions in the visible region enhances its intensity. The film with 5wt% Eu_2O_3 doping concentration show highest visible PL intensity. Europium doping in the ZnO lattice introduces defects in the lattice such as oxygen vacancies, zinc interstitials etc. The Raman spectrum of 5 wt% Eu_2O_3 doped ZnO film presents defect related vibrational mode at 288 cm^{-1}. The enhanced PL intensity in the visible region observed in Eu_2O_3 doped films can be due to the enhanced defects introduced in the ZnO lattice by Eu_2O_3 doping.

4 Conclusion

Highly c-axis oriented Eu_2O_3 doped ZnO films with good structural and optical quality are prepared using RF magnetron sputtering technique. The X-ray diffraction analysis shows that all the films present a hexagonal wurtzite structure of ZnO. The presence of an intense Raman band around 437 cm^{-1} corresponding to E_2 high mode in the micro-Raman spectra of undoped ZnO and Eu_2O_3 doped ZnO films supports the presence of wurtzite ZnO phase in the films. AFM images present a smooth surface morphology for all the films. Average transmittance of the films in the wavelength range 400 – 900 nm are calculated and found that all films show very high transmittance above 85%. The blue shift in the band gap energy observed in the Eu_2O_3 doped ZnO films due to Burstein-Moss effect. The enhanced PL intensity in the visible region observed in Eu_2O_3 doped films can be due to the enhanced defects introduced in the ZnO lattice by Eu_2O_3 doping.

5 . References

[1] Klingshirn 1975 Phys. Status Solidi B **71**547.

[2] Shan F K, Liu G X, Lee W J, Shin B C 2007 J. Appl. Phys **101** 053106.
[3] Lu J G, Fujita S, Kawaharamura T, Nishinaka H and Kamada Y 2008 Phys. Stat. Sol.(c) **5** 3088 .
[4] Sun X W, Kwok H S, 1999 J. Appl. Phys **86** 408.
[5] Tsoutsouva M G, Panagopoulos C N, Papadimitriou D, Fasakic I and Kompitsas M 2011 Materials Science and Engineering B **176** 480
[6] Heo Y W, Ip K, Pearton S J, Norton D P and Budai J D 2006 Applied Surface Science **252** 7442
[7] Smirnov M, Baban C and Rusu G I 2010 Applied Surface Science **256** 2405.
[8] Tan S T, Chen B J, Sun X W, Fan W J, Kwok H S, Zhang X H and Chua S J 2005 J.Appl. Phys.**98** 013505.
[9] Lehraki N, Aida M S, Abed S, Attaf N, Attaf A and Poulain M 2002 Curr. Appl. Phys. **12** 1283.
[10] Suvorova N A, Usov I O, Stan L, DePaula R F, Dattelbaum A M, Jia Q X and Suvorova A A 2008 Appl. Phys. Letters **92** 141911.
[11] Culity B D 1978 *Elements of X-Ray Diffraction* (Addison-Weseley) p.102.
[12] Deng H, Ressel J J, Lamb R N and Jiang B 2004 Thin solid Films **458** 43.
[13] Fujimura N, Nishihara T, Goto S, Xu J, Ito T 1993 J. Cryst. Growth **130** 269.
[14] Zhang Y Z, Liu Y P, Wu L H, Xie E Q and Chen J T 2009 J. Phys. D: Appl. Phys. 42085106.
[15] Ilican S, Caglar Y and Caglar M 2008 J Optoelectronics and Advanced Materials **10** 2578.
[16] Vinodkumar R, Navas I, Chalana S R, Gopchandran K G, Ganesan V, Reji Philip, Sudheer S K and Mahadevan Pillai V P 2010 Applied Surface Science **257** 708.
[17] Sharma S K and Exarhos G J 1997 Solid State Phenom. **55** 32.
[18] Soni B H, Deshpande M P, Bhatt S V, Chaki S H and Sathe V 2013 J.Appl.Spectros. **79 901**
[19] Scepanovic M, Srećkovic T, Vojisavljevic K and Ristic MM 2006 Sci. of Sint. **38** 169.
[20] Swanepoel R 1983 J. Phys E. Sci. Instrum.**16** 12.
[21] Moshfegh A Z, Azimirad R, Akhavan O 2005 Thin Solid Films **484** 124.
[22] Rambu A P, Sirbu D, Sandu A V, Prodanand G and Nica V 2013 Bull. Mater. Sci.**36** 231.
[23] Suchea M, Christoulakis S, Kartharakis M, Vidakis N and Koudoumas E 2009 Thin Solid Films **517** 4303
[24] Burstein E 1954 Phys. Rev. **93** 632.
[25] Chen Y, Bagnall D M, Koh H J, Park K T, Hiraga K, Zhu ZQ and Yao T 1998 J. Appl. Phys. 84 3912.
[26] Vinodkumar R, Navas I, Porsezian K, Ganesan V, Unnikrishnan N V and Mahadevan Pillai V P2014 Spectrochimica Acta Part A: **118** 724

Investigation of the impacts of selected nanotechnology products with view to their demand for raw materials and energy

M Möller[1], M Diesner[2], A Manhart[1], P Küppers[3], A Spieth-Achtnich[4], C Pistner[4]

[1]Senior Researcher, Öko-Institut e.V., Freiburg, Germany

[2]Researcher, Öko-Institut e.V., Freiburg, Germany

[3]Researcher, Öko-Institut e.V., Darmstadt, Germany

[4]Senior Researcher, Öko-Institut e.V., Darmstadt, Germany

E-mail: m.moeller@oeko.de

Abstract. In the study presented here qualitative and quantitative life-cycle considerations were employed to assess the potential material and energy savings that might be achieved through nano-enabled applications. Ten nanotechnology application fields with broad market coverage and immediate impact to either the generation of renewable energies or the use of critical resources were analyzed. Organic photovoltaic modules (solar cells that essentially consist of organic materials) and electronically dimmable windows (electrochromic laminated glass, which can be adjusted to conform to the ambient light conditions) as two very promising nano-enabled applications were quantitatively analyzed. Eight further products including neodymium magnets were evaluated on a qualitative basis. All assessments contain classical indicators such as energy efficiency, product carbon footprint, and resource consumption. In addition, pollutant aspects (exposure and toxicology) as well as other sustainability aspects (such as user benefits) were taken into account in the framework of a so-called "hot spot analysis". Furthermore, drivers behind the innovation as well as associated rebound effects were identified. The results highlight the importance of product specific analyses based on a life-cycle thinking approach.

1. Introduction to the subject

Nanotechnology is already being used in a wide variety of applications. The conservation of resources and energy play a major role in the decision to employ nanotechnology. Important examples of this are the generation of renewable energies, thermal insulation, power storage, new (building) materials and industrial production processes.

On the one hand the expected savings potentials are based on the new opportunities for the exploitation of quantum effects in the semiconductor sector and on the other hand on the large specific surface area of nanostructured materials, which allows for increased material efficiency. Thanks to entirely new or improved mechanical, electrical, optical or magnetic properties, a reduction in the consumption of raw materials and energy can be achieved throughout the entire lifecycle of nanotechnological applications.

2. Objective of the study

In this context, the general objective of the presented study [1] is to describe the anticipated future resource and energy requirements of particularly promising nanotechnological applications and products both in qualitative terms and, as far as possible, in quantitative terms, with a view to analyzing the efficiency in the use of resources. To this end, it is essential to characterise the respective nanoactive components underlying an application and to identify the specific drivers for a nanotechnological functionalization or for the substitution of well-established materials.

Finally, it will be investigated whether an increased use of nanotechnology in consumer products may result in rebound effects, which would counteract the goal of sustainability enhancement associated with these nanoproducts.

3. Methodological approach

For the survey, a total of ten practical examples of promising nanotechnology applications were analyzed in terms of the applications' raw materials and energy demand. The study "Blueprint Germany" [2] which depicted the most important key technology fields with respect to combating the greenhouse effect and the need to raise energy efficiency in industrial production was used as a basis for the selection of the most promising fields of application. Uniform criteria were used in the selection of the application fields. Hence, the respective applications had to be suitable for the purpose of mass production and, in addition, already exist on the market or at least be on the verge of entering the market. In addition, the existence of a suitable, non-nanotechnology-based reference product was necessary to obtain a reference point for comparative analyses. Eight out of the ten examples were evaluated in qualitative terms, while two applications underwent a quantitative analysis.

The eight case studies that were evaluated from a qualitative point of view on the basis of the list of criteria[1] set out by the German NanoDialogue. The two quantitative case studies were evaluated via selected key indicators outlined in the Öko-Institut's "Nano-Sustainability Check"[2]. The key indicators "energy efficiency", "product carbon footprint[3]" and "resource consumption" were chosen here. When considering the key indicator "resource consumption", special attention was given to the evaluation of the metals and rare earths[4] involved. In addition, toxicology as well as other sustainability aspects (such as user benefits) were taken into account in the framework of a so-called "hot spot analysis".

---

1 A tool elaborated by Issue Group 2 during the second phase of the NanoDialogue initiated by the Federal German Government (2009-2011), providing criteria for a qualitative analysis of the benefit and risk aspects of nanoproducts [3].

2 Instrument developed by the Öko-Institut, which is primarily intended for enterprises, offering a systematic raster for an integrated approach relative to sustainability aspects of nano-enabled applications based on a quantitative analysis [4].

3 Quantitative parameter that expresses the greenhouse gas potential of a product; including the balance of all climate-relevant emissions along the entire life cycle of the product.

4 Rare earth metals comprise the chemical elements in the 3rd row of the periodic system (with the exception of actinium), as well as the lanthanides; they are needed for many high-tech products (e.g. electro-mobility, information technology, renewable energies) and often provoke conflicts due to their limited reserves and the complex production process.

4. Neodymium magnets

Neodymium magnets are an interesting nano-enabled application, allowing a qualitative investigation of the effects on resources and energy demand and the identification of possible rebound effects. The results are highlighted here as a typical example for the eight qualitative application studies.

Neodymium magnets are extremely powerful magnets, which are primarily made of nanoscale neodymium and iron powder [5]. They are of critical importance for industry, since they are a key component in many electrotechnical applications. They are indispensable, inter alia, for the conversion of electrical energy into kinetic energy (electric motor) and vice versa (generator). These magnets are most commonly used in modern wind turbines; in addition, neodymium magnets are integrated into loudspeakers, hard drives, and electric vehicles among other applications.

The use of nanoscale metal powders can increase the magnetic performance (the so-called energy product) by 10-30 %. As a result, it is possible – in comparison to the reference product containing a conventional grain-size, metal powder - to save up to 30 % of the required neodymium (Nd). [6]

In addition to neodymium, further rare earths such as praseodymium, dysprosium or terbium are needed for Nd magnets. Currently, approximately 70 % of the worldwide Nd production is used for Nd magnets. The production of rare earths is characterized by complex technical measures, the release of toxic and radioactive substances (thorium, uranium, heavy metals, acids, etc.) as well as by particularly hazardous working conditions.

Due to the complex production process of rare earths and their limited reserves, it is of great economic and technological importance to further optimise the magnets to achieve either an enhancement of energy efficiency or a reduction of the weight of the rare earths used. The public acceptance and by extension the market share of electric mobility will have significant impact on the economic and technological development in this application sector. The electric motors in the automotive sectors could account for a significant portion of the neodymium used globally in the future.

Key drivers for switching to nano-scale Nd magnets are therefore the associated raw material savings in the field of rare earths and the enhancement of energy efficiency. Against the background that most of the world's neodymium is mined in China, raw material savings of up to 30 % are of particular relevance to industrial policy.

On the other hand, relevant rebound effects may arise from the production of Nd magnets, in particular taking into account the wide dispersal of the magnets used throughout the markets. With smaller and/or more efficient magnets it is possible to assemble small magnets into both cheaper products of equal efficiency, as well as into other, previously unattractive products. The development of new application areas is thus another key driver for nano-technological innovation. This might, however, result in direct rebound effects. Consumer demand for products with neodymium magnets might expand considerably, thereby boosting consumption of resources and favouring material dissipation. The savings potential that had been identified on the rare earths elements market could be negated; even the opposite of the desired effect might result from this development. In addition, the recycling of neodymium magnets – which is a complex matter anyway – is becoming less attractive because of the continuing size reduction of magnets and their broad-scale use.

5. Electronically dimmable windows

Electronically dimmable windows are made of electrochromic laminated glass, which dynamically, automatically or manually can be adjusted to conform to the ambient light conditions [7]. The electrochromic effect can be generated by using a thin film of nanostructured tungsten oxide, which - by applying a voltage - turns bluish as a result of a reaction with lithium ions. This glazing is suitable both for the private as well as for the commercial sector. The presented study, however, analyzed it in the context of its use in the commercial sector (such as in office buildings). On this market, it competes with mixed systems combining multiple glazing and shading devices which are often attached to the exterior of the building in the form of aluminum blinds (reference product).

The company EControl-Glas GmbH & Co. KG - which provided the necessary data for the application example - is the only supplier of electronically dimmable windows in Germany.

The detailed analysis of energy and resources aspects conducted by means of this data base comes to the conclusion that electronically switchable windows (nano-product) - as compared to the reference product - currently (scenario 2011) result in a significantly greater energy expenditure in production since the production facilities at EControl-Glas operate below their full capacity so far. It is not before the use phase that the new glazing becomes profitable in energetic terms. This is due to savings possible because of the lower energy consumption of air conditioning, lighting and the lack of necessity for external blinds that are also operated with the aid of electricity [8]. If the energy expenditure is converted into the CO_2 emissions resulting from the consumption, the following overall picture emerges:

Table 1: Comparison between nano- and reference product regarding the key indicator Product Carbon Footprint (PCF) for both scenarios (2011 and 2010)

	Nano-product	Reference Product
PCF [kg CO_2e/m^2] for the scenario 2011	**256**	**304**
thereof Production	153	42
Transportation	7	1
Use phase	96	262
End-of-life stage	0	-1
PCF [kg CO_{2e}/m^2] for the scenario 2020	**201**	**304**
thereof Production	99	42
Transportation	7	1
Use phase	95	262
End-of-life stage	0	-1

As can be seen from the above table, the results for the scenario 2020 show that the future situation is expected to change in favour of the nano-product. According to EControl-Glas, the energy efficiency in production will considerably improve with an increase in production volume by 2020.

Based on current information, a rebound effect emerges in the cold season only due to a slight increase in heating energy requirements as compared to the reference product. This adverse effect, however, does not nullify the savings from reduced cooling energy consumption, but slightly diminishes them.

With respect to resources, EControl glazing – in contrast to the reference product – requires the use of various metals that, in part, have been classified as critical in terms of their availability [9]. The above findings, however, should be seen in the light of the fact that the concentration levels of the relevant elements in EControl windows are typically very low. The annual production of EControl-Glas currently does not exceed 10 kg of tungsten, for example. This is about the tungsten amount needed for 65 sets of metal drills of a size between 1 and 10 mm, in which tungsten compounds are used as a curing agent. Even if the annual production of EControl-Glas would multiply itself per more than one hundred, this would not result in a significant rise in demand of any of the metals required.

Besides energy savings, an added value in terms of user benefits is also being achieved, which, however, is difficult to quantify: When a room is dimmed, the novel glazing selectively filters red and infrared light wavelengths out of the daylight spectrum. The resulting proportion of blue light in the room is demonstrably conductive to attention and concentration. Furthermore, the view from the windows will not be obstructed by blinds. Thus, this additional user benefits as well as the reduced energy consumption throughout the entire life cycle are the main drivers of this nano-innovation.

6. Organic photovoltaics

Organic photovoltaics (OPV) are solar cells that essentially consist of organic materials. As their plastic base materials are relatively easy to produce, OPV is considered to be an interesting alternative to conventional solar cells made up of inorganic semiconducting material (silicon) [10]. On the other hand, OPV modules - due to technological reasons - need fullerenes, i.e. nanomaterials, the production of which is particularly energy-intensive [11].

"Konarka PP 540", the product evaluated here, consists of a coated polyethylene terephthalate (PET), during the production of which several layers are printed in a continuous process. It is characterized by a high degree of flexibility, allowing a three-dimensional deformation and adaptation to the installation site. The reference product chosen was a model consisting of thin film solar cells based on amorphous silicon (aSi). The evaluation of further current analyses [12] on the subject of energy efficiency leads to the conclusion that the primary energy requirement of OPV modules - despite the fullerenes required to produce them – is below the value of aSi modules by a factor of around 3, the main reasons being the small material input required for the OPV modules and the low-temperature processing.

Furthermore, OPV modules offer excellent overall performance even in poor lighting conditions and in case of an unfavourable incidence angle of the light. For that reason and thanks to their flexible properties they can allow for solar power generation at locations, where this would not be technically feasible with conventional solar modules. The greater degree of freedom as compared to conventional PV modules represents one of the major drivers for the development of OPV modules.

In addition, the relatively high silver content (0.5-5 %) of OPV modules makes them attractive for the recycling industry (in particular copper and precious metals smelters). This promotes the recycling of the metals and materials contained in the modules. Moreover, it can be expected that the fullerenes contained in the modules will be destroyed this way.

7. Conclusions

When considering the total of ten application examples investigated, it becomes apparent that, in general, nanotechnological innovations can save significant amounts of resources and energy. However, the potentials described above, depend at least partially on boundary conditions that still have to become a reality in the future. As regards electrically dimmable windows, for example, it is assumed that production capacities that are currently not being fully utilized could be operated more efficiently in the future thanks to serial production on a large scale. Otherwise, the energy savings potentials of about 30 % as compared to conventional windows could not materialise.

Thanks above all to the qualitative analysis of several application fields, it was possible to determine the impact of new developments in nanotechnology on the consumption rate of critical metallic raw materials and the potential rebound effects that might occur. A good example of this are sintered neodymium magnets, where potential rebound effects that might occur as a result of increasing demand and broad-scale use are hindering the objective initially pursued, that is to reduce the specific consumption of rare earth resources.

With regard to the driving forces behind the investigated nano-enabled applications, it is found that, in addition to traditional drivers such as cost savings and the opening-up of novel application fields, issues such as material saving and energy efficiency in particular have recently become major drivers for innovation. This can be illustrated by the example of the Nd magnets, but is also obvious in the fields of energy technologies, industrial production processes, as well as (building) materials.

Against the background of this study, it again becomes particularly obvious that - with regard to the effects of nanotechnological products on the consumption rate of raw materials and energy, as well as for the determination of relevant rebound effects - the analysis of the entire life cycle of the application under consideration is an absolute prerequisite for obtaining reliable results.

References

[1] Möller M et al (2013). Untersuchung der Auswirkungen ausgewählter nanotechnischer Produkte auf den Rohstoff- und Energiebedarf. Final Report. UFOPLAN FKZ 3711 93 428

[2] Öko-Institut e.V. & Prognos AG. (2009). Blueprint Germany. A strategy for a climate-safe 2050. Online available at: http://www.wwf.de/fileadmin/fm-wwf/Publikationen-PDF/WWF_Blueprint_Germany.pdf

[3] BMU (2011). Final report of NanoKommission issue group 2. Guidelines for collecting data and comparing benefit and risk aspects of nanoproducts. Second dialogue phase 2009-2010. Online available at: http://www.bmub.bund.de/fileadmin/bmu-import/files/english/pdf/application/pdf/nano_finalreport2_bf.pdf

[4] Möller M et al (2012). Analyse und strategisches Management der Nachhaltigkeitspotenziale von Nanoprodukten. TEXTE 15/2012, Umweltbundesamt Dessau-Roßlau. Online available at: http://www.uba.de/uba-info-medien/4276.html

[5] Gutfleisch O et al (2011). Magnetic materials and devices for the 21st century: stronger, lighter, and more energy efficient. Advanced materials (Deerfield Beach, Fla.), 23(7), 821–42. doi:10.1002/adma.201002180

[6] Wecker J et al (1995). Nanostructured Nd–Fe–B magnets with enhanced remanence. Applied Physics Letters, 67(4), 563. doi:10.1063/1.115170

[7] Wittkopf, H. & Dittmar, M. (2012). Variable Sonnenschutzgläser – von den Grundlagen zur Praxis. Stahlbau, 81(S1), 279–290. doi:10.1002/stab.201290071

[8] Schlitzberger, S. & Kempkes, C. (2012). Durchführung von Simulationsrechnungen zur vergleichenden Bewertung von EControl. Kassel.

[9] EUCOM. (2010). Critical raw materials for the EU (pp. 1–84). European Commission: European Commission. Online available at: http://ec.europa.eu/enterprise/policies/raw-materials/files/docs/report-b_en.pdf

[10] Konarka Technologies Inc. (2011). Konarka Power Plastic ® 40 Series Product Specifications.

[11] Kushnir, D. & Sandén, B. A. (2008). Energy Requirements of Carbon Nanoparticle Production. Journal of Industrial Ecology, 12(3), 360–375. doi:10.1111/j.1530-9290.2008.00057.x

[12] Anctil, A., Babbitt, C., Landi, B. & Raffaelle, R. P. (2010). Life-cycle assessment of organic solar cell technologies. 2010 35th IEEE Photovoltaic Specialists Conference (pp. 000742–000747). IEEE. doi:10.1109/PVSC.2010.5617085

IOP Conf. Series: Materials Science and Engineering **64** (2014) 012029 doi:10.1088/1757-899X/64/1/012029

Nanotoxicity and Life Cycle Assessment: First attempt towards the determination of characterization factors for carbon nanotubes

Gonzalo Rodriguez-Garcia[1], Benedikt Zimmermann[2], Marcel Weil[1, 2]

[1]Helmholtz-Institute Ulm for Electrochemical Energy Storage (HIU), Albert-Einstein Alle 11, 89081 Ulm, Germany

[2]Institute for Technology Assessment and Systems Analysis (ITAS) at Karlsruhe Institute of Technology (KIT), Hermann-von-Helmholtz Platz 1, 76344 Eggenstein-Leopoldshafen, Germany

gonzalo.garcia@kit.edu

Abstract. Carbon materials, whether at macro, micro or at nanoscale, play an important role in the battery industry, as they can be used as electrodes, electrode enhancers, bipolar separators, or current collectors. When conducting a Life Cycle Assessment (LCA) of novel batteries manufacturing processes, we also need to consider the fate of potentially emitted carbon based nanomaterials. However, the knowledge generated in the last decade regarding the behavior of such materials in the environment and its toxicological effects has yet to be included in the Life Cycle Impact Assessment (LCIA) methodologies. Conventional databases of chemical products (e.g. ECHA, ECOTOX) offer little information regarding engineered nanomaterials (ENM). It is thus necessary to go one step further and compile physicochemical and toxicological data directly from scientific literature. Such studies do not only differ in their results, but also in their methodologies, and several calls have been made towards a more consistent approach that would allow us model the fate of ENM in the environment as well as their potentially harmful effects. Trying to overcome these limitations we have developed a tool based on Microsoft Excel® combining several methods for the estimation of physicochemical properties of carbon nanotubes (CNT). The information generated with this tool is combined with degradation rates and toxicological data consistent with the methods followed by the USEtox methodology. Thus, it is possible to calculate the characterization factors of CNTs and integrate them as a first proxy in future LCA of products including these ENM.

1. Introduction

Energy storage systems play a major role in the transition towards a more environmentally sustainable society, being critical for both increasing the production of electricity through renewable energies and to end a fossil fuel dependent mobility [1]. As such, a wide variety of technologies and an even wider range of material are now used or studied for the storage of electricity. One of those materials is carbon, used extensively in electrochemistry due to its good chemical stability and high electrical conductivity [2]. The importance of this element has increased in the last decades due to the apparition of carbon based nanomaterials. Carbon nanotubes (CNT) are rolled sheets of graphene with a diameter in the range of nanometers (nm) [3]. They can be used as electrodes, electrode enhancers, bipolar separators, or current collectors [2, 4–6]. Previous studies have shown the use of CNT on traction batteries can reduce the overall environmental profile of electric vehicles [7]. However, similarities between the structure of CNT and that of asbestos fibers have raised the awareness of the potential toxic effects of these substances and although there is ample literature on the subject [8], systematized information is still scarce.

Life Cycle Assessment is an ISO standardized methodology used to study the environmental effects of products and processes through their entire life cycle; this is, from the moment they are extracted from nature until they are disposed as waste (or return to a production cycle). In the Impact Assessment stage, all potential emissions which could cause a harmful impact are classified according to the impact categories they could potentially affect (Global Warming, Ozone Layer Depletion, Human Toxicity…) and relate them to a reference unit (e.g. kg CO_2 equivalents in the case of Global Warming) using characterization factors (CF) [8,9].

Due to the aforementioned relative scarcity of literature on CNT toxicity, there are no CF for CNT yet and, as a result of this, the effect of these nanomaterials in the environment is not included when conducting the LCA of their production process. Consequently, our objective was to test the capability of current impact assessment methodologies when assessing CNT and calculate the CF of single-walled carbon nanotubes (SWCNT) and multi-walled carbon nanotubes (MWCNT).

2. Materials and methods

Commercially available MWCNT, Baytubes 150HP, were chosen based on the availability of data regarding their behaviour in the environment and their toxicological effects [10,11]. For comparative purposes, SWCNT of the same dimensions as the MWCNT (Ø13nm, length 1 μm) were selected.

The CF were calculated with USEtox [13], a consensus methodology recommended by the ILCD handbook for the categories of Human Toxicity (HT) and Freshwater Ecotoxicity (FET) [10]. The parameters required for the CF are detailed in Table 1 and were introduced in USEtox using a spreadsheet tool previously developed by our research group [14]. For the calculation of the CF of MWCNT, the parameters found at the European Chemical Database (www.echa.eu) were used unless specified otherwise. Since no SWCNT are registered yet, the following assumptions and data were used:

- The molecular weight (MW) was calculated based on the diameter and length of the nanotube as in [15]. For the MWCNT, 27 layers were assumed based on its relative density.
- The octanol-water partition coefficient (K_{OW}) was calculated based on the chiral vector as suggested in [16].
- For both SWCNT and MWCNT the organic-carbon partition (K_{OC}) coefficient and the bioaccumulation factor in fish (BAF) were calculated based on their K_{OW} as recommended by the USEtox User's Manual [17].
- For both CNT types the vapour pressures were set to make their Henry Constants (K_H) equal to 10^{-20} as USEtox does for metals [17], since the vapour pressure of carbon at standard conditions is virtually none [18].
- For their two kinds of CNT the degradation rate in air was set to 10^{-20} as for metals in [17]. CNT are not degraded by free radicals, in fact, they induce the formation of free radicals and the oxidation of other substances [19].
- For both CNT the degradation rates in water and its derived degradation rates were calculated assuming they are recalcitrant substances [20].
- The aquatic ecotoxicity parameter was calculated based on the toxicity of these compounds for *Chlorella vulgaris* [21] and *Daphnia magna* [22].
- For HT, the ingestion non-cancer parameter was taken from a study conducted on mice [23] and the inhalation non-cancer one was taken from a study on rats [24]. Since the potential carcinogenity of SWCNT has not been established yet [25], no values were introduced for ingestion and inhalation cancer effects. Thr same review indicated certain MWCNT may

induce malignant mesothelioma if inhaled. However, to the best of our knowledge, no NOEC[1] is available yet. As a result of this, no carcinogenic effects were assessed for MWCNT either.

Table 1. Parameters required for the calculation of the characterization factors

Physicochemical properties		Degradation rates		Ecotoxicity	
Molecular weight (MW)	g/mol	in air	s^{-1}	Aquatic ecotoxicity	EC_{50}
Octanol-water partition coefficient (K_{OW})	-	in water	s^{-1}		
Organic-carbon partition coefficient (K_{OC})	L/kg	in (aerobic) sediments	s^{-1}	**Human toxicity**	
Henry constant (K_{H25})	Pa.m^3/mol	in anaerobic sediments	s^{-1}	Ingestion, non-cancer	kg/kg/day kg/lifetime
Vapour pressure (P_{vap})	Pa	in soil	s^{-1}	Inhalation, non-cancer	kg/kg/day kg/lifetime
Solubility 25°C (SOL)	mg/L	**Human exposure**		Ingestion, cancer	kg/kg/day kg/lifetime
		Bioaccumulation factor in fish (BAF)	L/kg$_{fish}$	Inhalation, cancer	kg/kg/day kg/lifetime

3. Results interpretation and discussion

3.1. Human Toxicity

Table 2 presents the CF for the USEtox methodology. To facilitate the comparison between the assessed substances and other compounds, the CF of Cu(II) are also presented, being this element a common negative current collector [6] and thus, substitutable by CNT.

Table 2. Human Toxicity (non-cancer) characterization factor (cases/kg$_{emitted}$)

Emission to	Urban air	Rural air	Freshwater	Seawater	Natural soil	Agric. soil
SWCNT	7.5E-05	6.7E-05	6.6E-04	6.4E-04	2.8E-07	2.9E-07
MWCNT	2.5E-03	2.6E-03	2.7E-03	9.2E-06	8.3E-04	1.4E-03
Cu(II)	1.3E-05	1.4E-05	8.6E-07	2.2E-07	4.6E-07	3.7E-05

The factors calculated for USEtox suggest MWCNT to be more harmful to human health than SWCNT when emitted to all compartments but seawater. If we take into account that the precision of the USETox CF for HT is usually of 2-3 orders of magnitude [13], the CF of both CNT could be considered identical. Similarly, for all emission compartments, the CF of one or both CNT are within 2-3 orders of magnitude of those of C. Hence, it is possible CNT are only marginally more toxic than the alternative. In order to explain the differences, both between CNT and of them with the reference substance, it is necessary to see how the parameters affecting the CF change between substances. According to [13]:

$$CF = FF \times XF \times EF \text{ ,} \tag{1}$$

or $$CF = iF \times EF \text{ , for human toxicity,} \tag{2}$$

being $$iF = FF \times XF \tag{3}$$

Where *FF* is the fate factor, how the chemical behaves in the environment, *XF* the exposure factor, the probability of being exposed to the chemical found in the environment, and *EF* the effect factor,

[1] No-Observed-Effect-Concentration: it is the parameter required to estimate the carcinogenity of substances in USEtox.

how the chemical behaves once it has reached the target organism. *iF* is the intake factor, which is the probability of being exposed to a chemical released to the environment.

Based on the calculated EF, SWCNT have almost 4 times more chances of affecting someone when inhaled than MWCNT (Table 3) but they still have a lower CF. This is because for all these three substances most of the impacts from their airborne emissions come from indirect sources. As indicated in section 2, the air-water partition coefficient of CNT is, as for metals, assumed to be negligible. As such, airborne CNT and Cu are going to be quickly deposited on water and soil, entering the human organism through drinking water or food rather than being inhaled. If ingested, MWCNT are the ones with a greater chance of causing damage, which is the reason why MWCNT's CF are higher than those of SWCNT for air, freshwater and soil emissions. The lower CF of MWCNT for seawater emissions is explained by means of their intake factors. The transfer of SWCNT from seawater to fish (the most direct route of exposure for seawater emissions) is 6 orders of magnitude higher than the one for MWCNT. Therefore, even when the latter are more toxic on a mass basis, they have a higher EF, the SWCNT have an overall greater chance of affecting a human body if released to the sea.

Table 3. Selected factors affecting Human Toxicity (non-cancer) CF

	EF ($cases/kg_{intake}$)		**iF ($kg_{intake}/kg_{emitted}$)**						
	inh	ing	Uair-air	Uair-other[1]	Fwater-fish	Swater-fish	Nat. soil-fish	Ag. soil-prod[2]
SWCNT	5.3E-02	1.1E-03	2.6E-05	6.5E-02	5.8E-01	5.7E-01	2.5E-04	3.0E-11
MWCNT	1.4E-02	1.3E+01	2.6E-05	2.0E-04	2.3E-05	7.2E-07	7.0E-06	4.7E-05
Cu(II)	9.1E-03	9.1E-03	2.6E-05	1.4E-03	6.7E-05	2.4E-05	3.6E-05	3.9E-03

[1] drinkig water+expected products+ unexpected products+meat+dairy+fish
[2] exposed produce+unexposed produce

The differences between the EF of CNT and Cu do not justify entirely the lower CF the latter showed in Table 2. Cu ingestion's EF 8 times larger than the one for SWCNT, should play an important role in defining the toxicity of its water and soil emissions. However, waterborne SWCNT are expected to transfer more easily from to fish than Cu (4 orders of magnitude more, see Table 3), hence their higher CF. For both SWCNT and Cu, fish is also the most direct route of exposure for natural soil emissions. However, in this case Cu is transferred more easily than SWCNT (14 times more) and therefore is the one presenting a higher CF. For agricultural soil, produce are the main route of exposure for Cu, enough to make them more dangerous than SWCNT, whose main route of exposure continues to be fish.

3.2. Data dependency of Human Toxicity CF: Current limitations and recommendations for future work

As seen in the previous section, there are two main factors affecting the differences in the CF calculated for the CNT and of these two with Cu: 1) their ingestion EF, and 2) their iF from several media, namely water, to fish.

EF are a direct consequence of the HT parameters, higher in the case of SWCNT for inhalation and for MWCNT in case of ingestion. For both CNT these parameters were obtained directly from experimental data. However, the ingestion factor of SWCNT is based on a study where a single bolus was administered, assessing then acute and sub-chronic effects [23]. This could potentially underestimate the oral toxicity of SWCNT and thus, in order to increase the validity of this parameter, future studies need to administer repeated doses so the chronic effects of the ingestion of SWCNT can be better known. Regarding EF, it is also necessary to mention that the size of the SWCNT used to define the inhalation and ingestion toxicity factors are not the same and that none of them are consistent with the one chosen for the comparison with MWCNT. As for MWCNT [26], it could be possible the behaviour of SWCNT in the organism to change with size, increasing thus the uncertainty of the calculated CF.

The intake factors from fish might, on the other hand, overestimate the availability of SWCNT for human consumption. According to the calculated results, fish consumption would be the main route of exposure to SWCNT, irrespectively of where they are emitted. The transfer of a substance to fish is affected by several factors, including the exchange between the media where they are emitted and water. For SWCNT, the most significant parameter is the BAF. As mentioned in section 2, for both CNT, this parameter is not measured directly but calculated instead based on the $logK_{OW}$. This latter parameter is based on experimental data for MWCNT (2.42) but its value is estimated for SWCNT (12.42) which might be the reason for the the large differences between the two CNT. Due to the critical importance of these two parameters, BAF and $logK_{OW,}$ in defining the behaviour of CNT in the environment, we recommended that for future calculations of CF at least one of them is based on empirical data.

Finally, as previously mentioned, there are no reports of SWCNT being carcinogenic [25] and as such no carcinogenic effects have been included in the CF presented here. However, if SWCNT are ever confirmed to cause cancer, those factors would need to be recalculated. Regarding MWCNT, there carcinogenity is relatively well known but we will require more data, in the form of NOEC, before their CF can be calculated.

3.3. Freshwater Ecotoxicity of Single-walled and Multi-walled Carbon Nanotubes

Regarding the FET results presented in Table 4, MWCNT are clearly more dangerous than SWCNT irrespectively of where they are emitted, since the differences between them are much larger than the precision of 1-2 orders of magnitude given by USEtox for FET CF [13]. Following the same reasoning, MWCNT are probably less dangerous than Cu, since most of the differences between them are in the range of 2 orders of magnitude. The exceptions are seawater emissions, where Cu is clearly more toxic than any CNT.

Table 4. Freshwater ecotoxicity characterization factor ($PAF.m^3.day/kg_{emitted}$)

Emissions to	**Urban air**	**Rural air**	**Freshwater**	**Seawater**	**Natural soil**	**Agric. soil**
SWCNT	4.85E-03	3.04E-03	1.25E-01	2.83E-30	5.32E-05	5.32E-05
MWCNT	1.91E+02	1.88E+02	7.40E+02	2.44E-21	2.26E+02	2.26E+02
Cu(II)	2.31E+04	2.33E+04	5.52E+04	1.03E-16	2.92E+04	2.92E+04

In the same way as HT CF, EF alone does not explain the differences between the FET CFof CNT, being 2 orders of magnitude higher for SWCNT than for MWCNT (Table 5). It is the XF, the possibility of being taken in by the objective organism (algae, daphnia, fish, etc.), the factor affecting the final value of the CF the most, being 6 orders of magnitude higher for MWCNT than for SWCNT. EF is however, the main responsible for the differences between MWCNT and Cu since it is 3 orders of magnitude lower for the former than for the latter. Nevertheless, the FF play an important role in defining the CF for seawater emissions. Contrarily than for the other compartments, where the FF to freshwater present the same tendency as the XF (MWCNT>Cu>SWCNT), in this case Cu and MWCNT exchange positions (Cu>MWCNT>SWCNT), which can explain partially why Cu is more toxic than CNT when emitted to seawater. Another factor affecting this discrepancy is that for both SWCNT and Cu, the transfer from seawater to freshwater is not as direct as it is for MWCNT (data not shown). Most of the SWCNT and Cu emitted to seawater would not exchange directly to freshwater, but would transfer there after moving to a different compartment, i.e. air and soil.

Table 5. Selected factors affecting Freshwater Ecotoxicity CF

	EF (PAF.m^3/kg)	**XF (-)**	**FF to freshwater from (days)**				
			Air[1]	Fwater	Swater	Nat. soil	Ag. Soil
SWCNT	6.5E+02	6.5E-06	6.4E-01	2.9E+01	5.5E-28	1.2E-02	1.2E-02
MWCNT	8.0E+00	1.0E+00	2.1E+01	9.2E+01	2.0E-14	2.8E+01	2.8E+01
Cu(II)	4.5E+03	3.3E-01	1.4E+01	3.7E+01	3.6E-20	1.9E+01	1.9E+01

[1] FF for Urban air and Rural air emissions present the same value

3.4. Data dependency of Freshwater Ecotoxicity CF: Current limitations and recommendations for future work

As for HT, two factors are the ones affecting the most the FET of CNT: 1) the EF and the 2) XF. Contrarily to how the HT EF are calculated, for FET several species, and potentially several studies for each species, can be considered. As indicated in section 2, two studies were used to calculate the EF of SWCNT, one for *Daphnia mangna* and another one for *Chlorella vulgaris.* Although the literature regarding the toxicity of engineered nanomaterials in fish is extensive [27], to the best of our knowledge no studies assessing the mortality caused by SWCNT are available yet. SWCNT are known respiratory toxicants in trout, producing gill pathologies at concentrations of 0.1 mg/L [28] but, partially due to the low solubility of these compounds, no LC_{50} has been determined for SWCNT yet. This kind of study would be required to achieve more robust FET CF. Nevertheless, all EC_{50} used in this assessment suggest SWCNT to be more harmful to aquatic life than MWCNT. Thus we consider there is only a small margin for error regarding their relative CF.

For FET, the XF is calculated as the truly dissolved fraction of a substance [29]. According to [17]:

$$XF = \frac{1}{1 + \frac{(Kp \cdot SUSP + K_{doc} \cdot BCF_{fish} \cdot BIOmass)}{10^6}} \tag{4}$$

Where *Kp*, *Kdoc*, and BCF_{fish} increase with K_{OC}, which is calculated based on K_{OW} thus the 6 orders-of-magnitude between CNT shown in Table 5. Contrarily to what happened to HT, the high K_{OW} of SWCNT makes them less available for aquatic life. Regardless, K_{OW}, calculated for SWCNT based on their chirality, has proven to affect significantly both human and freshwater toxicity CF. Accordingly, it should be a priority for the calculation of future CF to use experimental K_{OW}, as well as using BAF and K_{OC} not based on the water-octanol partition coefficient.

4. Conclusions

CNT are a promising material group with excellent technical performed which can be applied in the in the field of energy storage as well as in many others. Regarding known effects, they could potentially cause less environmental harm than alternative materials. However, their potential effects on human health might discourage its implementation. Important differences were found between the two CNT assessed, suggesting that 1) CNT are not a homogeneous group in terms of toxicity, 2) whether or not CNT should substitute current materials might depend on which particular nanotube is used.

This work shows that, despite current limitations, it is possible to calculate the CF of CNT for Life Cycle Impact Assessment with the existing information and methodologies. However, these CF should be considered as interim due to their dependency on estimated parameters, namely K_{OW}, and the lack of toxicity studies regarding certain effects such as carcinogenity. Overall, more research, both in the field of LCA and outside, will be required before the impact of engineered nanomaterials can be fully assessed over their whole life cycle.

Nomenclature

CF	Characterization factor(s)	HT	Human Toxicity
CNT	Carbon Nanotube(s)	iF	Intake factor(s)
EF	Effect factor	MWCNT	Multi-walled Carbon Nanotube(s)
FET	Freshwater Ecotoxicity	SWCNT	Single-walled Carbon Nanotube(s)
FF	Fate factor(s)	XF	Exposure factor(s)

References

[1] BMU 2012 *Langfristszenarien und Strategien für den Ausbau der erneuerbaren Energien in Deutschland bei Berücksichtigung der Entwicklung in Europa und global (Long-term Scenarios and Strategies for the Development of Renewable Energy in Germany Considering Development in Europe and Globally)* Final Report for the Federal Ministry of the Environment (in German)
[2] Endo, M *et al* 2008 Potential Applications of Carbon Nanotubes *Carbon Nanotubes: Advanced Topics in the Synthesis, Structure, Properties and Applications* (*Topics in Applied Physics vol 111*) ed Jorio A, Dresselhaus M S, Dresselhaus G (Heidelberg: Springer) pp 13–61
[3] Murty, B. 2013 *Textbook of Nanoscience and Nanotechnology* (Bangalore, Springer)
[4] Takamura, T and Brodd, RJ 2006 Carbonaceous Materials for Batteries *New Carbon Based Materials for Electrochemical Energy Storage Systems: Batteries, Supercapacitors and Fuel Cells* (NATO Science Series II: Mathematics, Physics and Chemistry vol 229) ed Barsukov I V, Johnson C S, Doninger J E and Barsukov V Z (Amsterdam: Springer) pp 157–69
[5] Nunes-Pereira J *et al 2013 Solid State Ionics* **249–250** 63
[6] Whitehead A H and Schreiber M 2005 *J. Electrochem. Soc.* **152** A2105
[7] Zimmermann B and Weil M 2013 *IEA-HEV Task 19 LCA Meeting* (Argonne, IL)
[8] Uo M *et al 2011 Dental Materials Journal* **30** 245
[9] ISO 2006 ISO 14040:2006 *Life cycle assessment: Principles and framework*
[10] European Commission - Joint Research Centre - Institute for Environment and Sustainability 2010 *ILCD Handbook: General Guide for Life Cycle Assessment : Detailed Guidance* (Luxemburg: Publications Office of the European Union)
[11] Hocke, H 2010 *BAYTUBES®: Carbon Nanotubes at Bayer* Bayer MaterialScience - Carbon Nanotubes BaytubesR seminar (Tel Aviv)
[12] http://apps.echa.europa.eu/registered/data/dossiers/DISS-9e9d7f16-96ef-5932-e044-00144f67d031/DISS-9e9d7f16-96ef-5932-e044-00144f67d031_DISS-9e9d7f16-96ef-5932-e044-00144f67d031.html
[13] Rosenbaum R K *et al 2008 Int J Life Cycle Assess* **13** 532
[14] Rodriguez-Garcia, G *et al* 2013 *Proc. SETAC Europe 19th LCA Case Study Symposium* (Rome) p 138
[15] Laurent C *et al 2010 Carbon* **48** 2994
[16] Toropov A A *et al 2007 Computational Biology and Chemistry* **31** 127
[17] Huijbregts, MAJ, *et al* 2010 *USEtox™: User Manual*
[18] Leider H . *et al 1973 Carbon* **11** 555
[19] Shvedova A *et al 2003 Journal of Toxicology and Environmental Health, Part A* **66** 1909
[20] Kümmerer K *et al 2011 Chemosphere* **82** 1387
[21] Schwab F *et al 2011 Environ. Sci. Technol.* **45** 6136
[22] Zhu X *et al 2009 J Nanopart Res* **11** 67
[23] Kolosnjaj-Tabi J *et al 2010 ACS Nano* **4** 1481
[24] Kobayashi N *et al 2011 Inhal. Toxicol.* **23** 814
[25] Toyokuni S 2013 *Adv. Drug Deliv. Rev.* **65** 2098
[26] Rydman, E 2013 *Evaluation of the health effects of carbon nanotubes* (Helsinki, Finnish Institute of Occupational Health)
[27] Handy R D *et al 2011 Journal of Fish Biology* **79** 821
[28] Smith C J *et al 2007 Aquatic Toxicology* **82** 94
[29] Henderson A D *et al 2011 Int J Life Cycle Assess* **16** 701–709

IOP Conf. Series: Materials Science and Engineering **64** (2014) 012030 doi:10.1088/1757-899X/64/1/012030

On the experimental approaches for the assessment of the release of engineered nanomaterials from nanocomposites by physical degradation processes

M Blázquez[1*], A Egizabal[2, 3], I Unzueta[1]

[1]INKOA SISTEMAS, SL. Ribera de Axpe 11 – Edificio D1, Dpto 208. 48950 Erandio, Bizkaia Spain

[2]FUNDACIÓN TECNALIA Mikeletegi pasealekua, 2 E-20009, Donostia Spain

[3]CIBER – BBN Spain

*E-mail: maria@inkoa.com

Abstract. The LIFE+ Project SIRENA, Simulation of the release of nanomaterials from consumer products for environmental exposure assessment, (LIFE11 ENV/ES/596) has set up a Technological Surveillance System (TSS) to trace technical references at worldwide level related to nanocomposites and the release from nanocomposites. So far a total of seventy three items of different nature (from peer reviewed articles to presentations and contributions to congresses) have been selected and classified as "nanomaterials release simulation technologies". In present document, different approaches for the simulation of different life cycle stages through the physical degradation of polymer nanocomposites at laboratory scale are assessed. In absence of a reference methodology, the comparison of the different protocols used still remains a challenge.

1. Introduction

From 1991 when Toyota Motor Co introduced nylon6 /clay nanocomposites in the market, the commercialization of polymer nanocomposites is continuously growing.

Vehicle safety and the need for light-weight materials to improve vehicular performance have been the most important factors driving the market in the past few years. An increased consumption of nanocomposites in building and construction and their importance in the field of electronics and semiconductors have been other factors driving the market of late (Transparency Market Research; 2013). In fact, according to a recent market review by BBC Research (2014), it is expected that global consumption of nanocomposites will grow in unit terms from nearly 225,060 metric tons in 2014 to nearly 584,984 metric tons in 2019, at a compound annual growth rate (CAGR) of 21.1%.

The most commonly used filler materials are nanoclays, however, other nanomaterials such as carbon nanotubes or graphene are entering the market.

The main goal of the SIRENA LIFE Project is to demonstrate and validate a methodology to simulate the unintended release of engineered nanomaterials (ENMs) from consumer products by replicating different life cycle scenarios involving their physical degradation. To date, the release from nanocomposites has been examined under three main scenarios: mechanical, weathering and incineration; sometimes in a combined manner. In present report, we review thirty two studies that investigate the release from solid nanocomposites through mechanical degradation processes.

2. Methods

The following databases were searched for scientific literature with last search March 2014: Scopus, Science Direct, Web of Science (actually Web of Knowledge) and DOAG (Directory of Open Access Journals). A list of search terms was developed which included nano*, nanocomposite and release-simulation related terms (drilling, abrading…). Additionally, information of interest was captured using alert systems and controlling the updating of different webpages with the Web Site Watcher program.

Abstracts of all articles found were critically reviewed and those matching the scope of this review were selected. In addition to original research articles, documents of other nature have been evaluated including conference proceedings, presentations, thesis, documents generated by related projects and review articles. Information from blogs, distribution lists, e-portals, networks has also been evaluated.

All the items assessed in this review have been classified under Topic 3 of SIRENA's TSS: *Protocols and methodologies used by the different research groups in order to simulate the release to the environment of ENMs from solid polymeric matrixes throughout an item´s lifecycle.* Information compiled under present topic includes data related to the physical, chemical and toxicological characterization and quantification of particles released to the environment in the simulated processes. Simulated processes degrading nanocomposites in all possible configurations are targeted (encapsulated, compounded, cementious, thermoplastic and surface coatings are, therefore, included).

The simulation processes for particles release from drug therapy applications and from composite debris from medical devices are not covered. Migration studies from food composites and probabilistic methods and kinetic modelling are not covered.

Information about real processes or in situ measurements that may cause environmental release (caused by intentional or incidental mechanisms) of ENMs from solid polymeric matrixes through an item´s lifecycle is covered under Topic 2 (http://www.life-sirena.com/index.php/en/technological-surveillance).

For each of the different items selected the next relevant data have been extracted and summarized in short fiches (one per scenario, as one single publication can describe different nano-release scenarios):

- Simulated process (example: sanding)
- Reference sample
- Nanoadditivated test specimen
- Sample characterization pior ENMs release simulation
- Where is the simulation performed (example: specifically dedicated chamber)
- Approach used for laboratory scaled ENMs release simulation
- Approach used for the recovery of the released ENMs
- Relevant outcomes in terms of release and associated exposure

If necessary, complementary data have been extracted from the supporting information -when available- and/or direct contact has been established with the authors for additional information retrieval or doubts clearing.

3. Nanocomposites of interest

The majority of the items reviewed have assessed laboratory-scaled manufactured materials; in only a few cases the commercial reference of the nanocomposite material is provided (Raynor et al., 2012). In some cases the samples were provided by industry but no specific commercial reference is mentioned on the text (Koponen et al., 2011), (Huang, 2012).

In relation to the different applications of the materials tested, those are generally not indicated on the text, not being able correlate the possible nano-release pathways to a specific scenario for sample use.

A total of 10 items have been identified addressing nanomaterials used as coatings (paintings, surface treatments…) (Koponen et al., 2012); (Göhler et al., 2010); (Koponen et al., 2011); (Hsu and Chein, 2007); (Göhler et al., 2012); (Vorbau et al., 2009); (Göhler et al., 2013a); (Göhler et al.,

2013b); (Gheerardyn et al., 2010); (Koponen et al., 2009). In case of thermoplastic or thermoset materials, examples addressing epoxy, polypropylene, polyamide 6, polymethyl methacrylate (PMMA), polycarbonate, polyoxymethylene (POM), polystyrene and polyurethane samples have been found. As an example of non-plastic material the use of cement has been reported by Wohlleben et al. (2011) and Hirth et al. (2013).

In most cases samples used as a reference (when used) have consisted of the same matrixes without the nano-scaled reinforcement. The comparison of the emissions generated by nano-reinforced materials in therefore not compared with the traditional fillers –generally at the micrometric scale- but with the original non-reinforced matrix. The functionality of the original matrixes is generally not comparable to that of the micro or nano-reinforced samples in terms of mechanical properties (hardness, impact resistance), conductivity or fire resistance, amongst other.

4. Characterization of the nano-functionality of the selected test materials

Most authors do not characterize the properties that the engineered nanomaterials confer to the reference test samples; this fact is especially critical in those experiments that use laboratory-scaled manufactured nanocomposites since due to problems associated to the scaling up, experimental formulations do not always reach the market. A few exceptions to present issue are the publications by Göhler et al. (2010); Wohlleben et al. (2013) and Göhler et al. (2013a), where the nano-additivated test samples are evaluated in terms of performance.

5. Different approaches for the simulation at laboratory scale of the processes with an associated nano-release.

The different mechanical degradation processes have been classified according to the definition of each of these processes as described by Canady et al. (2013). The next table summarizes the number of scenarios that have been described in the items collected up to March 2014:

Process	Mechanical process definition according to Canady et al.	Number of scenarios	Examples of instrumentation used
Sanding	Higher-energy type of mechanical stress where shear forces of a rough surface act on the matrix.	14	Commercial handheld orbital sander (Koponen et al., 2011)
Drilling	Mechanical process where high-speed mechanical shear forces are used often to produce a hole.	7	Drill press (Bello et al., 2010)
Abrasion	Mechanical process describing the dynamic friction between two surfaces, one of which sometimes has a rough surface. It is supposed to simulate, for example, use of consumer products like cleaning, sliding, walking, and scratching. Hence, it is low-energy type of mechanical stress with a lower likeliness of heat production.	9	Taber Abraser (Wohlleben et al., 2011)
Cutting/Sawing	Relatively low-speed mechanical process with a limited contact area to the material. It is used to derive specific forms and pieces from composite materials. Wet cutting and sawing is sometimes used to reduce heat.	4	Diamond-bladed table saw (Methner et al., 2012)
Grinding / Shredding	Grinding is understood as a mixed process of milling and cutting of materials.	3	Granulator (Raynor et al., 2012)

Other (Scratching, Impact tests…)	Scratching is a special case of a low-speed mechanical process with a limited contact area to the material. Mechanical shock is another special case of low-energy mechanical process. Impact tests refer to high- energy mechanical processes. These processes simulate hits as they may occur when the product falls down to the ground or experiences sudden mechanical hits by accident.	7	Metallic comb (for scratching) (Golanski et al., 2012); drop impact tower (for impact tests) (Sachse et al., 2013)

The total number of identified scenarios goes up to 44.

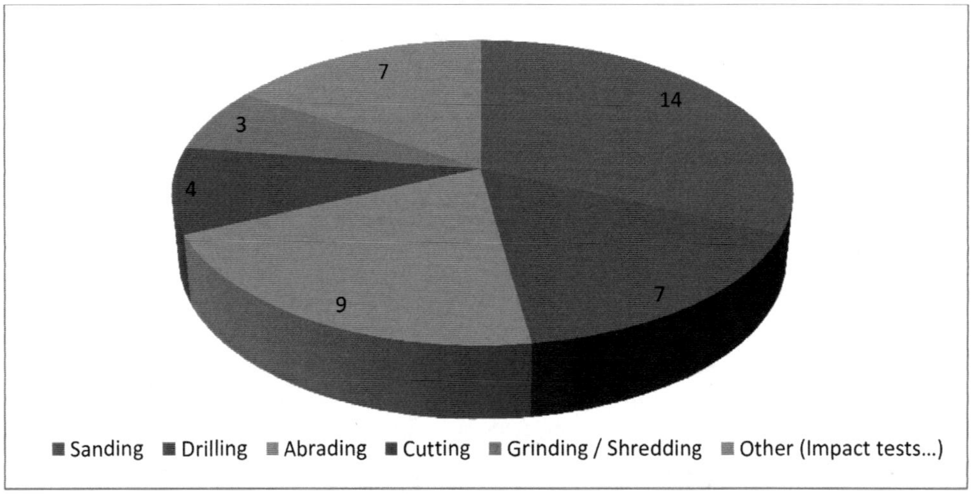

Figure 1. Number of scenarios at laboratory scale simulating different mechanical processes that degrade nanocomposites for the assessment of the release of engineered nanomaterials. As it can be noticed, *sanding* followed by *abrasion* are, to date, the most frequently simulated processes for nano-release assessment.

6. Equipment used and technical media

In relation to the experimental set-up there are basically two approaches: some authors have simulated the release in confined conditions (chamber, specifically designed enclosures, globe boxes…) whereas other authors have used "open" systems (Methner et al., 2012). In general, the lack of details provided prevents the replication of the experiments conducted.

For what refers to the instrumentation used, also, a variety of different instruments has been used. The selection of the instruments has a strong influence on the detection of released nano-objects. The Scanning Mobility Particle Sizer (SMPS) or the Fast Mobility Particle Sizer (FMPS) in combination with Aerodynamic Particle Sizer (APS) and the Condensation Particle Counter (CPC) are some of the most frequently used instruments. According to Schutz and Morris (2013), the measurement metrics used most frequently is that of particle number count. This technique is much more sensitive to ultra-fine particles than measurement by mass because nano-objects can occur in very large numbers without making a noticeable contribution to mass.

7. Assessment of the most outstanding results and conclusions.

There is no clear evidence on the differences of the emissions from the mechanical degradation of nanocomposites containing ENMs from the mechanical degradation of non-reinforced composites: no conclusions can be extracted from the existing data about how emissions vary with material composition. The occurrence of free ENMs from the samples of interest upon nano-release simulation experiments has been only occasionally reported and non-conclusive in terms of repeatability (Methner et al., 2012); (Schlangenhouf et al., 2012). The mechanical process and associated energy that is transferred to the samples, however, has a direct effect on the nature and quantity of emitted nano-objects.

8. Perspectives and future reflexions

The most relevant outcomes of present review are hereby summarized:

- In general, laboratory scaled materials are assessed but no data are available in relation to industrial materials or consumer nanocomposite- products integrating ENMs.
- The real potential of the laboratory scaled materials to replace traditional formulations (data related to increased performance) is generally not assessed.
- The absence of standard operating protocols to be used for scenarios simulation makes inter-assays comparison highly challenging (in some cases not even reference samples have been used).
- The instrumentation used is highly variable.
- Two major critical steps can be observed; the development of an appropriate set-up for nano-release assessment (confinement, clean air supply…) and the collection step in which released nano-objects are characterized.

In view of the state of the art in the emerging science of nano-release assessment, the need of standardized protocols is evident. In this sense, in February 2010 the Commission addressed a mandate to CEN, CENELEC and ETSI –Mandate M461 (http://ec.europa.eu/enterprise/standards_policy/mandates/database/index.cfm?fuseaction=refSearch.search#)- asking these bodies to elaborate with respect to nanotechnologies and nanomaterials a series of standardisation deliverables. Particularly, in relation to the objectives of the SIRENA LIFE Project, Annex I of Mandate M461 invited the above mentioned parties to develop as final outputs European standards for methods to simulate exposures to nanomaterials. On Annex II the need to develop a standard method to assess emissions from handling or machining of nanomaterial containing products is outlined. The demonstration actions targeted within SIRENA provide a pathway to complete these standardisation deliverables in order to assure the eco-responsible introduction of nanotechnologies into consumer products.

Acknowledgments

Present work has been co-funded by the LIFE+ Programme under Contract LIFE11 ENV/ES/596 SIRENA

IOP Conf. Series: Materials Science and Engineering **64** (2014) 012030 doi:10.1088/1757-899X/64/1/012030

9. Supporting Information

List of items reviewed and classification made on the simulations addressing mechanical degradation of polymer nanocomposites.

Title	SIMULATED DEGRADATION PROCESS	SPECIFIC MECHANICAL PROCESS	Authors	Year	Type of publication	Source
Aerosol particle emissions from sanding coated plates: effect of the sanding paper and filtering method to the human exposure	MECHANICAL	SANDING	Koponen I.K., Levin M., Jensen K.A.	2012	Contribution	EAC 12 (european Aerosol Conference)
Assessing potential nanoparticle release during nanocomposite shredding using direct-reading instruments.	MECHANICAL	GRINDING (SHREDDING)	Raynor P. C., Cebula J. I., Spangenberger J. S., Olson B. A. , Dasch J. M., D´Arcy J. B.	2012	Article	J Occup Environ Hyg. 2012;9(1):
Assessment of nanoparticle release and associated health effect of polymer-silicon composites	MECHANICAL	DRILLING	Zhu H., Irfan A., Sachse S., Njuguna J.	2012	Article	NANOSTRUC2012 IOP Conf series:Materials Science and Engineering 40 (2012)
Characterization and control of airborne particles emitted during production of epoxy/carbon nanotube nanocomposites.	MECHANICAL	SANDING	Cena L. G., Peters T. M.	2011	Article	J Occup Environ Hyg. 8(2):
Characterization of abrasion-induced nanoparticle release from paints into liquids and air	MECHANICAL	ABRASION	Golanski L.,Gaborieau A., Guiot A., Uzu G., Chatenet J., Tardif F.	2011	Article	J. Phys.: Conf. Ser. 304
Characterization of exposures to nanoscale particles and fibers during solid core drilling of hybrid carbon nanotube advanced composites.	MECHANICAL	DRILLING	Bello D., Wardle B. L., Zhang J., Yamamoto N., Santeufemio C., Hallock M., Virji M. A.	2010	Article	INT J OCCUP ENVIRON HEALTH VOL 16/NO 4,

Title	SIMULATED DEGRADATION PROCESS	SPECIFIC MECHANICAL PROCESS	Authors	Year	Type of publication	Source
Characterization of nanoparticle release from surface coatings by the simulation of a sanding process.	MECHANICAL	SANDING	Göhler D., Stintz M:, Hillemann L., Vorbau M.	2010	Article	Annals of Occupational Hygiene 08/2010; 54(6): 615–624
Comparison of dust released from sanding conventional and nanoparticle-doped wall and wood coatings	MECHANICAL	SANDING	Koponen I.K., Alstrup Jensen K., Schneider T.	2010	Article	Journal of Exposure Science and Environmental Epidemiology (2011) 21, 408–418
Elastic CNT-polyurethane nanocomposite: synthesis, performance and assessment of fragments released during use.	MECHANICAL + WEATHERING	SANDING + ABRASION	Wohlleben W., Meier M. W., Vogel S., Landsiedel R., Cox G., Hirth S., Tomovic Z.	2012	Article	Nanoscale. 2013 Jan 7;5(1): 369-380
Evaluation of Airborne Particle Emissions from Commercial Products Containing Carbon Nanotubes	MECHANICAL	SANDING	Huang G.	2012	Thesis	University of Iowa, 2012.
Evaluation of the potential airborne release of carbon nanofibers during the preparation, grinding, and cutting of epoxy-based nanocomposite material	MECHANICAL	CUTTING + GRINDING+ BELT SANDING + HAND SANDING	Methner M, Crawford C, Geraci C	2012	Article	J Occup Environ Hyg. 2012;9(5):
Exposure to nanoscale particles and fibers during machining of hybrid advanced composites containing carbon nanotubes.	MECHANICAL	CUTTING	Bello D., Wardle B. L., Yamamoto N., Guzman de Villoria R., Garcia E. J., Hart A. J., Ahn K., Ellenbecker M. J., Hallock M.	2008	Article	J Nanopart Res 14

Title	SIMULATED DEGRADATION PROCESS	SPECIFIC MECHANICAL PROCESS	Authors	Year	Type of publication	Source
Exposures to nanoscale particles and fibers during handling, porcessing, and machininng of nanocomposites and nanoengineered composites reinforced with alginged carbon nanotubes	MECHANICAL	DRILLING	Bello D., Wardle B. L., Yamamoto N., Guzman de Villoria R., Hallock M.	2009	Contribution	ICCM 17
In-Depth Survey Report: Case Study: Particle Emissions from the Processes of Machining Nanocomposites	MECHANICAL	CUTTING + SANDING	Garcia A., Lo L.M., Heitbrink W. A.	2013	Report	NIOSH
Investigation of the Release of Particles from a Nanocoated Product	MECHANICAL + WEATHERING	ABRASION + SANDING + GRINDING + DRILLING	Le Bihan O., Shandilya N., Gheerardyn L., Guillon O., Dore E., Morgeneyer M.	2013	Article	Advances in Nanoparticles, 2013, 2, 39-44
Investigations on nanoparticle release from artificially weathered coatings composites that experience different treatment processes	MECHANICAL + WEATHERING	Not specific (abstract)	Göhler D., Nogowski A., Fiala P., Stintz M.	2012	Contribution	European Aerosol Conference 2012
Measurement of nanoparticle removal by abrasion	MECHANICAL	ABRASION	Guiot A., Golanski L., Tardif F.	2009	Article	Journal of Physics: Conference Series 170 (2009) 012014
Method for the characterization of the abrasion induced nanoparticle release into air from surface coatings	MECHANICAL	ABRASION	Vorbau M., Hillemann L, Stintz M.	2008	Article	Journal of Aerosol Science 40(3):
Nanoparticle release from nanocomposites due to mechanical treatment at two stages of the life-cycle	MECHANICAL + WEATHERING	SANDING + OTHER (DIYNAMIC FRICTION AND WIND EROSION)	Göhler D., Nogowski A., Fiala P., Stintz M.	2012	Contribution	Nanosafe 2012

IOP Conf. Series: Materials Science and Engineering **64** (2014) 012030 doi:10.1088/1757-899X/64/1/012030

Title	SIMULATED DEGRADATION PROCESS	SPECIFIC MECHANICAL PROCESS	Authors	Year	Type of publication	Source
Nanoparticle release quantification of sanding-induced airborne particle emissions from artificially aged/weathered nanocomposites	MECHANICAL + WEATHERING	SANDING	Göhler D., Nogowski A., Stintz M.	2013	Contribution	NSTI-Nanotech 2013
New method for the characterization of abrassion induced nanoparticle release into air from nanomaterials	MECHANICAL	ABRASION	Golanski L., Guiot A., Braganza D., Tardif F.	2010	Article	Nanotech 2010, Vol. 1
On nanoparticle release from polymer nanocomposites for applications in lightweight automotive components	MECHANICAL	OTHER (IMPACT)	Sachse S., Gendre L., Silva F., Zhu H., Leszczynska A., Pielichowski K., Ermini V., Njuguna J.	2012	Contribution	Nanosafe 2012
On the lifecycle of nanocomposites: Comparing Released Fragments and their In-Vivo Hazards from Three Release Mechanisms and Four Nanocomposites	MECHANICAL + WEATHERING	SANDING + ABRASION	Wohlleben W., Brill S., Meier M. W., Mertler M., Cox G., Hirth S., Von Vacano B., Strauss V., Treumann S., Wiench K., Ma-Hock L., Landsiedel R.	2011	Article	SMALL 2011
Physical characteristics of nanoparticles emitted during drilling of silica based polyamide 6 nanocomposites	MECHANICAL	DRILLING	Sachse S., Silva F., Irfan A., Zhu H., Pielichowski K., Leszczynska A., Blazquez M., Kazmina O., Kuzmenko O., Njuguna J.	2012y	Contribution	Nanostruc 2012
Potential release of carbon nanotubes from their composites during grinding	MECHANICAL	GRINDING	Ogura I., Kotake M., Shigeta M., Uejima M., Saito K. ,Hashimoto N., Kishimoto A.	2013	Contribution	Nanosafe 2012
Release of carbon nanotubes from an epoxy-based nanocomposite during an abrasion process (poster)	MECHANICAL	ABRASION	Schlagenhauf L., Chu B. T. T., Buha J., Nüesch F., Wang J.	2012	Contribution	EAC 2012

Title	SIMULATED DEGRADATION PROCESS	SPECIFIC MECHANICAL PROCESS	Authors	Year	Type of publication	Source
Release-ability of nano fillers from different nanomaterials	MECHANICAL	ABRASION + OTHER (SCRATCHING AND MECHANICAL SHOCKS	Golanski L., Guiot A., Pras M., Malarde M , Tardif F.	2012	Article	J Nanopart Res (2012) 14
Reproducing the use of a nanostructured product within an emission chamber: study of nanoparticle release	MECHANICAL	SANDING + CUTTING + OTHER (SCRATCHING)	Gherardyn L., Le Bihan O., Morgeneyer M.	2010	Contribution	Nanosafe 2010
Sanding dust from nanoparticle-containing paints: Physical characterisation	MECHANICAL	SANDING	Koponen I.K., Jensen K. A., Schneider T.	2009	Article	Journal ofPhysics:ConferenceSeries 151 (2009)
Scenarios and methods that induce protruding or released CNTs after degradation of nanocomposite materials	MECHANICAL + WEATHERING	SANDING	Hirth S., Cena L., Cox G., Tomovic´ Z., Peters T., Wohlleben W.	2013	Article	J Nanopart Res (2013) 15:1504
The Effect of Nanoclay on Dust Generation during Drilling of PA6 Nanocomposites	MECHANICAL	DRILLING	Sachse S., Silva F., Zhu H., Irfan A., Leszczynska a., Pielichowski K., Ermini V., Blazquez M., Kuzmenko O., Njuguna J.	2012	Article	Hindawi Publishing Corporation. Journal of Nanomaterials. Volume 2012, Article ID 189386
The release of nanofilers from polymer matrix: Case studies on polypropylene, polyamide and polyurethane nanocomposites	MECHANICAL	DRILLING + OTHER (IMPACT)	Sachse S., Njuguna J., Michalowski S., Leszczynska A., Pielichowski K., Zhu H.	2013	Artcile	Nanoproduct Review

10. References

BELLO, D., WARDLE, B. L., ZHANG, J., YAMAMOTO, N., SANTEUFEMIO, C., HALLOCK, M. & VIRJI, M. A. 2010. Characterization of exposures to nanoscale particles and fibers during solid core drilling of hybrid carbon nanotube advanced composites. *Int J of Occup Environ Health,* 16, 434-450.

CANADY, R., KUHLBUSCH, T., RENKER, M., LEE, E. & TSYTSIKOVA, L. 2013. Multi-Walled Carbon Nanotubes in Polymer Matrices Phase 2.5 Report—Comparison of Existing Studies of Release Measurement for MWCNT-Polymer Composites. Center for Risk Science Innovation and Application of the ILSI Research Foundation

GHEERARDYN, L., BIHAN, O. L. & MORGENEYER, M. 16-18/11/2010. *Presentation: Reproducing the use of a nanostructured product within an emission chamber: study of nanoparticle release.* NanoSAFE 2010.

GÖHLER, D., NOGOWSKI, A. & STINTZ, M. Nanoparticle release quantification of sanding-induced airborne particle emissions from artificially aged/weathered nanocomposites. NSTI-Nanotech 2013, 2013.

GÖHLER, D., NOGOWSKI, A., FIALA, P. & STINTZ, M. Investigations on nanoparticle release from artificially weathered coatings composites that experience different treatment processes. EAC 2012, 2012 Granada, Spain.

GÖHLER, D., NOGOWSKI, A., FIALA, P. & STINTZ, M. 2013. Nanoparticle release from nanocomposites due to mechanical treatment at two stages of the life-cycle. *Journal of Physics: Conference Series,* 429.

GÖHLER, D., STINTZ, M., HILLEMANN, L. & VORBAU, M. 2010. Characterization of nanoparticle release from surface coatings by the simulation of a sanding process. *Ann. Occup. Hyg.,* 56, 615–624.

GOLANSKI, L., GUIOT, A., PRAS, M., MALARDE, M. & TARDIF, F. 2012. Release- ability of nano fillers from different nanomaterials (toward the acceptability of nanoproduct). *Journal of Nanoparticle Research,* 14.

HIRTH, S., CENA, L., COX, G., TOMOVIĆ, Z., PETERS, T. & WOHLLEBEN, W. 2013. Scenarios and methods that induce protruding or released CNTs after degradation of nanocomposites materials. *Journal of Nanoparticle Research,* 15.

HSU, L.-Y. & CHEIN, H.-M. 2007. Evaluation of nanoparticle emission for TiO2 nanopowder coating materials. *Journal of Nanoparticle Research,* 9, 157–163.

HTTP://EC.EUROPA.EU/ENTERPRISE/STANDARDS_POLICY/MANDATES/DATABASE/INDEX.CFM?FUSEACTION=REFSEARCH.SEARCH#.

HTTP://WWW.LIFE-SIRENA.COM/INDEX.PHP/EN/TECHNOLOGICAL-SURVEILLANCE.

HTTP://WWW.MARKETWATCH.COM/STORY/GLOBAL-NANOCOMPOSITES-MARKET-IS-EXPECTED-TO-REACH-USD-591-BILLION-BY-2018-TRANSPARENCY-MARKET-RESEARCH-2013-07-17.

HUANG, G. 2012. *Evaluation of airborne particle emissions from commercial products containing carbon nanotubes.* (Thesis)

KOPONEN, I. K., JENSEN, K. A. & SCHNEIDER, T. 2009. Sanding dust from nanoparticle-containing paints: physical characterization *Journal of Physics: Conference Series,* 151.

KOPONEN, I. K., JENSEN, K. A. & SCHNEIDER, T. 2011. Comparison of dust released from sanding conventional and nanoparticle-doped wall and wood coatings *Journal of Exposure Science and Environmental Epidemiology,* 21, 408–418.

KOPONEN, I. K., LEVIN, M. & JENSEN, K. A. Aerosol particle emissions from sanding coated plates: effect of the sanding paper and filtering method to the human exposure. EAC 2012, 2012 Granada (Spain).

METHNER, M., CRAWFORD, C. & GERACI, C. 2012. Evaluation of the Potential Airborne Release of Carbon Nanofibers During the Preparation, Grinding, and Cutting of Epoxy-Based Nanocomposite Material. *Journal of Occupational and Environmental Hygiene,* 9, 308-318.

RAYNOR, P. C., CEBULA, J. I., SPANGENBERGER, J. S., OLSON, B. A., DASCH, J. M. & D'ARCY, J. B. 2012. Assessing Potential Nanoparticle Release During Nanocomposite Shredding Using Direct-Reading Instruments. *Journal of Occupational and Environmental Hygiene,* 9**,** 1:13.

BBC RESEARCH. 2014. Global markets for nanocomposites, nanoparticles, nanoclays and nanotubes.

SACHSE, S., GENDRE, L., SILVA, F., ZHU, H., LESZCZYŃSKA, A., PIELICHOWSKI, K., ERMINI, V. & NJUGUNA, J. 2013. On Nanoparticle Release from Polymer Nanocomposites for Applications in Lightweight Automotive Components. *Journal of Physics: Conference Series,* 429.

SCHLANGENHOUF, L., CHU, B. T. T., BUHA, J., NÜESCH, F. & WANG, J. Release of carbon nanotubes from an epoxy-based nanocomposite during an abrasion process (Poster). EAC 2012, 2012 Granada (Spain).

SCHUTZ, J. A. & MORRIS, H. 2013. *REPORT: Investigating the Emissions of Nanomaterials from Composites and Other Solid Articles during Machining Processes.*

VORBAU, M., HILLEMANN, L. & STINTZ, M. 2009. Method for the characterization of the abrasion induced nanoparticle release into air form surface coatings. *Journal of Aerosol Science,* 40**,** 209-217.

WOHLLEBEN, W., BRILL, S., MEIER, M. V., MERTLER, M., COX, G., HIRTH, S., VACANO, B. V., STRAUSS, V., TREUMANN, S., WIENCH, K., MA-HOCK, L. & LANDSIEDEL, R. 2011. On the Lifecycle of Nanocomposites: Comparing Released Fragments and their In-Vivo Hazards from Three Release Mechanisms and Four Nanocomposites. *Small,* 7**,** 2384-2395.

WOHLLEBEN, W., MEIER, M., VOGEL, S., LANDSIEDEL, R., COX, G., HIRTH, S. & TOMOVIĆ, Ž. 2013. Elastic CNT–polyurethane nanocomposite: synthesis, performance and assessment of fragments released during use *Nanoscale,* 5**,** 369-380.

IOP Conf. Series: Materials Science and Engineering **64** (2014) 012031 doi:10.1088/1757-899X/64/1/012031

A review on the effect of mechanical drilling on polymer nanocomposites

Kristof Starost and James Njuguna

Institute for Innovation, Design and Sustainability, School of Engineering, Robert Gordon University, Aberdeen, AB10 7GJ, UK

E-mail: k.starost@rgu.ac.uk

Abstract. Over the past decade, polymer nanocomposites have undergone intensive research and development ensued by its increasing implementation within commercial applications. Consequently, the full life-cycle performance and any health risks associated with these materials have become of major interest. Throughout its use, a nanocomposite will undergo industrial machining where drilling can lead to material damage and/or exposure to the potentially toxic nanoparticles. This study assesses the existing and perspective research on nanocomposite drilling. Currently, although considerable amount of studies have investigated machining on conventional composite materials, there is a lack in knowledge on the effect of drilling on nanocomposites. The data underlines the various drilling parameters that will affect and influence the damage to the material and nano-sized particles released. Importantly, previous studies have identified potential mechanical damage caused by drilling and the release-ability of toxic nanoparticles from nanocomposites. It is therefore crucial to develop a full understanding and characterization on the effect of drilling on polymer nanocomposites.

1. Introduction

Considered as 21st century advanced materials, nanocomposites are still relatively new to industry. Throughout its life-cycle, a composite material will undergo various machining processes to fabricate and regulate to its corresponding application. Within industry, drilling is a fundamental and significant machining process used during assembly operations. Various industries, such as the automotive and aerospace industry, have already established the use of nanocomposites within their structures [1]. An example of this is with a leading automotive manufacturer using 300,000 kg of nanoclay composites annually for exterior automotive parts, and according to a recent report, global consumption of nanocomposites is expected to grow from its current amount of around 225,060 metric tons in 2014 to 585,984 metric tons in 2019 [2]. Due to the multiphase structure within composite materials, damages can be introduced during drilling. Around 60% of rejected parts are due to defects introduced in holes [3]. As these defects can lead to a significant decrease in performance, which in some cases can be critical (e.g. aerospace industry), it is vital that the full effect of drilling on nanocomposites is understood.

Conventional polymer composites are fabricated using a selection of material fillers to improve the properties of the constituent polymer matrix. A nanocomposite is distinct from conventional composites, in that at least one dimension of the reinforcing filler material is in the nano range: i.e. less than 100 nanometres [4-6]. Many studies have established that the introduction of nano-sized fillers can significantly further improve the properties of traditional polymer composite materials. Polymer nanocomposites are proven to potentially enhance mechanical [4], optical [6], electrical [6] and thermal [7] characteristics. These enhancements are due to the reinforcing phase having a particularly high surface to volume ratio and its exceptionally high aspect ratio [8]. These many advantages of nanocomposite materials have led to considerable research and as mentioned, the establishment within a variety of industrial applications; commonly for lightweight purposes.

Although considerable amount of studies have investigated composite drilling, there is still however very few studies on the effect of drilling on nanocomposites. This is due to the materials having only recently being established within the commercial industry. As stated, nanocomposites have different properties to conventional composite materials, and will consequently also perform differently under mechanical drilling. This work will present an overview on the work carried out on the effects of drilling on composite materials, followed by the existing studies on nanocomposite drilling and nanocomposite health risks.

2. Composite drilling

Drilling mechanics is a recognised discipline on its own, and has been thoroughly investigated and studied. The research so far however has been on more conventional materials that have been established within industries. Work done on traditional composite materials has identified several parameters during the drilling process that can affect the magnitude and amount of damage to the material. Damage distinctively associated to drilling on composites mainly investigates delamination mechanisms, as this is the leading and major cause of material failure [9]. Other damage modes induced by drilling include intralaminar cracking, fibre pull-out and fuzzing, matrix catering and thermal alterations [10].

A technique commonly used within literature on drilling assessment and optimization is to employ Taguchi's method to study the influence of process parameters such as the drill feed rate and speed on the thrust force, torque and damage factor [11-13]. In work done by Davim et al [13], through the use of Taguchi's method and the analysis of variance (ANOVA), the relationship between the feed rate and cutting speed with the delamination of a CFRP laminate was established. Other studies such as Enemouh et al [14] have demonstrated similar correlations for optimum drilling conditions on composite laminates.

Reviewing some of the studies carried out on the effect of drilling on polymer composites, a table has been produced to identify the parameters most commonly assessed in the drilling process when investigating composite damage. This is shown in Table 1.

Table 1. The drilling process parameters assessed for damage effects on composite materials from a sample selection of literature available

Composite tested	Drill Speed	Feed rate	Drill type	Drill diameter	Drill angle	Reference
GFRP Epoxy	x	x				[15]
CFRP Epoxy	x	x		x		[16]
GFRP Epoxy	x	x				[11]
CFRP Epoxy	x	x		x		[12]
CFRP Epoxy	x	x	x			[13]
GFRP Epoxy	x	x			x	[17]
CFRP Epoxy	x	x	x		x	[18]
GFRP Polyester	x	x		x		[19]
GFRP Polyester & GFRP Epoxy	x	x				[20]

The sample data presented in table 1 show a dominant focus on investigating the drill speed and feed rate as parameters for drilling operations on composite materials. From the chosen studies, the feed rate was observed to have the greatest effect on the damage caused to the mechanical structure of the material. With an increase in feed rate, an increase in delamination damage at any given speed was cited throughout all of the studies. However, the drill speed also observed to have an incremental effect on delamination damage, except for one of the composites tested by Kashaba et al [20], where the inverse effect was evident.

Importantly, the experimental observations have shown that careful consideration is necessary when drilling on composite materials. The selection of drill parameters will have major implications on the damage, surface finish and degradation qualities of the holes [11-20].

3. Nanocomposite drilling

As nanocomposites are a relatively newly established material, few studies have been produced on the drilling on polymer nanocomposites. As shown in the previous section however, mechanical drilling has demonstrated to undesirably damage composite materials, thus it is therefore necessary to evaluate the effect of drilling on nanocomposites.

From the literature available, only three studies have investigated drilling on nanocomposites. The generic development and assessment of polymer nanocomposite properties is also considerably scarce within literature. Baker et al [21] utilised drilling to evaluate coating wear resistance, but there is a lack in other studies assessing the damage caused by mechanical drilling on nanocomposites. Several studies have investigated damage caused to nanocomposites from other machining methods [22]. Previous studies that have evaluated the effect of mechanical drilling on nanocomposites are summarized in Table 2.

Table 2: Overview of parameters assessed in current studies on the effect of drilling on nanocomposites

Reference	Material Damage	Nanoparticle Release		
		Number of Nanoparticles	Nanoparticle Concentration	Nanoparticle Characterization
Sachse et al. [23]		x	x	
Bello et al [24]		x	x	x
Sachse et al. [25]		x	x	x

From the limited studies on nanocomposite drilling, the research has solely focused on investigating nanoparticles release due to the drilling. This highlights that there are currently no studies assessing the material damage due to drilling on nanocomposites. The three works appeared to have no common or standardized set of parameters used to assess the drilling effect. In the drilling studies carried out by Sachse et al [23] [25], a manually controlled 10mm diameter angle drill was used with a maximum speed of 1800 min $^{-1}$. The setup for these experiments is shown in Figure 1.

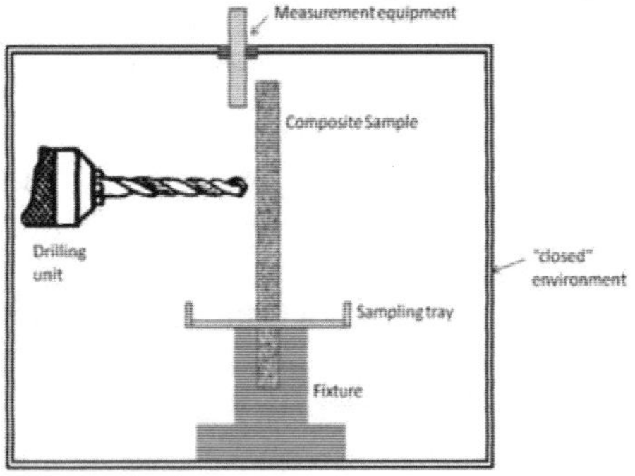

Figure 1. Apparatus and setup used for chamber experiments on investigating nanoparticle release due to drilling on nanocomposites [25]

IOP Conf. Series: Materials Science and Engineering **64** (2014) 012031 doi:10.1088/1757-899X/64/1/012031

Bello et al [24] specified two diamond drill diameters, ¼'' (6.35mm) and 3/8'' (9.525mm) along with investigating two drill speeds of 725min-1 and 1355min-1. This study investigated the use of CNT, alumina and graphite-epoxy composites, whereas the studies by Sachse [23][25] investigated PA6 and PP-silica nanocomposites. Apart from diverse nanocomposites being investigated, the differences in setups for the two studies demonstrate the need for a standardized set of drilling parameters to allow for a feasible comparison.

4. Why study nanoparticle release due to drilling?

As mentioned, due to the use of nano-sized particles, nanocomposites introduce a potential toxicological and eco-toxicological hazard. Research is comprehensively investigating the potential nanoparticle release and exposure to humans and the environment. Drilling on nanocomposites has the ability of unintentionally releasing these nanoparticles into the environment.

Studies in toxicology have shown that particles within the nano scale are potentially hazardous to humans. The advantages of the nanoparticle physiochemical properties employed for the use within materials also render potential unique toxic effects within biological systems [4]. Due to their small size, nanoparticles have the ability of penetrating the bloodstream and translocate throughout the body at which larger particles could not. Studies show that nanoparticles within the body can cross biological barriers such as the blood-brain and the blood-testis barrier and therefore treacherously reach highly protected organs [26]. Published data from in vitro studies, suggests that nanoparticles have the aptitude to induce cell growth inhibition [27], cell death [28], inflammatory response [29], DNA damage [30] and the generation of free radicals and reactive oxygen species. The toxicological effect is different for each type of nanoparticle, although the specific relationship between the chemical structure of every particle and the toxic implementations are still unknown [4]. It is also important to note that some nanoparticles appear to be non-toxic [31], and others are may even possess health benefits [32].

The type and characterization of the nanoparticles measured from release scenarios varies in most studies. In work by Bello et al [33], epoxy based nanocomposites were considered during laboratory cutting. Particle release measurements were taken and compared for both an unmodified and a carbon nanotube reinforced epoxy. The studies observed that nanoparticles were detected regardless of the composite type and presence of the carbon nanotube fillers. 1-10% of particles released were shown to be within the nanometre range (<100nm), whilst 71-89% were in the 1-10µm range. When examined using scanning electron microscopy (SEM) and transmission electron microscope (TEM), the released nanoparticles exposed no clearly distinguishable carbon nanotubes. The data suggests the nano-fillers continue to be merged within the polymer resin. However, in a different study by Huang et al, multi-walled carbon nanotubes were detected in the released nanoparticles from the composite matrix during sanding [34]. Multiple other studies have investigated the release of nanoparticles from carbon nanotube reinforced polymer nanocomposites due to mechanical influences. In most cases, nanoparticles have been observed to be released, but there is no universal conclusion on the quantity of exposure [35].

In the work carried out by Sachse et al [25], on drilling on PA6 and PP based nanocomposites, the data showed that the silica reinforced nanocomposite introduced fifty-six times more nanoparticles released than the neat polyamide. In a similar study, but with the use of montmorillonite as a nanofiller for the polyamide 6 resin, reverse data was observed. The reinforced nanocomposite was seen to release twenty times fewer nanoparticles than the neat polyamide [23].

In the study on the effect of drilling performed by Bello et al [25], the research appeared to exhibit aggregates of the carbon nanotubes in the released nanoparticles. Whereas in a comparative study also done by Bello et al [33] the same nanocomposites undergoing cutting did not show the aggregates of the carbon nanotube. This study highlights the effect of the machining process rather than the nanocomposite composition in production of nanoparticles released from nanocomposites.

The release of nano-sized particles from nanocomposites has been broadly researched [36]. So far however there is no standardized method established to simulate the release scenarios. From the literature available, including the studies performed on nanocomposite drilling, it can be said that most studies illustrated that nano-sized particles are released during some of the scenarios, but to a certain extent while laboratory experiments repeatability remains challenging.

5. Conclusions

The benefit of the nanoparticle fillers has caused a surge in investment and research across the world. From the literature available however, there is currently a lack of knowledge on the effect mechanical drilling has on polymer nanocomposites. Although considerable amount of studies have demonstrated that drilling can structurally damage conventional composite materials, no work has investigated the effects on nanocomposites. Along with assessing the damage drilling can cause on various polymer nanocomposites, it is crucial that any potential health or environmental risks associated with the materials are fully understood and characterized.

6. Acknowledgement

This work was funded by the European Commission under the ELIFE+ Programme within SIRENA Project No. ENV/ES/596.

7. References

[1] Njuguna, J, Pielichowski, K, and J Fan. "Polymer nanocomposites for aerospace applications." *Advances in Polymer Nanocomposites: Types and Applications* (2012): 472

[2] BCCResearch. Global Markets for Nanocomposites, Nanoparticles, Nanoclays, and Nanotubes. Technical Report NAN021F, BCC Research 2014

[3] Zitoune, Redouane, Vijayan Krishnaraj, and Francis Collombet. "Study of drilling of composite material and aluminium stack." *Composite Structures* **92**.5 (2010): 1246-1255

[4] Njuguna, J, Ansari, F, Sachse, S, Rodriguez, V M, and. "Nanomaterials, nanofillers, and nanocomposites: types and properties." Health and Environmental Safety of Nanomaterials: *Polymer Nancomposites and Other Materials Containing Nanoparticles* (2014): 1-27

[5] Njuguna, J, and Pielichowski, K. "Polymer nanocomposites for aerospace applications: properties." *Advanced Engineering Materials* **5**.11 (2003): 769-778

[6] Chapman, R, and P Mulvaney. "Electro-optical shifts in silver nanoparticle films." *Chemical physics letters* **349**.5 (2001): 358-362.

[7] Leszczyńska, A, Njuguna, J, Pielichowski, K, & Banerjee, J R "Polymer/montmorillonite nanocomposites with improved thermal properties: Part II. Thermal stability of montmorillonite nanocomposites based on different polymeric matrixes." *Thermochimica Acta* **454**.1 (2007): 1-22.

[8] Schmidt, G, and M M Malwitz. "Properties of polymer–nanoparticle composites." *Current opinion in colloid & interface science* **8**.1 (2003): 103-108

[9] Ho-Cheng, H, and C K H Dharan. "Delamination during drilling in composite laminates." *Journal of Manufacturing Science and Engineering* **112**.3 (1990): 236-239.

[10] Tagliaferri, V, G Caprino, and A Diterlizzi. "Effect of drilling parameters on the finish and mechanical properties of GFRP composites." *International Journal of Machine Tools and Manufacture* **30**.1 (1990): 77-84

[11] Ramesh, B, S J C Sharan, and R Kavialagan. "Experimental investigation and optimization in drilling GFRP polymeric composites using Taguchi and ANOVA." *International Journal of Mechanical and Production Engineering* **2**.1 (2013): 52-60

[12] Tsao, C C, and H Hocheng. "Taguchi analysis of delamination associated with various drill bits in drilling of composite material." *International Journal of Machine Tools and Manufacture* **44**.10 (2004): 1085-1090

[13] Davim, J P, and Pe Reis. "Study of delamination in drilling carbon fiber reinforced plastics (CFRP) using design experiments." *Composite structures* **59**.4 (2003): 481-487

[14] Enemuoh, E U, A Sherif El-Gizawy, and A Chukwujekwu Okafor. "An approach for development of damage-free drilling of carbon fiber reinforced thermosets." *International Journal of Machine Tools and Manufacture* **41**.12 (2001): 1795-1814.

[15] Tagliaferri, V, G Caprino, and A Diterlizzi. "Effect of drilling parameters on the finish and mechanical properties of GFRP composites." *International Journal of Machine Tools and Manufacture* **30**.1 (1990): 77-84.

[16] Lee, Seung-Chul, S-T Jeong, J-N Park, S-J Kim, and G-J Cho "Study on drilling characteristics and mechanical properties of CFRP composites." *Acta Mechanica Solida Sinica* **21**.4 (2008): 364-368

[17] Kilickap, E,"Optimization of cutting parameters on delamination based on Taguchi method during drilling of GFRP composite." *Expert Systems with Applications* **37**.8 (2010): 6116-6122.

[18] Shyha, I S, D K Aspinwall, Sein Leung Soo, and S Bradley. "Drill geometry and operating effects when cutting small diameter holes in CFRP." *International journal of machine tools and manufacture* **49**.12 (2009): 1008-1014.

[19] Mohan, N S, S M. Kulkarni, and A Ramachandra. "Delamination analysis in drilling process of glass fiber reinforced plastic (GFRP) composite materials." *Journal of Materials Processing Technology* **186**.1 (2007): 265-271

[20] Khashaba, U A "Delamination in drilling GFR-thermoset composites." *Composite Structures* **63**.3 (2004): 313-327.

[21] Baker, M A, et al. "Evaluating the microstructure and performance of nanocomposite PVD TiAlBN coatings." *Surface and Coatings Technology* **151** (2002): 338-343.

[22] Silva, F, Njuguna, J, Sachse, S, Pielichowski, K, Leszczynska, A, & Giacomelli, M, "The influence of multiscale fillers reinforcement into impact resistance and energy absorption properties of polyamide 6 and polypropylene nanocomposite structures." *Materials & Design* **50** (2013): 244-252.

[23] Sachse, S, et al. "The effect of nanoclay on dust generation during drilling of PA6 nanocomposites." *Journal of Nanomaterials* 2012 (2012): 26

[24] Bello, D, et al. "Characterization of exposures to nanoscale particles and fibers during solid core drilling of hybrid carbon nanotube advanced composites." *International journal of occupational and environmental health* **16**.4 (2010): 434-450

[25] Irfan, A, Sachse, S, Njuguna, J, Pielichowski, K, Silva, F, & Zhu, H, "Assessment of Nanoparticle Release from Polyamide 6-and Polypropylene-Silicon Composites and Cytotoxicity in Human Lung A549 Cells." *Journal of Inorganic and Organometallic Polymers and Materials* 23.4 (2013): 861-870.

[26] Pietroiusti, A. "Health implications of engineered nanomaterials." *Nanoscale* **4**.4 (2012): 1231-1247

[27] Zhu, Lin, et al. "DNA damage induced by multiwalled carbon nanotubes in mouse embryonic stem cells." *Nano letters* **7**.12 (2007): 3592-3597.

[28] Magrez, A, et al. "Cellular toxicity of carbon-based nanomaterials." *Nano letters* **6**.6 (2006): 1121-1125.

[29] Ding, L, et al. "Molecular characterization of the cytotoxic mechanism of multiwall carbon nanotubes and nano-onions on human skin fibroblast." *Nano Letters* **5**.12 (2005): 2448-2464.

[30] Kisin, E R., et al. "Single-walled carbon nanotubes: geno-and cytotoxic effects in lung fibroblast V79 cells." *Journal of Toxicology and Environmental Health*, Part A **70**.24 (2007): 2071-2079.

[31] Connor, E E, et al. "Gold nanoparticles are taken up by human cells but do not cause acute cytotoxicity." *Small* **1**.3 (2005): 325-327

[32] Schubert, D, et al. "Cerium and yttrium oxide nanoparticles are neuroprotective." *Biochemical and biophysical research communications* **342**.1 (2006): 86-91.

[33] Bello, D et al. "Exposure to nanoscale particles and fiber during abrasion of hybrid advanced composites containing carbon nanotubes." *Journal of Nanoparticle Research*, **11** (2009): 231-249.

IOP Conf. Series: Materials Science and Engineering **64** (2014) 012031 doi:10.1088/1757-899X/64/1/012031

[34] Huang, G, et al. "Evaluation of airborne particle emissions from commercial products containing carbon nanotubes." *Journal of Nanoparticle Research* **14**.11 (2012): 1-13.

[35] Schlagenhauf, L, F Nüesch, and J Wang. "Release of Carbon Nanotubes from Polymer Nanocomposites." *Fibers* **2**.2 (2014): 108-127.

[36] Njuguna, J., Pielichowski, K., and Zhu, H. (Eds.). Health and Environmental Safety of Nanomaterials: Polymer Nancomposites and Other Materials Containing Nanoparticles. Elsevier (2014).

IOP Conf. Series: Materials Science and Engineering **64** (2014) 012032 doi:10.1088/1757-899X/64/1/012032

Functional polymer - clay nanotube biocomposites with sustained drug release

Y Lvov and J Tully
Institute for Micromanufacturing, Louisiana Tech University, Ruston, LA 71272, USA
Email: Joshua Tully (jtu006@latech.edu), Yuri Lvov (ylvov@latech.edu)

Abstract. By adding 5% (w/w) of halloysite nanotubes that have been modified (loaded) with proteins or drugs it is possible to produce strong and functional biocomposites. Materials loaded with both types of materials were investigated using ultraviolet-visible spectrophotometry and thermogravimetric analysis to determine their release kinetics and overall loading efficiency. It was found that both released over a period of 5-20 hours with two distinct phases being present. An initial "burst stage" of release followed by a period of sustained release. Specifically, for proteins it has been shown that a significant amount (50-75%) remain immobilized even after being dispersed. The typical loading efficiency for both classes of molecules was 10-15%. These modified nanotubes can both strengthen a material and give it unique functionality and possible uses include more effective externally applied antibiotics and immobilized proteins with enhanced stability and reusability.

1. Introduction

Halloysite nanotubes (HNT) of 50 nm diameter and 1-1.5 μm length are a perspective dopant for stronger and functional polymer composites. We exploit the advantages of halloysite nanotubes for polymeric composites and its unique ability to act as a container for macromolecules (proteins and drugs). It was found that the addition of 3-5 % halloysite to polypropylene increases its mechanical strength by ca. 50 %. Besides, it is possible to take advantage of the unique chemical storage ability of halloysite to add needed functionality to polymers. Halloysite nanotubes are capable of storing biomolecules that are negatively charged inside of its lumen at 10 -15 wt. percent. Anti-septic chemicals[1], anti-corrosion chemicals[2], proteins, and a subclass of biocatalytic proteins called enzymes may also be stored and slowly (5-50 hrs) released from these natural biocompatible nanotubes.

By simply mixing HNTs and a protein, or drug, together it is possible to immobilize the protein or drug inside of the tubes. We found that loaded proteins and are split into two essential categories: protein molecules that are permanently attached inside of the nanotube lumen and proteins that will be released back into solution during 4-5 hours. Having two different categories and therefore two different "lifetimes" for a selected protein immobilized with halloysite allowed for the study of a burst phase of activity and a sustained phase. These phases correlate to the proteins that do not reside permanently and those that are tightly bound in the tube lumen, respectively. Drugs also show a similar phenomenon of a two-phase release pattern. Therefore, by taking halloysite, a naturally abundant nanocontainer, and utilizing the unique catalytic properties of enzymes and the potent effects of drugs it is possible to create strong and biofunctional composite materials.

2. Experimental

The first necessary step in our research was to study the loading efficiency and release kinetics of the selected proteins loaded inside of halloysite nanotubes. In Table 1 we summarize some proteins that were used in our experiments. An essential feature of these protein group is that they may be overall negative or positive depending on the pH at which we are making loading and release (e.g. above or below their isoelectric point).

There is a standard procedure that can be used to load any molecule inside of halloysite and can be found in various pieces of literature[3]. Proteins are very sensitive to their environment and in particular to pH and temperature. Temperatures above 60C often cause irreversible thermal denaturation, effectively destroying the protein. Therefore it is suggested to use ice during sonication to neutralize heat generation from this technique. Also, water based buffers are the preferred solvent for proteins, and although it is possible to use organic solvents like acetone it's not suggested. After performing the loading procedure the enzymes release kinetics were studied by studying the concentrations of the protein after halloysite being dispersed in water by measuring the UV absorbance of a sample at 280 nm. Thermogravimetric analysis (TGA) was employed to measure the tube loading efficiency and to elaborate on the two categories of proteins (permanently immobilized and capable of release). Some of the enzymes, glucose oxidase and horseradish peroxidase, had their enzymatic activity qualitatively tested. The results from those experiments confirm that the enzymes retain their enzymatic activity after loading and drying.

Table 1. The isoelectric point of a protein can be used to manipulate its charge. At a pH above the isoelectric point the protein will have a net negative charge and at a pH below the pI the protein will be positively charged.

Enzyme	EC Number	Molecular mass (kDa)	pI (Isoelectric point)
Bovine Serum Albumin	N/A	70 (monomer)	5.0
Glucose Oxidase	1.1.3.4	65 (dimer) = 130	4.2
Lipase	3.1.1.3	50 (dimer) = 100	5.0
Myoglobin	N/A	17 (dimer) = 34	7.0
Pepsin	3.4.23.1	35 (monomer)	1.0
Urease	3.5.1.5	90 (hexamer) = 540	5.0

3. Results and Discussion

The first data retrieved from the experiments was release kinetics of the enzymes. A large compilation of data for a wide variety of enzymes can be seen in Figure 1. From this figure several observations can be made about how proteins interact with halloysite nanotubes. First, is that in comparison to other molecules proteins have a relatively short release time frame. Typically, within the first hour over 75 % of the total amount of protein found was already in solution. This graph is somewhat misleading however in that 100 % of release is shown. The reality is that it is impossible to have every molecule of protein released from the halloysite nanotubes. However, after many experiments it was found that after 6 hours there is not a significant increase release. This is exemplified in Figure 2 where at 6 hours there is 86 % release and at 50 hours when the experiment stopped there was only 15 % more. When converted to mass this 15 % is in the microgram range.

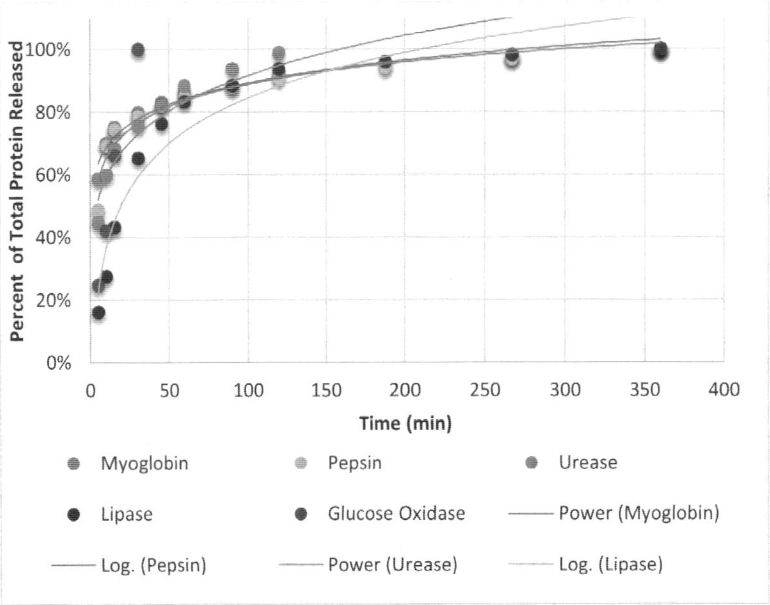

Figure 1. In these experiments the proteins were loaded at a 1:1 ratio (w/w) of protein to halloysite in water. The release study was also done using water. Absorbance was read at 280nm.

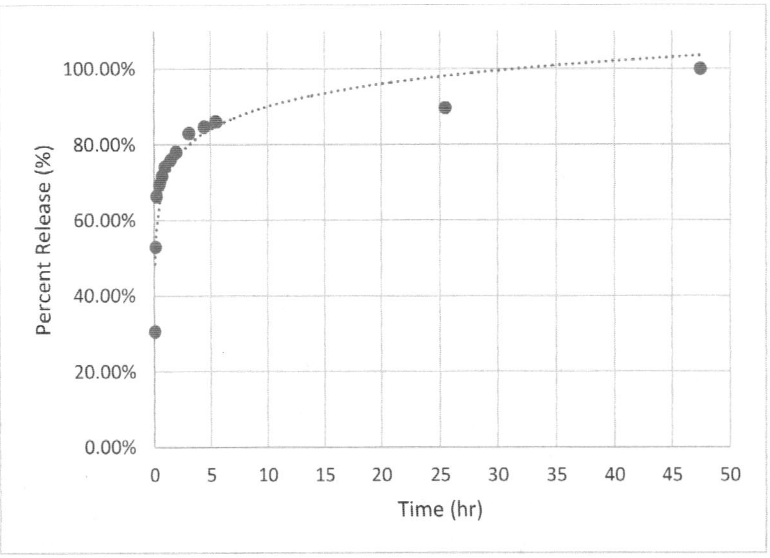

Figure 2. The same procedures as in Figure 1 were applied to this experiment. However, in this experiment the protein was positively charged and the ratio was 2:1 protein to HNT.

Ideally we would want the release of the protein to be extended over a much larger period of time. Somewhere along the lines of days to weeks would correlate better with other results from other chemicals. Despite this seemingly short release we were able to come across an interesting observation while using TGA to determine the loading efficiency. Figure 3 is an example of what would be constructed for every protein that was attempted to be loaded inside of the nanotubes.

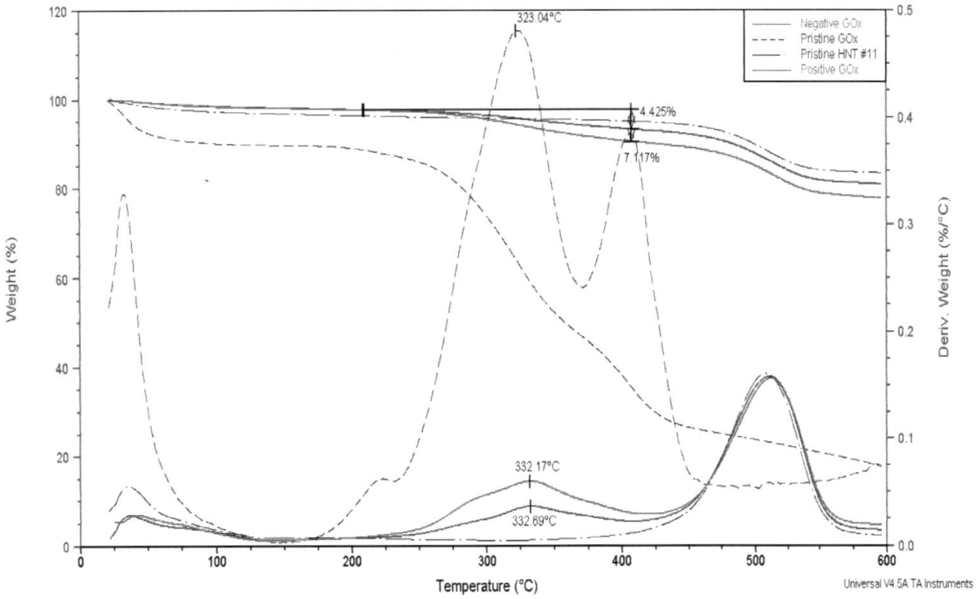

Figure 3. In this particular experiment glucose oxidase was loaded with halloysite when the protein had both a positive and negative charge. The positively charged sample shows a 7.11% decrease in mass while the negatively charged sample showed a 4.425% decrease.

There are couple meaningful observations of this TGA data. First the positively charged sample of glucose oxidase showed a significantly higher loss of mass over the same temperature range as the negatively charged sample. This can be attributed to the increase surface area on the outside of the nanotube which also happens to be negatively charged. The next is that both samples show almost the exact same maxima of degradation. This is indicative that the samples are stabilized in a similar fashion. For example, they may adsorb in the same orientation on both the outer surface and the inner surface. As mentioned earlier an interesting observation was made when a sample of protein loaded halloysite nanotubes was analyzed in this fashion after going through a release study. Figure 4 shows the result of this experiment.

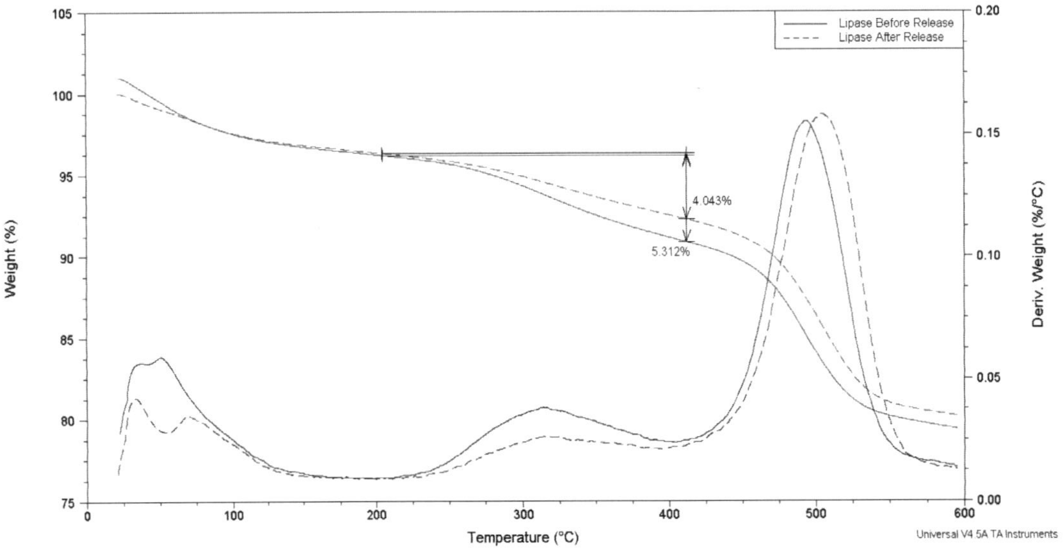

Figure 4. This sample of lipase loaded halloysite was prepare and release in deionized water. The ratio at loading was 1:1. The weight (%) curve for lipase before release was offset +1 unit to account for a larger loss of water in the first 100C.

It can be seen that 76 % remained after the release study had been completed. This led to the development of the idea that there may be two "categories" of proteins and essentially two phases of enzymatic activity that could be expected. The first phase is the burst phase as mentioned before. In this phase the enzyme that is release from the tube is allowed to react freely with any available substrate while at the same time the protein that is immobilized inside of the tube is also free to react albeit likely with less efficiency due to diffusion constraints or reduction in activity due to immobilization. Typically proteins and enzymes degrade over time and the free protein will eventually stop working. This is where the immobilized protein, which is likely to be much more stable will be able to continue to act as a catalyst, hence the sustained phase. There has been a lot of work done on immobilized protein catalysts and they have found important uses in industry[4] hence exploring this phenomena could find potential large scale use.

We must continue to work to establish the parameters that control the loading efficiency of proteins. As of now the mass, charge, and secondary structure of the protein seem to be the main factors that influence the loading efficiency. Big heavy proteins seem to be loaded less efficiently, while proteins with a condensed β-sheet structure seem to load well regardless of weight as opposed to those with mostly α-helixes. Charge is not as simple as mass structure to state a correlation. In general the loading efficiency for positively charged proteins is 40 % higher than that of negative proteins. However, it is likely they these proteins are adsorbing to the surface of the tubes and not on the inside. Finally, we need to produce quantitative data on enzymatic activity after loading.

Loading and release of different drugs was also studied. In Figure 5 we show results on sustain release of three drugs (dexamethasone, furosemide and nifedipine) from halloysite nanotubes. These drugs loading rate was 6-7 wt. % and release time increase from 5-10 min in non-encapsulated state to 10-20 hours for the drugs encased in the clay nanotubes.

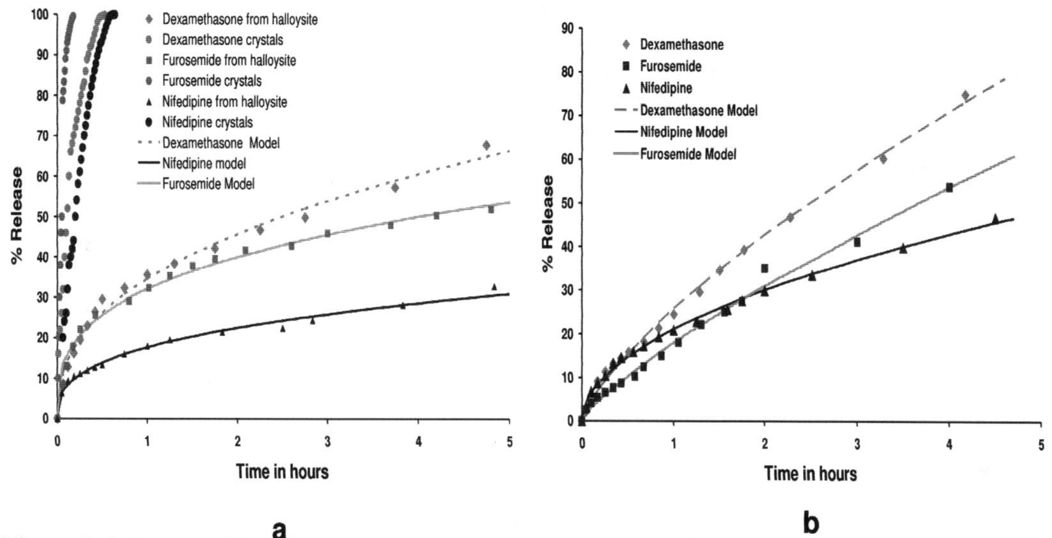

Figure 5. Release profiles of the drugs from halloysite nanotubules in water at pH 7.4: a-loading from 10 % alcohol; b- loading from 50 % ethanol / aqueous solvent.

Using for the loading the more concentrated drug solution in 1:1 mixture of water and ethanol we reach higher 9-10 wt. % loading efficiency and more linear though shorter time release characteristics (Figure 5 b).

4. Conclusion

Natural biocompatible halloysite clay nanotubes may be used for storage and sustained release of bioactive compounds, such as drugs and proteins. Typical release time is between 5 and 20 hours which is perspective for biomedical applications. However, one has to mind that these nanotubes are not biodegradable and some mechanisms of the nanotube removal from organisms have to be considered or such formulation can only be used for external (non-injective) applications.

We found that negatively charged proteins may be loaded into positively charged clay nanotube lumens at 5-8 wt. % and positively charged proteins may be loaded both inside and outside tubes at 10-15 wt. percent. Both type of immobilized proteins may be release within 4-5 hours, however in the case of negative proteins loaded into the tube lumen, approximately half of the load (3-4 % are strongly bound and remain immobilized for much longer time.

5. References

[1] Wei W, Minullina R, Abdullayev E, Fakhrullin R, Mills D and Lvov Y (2014) Enhanced efficiency of antiseptics with sustained release from clay nanotubes. *RSC Advances* **4** (1):488. doi:10.1039/c3ra45011b

[2] Abdullayev E, Price R, Shchukin D and Lvov Y (2009) Halloysite Tubes as Nanocontainers for Anticorrosion Coating with Benzotriazole. *ACS Applied Materials & Interfaces* **1** (7):1437-1443. doi:10.1021/am9002028

[3] Abdullayev E and Lvov Y (2013) Halloysite clay nanotubes as a ceramic "skeleton" for functional biopolymer composites with sustained drug release. *Journal of Materials Chemistry B* **1** (23):2894-2903. doi:10.1039/C3TB20059K

[4] DiCosimo R, McAuliffe J, Poulose AJ and Bohlmann G (2013) Industrial use of immobilized enzymes. *Chemical Society Reviews* **42** (15):6437. doi:10.1039/c3cs35506c

IOP Conf. Series: Materials Science and Engineering **64** (2014) 012033 doi:10.1088/1757-899X/64/1/012033

Investigations on surface wettability of ZnO nanowires using UV LEDs for biosensing applications

K Bhavsar, R Prabhu and P Pollard

IDEAS Research Institute, Robert Gordon University, Aberdeen, AB10 7GJ, UK.

Corresponding Author E-mail: r.prabhu@rgu.ac.uk

Abstract. ZnO is a biocompatible material suitable for biosensors and microfluidic devices. Nanowires of ZnO tend to show hydrophobic nature which decelerates the adhesion/adsorption of biomolecules on the surface. This paper discusses the investigations on tuning the wettability of ZnO nanowires using UV LEDs. The spectral effect of LED emission on ZnO nanowires wettability has been studied. Results indicate that UV LEDs offer an advanced control on tuning the wettability of ZnO nanowires. The spectral investigations have provided significant insight into the role of irradiating wavelength of light on the wettability.

1. Introduction

Nanowires have become one of the hot research topics due to their special properties and huge application potential. Semiconductor based nanowires, especially based on ZnO have been explored for photonics, electronics, biomedical and energy related applications [1]. ZnO is a wide bandgap (3.37 eV) semiconducting biocompatible and biodegradable material [2]. The properties of nanowires depend highly on their surface and structural properties and are also linked to the synthesizing method [3]. Various nanostructures of ZnO can be synthesized using a low cost and scalable hydrothermal method [4]. The surface properties (hydrophilic or hydrophobic nature) of ZnO play a crucial role in biosensing, biomedical and microfluidic applications [5]. Adsorption/adhesion of the cells, biomolecules (such as protein, RNA, DNA) on the surface and their reaction depends on the surface energy and it closely relates to the wetting nature of the surface, and strongly correlates with the biological interactions [6].

ZnO nanowires tend to show hydrophobic nature [7] and therefore, presumably decelerate the biological surface interactions in the aqueous environments. Also, it has been reported that a biomaterial surface with moderate hydrophilic nature improves the biological interactions [5]. Modification in the wetting nature (hydrophobicity) shown by the ZnO nanowires can lead to improved surface interactions. However, the major problem arises in the control over surface wettability which is significant for improving the biological interaction between the surface and surrounding medium. Therefore, it is important to have a well-defined control over the surface energy of ZnO nanowires. UV light irradiation can be used to control the surface wettability of ZnO nanowires [8] and the intensity and wavelength irradiation can be used to control the wettability. Several papers have reported on the tunable wettability shown by ZnO nanowires exposed to UV lamps [8-10]. However, these lamps are relatively slow to switch on, stabilize and also, generate a considerable amount of heat. Further, they are not eco-friendly, they are a potential health hazard due to the UV exposure and limit the miniaturization of the devices. A

UV lamp imposes the problem of destroying the biomolecules/cells on the surface because of deep UV light (UV-C) irradiation [11]. Also, the major problem arises in the control over surface wettability which is significant for improving the biological interaction between surface and surrounding medium. Therefore, it is important to have well defined control over the light intensity and wavelength which modulates the surface energy of the ZnO nanowires. UV LEDs are inexpensive, energy efficient, have a narrow emission spectrum and offer better tunability of various control parameters leading to controlled surface wettability shown by ZnO nanowires.

Investigations on the surface morphology of ZnO nanostructures, ZnO nanocomposites and the effect of UV light irradiation time on their surface wettability have been reported [9,10,12]. However, the role of UV wavelength in controlling the surface wettability of ZnO nanowires is unclear. So far, to our knowledge, this is the first study on the spectral effect of UV light on the wettability of ZnO nanowires. Initial investigations on the spectral effects of narrow band UV light irradiation, using UV LEDs (λ_p = 365 nm, 385 nm), on the hydrophobicity of ZnO nanowire films have been carried out and discussed in this paper. Also, the effects of UV light irradiation time and transition rate (rate of change in contact angle) have been studied. Results show that spectral overlap between LED emission spectra and ZnO nanowire absorption spectra defines the rate of change in hydrophobicity of ZnO nanowires. LED with λ_p = 365 nm have shown very high transition rate compared to the λ_p = 385 nm.

2. Materials and methods

2.1 Materials

All the materials used in the experiment were analytical grade. Zinc acetate ($Zn(CH_3COO)_2$), zinc nitrate hexahydrate ($Zn(NO_3)_2 \bullet 6H_2O$) and ammonium hydroxide (NH_4OH) were purchased from the Sigma Aldrich, UK. All glassware was cleaned in deionized water prior to the experiment.

2.2 Synthesis of ZnO nanowire films

ZnO nanowires were grown on glass substrates using hydrothermal synthesis method [13]. 10 mM Zinc acetate was drop casted on the glass substrates and annealed on a hotplate for one min at 100°C for better adhesion. The process was repeated 5 times to get a uniform thick seed layer. ZnO nanowires were grown by immersing the seeded substrates into the mixture of 125 ml 20 mM zinc nitrate and 7 ml 28-30% ammonium hydroxide and annealed on a hotplate at 95° C for 5 hours. Thereafter, substrates were removed from the growth solution and rinsed with de-ionized water several times. The substrates were allowed to dry at room temperature and atmospheric pressure before characterization.

2.3 Characterizations

Surface analysis of the synthesized ZnO nanowire films has been carried out using a Scanning Electron Microscope from ZEISS (EVO LS10). Surface wettability has been characterized using contact angle measurement system from the First Ten Angstrom (FTA 200) at room temperature (RH - 8%).

3. Experiments

A pictorial view of experimental setup for irradiating the sample with UV LED is shown in Figure 1. UV light from LED was coupled through a bi-convex lens and irradiated over the sample placed on a sample stage. A dichroic mirror was used in the path to tap a fraction of the UV light for monitoring the power level with a Coherent (FieldMaster GS) power meter and/or LED spectra with Stellernet (EPP2000C) spectrometer. Surface wettability of the developed ZnO nanowires has been studied using the contact angle measurement system, by measuring the water contact angle before and after the UV light irradiation for different irradiation times. All experiments were performed using 2 μl water drop under ambient conditions.

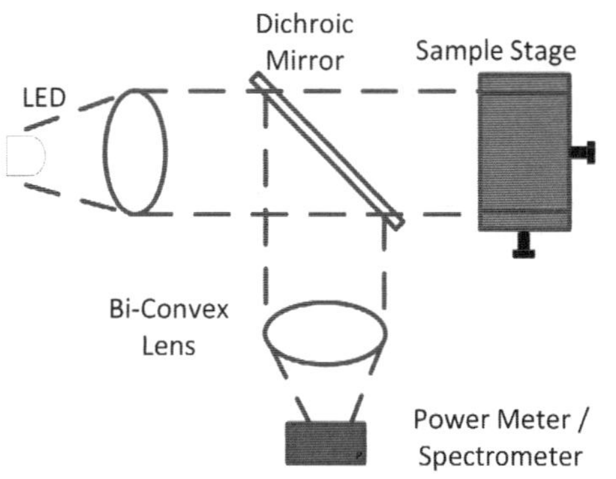

Figure 1. A pictorial view of experimental set-up.

4. Results and discussions

Figure 2 shows the SEM image of ZnO nanowires at different magnifications. From the SEM image, it can be seen that the synthesized ZnO nanowires are preferentially oriented towards the c-axis perpendicular to the glass substrate. Hexagonal structure of the ZnO nanowires can be seen in the magnified SEM image. Also, the magnified view of the nanowire tip reveals the growth of layered hexagonal nanostructures on the tip. This nanoscale roughness on the tip reduces the contact area between solid and the droplet which allows the entrapping of a large amount of air below the droplet, and therefore, enhances the hydrophobicity of the surface [7]. Length and diameter of the synthesized ZnO nanowires were about 11-12 μm and 400-500 nm respectively.

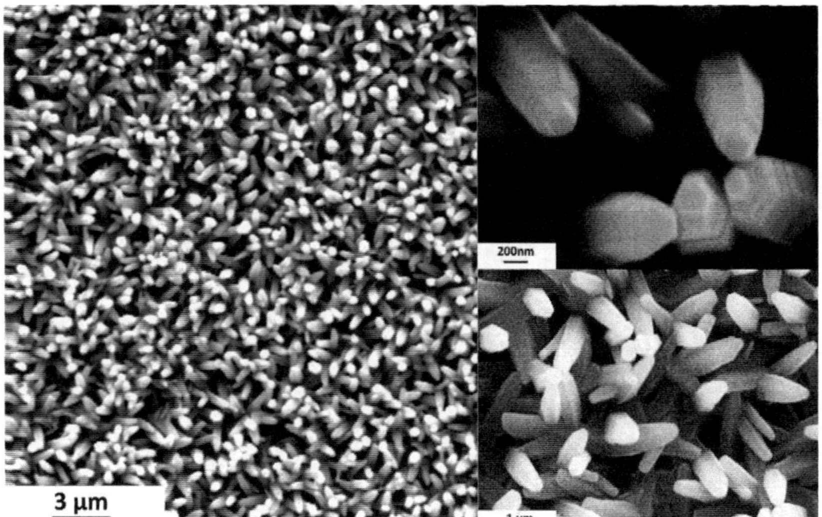

Figure 2. SEM image of ZnO nanowires with different magnifications.

The water contact angle was measured immediately after the synthesis of ZnO nanowires and found to be about 0°. This super hydrophilic nature shown by the synthesized ZnO nanowires using hydrothermal method can be attributed to the hydroxyl groups on the surface. Thereafter, substrates were left to dry under ambient conditions for 72 hours. ZnO nanowires have shown the contact angle of 125°.

Surface wettability of the synthesized ZnO nanowires have been analysed by measuring the static contact angle. ZnO nanowires surface wettability can be considered conferring to the Cassie-Baxter [14] or Wenzel [15] models.

A Young contact angle, θ_y, on the smooth surface can be defined in terms of surface energy as [16]:

$$\cos\theta_y = (\gamma_{sv} - \gamma_{sl}) / \gamma_{lv}$$

where, γ_{sv}, γ_{sl} and γ_{lv} are surface tension of solid-vapour, solid-liquid and liquid-vapour, respectively. Considering the surface roughness, the Wenzel model defines the apparent contact angle θ, defined as:

$$\cos\theta = r \cos\theta_y$$

where, r is a surface roughness factor. This suggests the effect of change in surface roughness on the hydrophobicity (contact angle). But the Wenzel model assumes that the liquid penetrates inside the rough surface which is the case after irradiation of UV light.

Under unexposed condition, ZnO nanowires tend to show a hydrophobic surface where liquid does not penetrate inside the rough surface (nanowire groves). Therefore, this condition refers to the Cassie-Baxter state [14].

$$\cos\theta = f(\cos\theta_y + 1) - 1$$

where, f is the area fraction of actual solid surface area to the projected solid surface area. Therefore, upon UV exposure, ZnO nanowires follow the wettability transition from Cassie-Baxter state to the Wenzel state.

UV light irradiation has been carried out on the ZnO nanowires using UV LEDs with a peak emission wavelength λ_p = 365 nm and 385 nm. Static water contact angles (WCA) have been measured on each sample before and after UV light irradiation, WCA_0 and WCA respectively. Figure 3 (a) and (b) shows the relative change in measured water contact angle and gradient with irradiation time, respectively. Rapid change in wetting angle has been observed for the LED with λ_p = 365 nm. Also, the gradient over the irradiation time indicates that the change in wetting angle is very high for the first 30 mins, thereafter, it slows down gradually. The reason behind this observed change in wettability by irradiating UV light on the ZnO nanowires can be explained as follows:

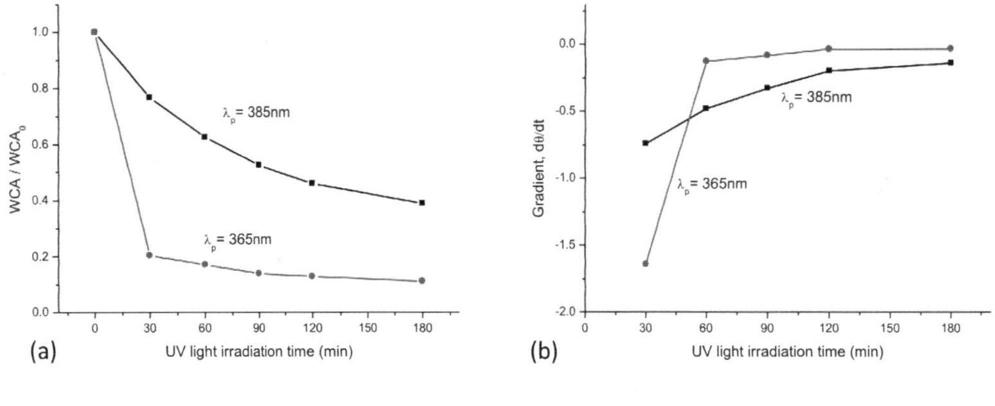

Figure 3. (a) Measured relative contact angle and (b) the gradient with irradiation time for the LED emission at λp = 365nm and 385nm.

IOP Conf. Series: Materials Science and Engineering **64** (2014) 012033 doi:10.1088/1757-899X/64/1/012033

UV light irradiation generates electron-hole pairs in the ZnO surface. These electron-hole pairs will create the surface defects/composition change leading to different wettability of the surface. The process which leads to the surface composition change can be expressed as [9]:

$ZnO + 2hv \rightarrow 2e^{-} + 2h^{+}$

$Zn^{2+} + e^{-} \rightarrow Zn_s^{+}$ (surface trapped electron)

$O^{2-} + h^{+} \rightarrow O^{-}$ (surface trapped hole)

$O^{-} + h^{+} \rightarrow 1/2O_2 + V_o$ (oxygen vacancy)

The surface trapped electrons (Zn_s^{+}) tend to react with oxygen molecules adsorbed on the surface:

$Zn_s^{+} + O_2 \rightarrow Zn_s^{2+} + O_2^{-}$

At the same time water molecules may adsorb at the oxygen vacancy (Vo) site. These defect sites are kinetically more favourable for hydroxyl groups (OH^{-}) adsorption than oxygen adsorption, hence promotes increased water adsorption at the UV light irradiated areas.

Further, to understand the observed spectral effect on the wettability, the absorption spectrum of the synthesized ZnO nanowires has been collected and plotted along with the UV LEDs emission spectra as shown in figure 4. Better spectral overlap indicates that ZnO nanowires absorb more light at 365 nm than 385 nm. Therefore, the observed change on the surface wettability of ZnO nanowires is high when irradiated with λ_p = 365 nm. This indicates that the wettability of the ZnO nanowires can be controlled by tuning the wavelength of irradiating light.

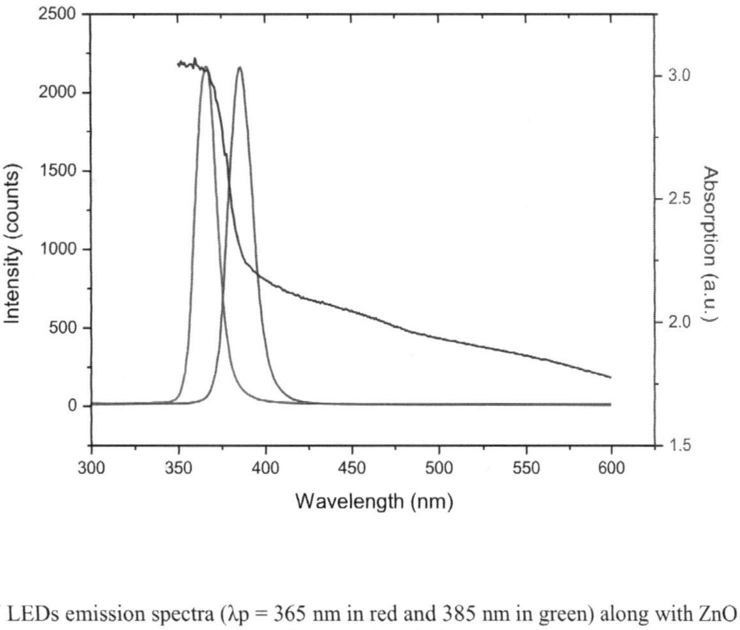

Figure 4. UV LEDs emission spectra (λp = 365 nm in red and 385 nm in green) along with ZnO nanowire absorption spectrum in blue.

5. Conclusion

The effect of irradiating UV light wavelength and irradiation time on tuning the wettability of ZnO nanwires has been studied. Results show that the wavelength of irradiating light has significant impact on tuning the surface wetting nature of ZnO nanowires. Developed ZnO nanowires have exhibited more advanced controllable wettability including a faster hydrophobic-hydrophilic switching upon UV light irradiation using LEDs. The UV LEDs offer better control over the wettability of the ZnO nanowires as well as the miniaturization of microfluidic devices and biosensors based on ZnO nanowires.

6. References

[1] Weintraub B, Zhou Z, Li Y and Deng Y 2010 Solution synthesis of one-dimensional ZnO nanomaterials and their applications *Nanoscale* **2** 1573
[2] Jun Zhou, Xu N and Zhong L. Wang 2006 Dissolving behavior and stability of ZnO wires iin biofluids: A study on biodegradability and biocompatibility of ZnO nanostructures *Advanced Materials* **18** 2432
[3] Hu X, Li G and Yu JC 2010 Design, Fabrication, and Modification of Nanostructured Semiconductor Materials for Environmental and Energy Applications *Langmuir* **26** 3031
[4] Zhang Y, Ram M, Stefanakos E and Goswami Y 2012 Synthesis, Characterization, and Applications of ZnO Nanowires *Journal of Nanomaterials* **2012** 1
[5] Janssen MI, van Leeuwen MBM, van Kooten TG, de Vries J, Dijkhuizen L and Wosten HAB 2004 Promotion of fibroblast activity by coating with hydrophobins in the beta-sheet end state *Biomaterials* **25** 2731
[6] Adamson AW and Gast AP 1997 *Physical Chemistry of Surfaces* 6th ed. (New York : Wiley)
[7] Badre C, Pauporte T, Turmine M and Lincot D 2007 A ZnO nanowire array film with stable highly water-repellent properties *Nanotechnology* **18**
[8] Lv J, Zhu J, Huang K, Meng F, Song X and Sun Z 2011 Tunable surface wettability of ZnO nanorods prepared by two-step method *Applied Surface Science* **257** 7534
[9] Khranovskyy V, Ekblad T, Yakimova R and Hultman L 2012 Surface morphology effects on the light-controlled wettability of ZnO nanostructures *Applied Surface Science* **258** 8146
[10] Li H, Zheng M, Liu S, Ma L, Zhu C and Xiong Z 2013 Reversible surface wettability transition between superhydrophobicity and superhydrophilicity on hierarchical micro/nanostructure ZnO mesh films *Surface & Coatings Technology* **224** 88
[11] Nakajima S, Lan L, Kanno S, Takao M, Yamamoto K, Eker A and Yasui A 2004 Light-induced DNA damage and tolerance for the survival of nucleotide excision repair-deficient human cells *Journal of Biological Chemistry* **279** 46674
[12] Liu Y, Lin Z, Lin W, Moon KS and Wong CP 2012 Reversible superhydrophobic-superhydrophilic transition of ZnO nanorod/epoxy composite films *ACS Applied Materials and Interfaces* **4** 3959
[13] Wen X, Wu W, Ding Y and Wang ZL 2012 Seedless synthesis of patterned ZnO nanowire arrays on metal thin films (Au, Ag, Cu, Sn) and their application for flexible electromechanical sensing *Journal of Materials Chemistry* **22** 9469
[14] Cassie ABD and Baxter S 1944 Wettability of porous surfaces.*Transactions of the Faraday Society* **40** 0546
[15] Wenzel RN 1936 Resistance of solid surfaces to wetting by water *Industrial and Engineering Chemistry* **28** 988
[16] Young T 1805 An Essay on the Cohesion of Fluids *Philosophical Transactions of the Royal Society of London* **95** 65

IOP Conf. Series: Materials Science and Engineering **64** (2014) 012034 doi:10.1088/1757-899X/64/1/012034

Structure and properties of PP/organoclay composite fibres

A Marcincin, M Hricova and A Ujhelyiova

Slovak University of Technology in Bratislava, Faculty of Chemical and Food Technology, Department of Fibres and Textile Chemistry, Radlinskeho 9, 812 37 Bratislava, Slovak Republic

anton.marcincin@stuba.sk

Abstract. In this paper, the effect of uniaxial deformation of PP/organoclay composite fibers in spinning and drawing on their supermolecular structure, thermal and mechanical properties is presented. The commercial organoclays Cloisite C15A and Cloisite C30B, both based on montmorillonite (MMT) were used in experimental work as inorganic fillers. The supermolecular structure of fibers was investigated by DSC analysis and X-ray diffraction (WAXS). The DSC measurements were carried out using conventional method (CM) and constant length method (CLM) in which the fibers with constant length during measurement were assured. Intercalation of polypropylene in the interlayer galleries of organoclay was evaluated by SAXS method. Tensile strength and Young's modulus of composite fibers are discussed in the paper with regard to their thermal properties and supermolecular structure as well as intercalation and exfoliation of (nano)filler in polymer matrix.

1. Introduction

The incorporation of the organoclay into the semicrystalline polymers affects their crystalline properties such as crystallization kinetics, crystal structures and total crystallinity [1, 2]. The layered silicates in the polypropylene (PP) matrix behave as effective heterogeneous nucleating agents. They enhance the crystallization rate [2], and at the same time decrease the spherulite size, perfection of the crystals and shift the cooling crystallization temperature to higher values [3]. Besides, the silicate nanoplatelets affect the shear flow induced crystallization and retardancy of the crystallisation kinetic of matrix at the quiescent crystallization of the PP/montmorillonite (MMT) composites [4]. The preferential orientation of PP lamellae perpendicular to the surface of the organoclay layers was found for PP/talc composite by TEM observation. The lamellar orientation on the clay layers was ascribed to nucleation and crystallization at the surface of the silicate platelets. Moreover, the asymmetric morphology of crystals was observed in PP/organoclay nanocomposites at higher temperature and at shear induced formation of the oriented threadlike crystallites [1].

Simultaneously with the development of highly oriented polymers the suitable methods for investigation of the oriented polymer structure were developed or modified. The special method of DSC analysis of fibers, known as "constrain method" or method of "constant length of fibers" (DSC-CLM), has been described in several papers. This DSC method is particularly suitable for evaluation of oriented structure elements of composite fibers [5].

In our previous papers the effect of organoclay on mechanical properties of PP fibers was studied [6, 7]. The positive effect of organoclay on mechanical properties of fibers was found for very low concentration of solid particles, only. In this case, the effect of nanoparticles as hard segments on mechanical properties is negligible. Thus, enhancement of the mechanical properties of composite fibers is affected indirectly by suitable changes in the structure of fibers. Some convenient methods were used for characterization of structure and morphology of PP composite fibers in our experimental work to explain their mechanical properties. The results are presented in this paper.

2. Experimental

2.1. Materials used

Polymers: Commercially available types of PP were used in the experimental work: polypropylene HPF (PP HP) with melt flow index (MFI) 8.0 g/10 min, in powder form and fiber-grade polypropylene TG 920 (PP) with MFI 10.5 g/10 min. Both polypropylenes were produced by Slovnaft a. s., Bratislava, Slovak Republic.
Fillers: Two kinds of organoclays (OC) used in this work were Cloisite 15A (C15A) and Cloisite 30B (C30B), both produced by Southern Clay Product, Inc, Gonzales, TX 78629, USA.

Compatibilisers: Two non-reactive compatibilisers-dispersants were used for the preparation of concentrated dispersion of organoclay in the PP HP. Commercially available Slovacid 44P (S44P) based on the ester of stearic acid and polypropylene glycol, produced by Sasol GmbH, Brunsbuettel, Germany and Tegopren 6875 (TEG) based on poly (alkylsiloxane) produced by Degussa Co., Düsseldorf, Germany.T

2.2. Preparation of PP/MWCNT composite fibres

1. The PP HP, organoclay and compatibiliser were mixed in a high rpm mixer for a defined time. Powder mixture was melted and kneaded using twin screw corotating extruder ϕ 28 mm. The temperature of the extruded melt was 229°C. The concentration of organoclay in PP HP was 10.0 wt.%. Content of compatibiliser was 4.0 wt.%.
2. The chips of organoclay concentrate dispersion and PP were mixed and spun using a single screw extruder ϕ 30 mm and a spinneret with 40 orifices. The spinning temperature was 280°C, metering of melt 30 g/min, spinning speed 360 m/min and the fineness of spun fibers (drawn ratio 1:1, λ_1) was 840 dtex. Fibers were drawn using laboratory drawing machine for draw ratio 1:3 (λ_3) and for maximum drawn ratio (λ_{max}) at drawing temperature 120°C.

2.3. Methods used

DSC measurement:The measurements were performed using Perkin Elmer DSC 7 in the temperature range 30-200°C. The standard heating rate was 10°C/min. The measurements were carried out using conventional method (DSC-CM) in which the cut fibers with length 1-2 mm were used, and constant length method (DSC-CLM) in which the constant length of fibers during measurement was assured. The constant length of the fibers was achieved through winding and fixation of fibers on a wire. The melting peak temperature T_m and melting enthalpy ΔH_m were evaluated.

X-ray scattering method: Wide-angle scattering (WAXS) method: The structure of fibers was evaluated by a wide-angle X-ray scattering method. The investigations were carried out using a Seifert X - ray diffractometer. On the basis of the WAXS patterns the parameters characterising the fiber structure were determined. The total crystallinity X_α and the content of the mesophase X_m were calculated as a ratio of the area under crystalline or mesophase peaks to the total area [18].
Small-angle scattering (SAXS) method: All SAXS patterns were recorded with an MBraun SWAX camera with the Kratky collimation system equipped with a PSD 50 MBraun linear position sensitive detector.

Mechanical properties: The Instron (3343 model) was used for the measurements of mechanical properties of fibers (according to ISO 2062:1993). They were evaluated from 15 measurements.

3. Results and discussion

3.1.Thermal properties and supermolecular structure of fibers

The DSC-CM method provides simple one-peak thermogram for undrawn fibers and characteristic double-peak thermogram for middle draw ratio of fibers (Figure 1a).

The radically different shapes of melting endotherms were observed for PP fibers in dependence on the draw ratio using DSC-CLM method. The endotherm peak is shifted to higher temperature gradually when the draw ratio increases (Figure 1b, 1c). One-peak endotherm becomes wider with the increase of draw ratio.

The melting temperature (T_m) for PP fibers and for PP/C15A composite fibers increase proportionally with the draw ratio up to maximum values (Figure 2a). The dependence of melting enthalpy (crystallinity) on organoclay concentration passes through clear maximum at 0.1 wt.% and practically does not change with the presence of compatibiliser-dispersant (Figure 2b).

Generally, DSC-CLM method provides the higher melting enthalpy for PP/organoclay fibers (higher crystallinity) in contrast to DSC-CM method, particularly for lower draw ratio.

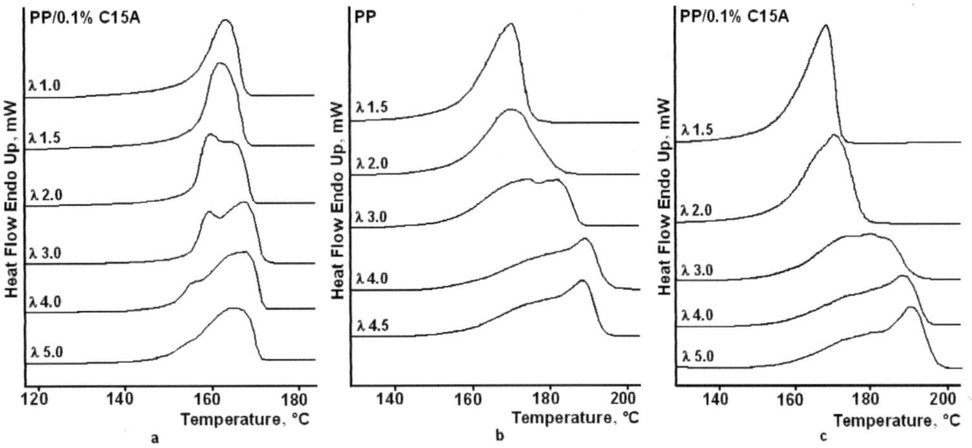

Figure 1 DSC-CM (a) and DSC-CLM (b, c) melting thermograms of PP and PP/0.1%C15A composite fibres in dependence on draw ratio

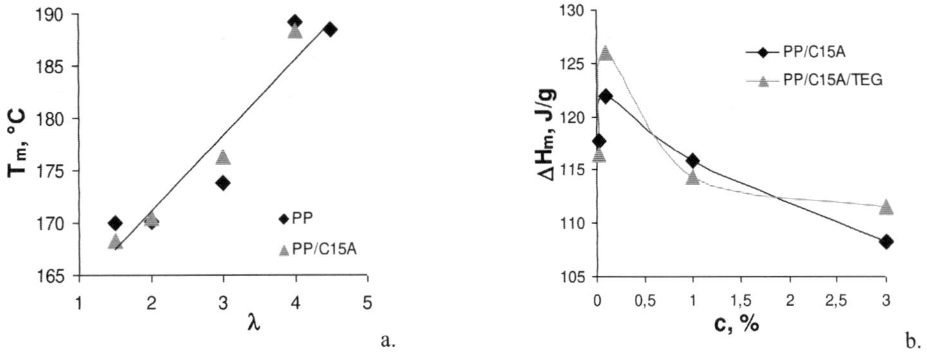

Figure 2 DSC-CLM melting temperature T_m (a) versus draw ratio and melting enthalpy ΔH_m (b) versus concentration of organoclay in the PP and PP composite fibres

3.2. WAXS and SAXS measurements

The total crystallinity of the oriented PP matrix of fibers consists of the fraction of α-form modification X_α and fraction of mesophase X_m (Table 1). On the basis of WAXS measurements the maximum total crystallinity was found for low concentration of organoclays, and at the same time for maximum draw ratio. Results are consistent with these obtained from DSC-CLM method. With higher concentration of C15A organoclay the total crystallinity of the PP matrix gradually decreases mainly due to significant decrease of the mesophase fraction X_m. The higher content of α-modification was found for drawn fibers and for higher content of C15A organoclay. The sizes of α-form crystallites passes through minimum at 0.1% C15A in PP fibers.

Table 1 Results of WAXS measurements of PP/organoclay composite fibres

Sample	Draw ratio λ	Total crystallinity x_C	Fraction of α form x_α	Fraction of mesophase x_m	Sizes of crystallites of α form $D_{(110)}$, nm	Sizes of mesophase crystallites, nm
PP	3.0	0.621	0.338	0.283	6.5	3.4
PP	4.5	0.625	0.311	0.314	5.4	3.1
PP/1.0% C15A	1.0	0.620	0.272	0.348	8.4	3.3
PP/1.0% C15A	3.0	0.668	0.407	0.261	7.1	3.4
PP/0.02% C15A	5.0	0.714	0.338	0.376	6.2	2.7
PP/0.1% C15A	5.0	0.688	0.347	0.341	4.6	2.8
PP/1.0% C15A	4.5	0.627	0.311	0.316	7.0	3.1
PP/3.0% C15A	3.0	0.571	0.388	0.183	11.6	3.5
PP/1.0% C30B	3.0	0.616	0.409	0.207	7.7	3.7
PP/1.0% C30B	4.5	0.658	0.389	0.269	6.3	3.1
PP/1.0% C30B/TEG	3.0	0.636	0.396	0.240	6.6	3.8
PP/1.0% C30B/TEG	4.5	0.636	0.435	0.201	5.2	4.0

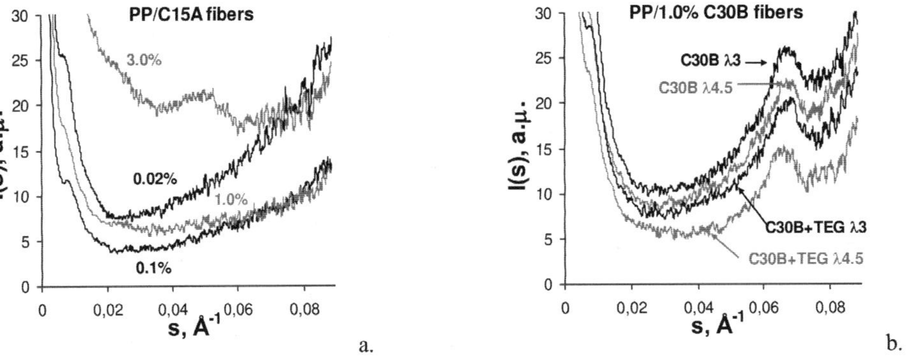

Figure 3 SAXS curves for evaluation the d-spacing of PP/organoclay composite fibres in dependence on concentration of C15A for λ_{max} (a) and draw ratio of PP/C30B 1.0 wt.% fibres (b).

SAXS measurement was used for evaluation of d-spacing between silicate layers in the C15A and C30B organoclays (Figures 3). The peak for C15A in PP fibers is weak and broad. It could not be resolved for high draw ratio and lower concentration (0.02-0.1%), (Figure 3a). The results show that cold drawing of the fibers contributes to intercalation and exfoliation of C15A organoclay in the matrix of PP fibers. Based on SAXS results the d-spacing of C30B collapsed and layers distance decreased from 1.9 to 1.5 nm. The organic phase partially removes from the layers space in the PP matrix. The SAXS results have shown that d-spacing of C15A organoclay in PP fibers grows during spinning and drawing mainly at low concentration and maximum draw ratio. In contrast, no positive change regarding to intercalation and exfoliation of C30B organoclay was observed.

3.3.Mechanical properties

Both organoclays positively affect the tensile strength and Young's modulus of PP composite fibers only at a low concentration, up to 1.0 wt% (Tables 2, 3). A lower tensile strength was found for PP fibers containing C30B without compatibiliser and with S44P one.

Table 2 Tensile strength σ, elongation at break ε and Young's modulus E of PP/C15A nanocomposite fibres

Fibres	**C_{OC} in fibre, %**	**Draw ratio λ_{max}**		
		σ, cN/tex	**ε, %**	**E, N/tex**
-	-	41.6 ± 0.7	33.8 ± 4.1	4.73 ± 0.17
PP/C15A	0.02	54.2 ± 3.5	27.7 ± 4.1	4.89 ± 0.25
	0.1	55.0 ± 1.9	25.6 ± 2.0	4.54 ± 0.42
	1.0	41.7 ± 2.4	36.1 ± 5.6	4.30 ± 0.28
	3.0	24.3 ± 0.8	98.5 ± 7.9	2.34 ± 0.08
PP/C15A/S44P	0.02	51.7 ± 2.9	27.2 ± 4.4	4.65 ± 0.30
	0.1	53.1 ± 1.0	26.0 ± 1.3	4.57 ± 0.26
	1.0	43.2 ± 1.8	34.2 ± 9.4	4.58 ± 0.24
	3.0	38.3 ± 2.0	41.0 ± 12.4	3.76 ± 0.16
PP/C15A/TEG	0.02	52.2 ± 1.9	25.3 ± 2.0	4.92 ± 0.16
	0.1	51.4 ± 1.3	24.1 ± 1.6	4.99 ± 0.11
	1.0	43.8 ± 1.4	44.7 ± 14.2	4.75 ± 0.23
	3.0	45.8 ± 2.9	25.8 ± 4.0	4.46 ± 0.17

Table 3 Tensile strength □ , elongation at breakε and Young's modulus E of PP/C30B nanocomposite fibres

Fibres	**C_{OC} in fibre, %**	**Draw ratio λ_{max}**		
		σ, cN/tex	**ε, %**	**E, N/tex**
-	-	41.6 ± 0.7	33.8 ± 4.1	4.73 ± 0.17
PP/C30B	0.02	52.5 ± 1.9	29.4 ± 4.0	4.99 ± 0.19
	0.1	49.4 ± 2.3	27.2 ± 3.3	4.83 ± 0.16
	1.0	43.9 ± 1.6	20.3 ± 1.3	4.78 ± 0.18
	3.0	-	-	-
PP/C30B/S44P	0.02	44.4 ± 1.7	43.4 ± 11.9	4.56 ± 0.16
	0.1	56.7 ± 0.8	24.6 ± 1.7	5.38 ± 0.14
	1.0	38.1 ± 3.4	24.7 ± 6.2	4.57 ± 0.27
	3.0	12.3 ± 0.7	58.5 ± 13.3	1.49 ± 0.10
PP/C30B/TEG	0.02	56.3 ± 0.8	25.2 ± 1.0	5.31 ± 0.22
	0.1	55.1 ± 0.9	26.0 ± 1.4	4.73 ± 0.35
	1.0	44.0 ± 2.4	28.3 ± 3.2	4.68 ± 0.18
	3.0	22.4 ± 1.5	81.8 ± 11.2	2.25 ± 0.14

Analysis of the mechanical properties of fibers shows predominant effect of supermolecular structure and orientation of fibers on both tensile strength and Young's modulus (Tables 1, Figure 2). The orientation of fibers results from the deformation of fibers in drawing. The positive effect of low concentration of organoclay and compatibilisers on mechanical properties of PP composite fibers resides in the favourable deformation during the drawing process. The higher λ_{max} at drawing of fibers was achieved in this case.

4. Conclusions

The analysis of the thermal properties, supermolecular structure, orientation and mechanical properties of PP/organoclay composite fibers obtained in the experimental work allow to draw the following conclusions:

- Tensile strength and Young's modulus of the PP/organoclay composite fibers correspond with their DSC-CLM melting temperature (orientation of non-crystalline regions) and melting enthalpy (total crystallinity).
- Tensile strength and Young's modulus of PP/organoclay fibers increase gradually with higher total crystallinity obtained by WAXS method, lower α-form fraction, higher fraction of mesophase crystallites and lower size of both the α-form and mesophase crystallites. These parameters were found for PP/organoclay composite fibers containing 0.02-1.0 wt% of organoclay C15A (C30B with TEG dispersant, only).

Acknowledgement: This work was supported by the Slovak Research and Development Agency under the contract No: VMSP-P-0007-09.

References

[1] Wang K *et al* 2005 Shear amplification and re-crystallization of isotactic polypropylene from an oriented melt in presence of oriented clay platelets *Polymer* **46** pp 9022-9032
[2] Ma J *et al* 2002 Crystallization behaviors of polypropylene/ montmorillonite nanocomposite *J. Appl. Polym. Sci.* **83** pp 1978-1985
[3] Zhang QX *et al* 2003 Multiple melting and crystallization of nylon-66/montmorillonite nanocomposites *J. Polym. Sci. Part B: Pol. Phys.* **41** pp 2861-2869
[4] Nowacki R *et al* 2004 Spherulite nucleation in isotactic polypropylene based nanocomposites with montmorillonite under shear *Polymer* **45** pp 4877-4892
[5] Grebowicz JS *et al* 2001 Deformation of undrawn poly(trimethylene terephthalate) (PTT) fibers *Polymer* **42** pp 7153-7160
[6] Marcinčin A *et al* 2008 Spinning of Polypropylene/Organoclay Composites, Thermal and Mechanical Properties of Fibres *Tekstil* **57** pp 141-148
[7] Marcinčin A *et al* 2009 Study of Rheological, Thermal and Mechanical Properties of Polypropylene/Organoclay Composites and Fibres *Fib. and Tex. in EE* **17** pp 22-28

IOP Conf. Series: Materials Science and Engineering **64** (2014) 012035 doi:10.1088/1757-899X/64/1/012035

Effect of nanoclay loading on the thermal decomposition of nanoclay polyurethane elastomers obtained by bulk polymerization

Javier Quagliano[1] and Javier Bocchio[2]

[1]Organic Synthesis Division and [2]Analytical Division, Applied Chemistry Department (DQA), Argentine R&D Institute for the Defense (CITEDEF), Buenos Aires, Argentina.

E-mail: jquagliano@citedef.gob.ar

Abstract. Thermoplastic urethane (TPU) nanocomposite was prepared successfully by dispersion at high shear stress of the nanoclay in polyol and further bulk polymerization. Our results from DSC studies showed an increase in decomposition temperature when nanoclay was loaded at 3,5% on elastomeric PU made from TDI, PTMEG and BDO, while not when nanoclay content was lower (1,5%). The exotherms at 370-375°C could be adscribed to the decomposition of the hard segments according to previous work.

1. Introduction

Nanocomposites of polyurethane exhibit improvement in various properties compared to the conventional microcomposites such as outdoor resistance of coatings [1], heat resistance, gas permeability, and flammability [2,3,4,5]. Mineral fillers as high-modulus additives are used as reinforcing agents in polymeric materials. In particular, polyurethane/clay nanocomposites have attracted great interest in recent years. They have gradually become more widely accepted in applications such as automobile parts [4,6].

On a broad basis, the preparation of polyurethanes (PU) is typically classified by the sequence of addition of the reactants (one-step, two-step or prepolymer process) or by the number of components, which the user has to mix together (a one-component or a two-component system).

Thermoplastic polyurethanes (TPU) have a two phase structure because of the thermodynamic incompatibility of the hard and soft segments of PU chain. Phase separation of hard segment from polyol soft segment determines elastomeric properties. Phase separation, leading to domain microstructure, has been postulated to explain the unusual viscoelastic properties of segmented PU [7].

In general, PU are formulated with an excess isocyanate to improve elastomeric properties during post-curing reactions. Mechanical properties of PU elastomers could be controlled by changing the molar ratio of two different monomer components [8]. The NCO/OH ratio determines the physical properties of PU, being generally a ratio of 1 the better for improving physical properties. The type of polyol, isocyanate, chain extender and synthesis conditions determines the physical and mechanical properties of PU, rendering a wide variety of materials with properties ranging from a soft foam to extremely hard and impact/tear-resistant materials.

Additionally, the way of incorporation of nanoclay into the PU matrix greatly affects its physical properties. The simplest approach of mixing nanoclay and PU is to physically mix them both, resulting in physical trap forces such as polar, hydrogen bonding and shear between the clay and the polymer [9]. Other mixing methods were reported in the state of the art. As an example, in one method for rigid PU foams, organoclay was dispersed first in the isocyanate component ultrasonically. It was found that using toluene as common solvent enhanced dispersion significantly [10]. Bulk and melt processing polymerization were compared as two methods of incorporation of nanoclay into thermoplastic polyurethane matrix [11].

The glass transition and decomposition temperatures of nanoclay PU increase with increasing clay content, due to the restricted motion of chains and barrier property of the clay platelets. In PU/clay nanocomposites, hard segments are highly attracted to the silicate surface but soft segments are not attracted in order to push the platelets apart to regain entropy [12,13].
Functionalized nanoclays were reported to increase the T_g gradually to the range of 60 to 62°C for 5 wt% loading in PU based thermoset adhesive [14]. On the other side, Thirumal and coworkers [16] observed that the glass transition temperature (T_g) decreases on loading of organically-modified nanoclay. These findings highlights that thermal behaviour depends on several factors such as type of TPU, clay incorporation method and how platelets are dispersed within the PU matrix.

Introduction of nanoclay limits the motion of TPU molecule and leads the nanocomposites to exhibit higher thermal stability [20]. Decomposition temperatures for 10% weight loss was at least 10ºC higher than control sample for nanoclay-reinforced PU foams up to 2% pphp (parts percent polyol basis) and were between 350 and 380ºC [17]. For thermoplastic PU (TPU) based on HMDI, PTMEG and BDO and loaded with nanoclay from 1 to 9% (pphp), decomposition temperatures were increased from 400 ºC for control TPU to almost 450 ºC for 5% nanoclay filled. The authors found that all the specimens of PU nanocomposites showed two stages of thermal degradation. The first stage is dominated by the degradation of the hard segment and the second stage correlated well with the thermal dissociation of the soft segment [18]. It is common knowledge that strong dipole-dipole, hydrogen bond interactions and crystallinity that nanoclay addition imparts improve heat resistance. Also for organoclay-modified PU foams enhancement degree of thermal stability and flame retardancy of composites was reported to coincide well with the sequences of gallery spacing of organoclay in the PU matrix [15].

In this contribution we report our preliminary results on the effect of nanoclay incorporation over the decomposition temperature of an elastomeric PU based on poly(tetramethylene)glycol (PTMEG), toluene diisocyanate (TDI) as curing agents and 1,4-butanediol (BDO) as chain extender.

2. Materials and Methods

Reactives: a polyether polyol poly(tetramethylene)glycol (Voranol®, mean molecular weight 2000 g/mol, hydroxyl index 56), was obtained from Mentvil Argentina S.A, Buenos Aires. An isomeric mixture of tolydene diisocyanate (TDI) was utilized as received from Petroquímica Río Tercero S.A, Córdoba, Argentina. The reactive nanoclay Cloisite C30B containing 90 mequiv./100 g clay of quaternary ammonium ions was donated by Southern Clay Products, US. The quaternary ammonium ion has the structure $N^+(CH_2CH_2OH)_2(CH_3)T$, with T representing an alkyl group of approximately 65% $C_{18}H_{37}$, 30%$C_{16}H_{33}$, and 5% $C_{14}H_{29}$.

The soft segment of elastomeric PU was synthesized from poly(tetramethylene)glycol (PTMEG), while the hard segment was synthetized from TDI and 1,4-butanediol (BDO, Sigma Aldrich, US) as chain extender.

OCT (organoclay)-polyurethane nanocomposite: OCT-polyol nanocomposites were prepared by two-step or prepolymer bulk polymerization. Increasing quantities of OCT (nanoclay) were added to the polyol, while stirring at 2000 rpm for 10 minutes. The resulting OCT-polyol nanocomposites were reacted with TDI for 120 minutes at 60ºC to form the prepolymer under vaccum (4 mm Hg). Then BDO was added as chain extender for 30-60 minutes the same temperature until elastomeric material developed, evidenced by the increased thickness and final formation of a gummous and soft material. Post-cure was made at ambient temperature for at least 7 days.

Differential scanning calorimetry (DSC) studies were performed in a TA model Q20 apparatus belonging to the Department. Samples were heated at 5°C/min under a nitrogen atmosphere.

3. Results

We obtained PU elastomers adding TDI to the polyol at 60ºC and adding BDO as chain extender in a second stage. The NCO/OH initial ratio was set to 1. When BDO was added, the reaction crude increased its viscosity until a hard elastic material was formed. For obtaining the nanocomposites, nanoclay was incorporated in the polyol fraction at high shear (2000 rpm) for 10 minutes.

In the following Table, we summarized the temperatures at with thermal peaks were obtained for different elastomeric PU.

Table 1: Decomposition temperatures of nanoclay PU elastomers based on PTMEG polyol, TDI and BDO as chain extender

OC (pphp, %)	Hard domain	Soft domain	Main temperature of peaks in DSC (°C)	Decomposition temperature (°C)
0	TDI	PTMEG	330/390	390
0	TDI-BDO	PTMEG	287/375/422	422
1,5	TDI-BDO	PTMEG	350/370/412	412
3,5	TDI-BDO	PTMEG	320/370/427	427

OC: organoclay. pphp (parts of OC on a percent polyol basis).

Linear PU made from TDI and PTMEG only showed less heat resistance than those branched with BDO, evidenced by a 20-30 °C reduction in decomposition temperature. Decomposition temperatures were 10ºC lower for 1,5% loaded nanoclay PU than for control sample. However, 3,5% loaded sample decomposed 5 °C after the control, coincidently with the majority of the reports [1,2,3,4]. The exotherms at 370-375ºC could be adscribed to the decomposition of the hard segments, according to previous reports [18, 20]. Saiani and coworkers [21] assigned the observed high-temperature endothermic transitions to the disruption of an ordered structure appearing in the hard phase under certain annealing conditions and to the microphase mixing of the soft and hard segments. Overall, decomposition temperatures were in the same range as for other elastomeric PU systems. For example, it was around 440ºC when poly(caprolactone) was the soft segment, and 4,4-methylene bis (cyclohexyl isocyanate) and BDO conformed the hard segment [19] and around 420ºC for a similar PU system cured with the aliphatic isophorone diisocyanate (IPDI) instead of aromatic TDI [22]. The thermal stabilization effect of nanoclay can be explained because nanoclay hinders the diffusion of the volatile degradation products (carbon dioxide, carbon monoxide, water molecules, etc.) from the bulk of the polymer matrix to the gaseous phase [20].

4. Conclusions

TPU nanocomposite was prepared successfully by dispersion at high shear stress of the nanoclay in the polyol and further bulk polymerization. Our results from DSC studies showed an increase in decomposition temperature when nanoclay was loaded at 3,5% on elastomeric PU made from TDI, PTMEG and BDO, while not when nanoclay content was lower (1,5%). The exotherms at 370-375ºC could be adscribed to the decomposition of the hard segments according to previous work. Our results justify further research with this PU system to determine specifically which nanoclay content improves thermal resistance without affecting general properties of this system.

References

[1] Dhoke S, Rajgolapan N and Khanna A 2012 *International Journal of Material Science*, **2**, 2, 47-55.

[2] Alexandre M and Dubois P 2000 *Mater. Sci. Eng-R*, **28** (1-2), 1-63.

[3] Wilkie C and Morgan A 2010 Fire Retardance of Polymeric Materials. Second Edition. CRC Press, Taylor and Francis Group, Florida, US.

[4] Quagliano J , Wittemberg V and Gavilán García I 2013 Recent advances on the utilization of nanoclays and organophosphorus compounds in polyurethane foams for increasing flame retardancy. *Structural Nanocomposites. Perspectives for Future Applications*, ed. J Njuguna (Springer Berlin Heidelberg)chapter 11, pp. 249-258

[5] Galimberti M 2011 *Rubber-clay nanocomposites Science, Technology, and Applications*. (US New Jersey: John Wiley & Sons).

[6] Usuki A, Hasegawa N and Kato M 2005 *Inorganic Polymeric Nanocomposites and Membranes, Advances in Polymer Sciences*, 179, ed. O Becker. pp135-195.

[7] Seymour R and Cooper S 1971 *Polymer Letters*, **9**, 689-694.

[8] Kojio K, Furukawa M, Nonaka Y and Nakamura S 2010 *Materials*, **3**, 5097-5011.

[9] Maji P, Gudchhait P and Bowmick A 2009 *J. Material Sciences*, **44**, 5861-5871.

[10] Widya T and Macosko C 2005 *Journal of Macromolecular Science, Part B: Physics*, **44**, 6.

[11] Pizzatto L, Lizot A, Fiorio R, Amorim C, Machado A, Giovanella M, and Crespo L 2009 *Materials Science and Engineering*: C **29**, 2, 1, 474–478.

[13] LaShanda T, Korley J, Liff S, Kumar N, McKinley G and Hammond P 2006 *Hastopoulos Microporous Laboratories*, HML 06-P-07.

[14] Dodiuk H, Belinski I, Dotan A and Kenig S 2006 *Journal of Adhesion Science and Technology* **20**, 12, 1345-1355.

IOP Conf. Series: Materials Science and Engineering **64** (2014) 012035 doi:10.1088/1757-899X/64/1/012035

[15] Xu Z, Tang X and Zheng J 2008 *Polymer-Plastics Technology and Engineering*, **47**, 11.

[16] Thirumala M., Dipak Khastgira D., Singhaa, N. Manjunathb and Napbi Y. 2009 Effect of a Nanoclay on the Mechanical, Thermal and Flame Retardant Properties of Rigid Polyurethane Foam *Journal of Macromolecular Science, Part A: Pure and Applied Chemistry*, **46**,7.

[17] Kim S, Lee M, Kim H, Park H, Jeong H, Yoon K and King B 2010 *Journal of Applied Polymer Science*,**117**,4, 1992-1997.

[18] Barick A and Tripathy D 2010 *Polymer Engineering & Science* **50**, 3, 484-498.

[19] Chan-Chan J, Solis-Correa R, Vargas-Coronado R, Cervantes J, Cauich-Rodríguez J, Quintana P. and Bartolo-Pérez P 2010 *Acta Biomaterialia* **6,** 2035–2044.

[20] Barick A and Trepathy D 2010 *Polymer Science, Engineering, and Technology,* 1471-1482.

[21] Saiani A, Daunch W, Verbeke A, Leenslag J and Higgins J 2001. Origin of Multiple Melting Endotherms in a High Hard Block Content Polyurethane. 1. Thermodynamic Investigation *Macromolecules*, **34**, (26), 9059–9068.

[22] Saha C, Tapan K, Chaki K and Singha N 2013 *J. Appl. Polymer Sci.* **130**, 5, 1-7.

IOP Conf. Series: Materials Science and Engineering **64** (2014) 012036 doi:10.1088/1757-899X/64/1/012036

Synthesis of PVDF ultrafiltration membranes supported on polyester fabrics for separation of organic matter from water

Sabelo D Mhlanga*, Tumelo G Tshabalala, Edward N Nxumalo and Bhekie B Mamba

Department of Applied Chemistry and the DST/MINTEK Nanotechnology Innovation Centre - Water Research Node, University of Johannesburg, P.O. Box 17011, Doornfontein, 2028, Johannesburg, South Africa

*Email: sdmhlanga@uj.ac.za; Tel: (+2711) 559 6187; Fax: (+2711) 559 6425

Abstract: Polyvinylidene flouride (PVDF) membranes supported on non-woven fabrics (NWF) of polyester are reported. The PVDF membranes were fabricated using the phase inversion method followed by modification of the active top layer of the PVDF thin film by adding polyvinylpyrolidone (PVP) into the cast solution. A PVDF resin was used with *N*-methyl-2-pyrrolidone (NMP) as a solvent. Sessile drop contact angle measurements and scanning electron microscopy (SEM) were used to study the physical properties of the membranes. Membrane rejection of humic acid was studied using a cross-flow membrane testing unit. The contact angle results revealed that the hydrophilicity of PVDF membranes increased as the PVP concentration was increased from 3 to 10 wt%. SEM analysis of the membranes revealed that the membrane pore sizes increased when PVP was added. AFM analysis also showed that membrane roughness changed when PVP was added. Total organic carbon (TOC) analysis of water samples spiked with humic acid was performed to test the rejection capacity of the membranes. Rejections of up to 97% were achieved for PVDF membranes supported on polyester NWF1, which had smaller thickness and higher permeability compared to polyester NWF2. The NWFs provided the high strength required for the membranes despite the modifications done on the PDVF surface and microstructure.

1. Introduction

Natural organic matter (NOM) is known to exist naturally in surface and ground waters [1]. It can react with a number of disinfectants to generate disinfection by-products (DBPs) such as trihalomethanes, halonitromethanes, haloacetonitriles, haloamides, halofuranones, iodo-acids and others which are known to be carcinogenic and having the potential of being highly detrimental to human life [2-7]. The effective removal of NOM is extremely important in order to prevent the formation of the DBPs and provision of safe drinking water. Moreover, NOM is known to cause fouling of membranes thus requiring high pressures to force water across the membrane.

In order to reduce NOM before disinfection, a direct correlation between TOC and NOM was established as per Standard Methods 5310 and the USEPA's disinfectants/disinfection by-products rule. Therefore, TOC analysis was chosen for measuring NOM in this study. The TOC analysis permitted the determination of the effectiveness of the NWF supported PVDF membranes in the removal of NOM, particularly humic acid (HA) from water [8,9].

---

* To whom any correspondence should be addressed.

In this study, PVDF membranes supported on two different non-woven fabrics (NWFs) of polyester (Table 1) were prepared using the phase inversion method [10]. We demonstrate the advantages of using the support materials compared to unsupported PVDF. It is well known that NWF materials can effectively be used for the removal of particles larger than 1 μm in the decontamination process in both air and water filtration [12]. Recent studies have shown that NWFs can be used in membrane bioreactors (MBR) to treat waste water [13]. It is also known that modification of polymeric membranes either through chemical functionalization of the surface (active layer) or through addition of nanoparticles compromises the strength of the membranes due to defects caused by these modifications. NWFs were used as support layers for the thin film PVDF membranes, which allowed us to modify the membranes while maintaining the desired strength provided by the intact NWF used as a support layer.

Table 1. Physical properties and chemical structures of the NWFs used as support materials for PVDF membranes.

Non-woven Fabric	**Material**	**Thickness/ μm**	**Permeability (μms^{-1})**	**Repeating unit of the polymer structure**
NWF 1	Polyester	95	162000	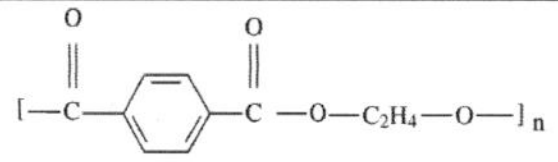
NWF 2	Polyester	134	2500	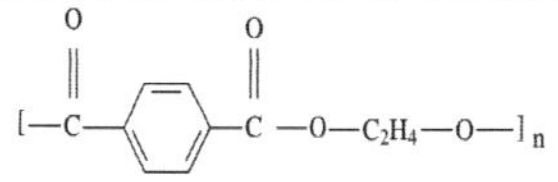

2. Experimental

2.1 Materials

PVDF resin, *N*-methyl-2-pyrrolidone (NMP) solvent, polyvinylpyrrolidone (PVP) with molecular weight of 29 KDa and HA powder were purchased from Sigma Aldrich, South Africa. The non-woven fabrics were purchased from Ahlstrom Co and Hirose Co in the USA. All chemicals were used without further purification. The physical-chemical properties of the NWFs are given in Table 1.

2.2 Membrane preparation

A 17 wt % of PVDF resin was dissolved in NMP with stirring over a period of 24 h to form a PVDF viscous homogenous solution. The PVP additive was varied between 3 wt %, 5 wt % and 10 wt % respectively. The resulting solution was allowed to stand for 24 h in order to allow the gas bubbles generated during the reaction to settle. The solutions were subsequently cast on the NWFs using a casting knife with a blade height set at 200 μm.

2.3 Membrane characterization

Water contact angles of the membranes were measured using a sessile drop method using a Data Physics optical goniometer, which is a contact angle measuring instrument with the droplet size controlled using a Gilmont syringe. Deionised water was used for analysis. The advancing angle was measured as water droplet was added spreading over the membrane surface. Droplets from the syringe were directed to come into contact with the membrane at several different points on each membrane sample to obtain a series of contact angle pairs. All measurements were executed at room temperature. The surface and cross-sectional analysis of the membranes were carried out using a JEOL–JSM-7500F scanning electron microscope using an energy source of 200 kV. The samples were first cryogenically fractured in liquid nitrogen. They were then securely mounted on aluminium stubs and coated with gold, in order to induce their capacity to conduct. Membranes were ultimately studied in order to understand the morphology of the bare membranes relative to the modified ones.

2.4 Membrane rejection studies

The fabricated PVDF NWF supported membranes were assembled in the cells of a cross flow unit. The membranes were then compacted at a pressure of 344.7 KPa using deionised water for 4 h. The water flux was then determined for all the membranes in the cross flow system. In this work, a 5 litre of 20 mg/L solutions of humic acid (HA) was prepared to be used in rejection studies. The pH of the solution was adjusted to be between 8 and 9 using NaOH to ensure complete dissolution of HA (HA is insoluble at lower pH values). The concentration of HA was kept at 20 mg/L with the aim of simulating the HA concentration found in natural surface waters. The flux of the membranes was determined at a pressure of 137.9 KPa by noting the time taken to collect 20 ml of the permeate volume, which was varied over 30 min intervals. The low pressure was used to conserve energy. A typical sea water reverse osmosis plant uses pressure in the region of 6000 KPa to 8000 KPa [14]. As membranes begin to foul the high pressure demand increases in order in to drive water across the membranes.

The concentration of the collected feed and permeate for each of the membranes was determined quantitatively using a An Apollo 9000 TOC Analyzer by TEKMAR DOHRMANN. The instrument uses the High Temperature Combustion–Infrared Method–5310B, which is more accurate than the UV-Persulfate method. The instrument was calibrated using a series of Potassium Hydrogen Phthalate (KHP) standard samples. The calibration curve was used to calculate the TOC values of unknown samples. These concentrations were then used to calculate percentage rejections (% R) of each membrane.

3. Results and Discussion

3.1 Contact angle measurements

The contact angles of the PVDF membranes supported on two NWFs are given in figure 1. A declining trend in contact angles was observed with the addition of PVP. Membrane contact angles declined sharply as the concentration of PVP was increased from 3 wt % to 10 wt % which suggested improved hydrophilicity of the membranes, thus improving their water flux properties. There were no noticeable differences between the supported membranes with those not supported by the NWFs. There effect of PVP addition was similar with all the membranes. An explanation to this observation was that the contact angle measurements were taken on the membrane surface. Since the NWF was located below the membrane surface, it did not make a contribution to the contact angle values. The contact angle graphs of the unsupported PVDF membranes (figure 1 b) illustrate the declining contact angle values, which showed that even the unsupported membranes became more hydrophilic with the increase in PVP concentration. The incorporation of PVP has a similar effect of making the membrane hydrophilic even if the membrane was not supported by NWFs.

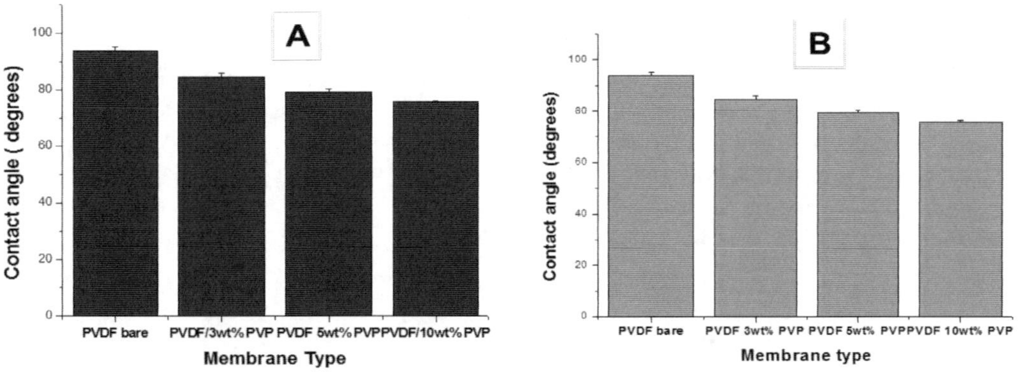

Figure 1. Surface water contact angles of PVDF membranes supported on NWF 1 (A) and unsupported PVDF membranes (B).

3.2 Scanning electron microscopy analysis

SEM surface and cross-sectional morphologies of the PVDF membranes supported on NWF 1 are given in figure 2. The SEM micrographs gave information about size, shape, and orientation of the pores present in the synthesised PVDF membranes. The bare PVDF membranes (i.e. without PVP) on the NWFs were found to comprise considerably smaller surface pores of about 10 nm and their cross-sectional view possessed a broad "finger-like" shape (figure 2 a and b). The bare membranes could be distinguished from the PVP modified membranes due to their increase pore sizes and asymmetric properties. When the concentration of the PVP was increased to 10 wt% in the casting solution, the size of the pores on the membrane surface increased to about 35 nm and the finger-like pores (macrovoids) on the membrane cross-section increased correspondingly and appeared to form asymmetric orientation when compared to relatively small macrovoids in the bare membranes (figure 2 c and d). Similar results were obtained when NWF2 was used as a support.

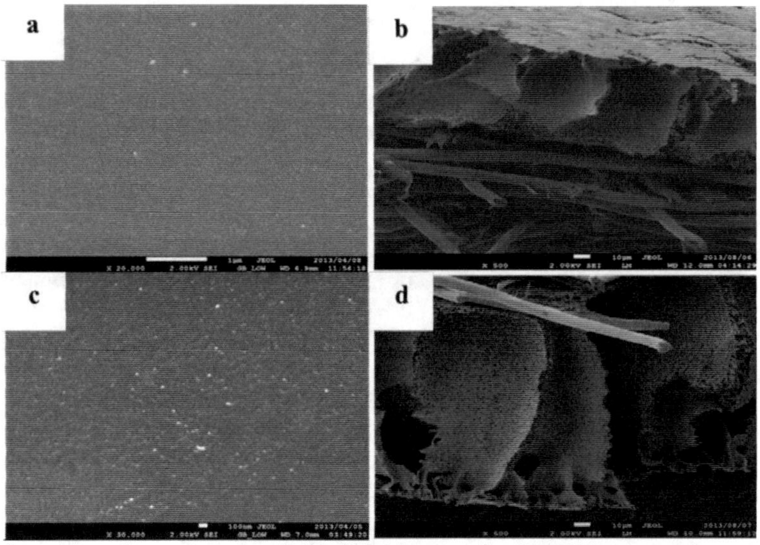

Figure 2. Surface (a) and cross-sectional (b) SEM images for PVDF bare membranes supported on NWF1 and surface (c) and cross-sectional (d) SEM images for PVDF membranes with 10 wt% PVP supported on NWF 1.

IOP Conf. Series: Materials Science and Engineering **64** (2014) 012036 doi:10.1088/1757-899X/64/1/012036

3.3 HA membrane rejection studies

The PVDF membranes supported on the polyester NWFs were all found to effectively reject HA from water. The highest rejection recorded for was 97% for PVDF membranes supported on NWF1 (figure 1) compared to the 95% for membranes supported on NWF 2 (data not shown). Incremental addition of PVP was found not to be consistent with rejections for most of the membranes, as shown in figure 1. It was presumed that this variation could be due to the fact that reject was not only influenced by pore size but also by interaction of the HA with the membrane surface.

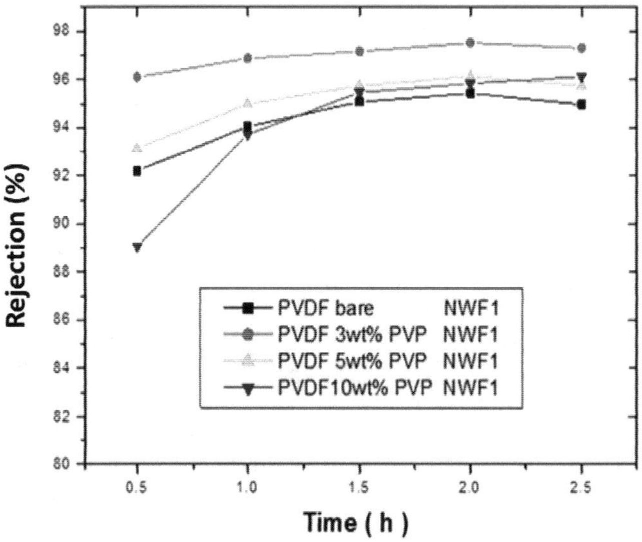

Figure 3. HA rejection for PVDF bare and PVP modified membranes supported on NWF1.

4. Conclusion

The effect of using NWFs as supports for PVDF membranes fabricated using the phase inversion method has been studied. Contact angle analysis of the membrane surfaces showed that the addition of PVP enhances hydrophilicity. The pore sizes of the membranes also increased with increase in PVP concentration. The PVDF membranes supported on NWF 1 all gave higher humic acid rejections compared to NWF 2 due to the permeability and thickness properties of NWF1. However the key results on the tensile strength of the membranes demonstrated that the use of NWF 1 and NWF 2 as supports provided advantages that could allow modification of the thin film PDVF membranes without compromising the strength. The tensile strength of the unsupported PDVF membranes decreased with increase in PVP concentration while the strength was not affected when the membranes were supported on the NWFs. The interactions between the PVDF membranes and NWFs were found to be good thus providing materials with robustness and flexibility that is desired for membrane materials.

Acknowledgement

The authors would like to acknowledge the DST/MINTEK Nanotechnology Innovation Centre in South Africa and the Centre for Nanomaterials Science Research (CNSR) at the University of Johannesburg for financial support.

IOP Conf. Series: Materials Science and Engineering **64** (2014) 012036 doi:10.1088/1757-899X/64/1/012036

References

1. Kilduff J E, Karanfil T and Weber W J 1996 *Environ. Sci. Technol.* **30** 1344
2. Krasner S W and Weinberg H S 2006 *Environ. Sci.Technol.* **40** 7175
3. Zularisam A W, Ismail A F and Salim R 2006 *Desalination* **194** 211
4. Domany Z, Galambos I, Vatai G and Bekassy-Molnar E 2002 *Desalination* **145** 333
5. LaKind J S, Richardson S D and Blount B C 2010 *Environ. Sci. Technol.* **44** 3205
6. Bellar T A, Lichtenberg J J, Kroner R C 1974 *JAWWA* **66** 703.
7. Krasner S W, McGuire M J, Jacangelo J G, Patania N L, Reagan K M and Aieta E M 1989 *JAWWA* **81** 41
8. Furlong J and Wallace B 2003 Application Note, Tekmar-Dohrmann, Mason, OH, USA
9. Wallace B 2003 Application Note - Teledyne Instruments, Tekmar-Dohrmann, Mason, OH, USA
11. Adams F V, Nxumalo E N, Krause R W M, Hoek E M V and Mamba B B 2012 *J. Membr. Sci.* **405-406** 291
12. Hutten I M 2007 *Handbook of nonwoven filter media*, Butterworth-Heinemann, Oxford, UK
13. Ren X, Shon H K, Jang N, Lee Y, Bae M, Lee J, Cho K and Kim I S 2010 *Water Res.* **44** 751
14. Miyano T, Matsuura T and Sourirajan S 1993 *Chem. Eng. Commun.* **119** 23

IOP Conf. Series: Materials Science and Engineering **64** (2014) 012037 doi:10.1088/1757-899X/64/1/012037

Preparation and characterization of ultra-hydrophobic calcium carbonate nanoparticles

A Barhoum[1,2]*, S M El-Sheikh[3], F Morsy[2], S El-Sherbiny[2], F Reniers[4], T Dufour[4], M P Delplancke[5], G Van Assche[1], H Rahier[1]

[1]Department of Materials and Chemistry, Vrije Universiteit Brussel, Pleinlaan 2, 1050 Brussels, Belgium
[2]Department of Chemistry, Helwan University, 11795 Helwan, Egypt
[3]Nanostructured Materials Division, Advanced Materials Department, CMRDI, P. O. Box 87, Helwan, 11421, Cairo, Egypt
[4]Analytical and Interfacial Chemistry, Université Libre de Bruxelles (ULB), Brussels, Belgium
[5]Department 4MAT, Universite Libre de Bruxelles, 50 avenue F.D. Roosevelt, 1050 Bruxelles, Belgium

E-mail: ahmed.abdelrasoul@vub.ac.be

Abstract. Anionic surfactants based on fatty acids are usually used to modify the particle surface properties of $CaCO_3$ with the aim to enhance its dispersion and compatibility with polymer matrices. In this study sodium oleate was used for the preparation of ultra-hydrophobic $CaCO_3$ nanoparticles using a wet carbonation route. The effect of sodium oleate on the characteristics, particle size, morphology, surface potential, thermal decomposition and hydrophobicity of $CaCO_3$, was investigated using X-ray diffraction (XRD), Fourier transform infrared spectroscopy (FTIR), transmission electron microscopy (TEM), Zeta potential, thermogravimetric analysis (TGA) and water contact angle measurement (WCA). The results showed that the addition of 2 wt% sodium oleate helps in reducing the particle size from 2 μm length scalenohedral particles to 45 nm rhombohedral particles and modifying of the hydrophobic property of $CaCO_3$.

1. Introduction

Calcium carbonate nanoparticles have received much attention owing to its wide industrial applications such as paper, rubber, plastics, and paints industries. $CaCO_3$ produces at industrial-scale via wet carbonation route by bubbling CO_2 gas through an aqueous $Ca(OH)_2$ suspension [1]. This rout is industrially used because of the availability of its raw materials, high yield as well as simplicity and low cost of production. Ultra-fine $CaCO_3$ could also be produced using the wet carbonation by adding surfactants and controlling the precipitation conditions, such as initial $Ca(OH)_2$ concentration, CO_2 flow rate, temperature, pH and stirring rate [2]. Anionic surfactants such as fatty acids and their sodium salts are widely used in industry for economic reasons. Such surfactants influence the $CaCO_3$ nucleation, crystal growth, grain shapes, and consequently, control the formation of crystal phases that are not usually stabilized under natural environments. They can bind with certain crystal planes during crystal growth, thereby modifying the particles morphology [3, 4]. The knowledge on the minimum

quantity of surfactant that could produce the desired precipitate is of importance, because the excess of surfactant could deteriorate the physical and chemical properties of the hosting matrix [5].
The goal of the present study is to prepare ultra-hydrophobic $CaCO_3$ nanoparticles using adjusted wet carbonation rout. Sodium oleate was used during preparation to inhibit crystal growth and modify the hydrophobic property of the $CaCO_3$. The effect of sodium oleate on the particle size, morphology, surface potential, thermal decomposition and hydrophobicity were investigated.

2. Experimental

2.1. Materials

Analytical grade sodium oleate ($C_{18}H_{33}NaO_2$, 82+ % oleic acid, Sigma), calcium oxide (CaO, 97+ % on dry substance, Acros Organics), carbon dioxide gas (CO_2 gas, 99+ %, Air Liquide) and monodistilled water were used to prepare $CaCO_3$ particles.

2.2. Method

Unmodified $CaCO_3$ was prepared from CaO lime and CO_2 gas via the wet carbonation rout. The preparation conditions, initial CaO concentration, CO_2 flow rate and temperature were adjusted at 1 M CaO, 100 $mL.min^{-1}$ CO_2 flow rate and 25°C, respectively. For oleate-modified $CaCO_3$, 2 wt% sodium oleate based on the theoretical $CaCO_3$ weight was added to the monodistilled water. In practices, the preparation of $CaCO_3$ was carried out in a plastic flask. The required amount of CaO reagent was slaked in the monodistilled water containing sodium oleate then the obtained lime was cooled to 25°C. After cooling, the pure CO_2 gas was blown into the lime milk from the bottom of the plastic bottle under vigorous stirring. The pH value of the reaction solution was monitored online using a pH meter (Jenway 3305). When the pH value decreased to 9 the reaction was completed and then the CO_2 flow was stopped. The slurry was washed, filtered and dried at 120°C in an oven for 24 hrs to obtain the $CaCO_3$ powder.

2.3. Characterization

The crystalline phase of the prepared $CaCO_3$ was characterized using X-ray diffraction (XRD, Bruker axs D8, Germany) with Cu-Kα(λ= 1.5406 Å) radiation. The morphology information was investigated using a high resolution transmission electron microscope (TEM, Jeol, JEM-2010, Japan). Bonding structures were analyzed using Fourier transform infrared spectrometer (FTIR-460plus, JASCO model 6100, Japan). Zeta surface potential of the $CaCO_3$ particles in suspension was measured using a Zeta meter (Malvern Instrument Zetasizer 2000). The thermal decomposition and weight loss was performed by thermogravimetric analysis (TA instrument, TGA Q5000, USA). The samples were dried isothermally at 55°C for 20 min before heating from 55 to 1000°C at a heating rate of 10 $°C.min^{-1}$ under Air atmosphere (50 $mL.min^{-1}$). The water contact angle (WCA) was measured with a Kruss DSA-100 contact angle analyzer. The measurements were performed on the prepared powders compressed into discs using 5 μL water droplet volume and the contacted angle was determined from the profile of the droplets that were fully separated from the pump syringe needle tip. The discs were prepared by compression under controlled conditions: 100 mg of the sample and a pressure of 107 Pa, in a typical IR die.

3. Results

X-ray diffraction patterns of the unmodified and oleate-modified $CaCO_3$ are shown in figure 1. XRD results indicate that the overall crystalline structure and phase purity of the $CaCO_3$ particles were obtained. XRD pattern exhibits the characteristic reflection of rhombohedral calcite (d-spacings/Å: 3.85, 3.04, 2.85, 2.49, 2.28; corresponding to hkl: 012, 104, 006, 110, 113, respectively) [6]. All the relative sharp peaks could be indexed as the typical calcite phase of $CaCO_3$ (JCPDS 88-1808). No characteristic peaks of other impurities were observed, which indicated that the products have high purity.

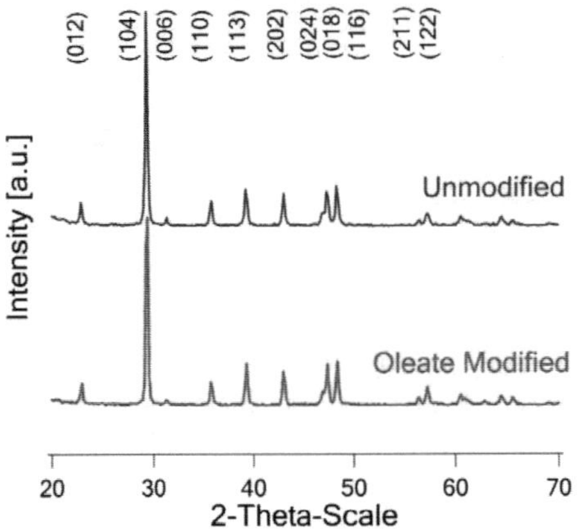

Figure 1. XRD patterns of the prepared calcium carbonates

FT-IR spectra of the unmodified and oleate-modified $CaCO_3$ are shown in figure 2. The characteristic absorption peaks of calcite $CaCO_3$ are stretching vibrations of the C–O approximately at 1420 cm^{-1} and bending vibrations of the C–O approximately at 872 and 710 cm^{-1}. The combination of the three peaks at 1420, 872 and 710 cm^{-1} appears at 2515 cm^{-1}. The broad absorption peaks at 3445 cm^{-1} are assigned to stretching vibration and asymmetric stretching vibration of O–H bond and it can be attributed to the presence of absorbed water and hydroxyl groups on the surface of $CaCO_3$ particles. The peaks around 2355 cm^{-1} are attributed to carbon dioxide in the atmosphere. Comparing with the literature [6] the characteristic IR spectra indicate that all the prepared samples are typically calcite crystals. Comparing the unmodified and oleate-modified samples, a shoulder at 1615 cm^{-1} and peaks at 2955, 2925 and 2885 cm^{-1}, are observed for oleate-modified $CaCO_3$. The shoulder at 1615 cm^{-1} corresponds to the appearance of a carboxylic salt, indicating that oleate has been attached to the surface of $CaCO_3$ via an ionic bond. The peaks at 2955, 2925 and 2885 cm^{-1}, are ascribed to the long alkyl chain of oleate (H–C–H stretching region) and prove the presence of oleate at the surface of $CaCO_3$ [7]. The XRD and FT-IR results indicate that no characteristic absorption of other phases is observed meaning that sodium oleate has no significant effect on $CaCO_3$ polymorphism.

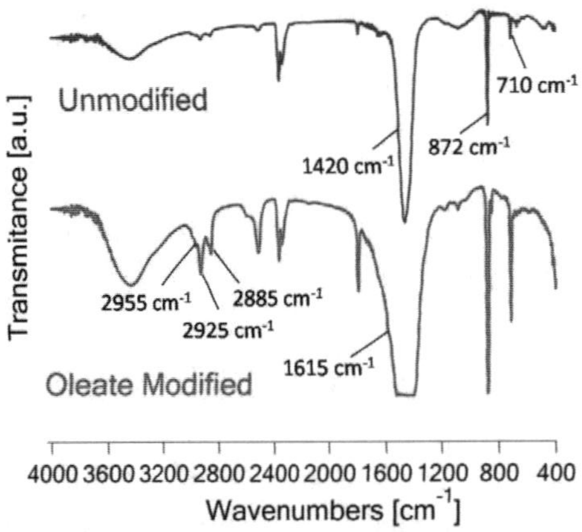

Figure 2. FT-IR spectra of the of the prepared calcium carbonates

Zeta potentials of unmodified and oleate-modified $CaCO_3$ were determined are shown in figure 3. The Zeta potential measurements show that the particle surface charge decreases from -14.5 to -22 mV with addition of 2 wt% sodium oleate. The change in surface potential of $CaCO_3$ is due to adsorption of oleate on the surface of the precipitated particles.

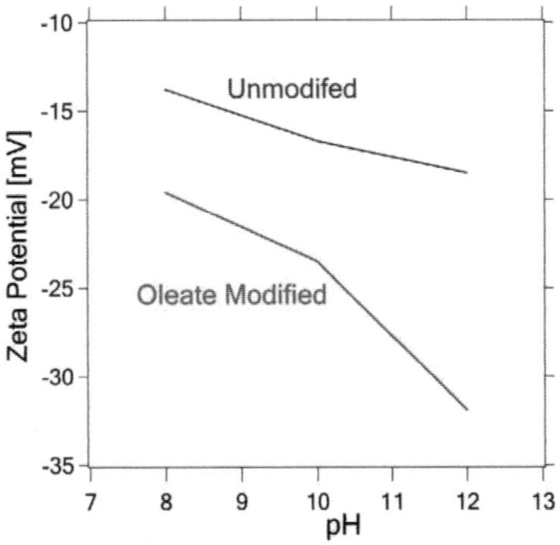

Figure 3. Zeta potentials of the prepared calcium carbonates

Figure 4 shows TEM and water contact angle images of unmodified and oleate-modified $CaCO_3$. The results show that the use of 1 M CaO concentration and 100 mL.min^{-1} forms 1.5-2.5 μm length scalenohedral particles. The average water contact angle (WCA) of these particles is about 25°. Upon addition of 2 wt% oleate, the average particle size of $CaCO_3$ reduced to 45 nm rhombohedral particles and the average water contact angle (WCA) increased to 130°.

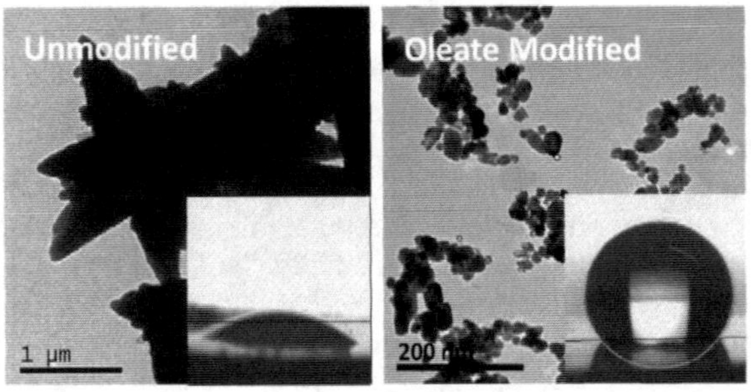

Figure 4. TEM and WCA images of the prepared calcium carbonates

Figure 5 shows the thermal decomposition curves of the unmodified and oleate-modified $CaCO_3$. The mass loss measurements revealed that the amount of adsorbed oleate reaches up to 75 % and the temperature at the maximum rate of decomposition of the $CaCO_3$ decreased from 757 to 715°C with addition of 2 wt% sodium oleate.

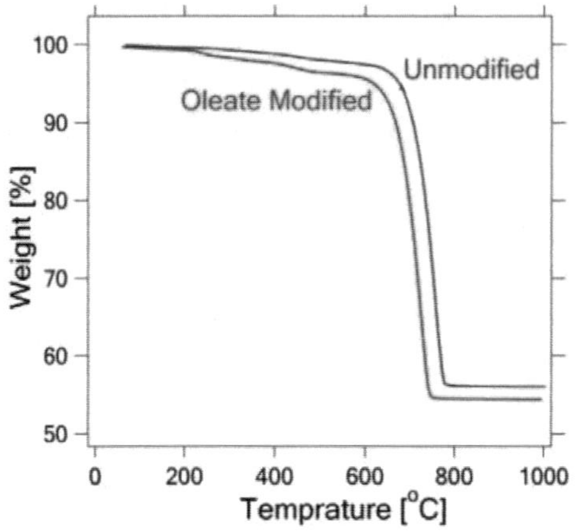

Figure 5. TGA curves of the prepared calcium carbonates

4. Discussion

Sodium oleate plays an important role on preparation and surface modification of the $CaCO_3$ particles. The data extracted from XRD and TEM reveal those oleates can successfully inhibit crystal growth and reducing the particle size of $CaCO_3$. The FT-IR, Zeta potential, TGA along with the WCA measurement confirm that sodium oleate can successfully adsorb on the $CaCO_3$ particles surface and alternates the particles surface properties from hydrophilic to ultra-hydrophobic. In absence of surfactant and the use of 1 M CaO, and low CO_2 flow rate of 100 mL.min^{-1} forms 2 μm length scalenohedral particles. Excess of Ca^{2+} in the bulk solution and a low CO_2 flow rate elongate the reaction and growth time of precipitated particles and lead to microsize particle. Up on addition of 2 wt% sodium oleate, the particle size is reduced. This could be explained by that sodium oleates can be situated at the gas–liquid interface. Consequently they increase the stability of CO_2 bubbles and prevent their aggregation. The maintenance of CO_2 bubbles enhances the mass transfer of CO_2 into solution and increases the CO_3^{2-}:Ca^{2+} ionic ratio and the nucleation rate. Moreover, the ability of oleate's carboxylic groups ($-COO^-$) to bind effectively to the Ca^{2+} ions on the crystals surface with ionic bonds can efficiently neutralize the positive charges on the crystals surface, decrease the surface charge to more negative values and inhibit the crystals growth [8]. The decrease of the particle size of the oleate-modified $CaCO_3$ causes the observed decrease of the thermal stability (temperature at maximum rate of decomposition) compared with unmodified $CaCO_3$ [9].

5. Conclusion

Ultra-Hydrophobic $CaCO_3$ particles were directly prepared via carbonation of CaO lime slurry in the presence of sodium oleate at room temperature. The sodium oleate solution was used in a precipitation process to control the particle size, morphology and to modify the surface characteristics of $CaCO_3$ particles simultaneously. The results from this work prove that coating of oleate on $CaCO_3$ particles was successfully done. Addition of oleate helps in reducing the particle size from 2 μm length scalenohedral particles to 45 nm rhombohedral particles and the water contact angle increases from 25° to 130° with addition of 2wt% oleate.

References

[1] Brecevic L and Kralj D 2007 On calcium carbonates: from fundamental research to application *Croat. Chem. Acta.* **80** 467.

[2] Ukrainczyk M, Kontrec J, Babic-Ivancic V, Brecevic L and Kralj D 2007 Experimental design approach to calcium carbonate precipitation in a semicontinuous process *Powder Technol.* **171** 192.

[3] Ukrainczyk M, Kontrec J and Kralj D 2009 Precipitation of different calcite crystal morphologies in the presence of sodium stearate *J. Colloid Interface Sci.* **329** 89.

[4] Wang Y L, Eli W M J, Zhang L T, Gao H Y, Liu Y F and Li P J 2010 A new method for surface modification of nano-$CaCO_3$ and nano-Al_2O_3 at room temperature *Adv. Powder Technol.* **21** 203.

[5] Sheng Y, Zhou B, Wang C Y, Zhao X, Deng Y H and Wang Z C 2006 In situ preparation of hydrophobic $CaCO_3$ in the presence of sodium oleate *Appl. Surf. Sci.* **253** 1983.

[6] Wu G H, Wang Y J, Zhu S L and Wang J D 2007 Preparation of ultrafine calcium carbonate particles with micropore dispersion method *Powder Technol.* **172** 82.

[7] Wang C Y, Piao C, Zhai X L, Hickman F N and Li J 2010 Synthesis and character of super-hydrophobic CaCO3 powder in situ *Powder Technol.* **200** 84.

[8] Lanzon M, Garrido A and Garcia-Ruiz P A 2011 Stabilization of sodium oleate as calcium oleate in cement-based mortars made with limestone fillers *Constr. Build. Mater.* **25** 1001.

[9] Chen X, Luo W and Liao Z 2006 Microstructure and thermal decomposition kinetics of nanocrystalline calcite *Nanoscience* **11** 311.

Synthesis of (Au)Ag core-shell nanocomposite in the water-ethanol mixture and its optical properties

A V Abakshonok[1], A Yu Panarin[2], V E Agabekov[1], A N Eryomin[1] and S N Terekhov[2]
[1] Institute of Chemistry of New Materials of NAS of Belarus, F. Skarina Sreet 36, 220141, Minsk, Belarus
[2] B.I. Stepanov Institute of Physics of NAS of Belarus, Nezalezhnasti Ave. 68, 220072, Minsk, Belarus

E-mail: nura2007@tut.by

Abstract. The technique of synthesis of (Au)Ag core-shell bimetallic nanocomposite was developed. Gold seed nanoparticles (NPs) were obtained by $HAuCl_4$ reduction with sodium citrate at ultrasonic treatment during 3 hours in a mixture of water - ethanol (1:1). Then, the surface of gold NPs was modified by silver. In the presence of polyvinylpyrrolidone (PVP) K30 (Mw ~ 24000) and K90 (Mw ~ 360000) the core-shell (Au)Ag NPs of spherical shape were formed. They are characterized by aggregate stability and well-defined absorption maximum at 400-514 nm. Composite (Au)Ag, prepared in the solution without a polymer or in the presence of carboxymethylcellulose (CMC), sodium polystyrene sulfonate (PSS), dextran T100 and T500, had a broad band plasmon resonance in the whole range of visible spectrum. The ability to use the (Au)Ag core-shell nanoparticles in absorption nanospectroscopy based on the phenomenon of plasmon resonance energy transfer (PRET) was evaluated. In the presence of 0,1-2,0 μM of water-soluble cationic Cu (II) -5,10,15,20-tetrakis (4-N-methyl pyridinium) porphyrin (CuTMPyP4) distinct dips due to plasmon quenching matched the absorption maximum of CuTMPyP4 were detected in the resonant scattering spectrum of (Au)Ag solution.

1. Introduction

Gold, silver and their composite nanoparticles are characterized by intense absorption and scattering due to surface plasmon resonance. The size- and shape-dependent unusual optical characteristics allow to use these particles in the various analytical methods [1-3]. The techniques can be based on shift of the plasmon resonance maximum depending on dielectric properties of environment, change the resonance light scattering in the dark-field microscopy as well as on the appearance of dips in the NPs resonant scattering spectra at a wavelength coinciding with optical absorption maximum of the test compound (the phenomenon of plasmon resonance energy transfer - PRET). Optical absorption spectroscopy requires relatively high concentrations of the compounds and has a low spatial resolution. In the case of PRET-based absorption nanospectroscopy, resonance energy migration is caused by a direct transfer, so PRET spectra can be recorded at lower concentrations of defined molecules [2, 3].

The purpose of the study is the development of method for the synthesis of (Au)Ag bimetallic core-shell NPs for their usage in absorption spectroscopy based on PRET.

2. Experimental section

Hydrogen tetrachloroaurate ($HAuCl_4$) and the polymers from Aldrich carboxymethylcellulose (CMC; ~ 250 kD), sodium polystyrene sulfonate (PSS; 70 kD), polyvinylpyrrolidone (PVP) K30 (Mw ~ 24000) and K90 (Mw ~ 360000), also dextrans T110 (Fluka, Switzerland) and T500 (Loba-Chemie, Austria) were used. Chemicals $AgNO_3$, $NaHCO_3$, sodium citrate, glucose and hydrogen peroxide ("Reahim", Russia), cationic Cu(II)-5,10,15,20-tetrakis (4-N-methyl pyridinium) porphyrin (CuTMPyP4) and anionic Fe tetra (sulfonatophenyl) porphyrin (FeTPPS) were of analytical grade and were used as received.

The ultrasonic bath Elmasonic S 30 H (Germany) with a frequency of 37 kHz, the effective power of 60 W and a maximum peak power of 240 watts was used throughout the experiments. Extinction spectra were recorded with a spectrofluorimeter CM2203 («SOLAR», Belarus). Diameter of gold NPs was calculated by the equation

$$d\ (\text{nm}) = 2.02X - 23.1,\ X = \lambda_{max}\ (\text{nm}) - 500 \geq 20 \qquad [4].$$

The size and shape of the gold NPs and composite (Au)Ag were assessed by transmission electron microscopy («JIM/100SX», Japan). The zeta potential and the z-Average size of the particles were determined using a Zetasizer Nano («Malvern Instruments», UK).

Resonant scattering spectra of gold NPs and the composite (Au)Ag colloidal solutions were recorded with a hand-made experimental setup [5]. To minimize the effect of reabsorption of the scattered light on spectra strongly deluted solutions were used. The optical path length of the exciting light was 5 mm. Also, the beam passed about 1mm near the wall of the cell. 90 ° scattering geometry was used. In this case, the reflected light did not enter into the detector system, and only scattered radiation was recorded.

Composite (Au)Ag was fabricated in two steps: at first seed gold NPs were obtained, and then they were modified with silver. Gold core NPs were synthesized in a mixture of water - ethanol (1:1) containing 0.4 mM $HAuCl_4$, 1.0 mM sodium citrate and 0.15 mg/ml PVP. Gold NPs were coated with silver by adding to the colloidal gold solution (the final dilution of gold NP 1.33 times) aliquots 10.0 mM $NaHCO_3$, 50.0 mM glucose, 5.0 mM hydrogen peroxide and 0.25 mM $AgNO_3$. Seed gold NPs and silver shell were synthesized at ultrasonic treatment during 3 hours at the temperature varied from 16 to 62 °C.

3. Results and discussion

Number of seed gold NPs rises proportionally to an increase of $HAuCl_4$ concentration, reaching a maximum at 0.4 mM (table 1). Size of gold NPs depends on the ratio of [sodium citrate] / [$HAuCl_4$]. It is well known that the concentration of sodium citrate influences on the size of gold NPs: the higher it is, the smaller the particles are formed [6]. This regularity is consistent with the obtained our data only for a concentration of $HAuCl_4$ no more than 0.2 mM. The probability of formation of smaller particles increases at a higher degree of supersaturation of the solution with $HAuCl_4$ and relatively low rate of germs condensation of the new phase. Gold sol prepared in a solution with 0.4 mM $HAuCl_4$ had the highest intensity of resonant light scattering. Therefore, the further synthesis of seed gold NPs was performed in an aqueous solution containing 0.4 mM $HAuCl_4$ and 1.0 mM sodium citrate. Gold NPs synthesized under these conditions had a spherical shape and formed aggregates about 40-60 nm (figure 1A).

The size of gold seeds and their tendency to agglomerate is influenced by the admixture of polymers in the reaction solution. In the presence of CMC gold NPs with diameter of 22-27 nm are obtained (figure 1B). In the case of PVP K30 (or K90) addition seeds of 7-10 nm occur (figure 1C). Gold sol formed both in the absence and in the presence of the aforementioned polymers, was characterized by colloidal stability and pronounced absorption maximum (figure 2). In the presence of dextran T110 (or T 500) a gray-violet gold sol was obtained, in which agglomeration of gold NPs was found. Its absorption spectrum did not show distinct bands of surface plasmon resonance (figure 2, spectrum 4). Furthermore, gold sols containing dextran (0.15 mg / ml) were unstable due to the poor solubility of these polymers in water – ethanol mixture.

Table 1. Characteristics of gold NPs obtained at different concentrations of $HAuCl_4$ in water - ethanol solution (1:1).

$[HAuCl_4]$, мМ	$\lambda_{макс}$, nm	$D_{макс}$	Diameter, nm
0.05	553	0.24	84
0.10	562	0.36	102
0.15	575	0.61	128
0.20	581	0.88	140
0.30	563	1.20	104
0.40	547	1.68	71

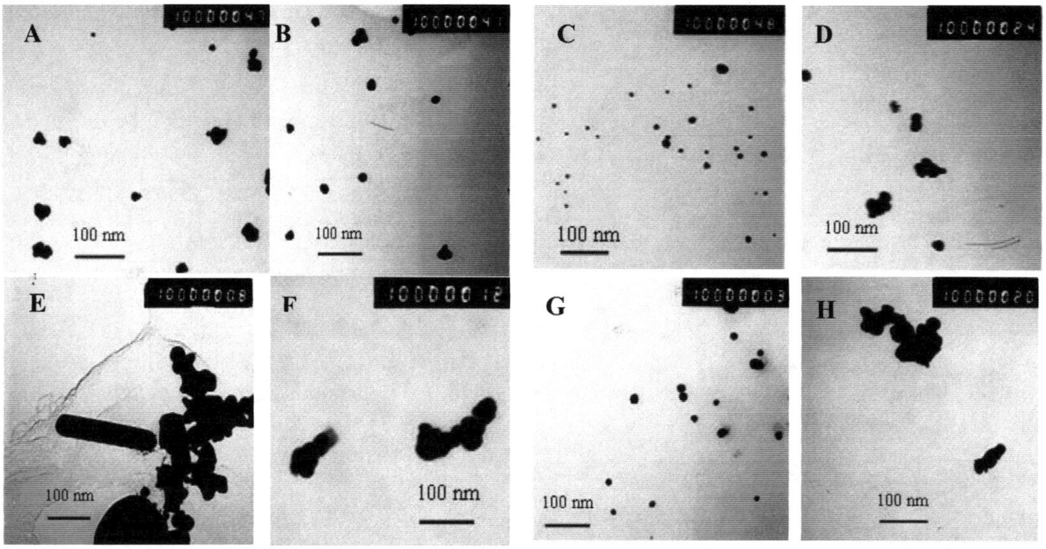

Figure 1. TEM micrographs of the gold NPs (A-D), synthesized in the absence of the polymer (A) and in the presence of CMC (B), PVP K30 (C), dextran T110 (D) and the composite (Au)Ag (E-H), prepared in medium without the polymeric additive (E), in the presence of CMC (F), PVP K30 (G), dextran T110 (H). Magnification is 100 000 times.

The surface of gold NPs was coated with silver in aqueous solution of 10 mM $NaHCO_3$, using glucose for the gradual reduction of $AgNO_3$, and also hydrogen peroxide to dissolve small clusters of silver. In the absence of polymer additives composite (Au)Ag is formed, which shows a broad absorption in the whole range of visible spectrum (300-1100 nm) without well-defined plasmon resonance band (figure 3, spectrum 1). Sol has a gray-violet color. It consists of the rod-shaped particles (figure 1E). Moreover, the particles agglomerate to form a ribbon like structures. Absorption of the composite increases in direct proportion to the concentration of $AgNO_3$ in the reaction solution. The (Au)Ag composite, obtained in a medium with 0.2 and 0.3 mM $AgNO_3$, had the highest intensity of light scattering.

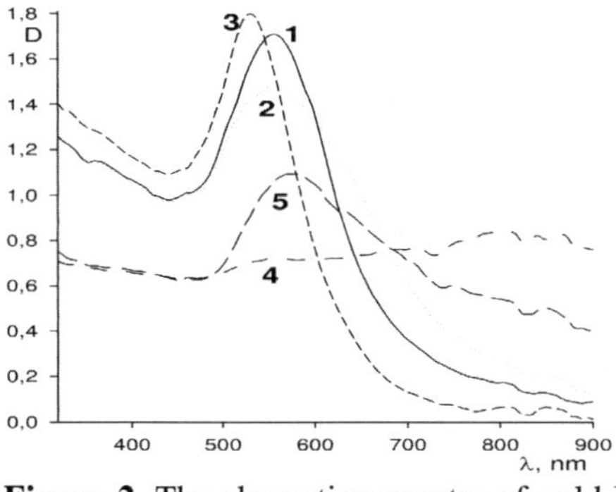

Figure 2. The absorption spectra of gold NPs synthesized in the medium without polymer (1) and in the presence of CMC (2), PVP K90 (3), dextran T110 (4) and PSS (5).

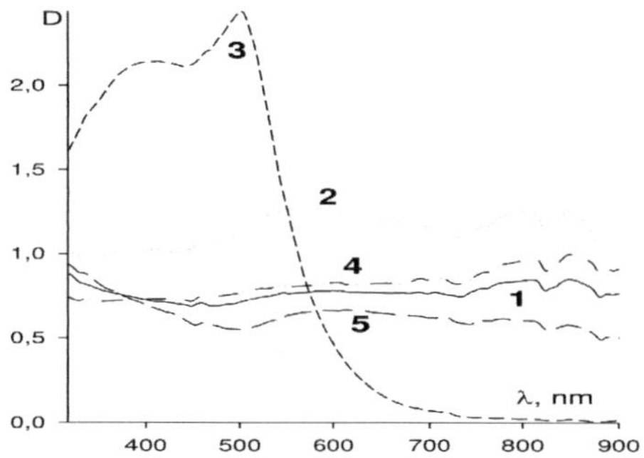

Figure 3. Absorption spectra of the composite (Au)Ag, synthesized in the medium without polymer (1) and in the presence of CMC (2), PVP K90 (3), dextran T110 (4) and PSS (5).

Ribbon like structures are formed in the presence in the reaction solution of CMC (figure 1F), PSS and dextran (figure 1H). Composite (Au)Ag prepared in the presence of these polymers has a low colloidal stability and is characterized by a broad absorption band in the visible region (figure 3, spectrum 2, 4 and 5). Only in the presence of PVP K30 or PVP K90 (0.11 mg / mL) spherical shape composite (Au)Ag is formed with particles size 20-26 nm (figure 1G). It is stable and has well - defined maximum of absorption (figure 3, spectrum 3). Increasing diameter of the composite particles as compared with germinal gold NPs and broad absorption band at 400 nm are indication of strong silver component in the composite (Au)Ag, which determines its optical properties.

Table 2. The hydrodynamic characteristics of the composite (Au)Ag, synthesized in the presence of different polymers (0.11 mg / ml).

Polymer	Zeta-potential, mV	Conductivity, mS/cm	Z-Average size, nm	Polydispersity
Without polymer	-15,1	0,021	767	0,392
CMC	-39,1	0,213	161	0,274
PVP K30	-33,8	0,117	88	0,262
PVP K90	-21,6	0,239	73	0,365
Dextran T110	-11,9	0,269	285	0,248
Dextran T500	-7,9	0,218	260	0,334
PSS	-29,2	0,285	237	0,297

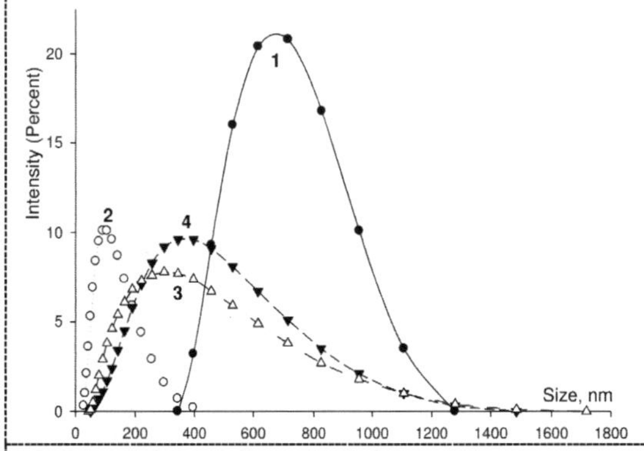

Figure 4. The distribution of the composite (Au)Ag on hydrodynamic size in the medium without the polymeric additives (1) and in the presence of PVP K30 (2), dextran T110 (3) and the PSS (4).

The zeta potential and the hydrodynamic size of the composite substantially depend on the presence of polymer in the reaction medium (table 2, figure 4). In general, the polymers reduce the hydrodynamic radius of the composite (Au)Ag. The largest size of the particles is inherent to the composite obtained in the medium without the polymeric additive and the least - in cases of PVP. This weak cationic polyelectrolyte does not reduce, but rather increases the zeta potential of the composite (Au)Ag as compared with the particles synthesized in a medium without the polymer (тable 2). Apparently, in the presence of PVP structures [PVP-(Au)Ag] are formed, in which polymer molecule forms the core, while negatively charged (Au)Ag particles – the sheath. Such kind of structure is very compact and has a well-defined surface plasmon resonance band at 400-514 nm caused by gold and silver components. Concentration of PVP equal to 0.15-0.20 mg / ml is optimal for the formation of [PVP-(Au)Ag]. The most intense spectrum of light scattering was observed for the composite (Au)Ag, prepared in the presence of PVP with a mass of 360 kDa.

To evaluate the usefulness of the (Au)Ag core-shell nanoparticles in absorption nanospectroscopy, based on PRET resonant scattering spectra of their mixture with cationic/anionic water-soluble porphyrins were measured. Cationic Cu (II) -5,10,15,20-tetrakis (4-N-methyl pyridinium) porphyrin (CuTMPyP4) has a maximum extinction in the visible region at 550 nm (figure 5, spectrum 1), which satisfactory overlaps with peak scattering of composite (Au)Ag (figure 5, spectrum 2) obtained in the medium with PVP (~ 360 kDa). Decrease in the intensity of band scattering at 530 nm and a dips at ~ 560 nm are detected for sol (Au)Ag depending on the concentration of porphyrin (figure 5). This behaviour is caused by the transfer of energy plasmon resonance to CuTMPyP4 molecules. After addition of FeTPPS in the composite sol (Au)Ag, synthesized in the presence of 0,11-0,15 mg / ml PVP of molecular weight ~ 360 kD, dip in the scattering spectrum at 413-415 nm, where porphyrin has an intensive Soret band [7], is found.

Apparently, the differences of light scattering spectra are due to various binding of these two porphyrins with [PVP-(Au) Ag]. Composite (Au)Ag, due to the odd negative charge (table 2), effectively binds cationic CuTMPyP4, but not anionic FeTPPS. Last one predominantly interacts with the cationic PVP, forming the core of [PVP-(Au) Ag], and only indirectly with the particles (Au)Ag, surrounding this core polymer.

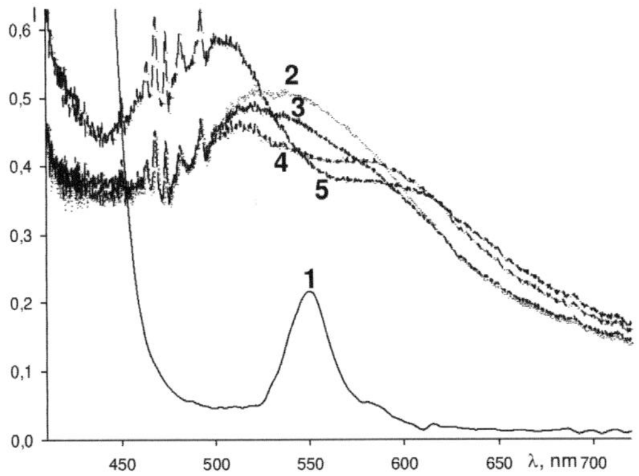

Figure 5. Absorption spectrum of CuTMPyP4 (1) and the scattering spectra of composite (Au)Ag (2), its complex with 0.1 (3), 0.4 (4) and 2.0 (5) μM CuTMPyP4.

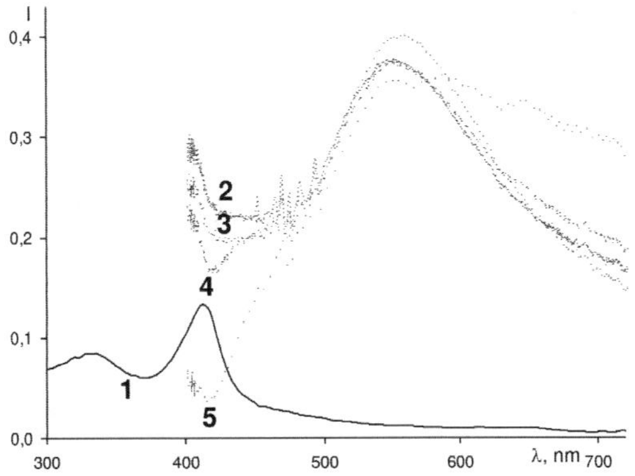

Figure 6. Absorption spectrum of 5 μM FeTPPS (1) and the scattering spectra of composite (Au)Ag (2), its complex with 1 (3), 10 (4) and 100 (5) μM FeTPPS.

4. Conclusion

In the water-ethanol mixture (1:1) (Au)Ag core-shell composite was synthesized in two steps, suitable for use in absorption spectroscopy, based on the effect of plasmon resonance energy transfer. For the course of synthesis of (Au)Ag the presence in the reaction solution of the polymer additive is important. It was established that the PVP - is the most effective additive, which not only promotes the formation (Au)Ag, but the structure [PVP-(Au)Ag], comprising a cationic polymer PVP and negatively charged nanoparticles (Au)Ag.

Colloidal solution of composite [PVP -(Au)Ag], comprising PVP having a molecular weight of ~ 360 kDa , is characterized by intense scattering of light in the 400-800 nm. With increasing concentration of the cationic CuTMPyP4 dip appears in the band of the light scattering, indicating that the porphyrin is directly bound with (Au)Ag. In the case of anionic FeTPPS - in the shortwave minimum of light scattering spectrum reflecting that the FeTPPS is directly bound with cationic PVP.

5. References

[1] Wang J, Kong L, Guo Z, Xua J and Liua J 2010 Synthesis of novel decorated one-dimensional gold nanoparticle and its application in ultrasensitive detection of insecticide *J. Mater. Chem.* **20** 5271–79

[2] Choi Y, Park Y, Kang T and Lee L P 2009 Selective and sensitive detection of metal ions by plasmonic resonance energy transfer-based nanospectroscopy *Nature nanotechnology* **4** 742–46

[3] Liu G L, Long Y, Choi Y, Kang T and Lee L P 2007 Quantized plasmon quenching dips nanospectroscopy via plasmon resonance energy transfer *Nat Method* **4** 1015-17

[4] Khlebtsov N G, Bogatyrev V A, Dykman L A and Khlebtsov B N 2007 Gold plasmon resonant nanostructures for biomedical applications *Russian Nanotechnologies* **2** 69-86

[5] Eryomin A N, Abakshonok A V, Agabekov V E, Zhavnerko G K, Panarin A Yu and Terekhov S N 2013 Spectra of resonant light scattering of gold nanoparticles *Proceedings of the academy of sciences of Belarus. Series of chemical sciences* 55-62

[6] Frens G 1973 Controlled nucleation for the regulation of the particle size in monodisperse gold suspensions *Nature Phys. Sci.* **241** 20-22

[7] Fleischer E B, Palmer J M, Srivastava T S and Chatterjee A 1971 Thermodynamic and Kinetic Properties of an Iron-Porphyrin System *J. Am. Chem Soc.* **93** 3162-67

Characterization of synthesized polyurethane/montmorillonite nanocomposites foams

Farahnaz Ansari [1], Sophia Sachse [2], S. Michalowski[3], Masoud Kavosh [2] Krzysztof Pielichowski [3] and James Njuguna*[1]

[1] Innovation, Design and Sustainability, School of Engineering, Robert Gordon University, Aberdeen, AB10 7GJ, UK
[2] Cranfield University, Bedfordshire, MK43 0AL, UK
[3] Department of Chemistry and Technology of Polymers, Cracow University of Technology, ul. Warszawska 24, 31-155 Kraków, Poland

* Correspondance aurthor E-mail: j.njuguna@rgu.ac.uk

Abstract. Nanophased hybrid composites based on polyurethane/montmorillonite (PU/MMT) have been fabricated. The nanocomposite which was formed by the addition of a polyol premix with 4,4'-diphenylmethane diisocyanate to obtain nanophased polyurethane foams which were then used for fabrication of nanocomposite panels has been shown to have raised strength, stiffness and thermal insulation properties. The nanophased polyurethane foam was characterized by means of scanning electron microscope (SEM), transmission electron microscope (TEM) measurements and X-ray diffraction (XRD). TEM and SEM analysis indicated that nanophased particles are dispersed homogeneously in the polyurethane matrix on the nanometer scale indicating that PU/MMT is an intercalated nanocomposite with a 2-3 nm nanolayer thickness.

1. Introduction

In the last 20 years, there has been a strong interest in the development of polymeric nanocomposites. Nanocomposite has been successfully produced with many polymers including polyamide, polystyrene, polycaprolactone, poly (methyl methacrylate), and epoxy resin and others [1–9]. The most versatile class of polymers is polyurethane (PU) due to its excellent physical properties, e.g. flexibility at low temperature, abrasion resistance, variable hardness, etc. However, PU has the drawbacks of low thermal stability and mechanical strength. A focus of recent efforts to improve its properties has been the addition of inorganic filler (including nanoclays) in order to increase the properties of the polymer matrix [10, 11]. With improvements in the mechanical, thermal, and gas barrier properties of the polymers [12-16], PU nanocomposites are attractive for various applications from aerospace components, automobiles, ships, ballistic vests, sporting goods and helmets through to providing heat shielding, structural stiffness and crash energy management for racing cars, among several others applications.

In general there are two properties of nanocomposites that are distinctly different from other materials: i) large surface area for a given volume and ii) quantum effects [17]. Since many important chemical and physical interactions are governed by surface properties [18, 19], a nanostructured material can have substantially different properties from a bulk scale material of the same

IOP Conf. Series: Materials Science and Engineering **64** (2014) 012039 doi:10.1088/1757-899X/64/1/012039

composition. Therefore this area requires fundamental studies of mechanical, electrical, thermal, optical, and chemical properties along with related research for real applications.

For the variety of practical applications we synthesized PU/MMT and developed nanocomposite preparations [20].As a preliminary to these, this study presents the characterization of the morphology of PU/MMT and the dispersion of nanoparticles by SEM and TEM. The structural and morphological data will be used to prepare well-defined lab materials with known composition that correspond to the currently used industrial materials.

2. Synthesis of nanophased foam

Material and preparation of nanocomposites Polyurethane foam with different weight percentages (up to 10%) of nanoclay was manufactured in four steps at the Department of Chemistry and Technology of Polymers (Cracow University of Technology, Poland). Polyol blend (polyether RF-551) and polyester (T-425R) mixture from Alfasystems, Brzeg Dolny, Poland, was stirred with powdered MMT (Optibent 987, Süd Chemie AG, Moosburg, Germany). Catalyst (N,N-dimethyl cyclohexylamine) water and surfactant (SR-321, Union Carbide, Marietta, GA) were added in order to prepare the polyol premix (component A). N-pentane as a physical blowing agent was added to component A. Component B was polymeric 4,4'- diphenylmethane diisocyanate (PM 200). It was added to component A and the mixture was stirred for 10 seconds with an overhead stirrer. Prepared mixtures were dropped into a mould. All the experiments were performed at ambient temperature of ca. 20 °C.

3. Characterization

3.1. X-ray diffraction

The Powder X-ray diffraction (XRD) θ-2θ studies were performed with scattering angle, θ ranging between 0° and 90° using nickel-filtered Cu Kα ($\lambda = 0.154$ nm) radiation at a scan rate of 1°/min. The XRD pattern of the PU/MMT indicates the presence of predominantly montmorillonite Na-Mg-Al-Si4O11 crystals.

3.2. Scanning electron microscopy (SEM) analysis

The morphology of the samples was investigated by using a FEI XL30 field emission scanning electron microscope. The operating voltage was in the range of 10–20 kV to minimize charging of the sample. The SEM images were not subjected to quantitative analysis, but merely served to gain a visual impression and it uses the reflection properties of surfaces for imaging it should be possible to visualize the binding of nanoparticles on the surface. In this attempt the fracture surface of nanocomposites were sputter coated with gold before observation.

3.3. Transmission electron microscopy (TEM) analysis

A JEOL–200CX transmission electron microscope was used to investigate the morphology of sample. For TEM studies samples were first washed by placing the nanocomposite into ultrapure water and sonicating for 2 hours. After centrifugation (6000g for 10 min), the final separated slurry was sonicated for approximately 5 min to better disperse the nanoparticles. A drop was placed on a carbon-coated copper TEM grid (200–300 mesh) and then left to dry in air.

3.4. Morphology Structure

Low impact tests were conducted using an instrumented falling weight impact device (drop tower). The device was equipped with data acquisition system to acquire force versus time data. Impact energy and velocity can be set as control factor by changing the mass and height of the dropping weight. The velocity of the falling striker is measured just before the striker hits the specimen. The specimen is hold in place by fixture at the bottom of the drop tower. The impact test was performed by placing PU/MMT samples (H 50mm, W 35mm, D 25mm) bottom of the drop tower and fixed by two beams.

IOP Conf. Series: Materials Science and Engineering **64** (2014) 012039 doi:10.1088/1757-899X/64/1/012039

4. Results and discussion

4.1. XRD

XRD analysis was performed in order to understand the chemical composition of the clay. Peaks identified on the θ-2θ scans were compared to those stored in the Joint Committee on Powder Diffraction Standards (JCPDS) database allowing identification of both crystal structure and specific crystal planes (Fig. 1).

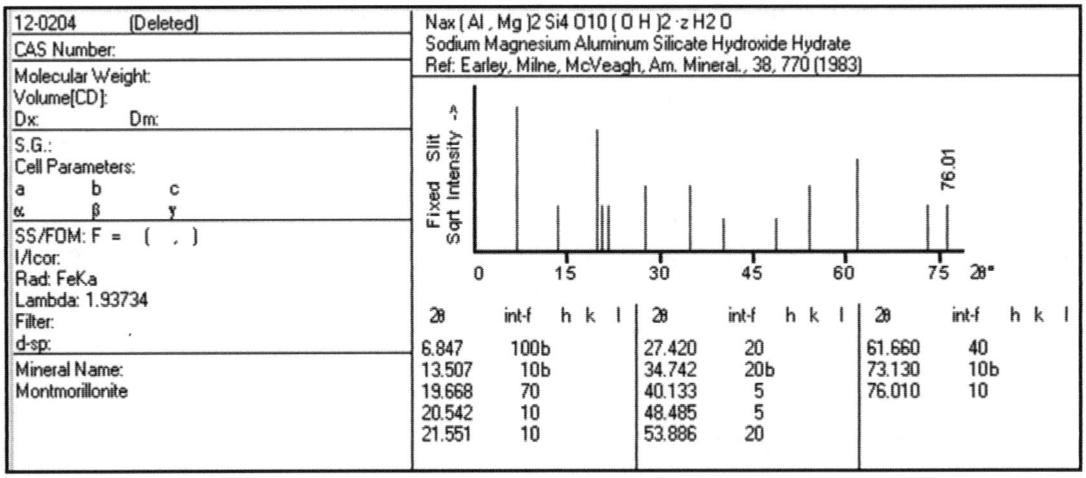
12-0204 (Deleted)
CAS Number:
Molecular Weight:
Volume[CD]:
Dx: Dm:
S.G.:
Cell Parameters:
a b c
α β γ
SS/FOM: F = (,)
I/Icor:
Rad: FeKa
Lambda: 1.93734
Filter:
d-sp:
Mineral Name:
Montmorillonite

Nax (Al , Mg)2 Si4 O10 (O H)2 ·z H2 O
Sodium Magnesium Aluminum Silicate Hydroxide Hydrate
Ref: Earley, Milne, McVeagh, Am. Mineral., 38, 770 (1983)

2θ	int-f	h k l	2θ	int-f	h k l	2θ	int-f	h k l
6.847	100b		27.420	20		61.660	40	
13.507	10b		34.742	20b		73.130	10b	
19.668	70		40.133	5		76.010	10	
20.542	10		48.485	5				
21.551	10		53.886	20				

Figure 1. XRD pattern of Montmorillonite (Clay), $Na_x (Al,Mg)_2Si_4O_{10}(OH)_2.zH_2O$.

The XRD curves of the PU/MMT are presented in Fig. 2. The discernible peaks which can be clearly identified in the scan in Fig. 2 can be matched to, the (6.9), (13.5), (19.7), (21.5), (27.0), (35.0), (48.5), (54.0), (61.76), (73.13), and (76.68) planes of a cubic MMT unit cell.

4.2. SEM

Fig. 3 shows a SEM microphotograph of the clay/polyurethane nanocomposite. The nanolayers (observed as small spots) were approximately 3-5 nm thick and well dispersed on the surface of the polyurethane. Although not indicated in the microphotographs the clay was distributed in the deep-dark region (the cavity of the polyurethane foam structure). The SEM results indicate that the MMT particles are completely disordered and dispersed relatively homogeneously on the nanoscale in the PU matrix.

IOP Conf. Series: Materials Science and Engineering **64** (2014) 012039 doi:10.1088/1757-899X/64/1/012039

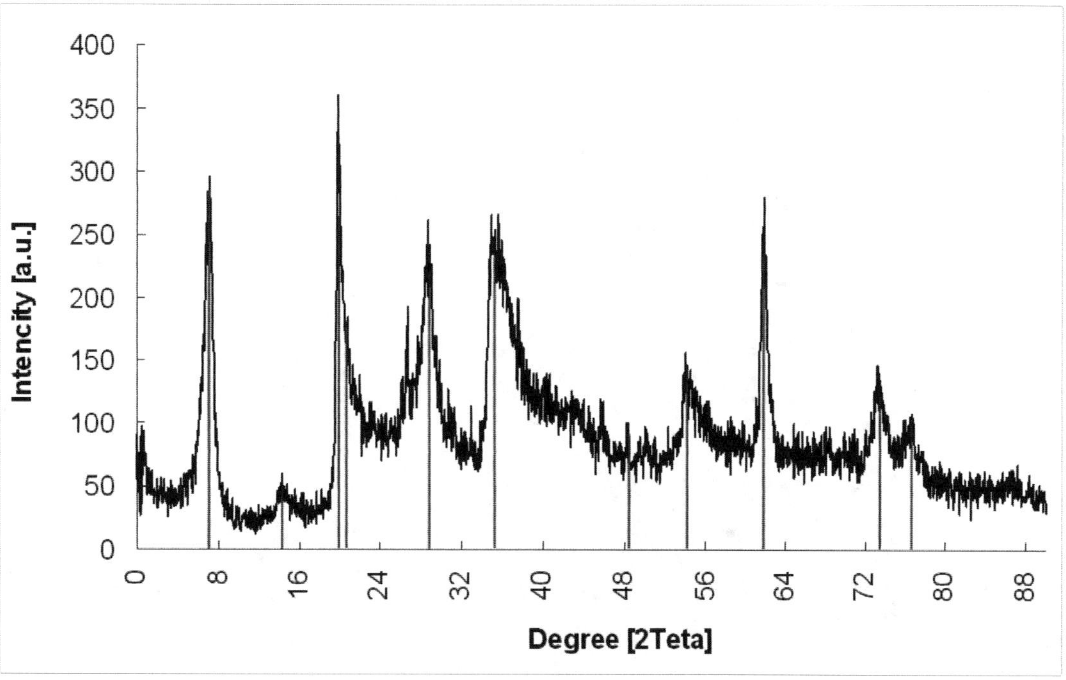

Figure 2. (a) The X-ray diffraction (XRD) pattern of (a) pure polyurethane (PU), (b) pure clay (MMT) and (c) PU/MMT. The numbers in parenthesis give the Miller indices of pure nanoclay (JCPDS19-0629) assigned to the observed peaks.

4.3. TEM

To obtain more accurate information on the morphology of the nanoclay TEM measurements were made. These showed that the MMT keeps its layered pattern inside the composite indicating that the PU penetrate through the MMT layers resulting in an intercalated structure. Pure clay and PU/MMT the result for which are shown in Fig. 4 exhibits a particle reinforced composite structure and in this case the particle surface area could be determined directly using the image analysis technique In this technique areas within a microscopic image are identified through their contract area and edge intensity level. It is normally restricted to subconfluent systems and consequently the visible projected nanoparticle area (A) is only an indicator of the actual area of contact between particle and matrix. Fig. 5 shows a bar chart of equivalent diameter calculated as $\sqrt{4A/\pi}$ of the nano size distribution.

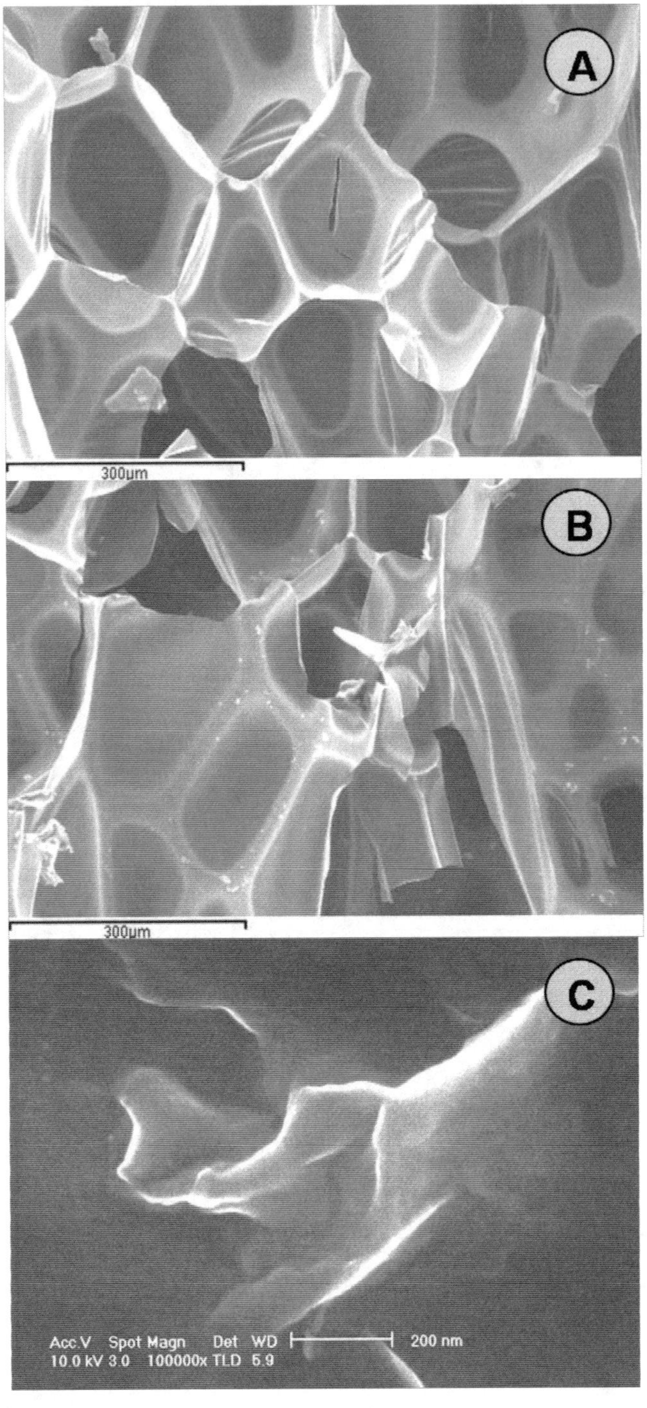

Figure 3. SEM photographs of: (a) pure PU; (b) PU/MMT; (c) pure nanoclay.

4.4. Morphology vs. mechanical Structure

The impact response of the nanocomposite with 2.5 and 10 wt. % MMT were evaluated. Several samples of each set were tested at different energy levels. Transient data including time, load, energy, velocity and deflection were collected for each sample as functions of time. Fig. 6 compares typical load versus time curves for the two specimen types at 19.5 J impact energy level. The peak loads were 70.64 kN for 2.5 wt. % MMT loaded composites and 72.84 kN for the wt. % MMT loaded sample.

Under impact loading, energy absorption in any material is mainly through elastic deformation in the initial stage with some energy absorbed through friction. Once the materials yield stress, the level of maximum elastic deformation, is reached the structure starts deforming irreversibly. The structure then dissipates the excess energy in the form of either plastic deformation in case of ductile materials or through various damage mechanisms in the case of brittle materials. It was further noted that samples with higher filling percentage had the highest energy intake at the failure point, in agreement with previous observations by Njuguna [11].

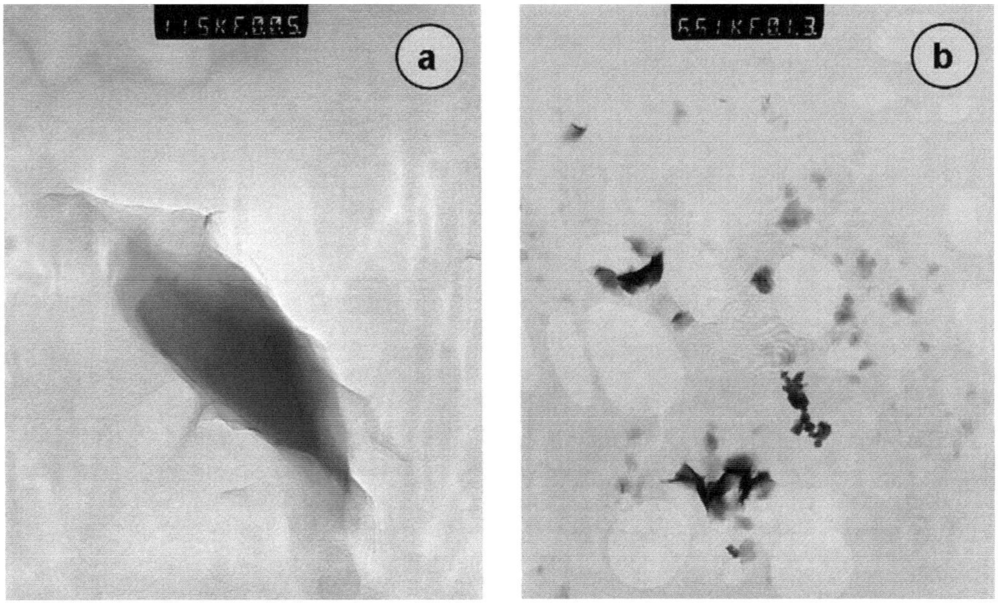

Figure 4. TEM microphotograph of: (a) pure nanoclay; (b) PU/MMT.

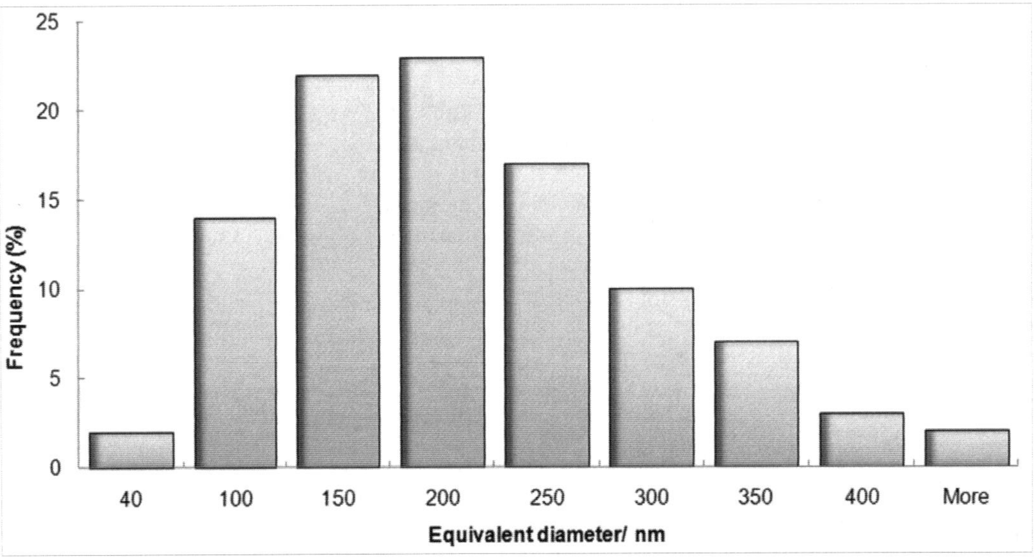

Figure 5. Bar chart of the equivalent diameter of MMT layers from a typical preparation determined from TEM images (a total of 100 particles were measured).

5. Conclusion

Fabricated PU/MMT nanocomposites have been subjected to XRD, SEM and TEM analyses and these show that the clay has the correct crystalline form and still retains a good dispersion of nanoparticles in the PU matrix. SEM examination of a clay/waterborne polyurethane nanocomposite showed that the clay was distributed in the region in which hard segment aggregation of polyurethane was located.

As demonstrated in a previous study [11] the incorporation of MMT increases the number of PU cells with smaller dimensions and higher anisotropy index. The reduction of pore size increased the composite density, so that composites with increasing amount of nanofiller were able to taking higher peak loads. Under impact loading, energy absorption in any material is mainly through elastic deformation in the initial stage with some energy absorbed through friction [20,21]. Once reached the materials yield stress, the level of maximum elastic deformation is reached the structure starts deforming irreversibly. The structure then dissipates the excess energy in the form of either plastic deformation in the case of ductile materials or through various damage mechanisms in the case of brittle materials. It was further noted that samples with higher filling percentage had the highest energy intake at failure point, in agreement with previous observations in the literature.

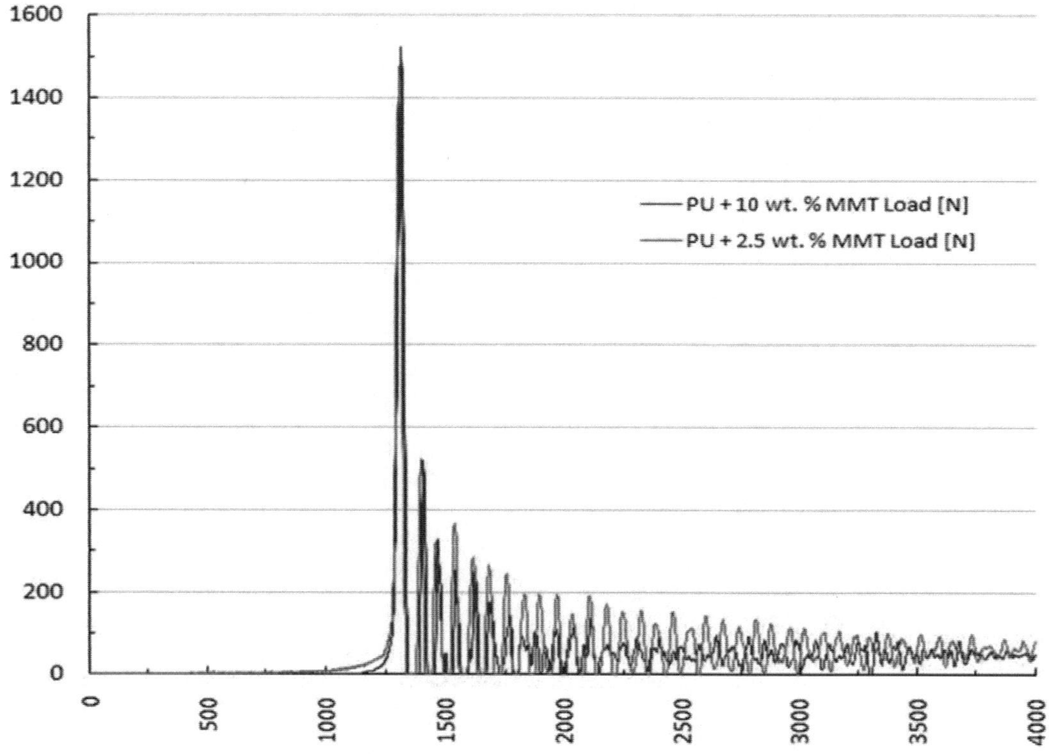

Figure 6. Comparison of load vs time for the two specimen types at 19.5 J impact energy level. The peak loads were 70.64 kN for 2.5 wt. % MMT loaded composites and 72.84 kN for the wt. % MMT loaded sample.

6. References

[1] L.P. Cheng, D.J. Lin, K.C. Yang, J. Membr. Sci. 2000, 172, 157-166.
[2] G. Jimenez, N. Ogata, H. Kawai, T. Ogihara, J. Appl. Polym. Sci. 1997, 64 (11), 2211-2215.
[3] E.P. Giannelis, Adv. Mater. 1996, 8, 29-35.
[4] P.C. LeBaron, Z. Wang, T.J. Pinnavaia, Appl. Clay Sci. 1999, 15, 11-29.
[5] M. Alexandre, P. Dubois, Mater. Sci. Eng. 2000, 28: 1-65.
[6] K.J. Yao, M. Song, D.J. Hourston, D.Z. Luo, Polym. 2002, 43 (3), 1017-1020.
[7] T.K. Chen, Y.I. Tien, K.H. Wei, J. Polym. Sci., Part A: Polym. Chem., 1999, 37, 2225-2233.
[8] Q. Wu, Z. Xue, Z. Qi, F. Wang, Polym. 2000, 41, 2029-2032.
[9] H. Dvorchak, J. Coatings Technology, 1997, 69, 47-52
[10] C. Hongxiang, Z. Maosheng, S. Hongying, J. Qingming, Mater. Scie. Eng. 2007, 725-730.
[11] J. Njuguna, K. Pielichowski, S. Desai, Polym. for App. Technologies 2008, 19(8), 947-959.
[12] L. Tie, J.P. Thomas, Chem. Mater. 1994, 6, 2216-2219.
[13] K. Masaya, J. Polym. Sci. Part A: Polym. Chem., 2004, 42, 819-824.
[14] A. Okada, A. Usuki, Mater. Sci. Eng. 1995, C3: 109-115.
[15] M. Alexandre, P. Dubois, Mater. Sci. Eng. 2000, 28, 1-63.

[16] M. Kurian, A. Dasgupta, M.E. Galvin, C.R. Ziegler, F.L.Beyer, Macromolecules, 2006, 39, 1864-1871.
[17] Nanoscience and Nanotechnologies, The Royal Society & the Royal Academy of Engineering, July 2004, (Reproduced in part with permission 2005).
[18] RTO Lecture Series, Nanotechnology aerospace applications. EN-AVT-129, May 2005.
[19] G. Chen, Y. Ma, Z. Qi, Scripta Mater. 2001, 44, 125-128.
[20]. J. Njuguna, S. Michalowski, K. Pielichowski,. K. Kayvantash, A. C. Walton, 2010, Polymer Composites, 2010: 32(1), 6–13.
[21] L. Sun, R. F. Gibson, F. Gordaninejad, J. Suhr, 2009, 69, 2392-2409.

IOP Conf. Series: Materials Science and Engineering **64** (2014) 012040 doi:10.1088/1757-899X/64/1/012040

Predictive Multiscale Modeling of Nanocellulose Based Materials and Systems

Andriy Kovalenko[1,2]

[1] National Institute for Nanotechnology, 11421 Saskatchewan Drive, Edmonton, AB, T6G 2M9, Canada

[2] Department of Mechanical Engineering, University of Alberta, Edmonton, AB, T6G 2G8, Canada

E-mail:andriy.kovalenko@nrc-cnrc.gc.ca
http://www.inrf.ualberta.ca/research/kovalenko

Abstract. Cellulose Nanocrysals (CNC) is a renewable biodegradable biopolymer with outstanding mechanical properties made from highly abundant natural source, and therefore is very attractive as reinforcing additive to replace petroleum-based plastics in biocomposite materials, foams, and gels. Large-scale applications of CNC are currently limited due to its low solubility in non-polar organic solvents used in existing polymerization technologies. The solvation properties of CNC can be improved by chemical modification of its surface. Development of effective surface modifications has been rather slow because extensive chemical modifications destabilize the hydrogen bonding network of cellulose and deteriorate the mechanical properties of CNC. We employ predictive multiscale theory, modeling, and simulation to gain a fundamental insight into the effect of CNC surface modifications on hydrogen bonding, CNC crystallinity, solvation thermodynamics, and CNC compatibilization with the existing polymerization technologies, so as to rationally design green nanomaterials with improved solubility in non-polar solvents, controlled liquid crystal ordering and optimized extrusion properties. An essential part of this multiscale modeling approach is the statistical-mechanical 3D-RISM-KH molecular theory of solvation, coupled with quantum mechanics, molecular mechanics, and multistep molecular dynamics simulation. The 3D-RISM-KH theory provides predictive modeling of both polar and non-polar solvents, solvent mixtures, and electrolyte solutions in a wide range of concentrations and thermodynamic states. It properly accounts for effective interactions in solution such as steric effects, hydrophobicity and hydrophilicity, hydrogen bonding, salt bridges, buffer, co-solvent, and successfully predicts solvation effects and processes in bulk liquids, solvation layers at solid surface, and in pockets and other inner spaces of macromolecules and supramolecular assemblies. This methodology enables rational design of CNC-based bionanocomposite materials and systems. Furthermore, the 3D-RISM-KH based multiscale modeling addresses the effect of hemicellulose and lignin composition on nanoscale forces that control cell wall strength towards overcoming plant biomass recalcitrance. It reveals molecular forces maintaining the cell wall structure and provides directions for genetic modulation of plants and pretreatment design to render biomass more amenable to processing. We envision integrated biomass valorization based on extracting and decomposing the non-cellulosic components to low molecular weight chemicals and utilizing the cellulose microfibrils to make CNC. This is an important alternative to approaches of full conversion of lignocellulose to biofuels that face challenges arising from the deleterious impact of cellulose crystallinity on enzymatic processing.

1. Applications of modified cellulose nanocrystals

Cellulose nanocrystals (CNC), the smallest building blocks of cellulose holding its physical properties, is a biocompatible and biodegradable material with mechanical properties comparable to these of reinforcement materials, such as carbon fibers, carbon nanotubes, and steel [1],[2]. The CNC can be produced in a sustainable and viable bulk process, unlike other nanomaterials (e.g., carbon nanotubes) that are typically produced in small amounts and at a very high cost [3]. The outstanding mechanical properties of CNC make it very attractive as reinforcing additive in nanocomposite materials [4]-[8], hydrogels [9]-[13] and foams [14],[15], as well as specialized security inks [16],[17]. Suspensions of CNC undergo phase separation into clear isotropic and structured chiral nematic phases above critical concentration [18] and upon drying from iridescent chiral nematic films [19]-[21].

Large-scale applications of CNC in nanocomposites require modifications to tune up its surface properties thus enhancing its solubility in nonpolar solvents and compatibility with polymer matrices [22]. The CNC have been extensively manipulated to provide a rich suite of new materials and platforms for further transformations [23]. Several methods for CNC surface modification have been developed, including grafting [24]-[26], self-assembly [27], and surfactants [28]. Effective surface modifications must provide sufficient solubility tune up without altering the CNC inter- and intramolecular hydrogen bonding network and deteriorating crystallinity and mechanical properties.

Despite the extensive experimental research effort, progress towards the development of effective surface modifications has been rather slow, mainly because (i) extensive chemical modifications destabilize the hydrogen bonding network of cellulose, causing deterioration of the mechanical properties, and (ii) relationship between surface modifications and solvation thermodynamics has not been explored. The 3D-RISM-KH molecular theory of solvation [32]-[36] provides an insight into CNC-water hydrogen bonding interactions relevant to relaxation and right-hand twist, and predicts the effective forces that control the dispersion of pristine and modified CNC in various solvents [29],[30].

2. 3D-RISM-KH integral equation theory of molecular liquids and solutions

Based on first principles of statistical mechanics [31], the 3D-RISM-KH molecular theory of solvation (three-dimensional reference interaction site model with the Kovalenko-Hirata closure approximation) [32]-[36] bridges the gap between molecular structure and effective forces on multiple length scales. It provides a firm platform to handle complex nanochemical and biomolecular systems in solution. As distinct from molecular simulations exploring the phase space by direct sampling, 3D-RISM-KH operates with spatial distributions in the statistical-mechanical ensemble rather than with individual trajectories of molecules. Analytical summation of the free energy diagrams beginning with molecular force field leads to the 3D-RISM-KH integral equations for 3D site correlation functions. Converging the latter gives the solvation structure in terms of 3D distribution maps of solvent interaction sites around a solute macromolecule (supramolecule) of arbitrary shape and then analytically the solvation thermodynamics in a simple closed form as a single integral of the 3D correlation functions obtained.

For simple and complex solvents and solutions of a given composition, including buffers, salts, polymers, ligands and other cofactors at a finite concentration, the 3D-RISM-KH molecular theory of solvation properly accounts for chemical functionalities by representing in a single formalism both electrostatic and non-polar features of solvation, such as hydrogen bonding, structural solvent molecules, salt bridges, solvophobicity, and other electrochemical, associative and steric effects. For real systems, solving the 3D-RISM-KH integral equations is far less computationally expensive than running molecular simulations which must be long enough to sample exchange and binding events. This enables handling complex systems and processes occurring on large space and long time scales, frequently not feasible for molecular simulations. The 3D-RISM-KH theory successfully describes both simple and complex associating liquids with different chemical functionalities [36]-[43], including ionic liquids [39] and polyelectrolyte gels [40], in a range of thermodynamic conditions and composition [41]-[43], and in various local environments such as interfaces with metal [32],[33] and metal oxide [44], zeolite [35],[45] and clay [46], and confinement of carbon nanotubes [38], synthetic organic rosette nanotubes [36],[47]-[49], and biomolecular systems [36],[50]-[62].

IOP Conf. Series: Materials Science and Engineering **64** (2014) 012040 doi:10.1088/1757-899X/64/1/012040

3. Structure and effective interactions of pristine and modified CNC in solution

A model for the computer simulation of CNC is based on the I_α crystal structure [63] containing 34 chains forming a three-layered cellulose fibril, as shown in Figure 1. The fibril cross section is approximately 4.3 nm in its longest direction and each chain in the fibril consists of 16 glucose units making 8 nm long particles.

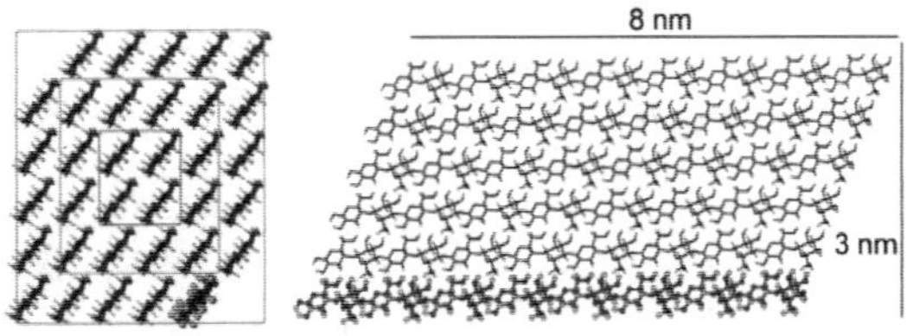

Figure 1. Cellulose fibril containing 34 glucan chains: (Left) cross section showing inner (red), intermediate (green) and outer (blue) layers; (Right) CNC particle dimensions.

For this model particle, structure relaxation is performed using MD simulation in a periodic box with dimensions 14.9 nm × 8.9 nm × 8.3 nm [30]. The distance between particle surface and the box edges is 2 nm and the space in the box around the particle is filled with water molecules. The aim of this simulation is to relax and equilibrate the model structure in the presence of solvent, mimicking as close as possible the experimental conditions of pristine CNC solvation in aqueous solution. In this neutral particle, the initial charge distribution in the cellulose fibril is inhomogeneous over different faces of the particle Figure 2(a) and (b). Cellulose layers in the crystal are stacked so that the positively charged groups face the negatively charged ones, giving a "crystal-like" charge distribution. The 3D density distributions of solvent species around an unrelaxed CNC particle clearly shows that anions (Cl^-) are concentrated around the particle surface with exposed positively charged groups, while positively charged ions (Na^+) are localized on the opposite side of the particle Figure 2(c).

On relaxing the CNC particle in explicit water solution for 6 ns and analyzing the structure after relaxation, we observe a smearing of the charge over the faces, Figure 2(d) and (e), which noticeably affects the distribution of the ions around the fibril Figure 2(f). The relaxed particle charge is weakly polarized, as evidenced by the rather even distribution of cations and anions around it. In both unrelaxed and relaxed particles, the ions distributions are calculated using the 3D-RISM-KH theory for a particle in aqueous solution of NaCl at concentration 0.040 mol/kg. One can hypothesize that the initial "crystal-like" highly polarized charge distribution in the CNC particle is one of the driving forces leading to the fibril twisting.

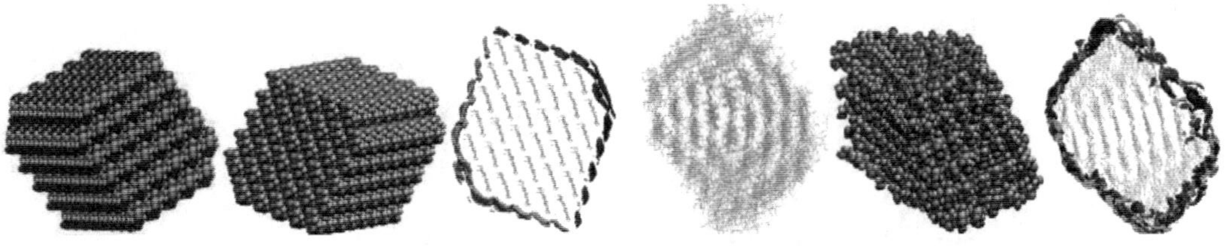

(a) (b) (c) (d) (e) (f)

Figure 2. Relaxation of a pristine CNC particle in explicit water MD simulation; positively (in blue) and negatively (in red) charged atoms. Surface charge distribution of the unrelaxed CNC particle:
(a) Faces exposing positively charged groups are predominantly blue colored;
(b) Faces exposing negatively charged groups are predominantly red colored;
(c) 3D distributions of Na^+ (blue) and Cl^- (green) around an unrelaxed fibril in NaCl aqueous solution.
Surface charge distribution of a relaxed CNC particle after 5 ns MD simulation:
(d) CNC particle cross section with right-hand twist; (e) Charge distribution in the relaxed particle;
(f) 3D distributions for Na^+ (blue) and Cl^- (green) around a relaxed fibril in NaCl aqueous solution.

The next step taken towards the development of representative CNC models is to build a charged sulfonated particle model, which is accomplished by grafting sulfate groups to some of the cellulose's surface chains at the surface density taken from elemental analysis [64]. The model contains 16 modified glucose units at the C6 position with sulfate groups so as to have 4 substitutions on each face. Each sulfonate group bears one unit of negative charge and the total CNC particle charge is -16*e*. The estimated surface charge density is 0.3 e/nm^2, in accordance with the experimental data for sulfonated CNC [65]. The MD setup for particle relaxation is similar to that of the neutral particle, the only difference being 16 Na^+ ions added to the simulation box to get overall electrostatic neutrality.

The 3D-RISM-KH theory has been applied to CNC surface modifications to tune-up the effective forces acting on CNC particles in solution. The Iβ pristine and esterified cellulose was considered in hydrophilic solvent (water), hydrophobic solvent (benzene), and ionic liquid ([mmim$^+$][Cl$^-$]). Ionic liquids (IL) are a novel class of non-volatile environmentally-friendly solvents that attracted interest of synthetic chemists due to the fundamentally different reaction thermodynamics and kinetics compared to typical molecular solvents [66]. The 3D-RISM-KH method predicts the properties of IL in a remarkable agreement with the results of Car-Parrinello high-level *ab initio* modeling [36],[39].

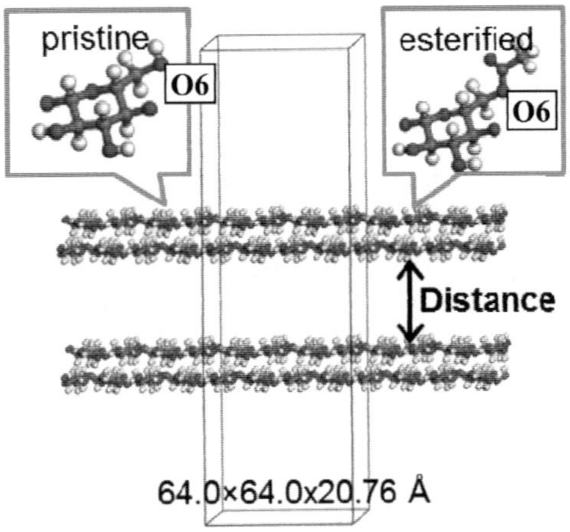

Figure 3. Periodically-extended model of pristine and esterified Iβ CNC.

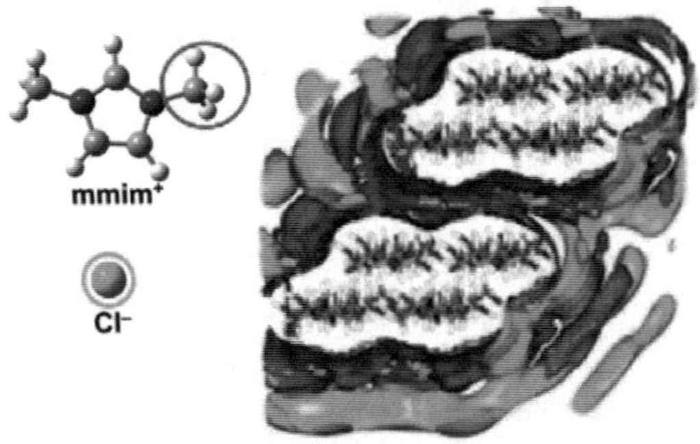

Figure 4. Solvation structure of [mmim][Cl] ionic liquid around pristine Iβ CNC particle.

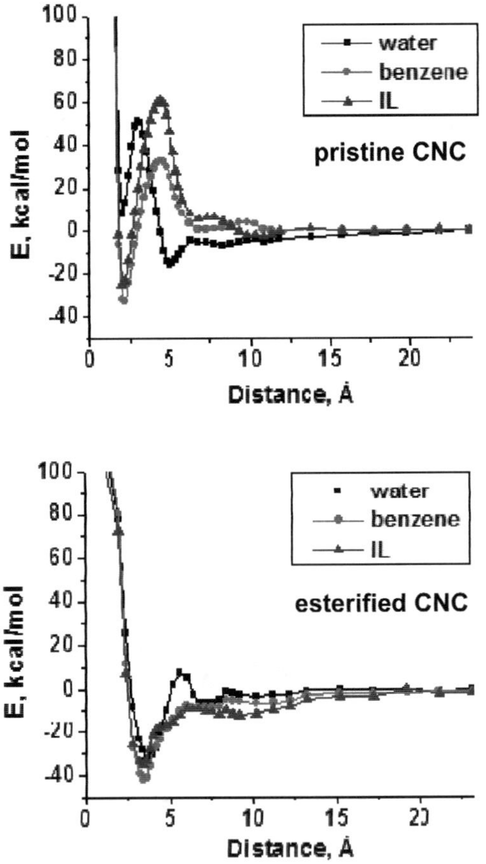

Figure 5. Potential of mean force between the pristine (upper panel) and esterified (lower panel) periodically-extended Iβ CNC rods immersed in water, benzene, and [mmim][Cl] ionic liquid (IL).

Figure 3 exhibits the periodically-extended Iβ CNC rod model that is particularly suitable for prediction of interaction among particles with uniform surface and large aspect ratio, using the pristine structure from Ref. [67]. The esterified structure (right-hand inset) containing acetate groups at every primary surface hydroxyl group (O6 atom, left-hand inset) is partially relaxed using molecular mechanics with constrained cellulose backbone. For every dispersion configuration, the 3D-RISM-KH theory predicts the solvent orientation based on the 3D solvent site density distributions, as shown in Figure 4 for the solvation structure of pristine CNC in methyl-methyl-imidazolium chloride IL.

Figure 5 presents the effective interaction (potential of mean force) plots for pristine and esterified CNC in different solvents. Similar to Iα cellulose, the Iβ cellulose is stabilized by the hydration shell. The pristine CNC disaggregation barrier increases in the order of solvents: water < benzene < IL, which indicates that dispersion is favorable in water but is not favorable in benzene and IL. This is typical for hydrophilic surfaces. For esterified CNC, the potential of mean force in benzene and IL have a lower disaggregation barrier than in water, leading to facilitated dispersion in hydrophobic solvents. The lack of significant solvent expulsion barriers in benzene and IL suggest weak solvent-solute interactions. This example demonstrates how surface modifications can be employed to tune dispersion properties of CNC in various solvents. In addition to addressing the type of the surface modification, the 3D-RISM-KH methodology allows prediction of the degree of surface modification necessary to achieve a certain dispersion effect. This is particularly important in industrial processing, where achieving high degrees of surface modification could be costly.

4. Effect of hemicellulose composition on nanoscale forces that control cell wall strength

Overcoming biomass recalcitrance constitutes the most fundamental unsolved problem of plant-based green technologies [68],[69]. Efficient conversion of lignocellulosic biomass to second-generation biofuels and valuable chemicals requires decomposition of resilient plant cell wall structure. Plants naturally evolved to withstand harsh external mechanical, thermal, chemical, and biological factors, and so are resistant to cell wall deconstruction. Plant secondary cell walls are composed of cellulose microfibrils embedded in a complex non-cellulosic matrix of mainly hemicellulose and lignin. These non-cellulosic components are responsible for the cohesive forces within the cell wall that entail structural support to plants [70]. It is well known that cell wall recalcitrance varies among plant species and even with different phenotypes of the same plant, depending on the chemical composition of the non-cellulosic matrix. Changing the amount and composition of branches attached to the hemicellulose backbone can significantly alter cell wall strength and microstructure [68],[70]-[72]. Therefore, full understanding of the chemical interactions within the cell walls is fundamental to gain control of the lignocellulosic biomass recalcitrance [69].

Glucuronoarabinoxylan – hemicellulose type most abundant in important lignocellulosic grasses used for biofuel production, such as sugar cane and corn – consists of a xylan backbone decorated with branches of mainly glucuronic acid and arabinose whose amount and ratio varies substantially with the plant genotype [73]. Genetic manipulation of glucuronic acid branching was shown to significantly improve xylan extractability from cell walls without impairing plant growth [70]. High-resolution imaging indicates a strong correlation between cell wall architecture and enzymatic digestibility [71].

Molecular dynamics simulations have been used to investigate decrystallization of cellulose [74],[75], the interactions between cellulose and non-cellulosic components of plant cell walls [76], and the structure and dynamics of lignin [77]. However, extremely long and costly simulations are required to get adequate statistical sampling addressing both solvation structure and thermodynamics of effective interactions in cell walls based on molecular forces. The 3D-RISM-KH molecular theory of solvation provides full statistical-mechanical sampling and molecular picture of hemicellulose arrangement around cellulose, and is uniquely capable of predicting the chemistry-driven effective interactions in plant cell walls as well as enzymatic and catalytic chemical deconstruction of biomass. It reveals molecular forces maintaining the cell wall structure and shows that hemicellulose branches of arabinose, glucuronic acid, and especially glucuronate strengthen the primary cell wall by strongly coordinating to hydrogen bond donor sites on the cellulose surface [78],[79].

IOP Conf. Series: Materials Science and Engineering **64** (2014) 012040 doi:10.1088/1757-899X/64/1/012040

Significant advances have been achieved lately for comprehension of the cell wall inner architecture and its correlation with the enzymatic digestibility. Nonetheless, due to the high complexity of the cell wall, there is still a lack of molecular level characterization of these structures. We construct a model of a primary cell wall fragment containing two 4-chain 8-glucose-long cellulose fragments immersed in aqueous solutions of the monomers arabinose, glucuronic acid and glucuronate at different concentrations [78]. Application of the 3D-RISM-KH molecular theory of solvation to this model to investigate the effects of hemicellulose composition on the effective interactions with cellulose fibrils shows the substantial role of hemicellulose in the stability of the primary cell wall. This reveals the relative contributions of different xylan substitutions to cell wall strength [78], and opens up an approach to the molecular basis of plant cell wall recalcitrance in an unprecedented level of details by considering more elaborate cell wall models and including lignin into account [79]. The effective interactions and the aggregation free energy of cellulose nanofibrils in hemicellulose hydrogel (Figure 6) show that the main effect of hemicellulose composition arises due to the presence of basic groups but not stereochemistry [78]. Extension to other classes of hemicelluloses such as galactoglucomannans suggests the effect of galactose, glucose and mannose would be rather weak and the largest contributions would arise from the carboxylate groups of random glucuronate substitutions.

Recent approaches to overcome biomass recalcitrance involve genetic modification of crops to control lignin contents or composition and xylan structure so as to enable less severe pretreatment and efficient enzymatic processing [70],[80]. Lignin extraction from transgenic plants at mild conditions is expected to enable enzyme access to the hydrophobic cellulose face while keeping the remaining cell wall structure intact and the polysaccharides least altered for effective digestion by enzymes [71]. We envision integrated biomass valorization based on extracting and decomposing noncellulosic components to low molecular weight chemicals and using cellulose microfibrils to make cellulose nanocrystals and nanofibrils. These nanocellulosic materials are extensively manipulated to provide a rich suite of new materials and platforms for further transformations [81],[82]. This is an important alternative to approaches of full conversion of lignocellulose to biofuels that face challenges arising from the deleterious impact of cellulose crystallinity on enzymatic processing [83].

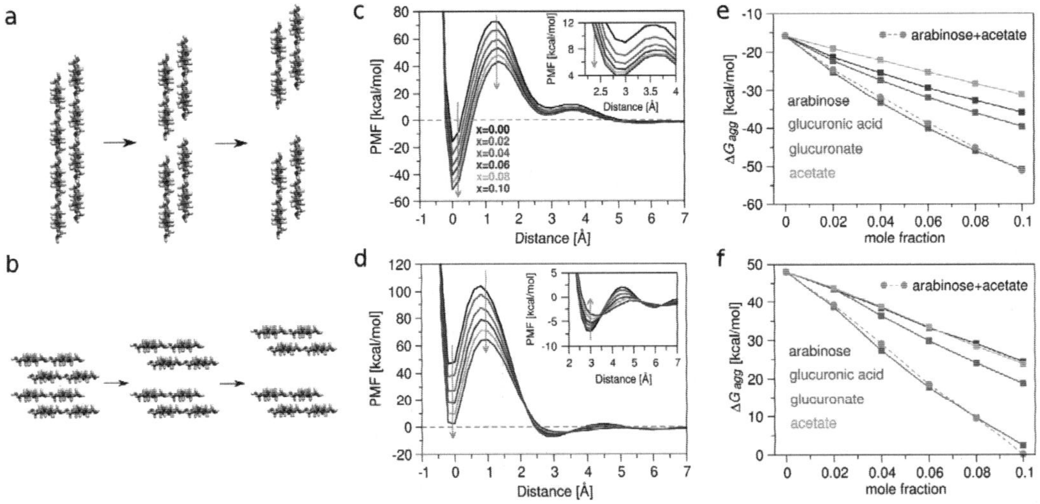

Figure 6. Potential of mean force (PMF) and aggregation free energy ΔG_{agg} of cellulose nanofibrils in hemicellulose hydrogels. Parts (a) and (b): Disaggregation by disrupting hydrophilic and hydrophobic contacts, respectively. Parts (c) and (d): PMF along disaggregation pathways (a) and (b), respectively, at glucuronate molar fractions x=0.0 – 0.1 [legend in part (c)]. Grey arrows show PMF change with glucuronate concentration. Parts (e) and (f): ΔG_{agg} for the hydrophilic and hydrophobic contacts (a) and (b) in hemicellulose hydrogels. Grey dotted line: Sum of the arabinose and acetate curves.

References

[1] Lahiji R R, Xu X, Reifenberger R, Raman A, Rudie A and Moon R J 2010 *Langmuir* **26** 4480
[2] Hamad W 2006 *Can. J. Chem. Eng.* **84** 513
[3] Postek M T, Vladar A, Dagata J, Farkas N and Ming B 2008 *Proc. SPIE* **7042** 70420D
[4] Siqueira G, Bras J and Dufresne A 2010 *Polymers* **2** 728.
[5] Zainuddin S Y Z, Ahmad I, Kargarzadeh H, Abdullah I and Dufresne A 2013 *Carbohydrate Polymers* **92** 2299
[6] de Mesquita J P, Donnici C L, Teixeira I F and Pereira FV 2012 *Carbohydrate Polymers* **90** 210
[7] Khan R A, Beck S, Dussault D, Salmieri S, Bouchard J and Lacroix M 2013 *J. Appl. Polymer Sci.* **129** 3038
[8] Chen D, Lawton D, Thompson M R and Liu Q 2012 *Carbohydrate Polymers* **90** 709
[9] Abitbol T, Johnstone T, Quinn T M and Gray D G 2011 *Soft Matter* **7** 2373
[10] Dorris A and Gray D G 2012 *Cellulose* **19** 687
[11] Kelly J A, Shukaliak A M, Cheung C C Y, Shopsowitz K E, Hamad W Y and MacLachlan M J 2013 *Angew. Chem. Int. Ed. Engl.* **52** 8912
[12] Lin N and Dufresne A 2013 *Biomacromolecules* **14** 871
[13] Yang J, Han C-R, Duan J-F, Xu F and Sun R-C 2013 *ACS Appl. Mater. Interfaces* **5** 3199
[14] Wik V M, Aranguren M I and Mosiewicki M A 2011 *Polymer Eng. & Sci.* **51** 1389
[15] Holt B L, Stoyanov S D, Pelan E and Paunov V N 2010 *J. Mater. Chem.* **20** 10058
[16] Zhang Y P, Chodavarapu V P, Kirk A G and Andrews M P 2012 *J. Nanophoton.* **6** 063516
[17] Picard G, Simon D, Kadiri Y, LeBreux J D and Ghozayel F 2012 *Langmuir* **28** 14799
[18] Dong X M, Kimura T, Revol J F and Gray D G 1996 *Langmuir* **12** 2076
[19] Beck S, Bouchard J and Berry R 2011 *Biomacromolecules* **12** 167
[20] Shopsowitz K E, Hamad W Y and MacLachlan M J 2012 *J. Am. Chem. Soc.* **134** 867
[21] Kelly JA, Shopsowitz KE, Ahn JM, Hamad WY and MacLachlan MJ 2012 *Langmuir* **28** 17256
[22] Hasani M, Cranston E D, Estman G and Gray D G 2008 *Soft Matter* **4** 2238
[23] Habibi Y, Lucia L A and Rojas O J 2010 *Chem. Rev.* **110** 3479
[24] Araki J, Wada M and Kuga S 2007 *Langmuir* **17** 21
[25] Montanari S, Roumani M, Heux L and Vignon M R 2005 *Macromolecules* **38** 1665
[26] Wang N, Ding E and Cheng R 2007 *Front. Chem. Eng China* **1** 228
[27] Cranston E D and Gray D G 2006 *Sci. Technol. Adv. Mater.* **7** 319
[28] Gousse C, Chanzy H, Excoffier G, Soubeyrand L and Fleury E 2002 *Polymer* **43** 2645
[29] Stoyanov S R, Gusarov S and Kovalenko A 2013 in: *Production and Application of Cellulose Nanoparticles*, Postek M T, Moon R J, Rudie A, Bilodeau M (Eds.) (TAPPI Press) 147-150.
[30] Stoyanov S R, Lyubimova O, Gusarov S and Kovalenko A 2014 *Nordic Pulp Paper Res. J.* (submitted).
[31] Hansen J-P, McDonald I 2006 *Theory of Simple Liquids,* 3rd ed. (Amsterdam: Elsevier)
[32] Kovalenko A 2003 In: *Molecular Theory of Solvation*, Hirata F (Ed.) (Dordrecht: Kluwer Academic) pp 169-275
[33] Kovalenko A and Hirata F 1999 *J. Chem. Phys.* **110** 10095
[34] Kovalenko A and Hirata F 2000 *J. Chem. Phys.* **112** 10391
[35] Gusarov S, Pujari B S and Kovalenko A 2012 *J. Comput. Chem.* **33** 1478
[36] Kovalenko A 2013 *Pure Appl. Chem.* **85** 159
[37] Gusarov S, Ziegler T and Kovalenko A 2006 *J. Phys. Chem. A* **110** 6083
[38] Casanova D, Gusarov S, Kovalenko A and Ziegler T 2007 *J. Chem. Theory Comput.* **3** 458
[39] Malvaldi M, Bruzzone S, Chiappe C, Gusarov S, Kovalenko A 2009 *J. Phys. Chem. B* **113** 3536
[40] Kovalenko A, Kobryn A E, Gusarov S, Lyubimova O, Liu X, Blinov N and Yoshida M 2012 *Soft Matter* **8** 1508
[41] Kovalenko A and Hirata F 2001 *Chem. Phys. Lett.* **349** 496
[42] Kovalenko A and Hirata F 2002 *J. Theor. Comput. Chem.* **1** 381
[43] Yoshida K, Yamaguchi T, Kovalenko A and Hirata F 2002 *J. Phys. Chem. B* **106** 5042

[44] Shapovalov V, Truong T N, Kovalenko A and Hirata F 2000 *Chem. Phys. Lett.* **320** 186
[45] Stoyanov S R, Gusarov S and Kovalenko A 2011 In: *Industrial Applications of Molecular Simulations*; Meunier M (Ed.) (Boca Raton: CRC Press) pp 203-230
[46] Fafard J, Lyubimova O, Stoyanov S, Dedzo K, Gusarov S, Kovalenko A and Detellier C 2013 *J. Phys. Chem. C* **117** 18556
[47] Moralez J G, Raez J, Yamazaki T, Motkuri R K, Kovalenko A and Fenniri H 2005 *J. Am. Chem. Soc.* **127** 8307
[48] Johnson R S, Yamazaki T, Kovalenko A and Fenniri H 2007 *J. Am. Chem. Soc.* **129** 5735
[49] Yamazaki T, Fenniri H and Kovalenko A 2010 *ChemPhysChem* **11** 361
[50] Yoshida N, Phongphanphanee S, Maruyama Y, Imai T and Hirata F 2006 *J. Am. Chem. Soc.* **128** 12042
[51] Yoshida N, Imai T, Phongphanphanee S, Kovalenko A and Hirata F 2009 *J. Phys. Chem. B Feature Article* **113** 873
[52] Phongphanphanee S, Yoshida N and Hirata F *J. Am. Chem. Soc.* **130** 1540
[53] Phongphanphanee S, Rungrotmongkol T, Yoshida N, Hannongbua S and Hirata F *J. Am. Chem. Soc.* **132** 9782
[54] Maruyama Y, Yoshida N and Hirata F *Interdiscip. Sci. Comput. Life Sci.* **3** 290
[55] Maruyama Y, Yoshida N and Hirata F *J Phys Chem B* **114** 6464
[56] Stumpe M C, Blinov N, Wishart D, Kovalenko A, Pande V S 2011 *J. Phys. Chem. B* **115** 319
[57] Yamazaki T, Blinov N, Wishart D and Kovalenko A *Biophys. J.* **95** 4540
[58] Blinov N, Dorosh L, Wishart D and Kovalenko A 2010 *Biophys. J.* **98** 282
[59] Blinov N, Dorosh L, Wishart D and Kovalenko A 2011 *Molec. Simul.* **37** 718
[60] Kovalenko A and Blinov N *J. Molec. Liq.* **164** 101
[61] Nikolic D, Blinov N, Wishart D and Kovalenko A *J. Chem. Theory Comput.* **8** 3356
[62] Imai T, Miyashita N, Sugita Y, Kovalenko A, Hirata F, Kidera A 2011 *J.Phys.Chem.B* **115** 8288
[63] Nishiyama Y, Sugiyama J, Chanzy H and Langan P 2003 *J. Am. Chem. Soc.* **125** 14300
[64] Boluk Y, Zhao L and Incani V 2012 *Langmuir* **28** 6114
[65] Jiang F, Esker A R and Roman M 2010 *Langmuir* **26** 17919
[66] Welton T and Wasserschield P 2007 *Ionic Liquids in Synthesis* (Weinheim: VCH-Wiley)
[67] Nishiyama Y, Langan P and Chanzy H 2002 *J. Am. Chem. Soc.* **124** 9074
[68] Himmel M E, Ding S, Johnson D K, Adney W S, Nimlos M R, Brady J W and Foust T D 2007 *Science* **315** 804
[69] Chundawat S P S, Beckham G T, Himmel M E and Dale B E 2011 *Annu. Rev. Chem. Biomol. Eng.* **2** 121
[70] Mortimer J C *et al* 2010 *Proc. Natl. Acad. Sci. USA* **107** 17409
[71] Ding S-Y, Liu Y-S, Zeng Y, Himmel M E, Baker J O and Bayer E A 2012 *Science* **338** 1055
[72] DeMartini J D, Pattathil S, Miller J S, Li H, Hahncde M G and Wyman C E 2013 *Energy Environ. Sci.* **6** 898
[73] Pauly M and Keegstra M 2008 *Plant J.* **54** 559
[74] Beckham G T, Matthews J F, Peters B, Bomble Y J, Himmel M E and Crowley M F 2011 *J. Phys. Chem. B* **115** 4118
[75] Payne C M, Himmel M E, Crowley M F and Beckham G T 2011 *J. Phys. Chem. Lett.* **2** 1546
[76] Linder, B.; Petridis, L.; Schulz, R.; Smith, J. C. Biomac-romolecules 2013, 14, 3390
[77] Petridis, L.; Schulz, R.; Smith, J. C. J. Am. Chem. Soc. 2011, 133, 20277
[78] Silveira R L, Stoyanov S R, Gusarov S, Skaf M S and Kovalenko A 2013 *J. Am. Chem. Soc.* **135** 19048
[79] Silveira R L, Stoyanov S R, Gusarov S, Skaf M S and Kovalenko A 2014 (to be published)
[80] Chen F and Dixon R A 2007 *Nat. Biotechnol.* **25** 759
[81] Habibi Y, Lucia L A and Rojas O J 2010 *Chem. Rev.* **110** 3479
[82] Nogi M, Iwamoto S, Nakagaito A N and Yano H 2009 *Adv. Mater.* **21** 1595
[83] Chundawat S P S *et al* 2011 *J. Am. Chem. Soc.* **133** 11163

IOP Conf. Series: Materials Science and Engineering **64** (2014) 012041 doi:10.1088/1757-899X/64/1/012041

Structure, entanglements and dynamics of polymer nanocomposites containing spherical nanoparticles

A Karatrantos[1],N Clarke[1], R J Composto[2], K I Winey[2]

[1] Department of Physics and Astronomy, University of Sheffield, Sheffield, S3 7RH, UK.
[2] Department of Materials Science and Engineering, University of Pennsylvania, Philadelphia, Pennsylvania, 19104, USA.

E-mail: a.karatrantos@sheffield.ac.uk

Abtract. We investigate the effect of nanoparticles on polymer structure, nanoparticle dynamics and topological constraints (entanglements) in polymer melts for nanoparticle loading above percolation threshold as high as 40.9% using stochastic molecular dynamics (MD) simulations. An increase in the number of entanglements (decrease of N_e with 40.9% volume fraction of nanoparticles dispersed in the polymer matrix) in the nanocomposites is observed as evidenced by larger contour lengths of the primitive paths. Attraction between polymers and nanoparticles affects the entanglements in the nanocomposites and alters the primitive path. The diffusivity of small sized nanoparticles deviates significantly from the Stokes- Einstein relation.

1. Introduction

The addition of nanoparticles to a polymer matrix can improve the electrical, thermal, mechanical, rheological and tribological properties [1]. We explore how nanoparticles can affect polymer structure when the polymer radius of gyration is larger than the nanoparticle radius (R). There is controversy as to whether the addition of nanoparticles to a polymer melt causes perturbed behaviour of polymers [2,3,4,5]. The quality of nanoparticle dispersion [3,4] can have an important effect on the polymer structure. However, we need to note that in the experimental area of polymer nanocomposites containing spherical nanoparticles, the role of monomer-nanoparticle interaction on polymer structure and nanoparticle dynamics is not clear yet.

The dynamics of long polymers is limited by entanglements, which are topological constraints imposed by the other chains. These can dramatically change the polymer's viscosity, mechanical and tribological properties. We investigate how the spherical nanoparticles also affect the primitive path network of polymers by using topological algorithms [6-9]. An increase in the entanglement polymer density originates mechanical reinforcement in nanocomposites [10].

2. Simulation Details

In the stochastic MD method [14] a set of interacting particles is considered, whose time evolution is governed by Newton's equation of motion. The Lennard-Jones potential V_{ij}^{LJ}, acting along the line between the centres of mass of two particles force [11] is given by:

$$V_{ij}^{LJ} = 4\varepsilon_{ij}\left(\frac{\sigma_{ij}^{12}}{r_{ij}^{12}} - \frac{\sigma_{ij}^{6}}{r_{ij}^{6}}\right) \qquad (1)$$

where $\varepsilon_{ij}=k_BT$ (k_B is the Boltzmann constant, and T is the absolute temperature) is the interaction energy between particle i and particle j (for monomers : $\varepsilon_m=\sigma_m=m_m=1$, for nanoparticles: $\varepsilon_p=1$, $\sigma_p=2R$, $m_p=0.85\cdot(4/3)\pi R^3$, r_{ij} represents the distance between particles i and j. For monomer (m)-

nanoparticle (p) interactions the combination rules: $\varepsilon_{mp} = (\varepsilon_m \varepsilon_p)^{1/2}$, $\sigma_{mp} = (\sigma_m + \sigma_p)/2$, were used. The monomers (or nanoparticles) are modelled with the repulsive only part of Equation 1, with a cut off radius at $r_c=2^{(1/6)}\sigma_m$ ($r_c=2^{(1/6)}\sigma_p$ for nanoparticles). For attractive monomer - nanoparticle interaction (attractive nanoparticles) the cut off radius is $r_c=2.4\ \sigma_{mp}$, and for repulsive monomer-nanoparticle interaction (repulsive nanoparticles) $r_c=2^{(1/6)}\sigma_{mp}$. Also, the monomers were connected using FENE potential [12]. The stiffness of the polymer chains is controlled by a cosine harmonic bending potential [13], which acts on three consecutive beads along the chain. By increasing the intramolecular stiffness of the polymer chain the entanglement length is decreased [14] to a value of $N_e \approx 58$.

$$V_{ij}^{LJ} = \frac{1}{2} k_\theta \left(cos(\theta_{ijk}) - cos(\theta_0) \right)^2 \qquad (2)$$

where θ_{ijk} is the bending angle between three consecutive bonds. The equilibrium value is $\theta_0 = 109.5$, and the bending constant $k_\theta=25$ [13].

The simulations of the polymer melt were performed at a monomer density $\rho=(N_m/V)=0.85$ in a simulation cell of 60000 particles (where N_m is the total number of monomeric units in the system and V is the volume of the system) using the NVT ensemble. The pressure calculated for the N=200 polymer melt was $P^*=P\cdot \sigma_m^3/\varepsilon_m=4.864$. That pressure was used to perform all the nanocomposite systems simulations in the NPT ensemble. To set the temperature at $T^*=k_BT/\varepsilon_m=1$ and pressure at $P^*=4.864$ the Langevin thermostat and Berendsen barostat were used respectively. The equations of motion were integrated using the Leap frog algorithm [15] with a time step equal to 0.005τ for polymer melts (a time step of 0.004τ was used for nanocomposite simulations), where $\tau=(m_m \sigma_m^2/k_BT)^{1/2}$ is the LJ time unit. The duration of the simulation runs were between $4\text{-}20\cdot 10^6$ steps depending on the length of the molecules. The equilibrium time is long enough for the polymer to move at least twice the radius of gyration of the polymer, R_g, which is the equilibration criterion usually used [16,17]. For nanocomposites, we consider systems of spherical nanoparticles in a dense polymer melt.

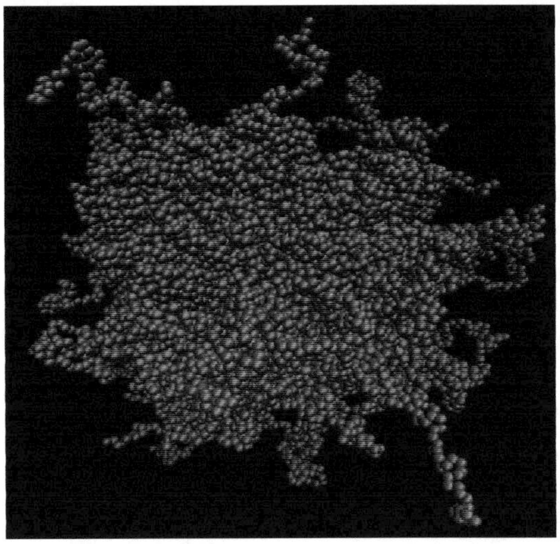

Figure 1. Spherical nanoparticles dispersed in polymer matrix (ϕ=31.6%).

In the nanocomposite systems (see Figure 1) studied, a total number of N_t=23600 monomers were used in a cubic cell. Details of the nanocomposite systems studied are summarized in the Table 1:

Table 1. volume fraction ϕ%, number of nanoparticles N_p of radius R

ϕ%	N_p of radius R=1	N_p of radius R=2
10.3	400	-
18.7	800	100
25.7	1200	-
31.6	1600	200
40.9	2400	300

3. Estimators for entanglement length N_e

In polymer melts of sufficiently long flexible chain molecules, neighbouring chains strongly interpenetrate and entangle with each other [18], thus, the motion of polymers whose degree of polymerization becomes larger than the ``entanglement length" N_e is confined to a tubelike region. The N_e can be determined by the estimator of Everaers (S-coil estimator) [19]. In addition, there are modified estimators that provide an upper bound for N_e, such as the modified S-coil [8], but they tend to overestimate N_e for weakly entangled chains. In order to eliminate the systematic errors that appeared in the previous estimators [7,9] and to obtain an accurate N-independent value, we use an ideal N_e estimator (M-coil) [8], which requires simulation of multiple chain lengths, using coil properties. The averages in our analysis are taken over the ensemble of all chains at each time step. Then the time average is taken for 250 saved configurations (at a time larger than the disentanglement time, t>τ).

4. Results and discussion

4.1 Polymer and nanoparticle structure

In Figure 2 we show the monomer-nanoparticle and nanoparticle-nanoparticle radial distribution functions g(r) for repulsive and attractive nanoparticles (R=1) respectively for different volume fractions. As can be seen the high monomer density of the layers establishes a well defined interface between the nanoparticles and polymer melt whose structure differs from that of the amorphous polymer melt. Monomer packing effects become important as monomers form layers around the nanoparticle, which is illustrated by the oscillations of g(r). A completely different behaviour of the nanoparticle structure due to the polymer nanoparticle interaction can be seen in Figures 2c, 2d.

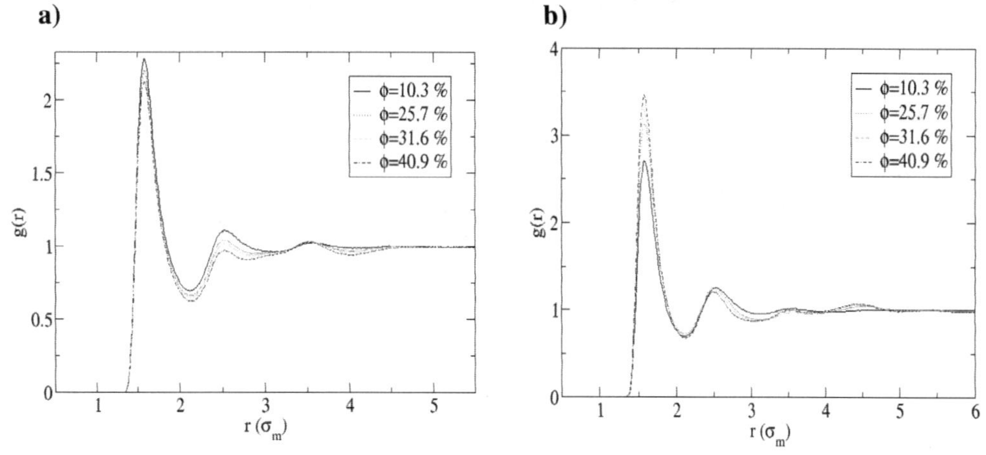

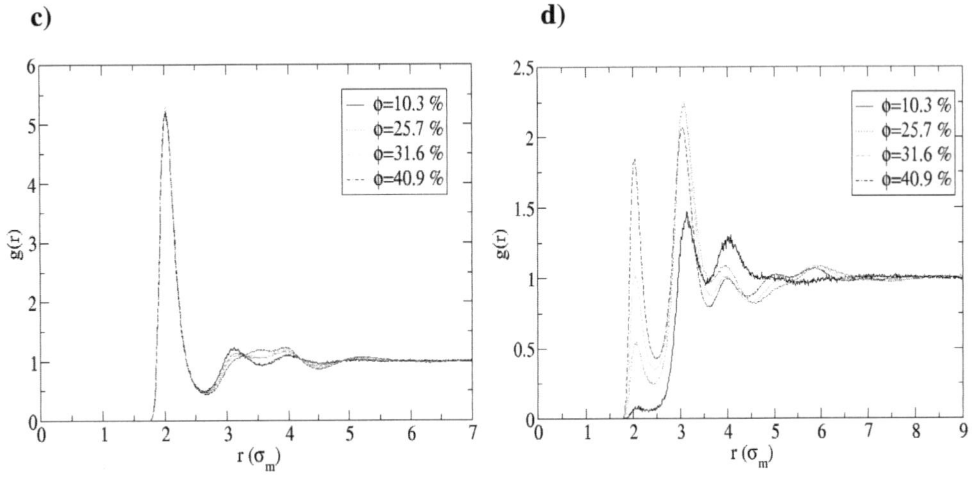

Figure 2. Radial distribution functions in the nanocomposite systems at different volume fractions ϕ for R=1. **a)** repulsive nanoparticles: monomer-nanoparticle, **b)** attractive nanoparticles: monomer-nanoparticle, **c)** repulsive nanoparticles: nanoparticle-nanoparticle, **d)** attractive nanoparticles: nanoparticle-nanoparticle. i) 10.3% (black line), (ii) 25.7% (dotted line), (iii) 31.6% (dashed line), (iv) 40.9% (dashed-dotted line).

4.2 Nanoparticle dynamics

It is known that the dynamics of spherical particles (colloids) in a fluid can be predicted by the classic continuum Stokes-Einstein relation. For a large spherical particle of radius R in a fluid of viscosity η the diffusion coefficient D of the particle is given by [20]:

$$D = \frac{k_B T}{4\pi\eta R} \quad (3)$$

However, in the case of nanoparticles dispersed in polymer melts and solutions, the validity of Equation (3) is questionable. The nanoparticle diffusivity of the simulated nanocomposite systems is calculated through the mean square displacement measurements and is given by [11]:

$$D = \lim_{t\to\infty} \frac{\langle \left(r_{cm}(t) - r_{cm}(0)\right)^2 \rangle}{6t} \quad (4)$$

It can be seen from Figure 3 that the nanoparticle diffusivity for different polymer matrices deviates from the Stokes – Einstein relation for small nanoparticles of R=1. This behaviour is obvious in the dilute regime (φ=10.3%) for polymers with N>20. However, the nanoparticle diffusivity follows the Stokes-Einstein relation in nanocomposites containing short chains at the dilute regime. In addition, in the regime above the percolation threshold (φ_c=31%) the nanoparticles diffusivity is decreased dramatically.

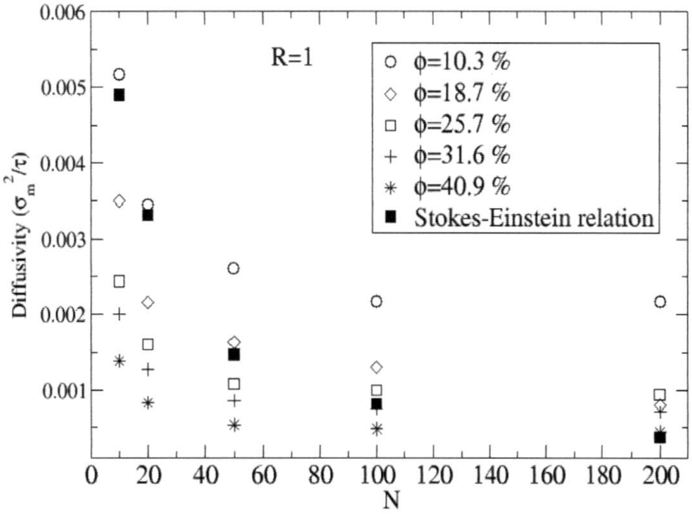

Figure 3. Nanoparticle (R=1) diffusivity estimated using Equation (4) for attractive spherical nanoparticles at different volume fractions ϕ: i) 10.3% (spheres), ii) 18.7% (diamonds), (iii) 25.7% (squares), (iv) 31.6%, (v) 40.9% (stars), (vi) Stokes-Einstein relation (equation 3) (filled circles).

4.3 *Entanglement length*

In Figure 4 we show the dependence of N_e as calculated by the M-coil estimator [8,21] of nanocomposite systems, for repulsive and attractive nanoparticles. The polymer-nanoparticle interaction alters the number of entanglements for the high nanoparticle volume fractions studied, as can be seen in Figure 4a due to an increase of the contour length of the primitive path (L_{pp}) (Figure 4b). The addition of nanoparticles decreases the N_e since there are more topological constraints that hinder the polymer chain diffusivity.

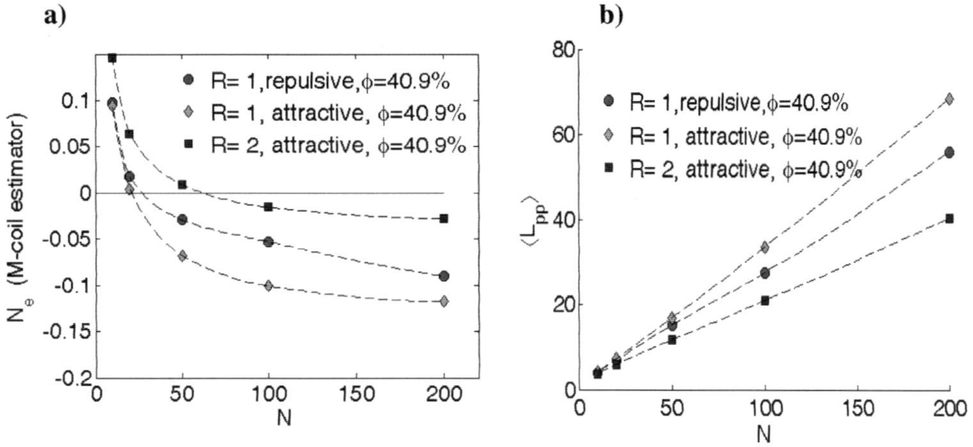

Figure 4. a) N_e estimated using the M-coil estimator for polymer nanocomposites: (i) repulsive nanoparticles: R=1 (circles), (ii) attractive nanoparticles: R=1 (diamonds), (iii) repulsive nanoparticles: R=2 (squares). **b)** Contour length of the primitive path L_{pp}: (i) repulsive nanoparticles: R=1 (circles), (ii) attractive nanoparticles: R=1 (diamonds), (iii) repulsive nanoparticles: R=2 (squares).

5. Conclusions

The polymer structure, nanoparticle dynamics and topological constraints (entanglements) of polymers in melts and nanocomposites containing spherical nanoparticles were investigated by means of stochastic molecular dynamics simulations. We show that small sized nanoparticles can diffuse much faster than that predicted from the Stokes-Einstein relation. Also, we applied an ideal N_e estimator for the calculation of the number of entanglements in our systems, and extracted the N--independent entanglement length N_e. We observe that the entanglement length decreases significantly with volume fraction of nanoparticle. This decrease of N_e in the polymer melt with nanoparticles originates from the polymer/nanoparticle entanglements, because of the increase in contour length of the primitive path due to the addition of nanoparticles.

Acknowledgements

We thank Professor Martin Kroger for providing the Z1 algorithm and for useful discussions. This research was funded by the EPSRC/NSF Materials Network program EP/5065373/1 (EPSRC: NC, AK) and DMR-0908449 (NSF: KIW, RJC).

References

[1] Kumar S K and Krishnamoorti R 2010 *Annu. Rev. Chem. Biomol. Eng.* **1** 37
[2] Tuteja A, Duxbury P, and Mackay M 2008 *Phys. Rev. Lett* **100** 077801
[3] Sen S, Xie Y, Kumar S K, Yang H, Bansal A, Ho D L, Hall L, Hooper J B, and Schweizer K S 2007 *Phys.Rev. Lett.* **98**, 128302
[4] Jouault N, Dalmas F, Said S, Di Cola E, Schweins R, Jestin J, and Boue F 2010 *Macromolecules* **43**, 9881
[5] Crawford M K, Smalley R J, Cohen G, Hogan B, Wood B, Kumar S K, Melnichenko Y B, He L, Guise W, and Hammouda B 2013 *Phys. Rev. Lett.* **110** 196001
[6] Kroger M 2005 *Computer Physics Communications* **168** 209
[7] Shanbhag S and Kroger M 2007 *Macromolecules* **40** 2897
[8] Hoy R S, Foteinopoulou K, and Kroger M 2009 *Phys. Rev. E.* **80** 031803
[9] Karayiannis N C and Kroger M 2009 *Int. J. Mol. Sci.* **10** 5054
[10] Jouault N, Dalmas F, Boue F, and Jestin J, 2012 *Polymer* **53** 761
[11] Allen M P and Tildesley D J 1987 *Computer Simulation of Liquids* (Oxford: Clarendon Press)
[12] Kremer K and Grest G S 1990 *J. Chem. Phys.* **92** 5057
[13] Bulacu M and van der Giessen E 2005 *J. Chem. Phys.* **123** 114901
[14] Uchida N, Grest G S, and Everaers R 2008 *J. Chem. Phys.* **128** 044902
[15] van Gunsteren W F and Berendsen H C J 1998 *Molecular Simulation* **1** 173
[16] Liu J, Wu Y, Shen J, Gao Y, Zhang L, and Cao D 2011 *Phys. Chem. Chem. Phys.* **13** 13058
[17] Desai T, Keblinski P, and Kumar S K 2005 *J. Chem. Phys.* **122** 134910
[18] Doi M and Edwards S F 1986 *The Theory of Polymer Dynamics* (Oxford: Clarendon Press)
[19] Everaers R, Sukumaran S K, Grest G S, Svaneborg C, Sivasubramanian A and Kremer K 2004 *Science* **303** 823
[20] Einstein A 1905 *Ann. Der Physik* **17** 549
[21] Karatrantos A, Clarke N, Composto R J, and Winey K I 2013 *Soft Matter* **9** 3877

IOP Conf. Series: Materials Science and Engineering **64** (2014) 012042 doi:10.1088/1757-899X/64/1/012042

Novel $ZnAl_2O_4$:SiO_2 Nano-composites for High Temperature Refractory Applications

Gunjan Sukul[1], P.V. Balaramakrishna[1]
[1]Research and Development, Hydrocarbon-IC, Larsen and Toubro Limited, Mumbai, India

E-mail: gunjan_sukul@lntnec.com

Abstract. High Temperature properties in nanocomposites have received significant attention in recent years due to its potential as an efficient refractory material. Nanoenabled spinel refractory products are finding widespread application owing to their enhanced temperature withstanding limits coupled with excellent abrasion resistance. Several synthesis techniques have been employed to develop these spinel nanopowders. Sol-Gel synthesis provides lower processing temperatures, control over purity, composition and easy introduction of doping elements. In this work an in-situ sol-gel route was adopted for preparation of novel nanocrytalline $ZnAl_2O_4$ dispersed in silica matrix. The gels of composition 5%ZnO–6%Al_2O_3–89%SiO_2 were developed by using tetraethyl ortho silicate, zinc nitrate, aluminium nitrate and ethyl alcohol as precursors. The transparent gels were converted to xero gel and subsequently to crystalline phase by controlled heat treatment. The structure and thermal behavior of these nanopowders was studied by utilizing various characterization techniques. Differential Scanning Calorimetry and Thermo Gravimetric Analysis were performed on the xero gel in inert argon atmosphere indicating the crystallization of spinel $ZnAl_2O_4$ and formation of oxide network. X-ray diffraction spectra were studied for samples heat treated at different temperatures in the range of 800 deg C to 1200 deg C confirming the formation of crystalline $ZnAl_2O_4$ phase. Fourier Transfer Infrared specta was recorded to understand the mechanism of development of glass from xero gel and the various bond-formations during the transformation. The morphology and crystallite size of nanocrystals were observed by Atomic Force Microscopy (AFM). The crystallite size measured by AFM was in the range 23 – 28 nm and the mean size calculated using Scherrer's equation was 29 nm. This approach may enable rapid and cost-efficient manufacturing of bulk refractory nanocomposites for supporting the industrial demands of stringent continuous processes with higher availability.

1. Introduction

Nano-enabled refractory materials are being developed in order to meet stringent demands in high temperature processes, which could offer improved service life with minimum shutdowns. Spinel presents itself as a promising structural material for mechanical and thermal applications because of its excellent high-temperature properties such as strength, hardness and chemical stability.

Synthesis of nanostructured spinel based composites, dispersed in silica glassy matrix have drawn great interest of research owing to their potential applications in refractories, high alumina cement, optical and as oxidation catalysts [1-4]. These materials are found to exhibit enhanced properties such as thermal shock resistance, wear resistance and increased hardness [5-6]. Synthesis of nanomaterials depends on the ability to control the size and distribution of the particles in the host matrix, the mixing modes of the reactants and drying techniques [7]. The advantages of Sol gel synthesis route are lower processing temperatures, control over purity, composition and easy introduction of doping elements [8-10]. In this paper, we are presenting a novel in-situ sol gel route for $ZnAl_2O_4$ nanocomposite and its characterization using x-ray diffraction, infra-red spectroscopy and atomic force microscopy techniques.

2. Experimental Procedure

In-situ sol-gel synthesis route was adopted for developing gel samples with the composition 5%ZnO–6%Al_2O_3–89%SiO_2. Two solutions were made simultaneously at a controlled temperature of 298 K with continuous stirring at 500 rpm for 60 minutes. First solution was made by mixing $Zn(NO_3)_2.6H_2O$ and $Al(NO_3)_3.9H_2O$ in distilled water according to their respective molecular weight percentages as mentioned above. Second solution was made by mixing TEOS, EtOH and distilled water in order to obtain the ratio of water to TEOH as 16:1. The two solutions were mixed and stirred at 500 rpm at 298 K for one hour and then allowed to age at the same temperature for 24 hours for gel formation. The gel was then dried at 313K for 6 weeks to form xero gel. Xero gel was calcined at 1073 K for 2 hours and then heat treated between 1173 – 1473 K for 5 hours at a heating rate of 1 K/min. The flow sheet for the synthesis process is shown in Fig 1.

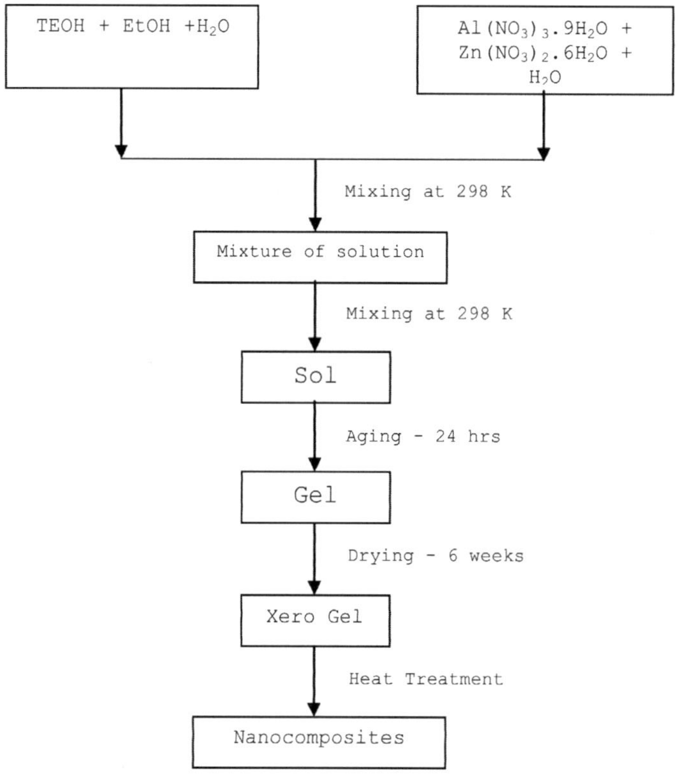

Figure 1: Flow Chart for Sol Gel Synthesis of Nanocomposites

Post synthesis, Differential Scanning Calorimetry (DSC) and Thermogravimetric Analysis (TGA) were performed on the xero gel in inert argon atmosphere with a purge rate of 100 ml/min and heating rate of 5K/min. X-ray diffraction spectra were studied for samples heat treated at different temperatures in the range 10° to 90° 2θ with Cu Kα radiation wavelength $\lambda = 1.540598$ Å in steps of 0.1^o. Fourier Transfer Infrared specta (FTIR) was recorded to understand the mechanism of development of glass from xero gel and the various bond-formations during the transformation. The morphology and crystallite size of nanocrystals were investigated by Atomic Force Microscopy (AFM).

3. Results & Discussion

DSC and TGA scan of dried gel sample shown in Figure 2 indicated an endothermic peak at 382 K corresponding to loss of water molecules present in the dried gel capillaries. The exothermic peak at 1180 K is for the crystallization of spinel $ZnAl_2O_4$. A step-like change at 1112 K corresponds to glass transition because of the formation of oxide network.

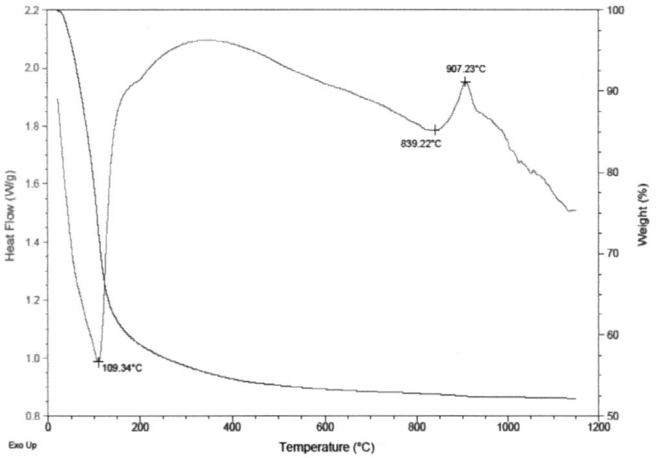

Figure 2: DSC and TGA Results for Dried Gel

The synthesis of sol to gel and insitu reaction for the formation of $ZnAl_2O_4/SiO_2$ is described as follows –

$$(C_2H_5O)_3SiOH + Zn(NO_3)_2 + Al(NO_3)_3 + H_2O \xrightarrow{298\ K} ZnO.Al_2O_3.SiO_2\ (Gel) \quad (1)$$

$$ZnO.Al_2O_3.SiO_2\ (Gel) \xrightarrow{1073\ K} ZnO.Al_2O_3.SiO_2\ (Glass) \quad (2)$$

$$ZnO.Al_2O_3.SiO_2\ (Glass) \xrightarrow{1273\ K} ZnAl_2O_4/SiO_2\ (Nanocrystals) \quad (3)$$

The total reduction of weight observed is 45%, which can primarily be attributed to the loss of water and ethanol.

X-ray diffraction pattern of the dried gels heat treated in the temperature range 1073 – 1473 K, shown in Figure 3, indicates that the sample, heat treated upto 1073 K is amorphous in nature. This is evident from the broad peak centered at 22°, which is characteristic of the diffraction pattern of the amorphous SiO_2 glassy matrix.

IOP Conf. Series: Materials Science and Engineering **64** (2014) 012042 doi:10.1088/1757-899X/64/1/012042

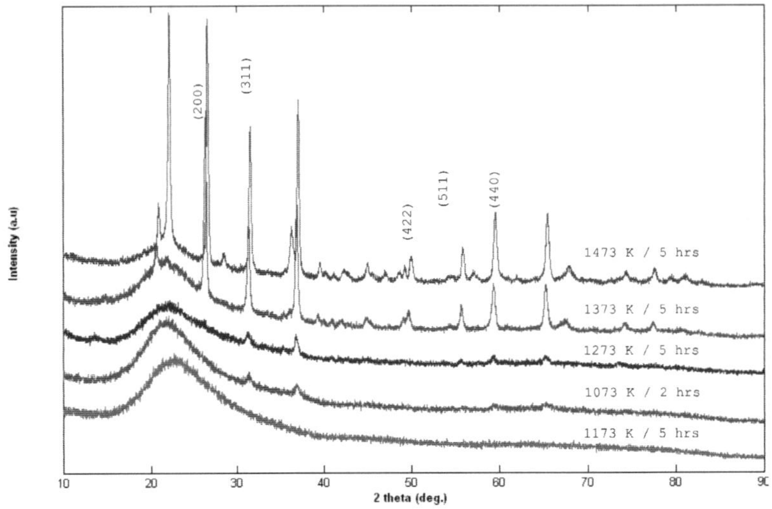

Figure 3: X-ray Diffraction Spectra for Heat Treated Samples

The diffraction pattern of sample heat treated at 1273 K for 5 hours shows clear diffraction peaks at 31.3°, 36.8°, 55.7°, 59.3° and 65.2° of 2θ value. On comparison with JCPDS data, these peaks confirm to the formation of crystalline $ZnAl_2O_4$ phase. It is also seen that on heating above 1273 K, the broad peak at 22° disappears, indicating rupture of this glassy matrix.

Scherrer's equation can be used to calculate the crystallite size:

$$D = K\lambda / (\beta \cos\theta) \qquad (4)$$

where,
λ - X-ray wavelength (1.540598 Å)
β - Full width half maximum (FWHM) of the peak
K - Shape factor (0.89)

Using the above equation, the calculated average crystallite size is 29 nm.

FTIR spectra of the gel sample and heat treated samples in the wave number range 3800 to 600 cm^{-1} is shown in Figure 4. The spectrum of xero gel has a strong and broad absorption peak between 3500 cm^{-1} and 3380 cm^{-1} indicating the vibration modes of metal hydroxyl groups. Xero gel spectrum also has peaks at 1376 cm^{-1} and 816 cm^{-1} corresponding to the presence of nitrate groups. The peaks of metal hydroxyl and nitrate groups are not present for heat treated samples. The peaks between 1650 to 1635 cm^{-1} and between 950 and 940 cm^{-1} may be attributed to stretching and bending vibrational modes of O-H of molecular water and the Si-OH stretching of surface silicon hydrogen bond to molecular water respectively[11,12].

IOP Conf. Series: Materials Science and Engineering **64** (2014) 012042 doi:10.1088/1757-899X/64/1/012042

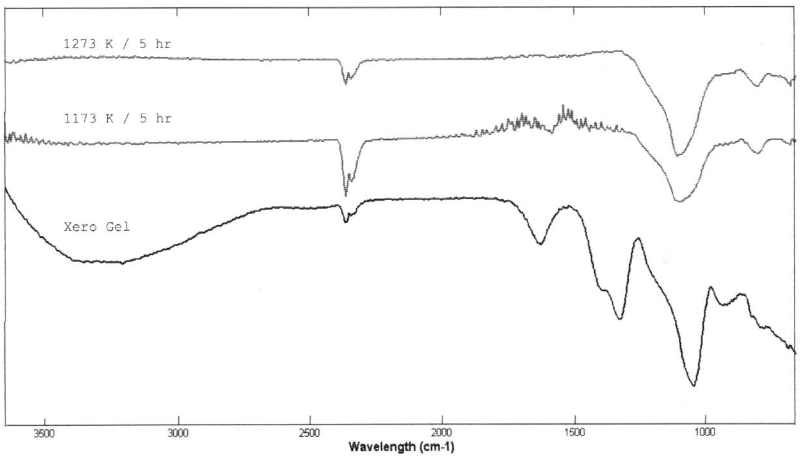

Figure 4: FTIR Pattern for Dried Gel and Heat Treated Samples

The peaks at 1105 cm^{-1} and 795 cm^{-1} signifies the stretching and bending vibrations of Si-O-Si bonds which are present in heat treated samples [13]. It is observed that with increase in heat treatment temperature, the metal hydroxide peaks get weaker, and the Si-O-Si peaks get stronger, confirming the formation of glass network. The appearance of absorption peaks at 670 cm^{-1} for sample heat treated at 1273 K is characteristic of the spinel structure [14]. The intensity of spinel peaks increases with increase in the heat treatment temperature of the samples.

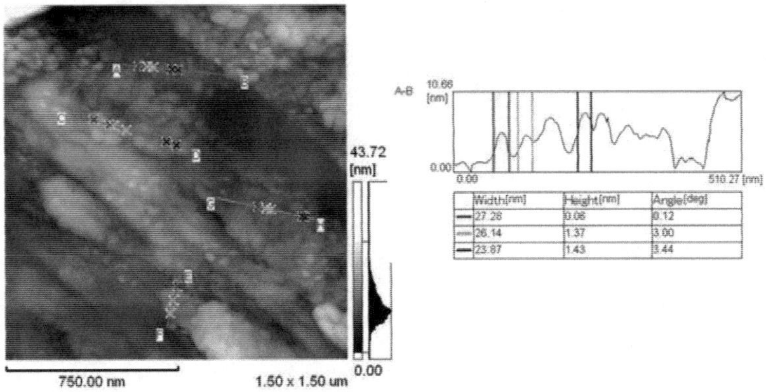

Figure 5: Atomic Force Micrograph and Analysis of Nano Powders

Atomic Force Microscopy (AFM) in non-contact mode was carried out on the sample heat treated at 1273 K for 5 hours. Powder sample was dispersed in acetone and water using ultrasonic mixer and analyzed on a mica plate by taking a drop of the dispersed powder. The typical micrographs are shown in Figure 5. It was observed that there was a tendency of the particles to agglomerate in the dispersion medium. The spherical particles in the image are $ZnAl_2O_4/SiO_2$ nanocomposites. The measured crystalline sizes were in the range of 23 – 28 nm.

4. Conclusions

A novel method for synthesis of $ZnAl_2O_4/SiO_2$ spinel nanocomposites was investigated using laboratory scale precursors. Calcination studies were carried out to arrive at nanocrystallaine phase which was verified by x-ray diffraction patterns. The phase and microstructural evolution was optimized and characterized by differential scanning calorimetry and fourier transform infra-red spectroscopy. Nano-size crystals were confirmed by atomic force microscopy and Scherrer's equation. The synthesis strategy adopted here might open a new route for development of refractory nanocomposites with enhanced performance for future applications.

References

[1] V. Ciupina, I. Carazeanu, G. Prodan, J. Opt. and Adv. Mater., Vol. 6, No. 4, 1317 – 1332 (2004).
[2] K.T. Jacob, K.P. Jayadevan, R.M. Mallaya, Y. Waseda, Adv. Mater.12 440, (2000).
[3] X. Duan, D. Yuan, X. Cheng, H. Sun, Z. Sun, X. Wang, Z.Wang, D. Xu, M. Lv, J. Am. Ceram. Soc., 88 [2] 399-403 (2005).
[4] X Zhao, L Wang, X Xu, X Lei, S Xu, F Zhang, AIChE Journal, Vol.58, No.2, 573-582 (2012).
[5] L Henkel, D Koch, G Grathwohl, J.Am.Ceram.Soc., 92 [4] 805–811 (2009).
[6] G Fan, F Li, AIChE Journal, Vol.58, No.9, 2639-2648 (2012).
[7] C.J. Brinker, G.W. Scherer, Sol Gel Science, Academic Press, New York, (1990)
[8] S.Kurajica, E. Tkalcec, J. Schmauch, J. Euro. Ceram. Soc., 27 951-958 (2007).
[9] D.A. Fumo, M.R. Morelli, A.M. Segades, Mater. Res. Bull. 31, 1243 (1996)
[10] M. Jafarzadeh, I.A. Rahman, C.S. Siaput, J. Sol-Gel Sci. Technol. 50, 328-336 (2009).
[11] E.A. Haryri, M. Greenblatt, T.Tsai, P.P. TsiaSolid State Ionics, 37, 233 (1990)
[12] S.W. Lee, R.A. Condrate Sr., J. Mater. Sci. 23, 2951 (1988)
[13] A. Adamczyk, Journal of Mol. Str. 596, 127 (2002)
[14] S. Mathur, M. Veith, M. Haas, H. Shen, N. Lecerf, V. Huch, J. Am. Ceram. Soc. 84 (9) 1921 (2001)

Nanotubular Halloysite Clay as Efficient Water Filtration System for Removal of Cationic and Anionic Dyes

Yafei Zhao, Elshad Abdullayev, and Yuri Lvov*

[1]Institute for Micromanufacturing, Louisiana Tech University, Ruston, LA 71270

E-mail: ylvov@latech.edu

Abstract. Halloysite nanotubes, chemically similar to kaolinite, are formed by rolling of kaolinite layers in tubes with diameter of 50 nm and length of ca. 1 μm. Halloysite has negative SiO_2 outermost and positive Al_2O_3 inner lumen surface, which enables it to be used as potential absorbent for both cationic and anionic dyes due to the efficient bivalent adsorbancy. An adsorption study using cationic Rhodamine 6G and anionic Chrome azurol S has shown approximately two times better dye removal for halloysite as compared to kaolinite. Halloysite filters have been effectively regenerated up to 50 times by burning the adsorbed dyes. Overall removal efficiency of anionic Chrome azurol S exceeded 99.9% for 5th regeneration cycle of halloysite. Chrome azurol S adsorption capacity decreases with the increase of ionic strength, temperature and pH. For cationic Rhodamine 6G, higher ionic strength, temperature and initial solution concentration were favorable to enhanced adsorption with optimal pH 8. These results indicate a potential to utilize halloysite for the removal of ionic dyes from environmental waters.

1. Introduction

Synthetic dyes are widely used in leather, paper, textile, etc. Discharge them into waters causes environmental various problems due to their carcinogenicity, toxicity to aquatic life and undesirable aesthetic aspect [1, 2]. Various methods including precipitation, coagulation, membrane separation, photo-degradation, ion exchange and adsorption have been extensively exploited for dyes-contaminated wastewater treatment [3-6]. Adsorption has been proved to be a highly efficient and relatively low-cost technique for the treatment of dye contaminated waters due to high efficiency, environmental friendliness, low cost, easy operation, insensitivity to toxic substances and the possibility of the materials recycling. Generally, zeolites, kaolinite and montmorillonite clays, banana peel, saw dust, fly ash, and other materials [1, 4, 7-11] have the disadvantages of high costs and resulting problems associated with their subsequent treatment and regeneration. Most important of all, none of these materials has high adsorption capacity for both positive and negative dyes.

Halloysite is a two-layered aluminosilicate clay mineral, consisting of one alumina octahedron sheet and one silica tetrahedron sheet in 1:1 stoichiometric ratio, which is available in abundance in China as well as other locations around the world. A comparative study on structure and morphology of halloysites from six different natural deposits have been presented in a recent paper and their potential as nanoadsorbents, fillers in polymer composites and containers for loading active agents was demonstrated [12]. Halloysite nanotubes with *ca.* 50 nm diameter, 10 nm lumen and 1 μm length are formed by rolling kaolinite sheets during natural hydrothermal process [13-18]. Halloysite tubes have multilayer walls with negatively charged Si-OH on the outer surface and positively charged Al-OH on the inner surface (at pH between 4 and 9) and don't require exfoliation, contrary to the kaolinite plate-like morphology. This unique bivalent morphology with spatially separated negative and positive surfaces makes halloysite tube a promising absorbent for variety of pollutants, both positive and negative [19-23]. Literature has studied the adsorption properties of halloysite with cationic dyes and heavy metal ions, but anionic dye adsorption was not studied though it may be specifically controlled with its positively charged lumen. Halloysite can be modified by silane coupling agents, surfactants, metal complexes and polyelectrolytes for higher adsorption capacity [24-28].

In this paper, we investigate the halloysite adsorption study of cationic dye Rhodamine 6G and anionic dye Chrome azurol S and its comparison with kaolinite clay adsorption. Comparison of adsorption isotherms for two mesoporous materials with similar chemistry but different morphology (tubule and plate) was performed in dependence on adsorbent dose, pH, temperature and initial concentration. Halloysite surface was modified with surfactants and polyelectrolytes to reveal the major forces involved in adsorption process.

2. Experimental and Methods

Halloysite from Applied Minerals Inc (USA) was used without further treatment. Kaolinite was purchased from Sigma (USA) and used without pretreatment. The batch experiments were carried out in 2 mL centrifuge tubes containing 1 mL of dye solutions and 2 mg adsorbent clay (except for the study of the adsorbent dose effect on dye adsorption, in which amount of clay indicated within the context). The tubes were left at 20-60 °C to reach equilibrium on dye adsorption. Then tubes were centrifuged for 5 min at 14000 rpm and the concentration of supernatant was determined using UV-Vis Spectrophotometer (Agilent, 8453). For the water filter system, 1 g of adsorbent clay was added to the water filter and cellucotton was used to block the adsorbent powder. The system was kept at 20 °C. 300 mg/L Rhodamine 6G and Chrome azurol S dyes solution were studied in the system and dyes concentration were determined after filtration, respectively. In order to study the reusability of halloysite, the used adsorbents were dried under 60 °C in the oven and burned at 300°C. After that, the burned samples were reused as adsorbents in the water filter system and dyes concentration were measured again after filtration. The adsorption-burning-adsorption cycle were repeated for five times.

3. Results and Discussion

3.1. Morphology characterization of halloysite and kaolinite

SEM is used to observe the morphological structures of HNTs and kaolinite. It can be seen from Figure 1a and 1b that HNTs has a cylindrical shape with 15-20 aluminosilicate layers rolled in the multilayer tubule walls with a layer spacing of 0.72 nm. It has hollow and open-ended lumen with a length of 0.5-1 mm. The inner and outer diameter of HNTs is in the range 20-30 nm and 40-50 nm, while the wall thickness is 15-20 nm. kaolinite has a base shape and lamellar structure. The particle size of kaolinite is estimated 1-15 μm.

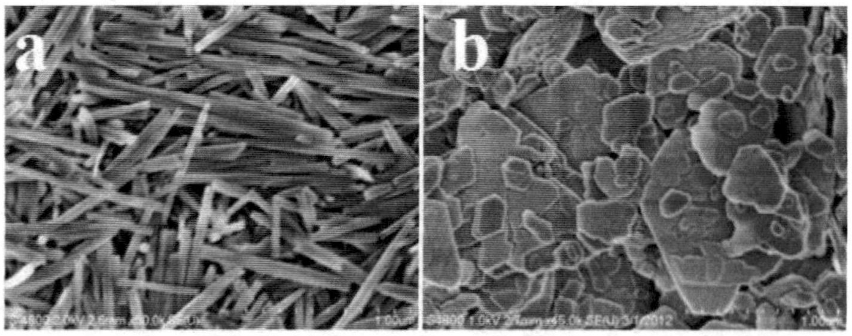

Figure 1. SEM images of halloysite nanotubes and kaolinite

3.2. Adsorption properties of halloysite and kaolinite

Adsorption behaviors of positively charged dye -Rhodamine 6G and negatively charged dye - Chrome azurol S (Fig. 1) were studied with halloysite and kaolinite powders using aqueous dyes of 0 – 0.05wt%. The maximum adsorption capacity calculated from Langmuir model is 43.6 mg/g for Rhodamine 6G and 38.7 mg/g for Chrome azurol S onto halloysite, and 21.4 mg/g for Rhodamine 6G and 36.7 mg/g for Chrome azurol S onto kaolinite.

It can be observed that halloysite has better adsorption for both of the dyes as compared to kaolinite. The larger surface area of halloysite, gives more sites for dye adsorption. Besides, both negative outer surface and positive inner surface of the halloysite are exposed to the solution, allowing adsorption of both positive and negative molecules. Water filtration system and regeneration of halloysite and kaolinite

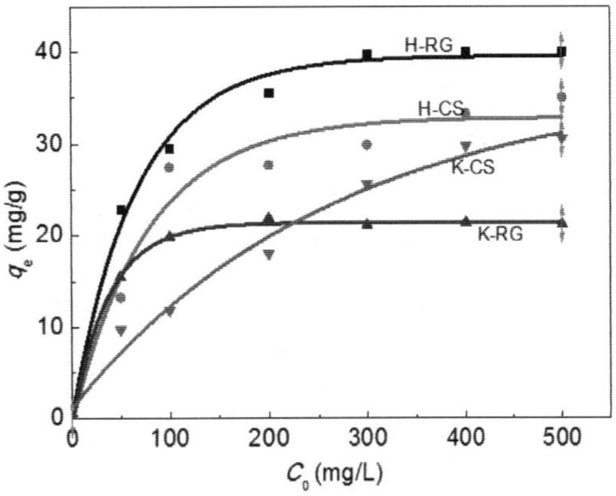

Figure 2. Adsorption of Rhodamine 6G and Chrome azurol S onto halloysite and kaolinite (H-halloysite, K-kaolinite, RG-Rhodamine 6G, CS-Chrome azurol S).

3.3. Water filter system and regeneration study

Water filters were prepared using halloysite and kaolinite clays (Figure 3). Filtration efficiency was tested by passing 30 mL of 300 mg/L dye solution. Chrome azurol S was completely removed from contaminated water by halloysite, as it is clearly evident from UV spectra of the filtered liquid samples. In spite of having similar chemical structure, kaolinite was shown to be poor filter. Probably this is associated with nanoporous structure of the halloysite. Small pores of *ca.* 5-10 nm diameter efficiently adsorb dyes from the solution before it passes from the filter. Similar tendency was observed for the case of Rhodamine 6G as well; over 99% of the dye was removed by the halloysite filter.

Dyes can be removed from adsorbed filters by burning at 300 °C. Second generation reusable filters were made and their efficiency was demonstrated using Chrome azurol S and Rhodamine 6G. Halloysite showed significantly better reusability compared to kaolinite. Filtrate from halloysite filters are almost completely transparent for both the dyes, indicating near filtration efficiency for both the negative and positive dyes. Their efficiency compares fairly well with the efficiency of first generation halloysite filters. With this we may conclude that halloysite almost completely preserved its dye removal efficiency after first reuse. On the other hand, significant amount of dye passed from kaolinite filters, indicating its poor filtration efficiency. Furthermore, halloysite filters have been effectively regenerated up to 5 times by burning the adsorbed dyes.

IOP Conf. Series: Materials Science and Engineering **64** (2014) 012043 doi:10.1088/1757-899X/64/1/012043

Figure 3. photographys of the water filters prepared by one gram of halloysite and kaolinite clays, tested on 300 mg/L concentration Chrome azurol S (a), 300 mg/L concentration Rhodamine 6G (b) and mixture of these dyes 150 and 150 mg/L (c) solutions with 300 mg/L concentration.

3.4. External factors affecting dye adsorption efficiency

Effect of initial concentration and temperature, adsorbent dose, pH and ionic strength onto HNTs were studied in detail and displayed in Figure 4. From Figure 4a, it shows the adsorption capacity increased with the increase of dyes concentration. This is due to higher concentration can provide higher driving force to overcome the resistance to the mass transfer of dyes between the aqueous and solid phases. Temperature also has important impact onto the adsorption process. The adsorption capacity of Rhodamine 6G onto HNTs was found to increase with increase in temperature while that of Chrome azurol S onto HNTs was decrease. These results indicate the adsorption process for Rhodamine 6G and Chrome azurol S is endothermic and exothermic, respectively. A possible mechanism of interaction is the reaction between the hydroxyl groups of HNTs and cationic/anionic group in the dye molecules. At initial pH 6, HNTs is negatively charged and thus such reaction is electrostatic attraction for cationic dyes and repulsion for anionic dyes. Either electrostatic attraction or repulsion effect could be enhanced at higher temperature. Thus, this results in the increase for Rhodamine 6G removal but the decrease for Chrome azurol S adsorption.

From Figure 4b, it was observed that the removal efficiency increased from 51.4% to 89.0% for Rhodamine 6G and from 43.9% to 72.8% for Chrome azurol S with an increase in adsorbent dose from 1 to 15 mg. This may be due to more adsorbent dose gives more surface area and thus results in more available adsorption sites.

It is well-known that pH is a crucial parameter because it strongly affects not only the surface charge of the adsorbent, the degree of ionization of the materials, the dissociation of functional groups on the active sites of the adsorbent, but also the structure of the dye molecule, Furthermore, it delineates the mechanism involved in adsorption process. Figure 4c illustrates dyes adsorption by HNTs as affected by pH from 3 to 12. The optimum pH for Rhodamine 6G is found to be around 8 and the adsorption capacity of Chrome azurol S decreases with the increase of pH. The reason for the strong dependence of adsorption capacity on pH can be explained that the surface charge of HNTs is significantly affected by solution pH. The pH_{zpc} of HNTs was found to be 2.75. The surface of HNTs is positively charged below pH 2.75, while it is negatively charged above 2.75. Besides, the zeta potential of HNTs is more negative with increase in pH in the range pH 3 to pH 12. The electrostatic interaction between the dye molecules and the adsorbent surfaces can be either attractive or repulsive, therefore the adsorption capacities should increase when the interactions are attractive and decrease when they are repulsive. With the increase in pH, either attraction or repulsion can increase. As discussed previously, Rhodamine 6G is a cationic dye and therefore experiences attraction with the negatively charged HNTs, while Chrome azurol S is an anionic dye and thus experiences repulsion with negative HNTs. Therefore, increasing electrostatic attraction between negatively charged adsorption sites and positively charged Rhodamine 6G causes an increase in Rhodamine 6G removal, whereas increasing electrostatic repulsion between negatively charged adsorption sites and negatively charged Chrome azurol S results in a decrease in Chrome azurol S removal. On the other hand, the poor adsorption at lower pH (pH<8) may be due to the excessive H^+ ions competing with Rhodamine 6G for adsorption sites. The adsorption for Rhodamine 6G decreases when pH above 8. This may be due to the deprotonation of the substance in alkaline solution (as one can see Rhodamine 6G (pKa=7.5) is a weak base and can easily lose its protons by the effect of strong bases, which leads to the partial neutralization of its positive charge. As a result, electrostatic interaction with negatively charged halloysite decreases and thus results in the decrease in the adsorption capacity.

Since industrial wastewater always contains various additives such as inorganic salts, it is necessary to investigate how these ions affect the adsorption behavior for dyes. 2 mg of HNTs were added into 1 mL of 100 mg/L dye solution with different NaCl concentration at room temperature. The results are depicted in Figure 4d. The adsorption capacities of Rhodamine 6G and Chrome azurol S on HNTs were significantly affected by the presence of NaCl. When NaCl concentration increased from 0 to 1 mol/L, the adsorption capacity of Rhodamine 6G by HNTs increased from 29.5 to 45.3 mg/g but the adsorption capacity of Chrome azurol S decreased from 20.4 to 8.3 mg/g. Ionic strength can influence zeta potential of HNTs. Zeta potential of HNTs is negative at natural condition and it becomes less negative when increase the concentration of NaCl. Accordingly, the electrostatic attraction effect between HNTs and Rhodamine 6G decreases whereas the electrostatic repulsion effect between HNTs and Chrome azurol S increases. Consequently, the adsorption capacity of HNTs for Rhodamine 6G decreases but for Chrome azurol S increases.

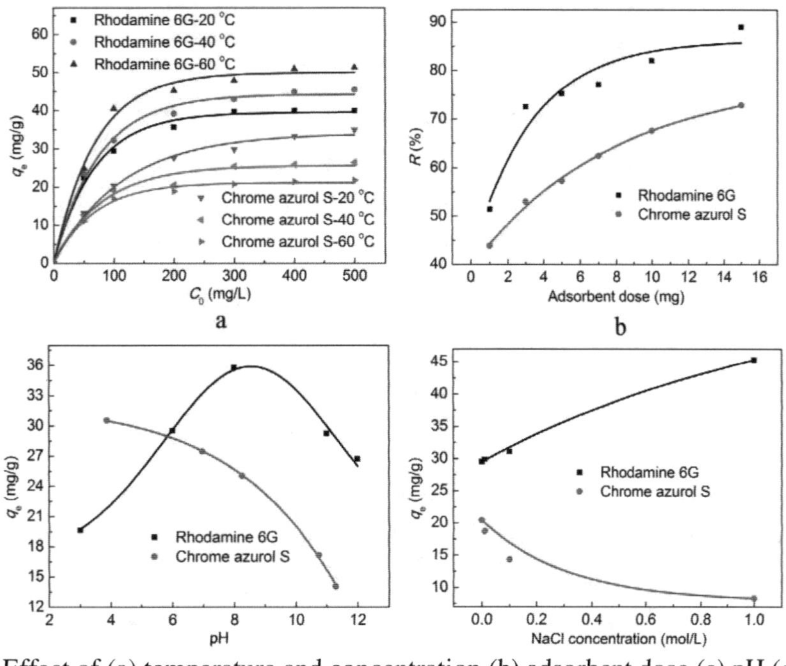

Figure 4. Effect of (a) temperature and concentration (b) adsorbent dose (c) pH (d) ionic strength

4. Conclusion

Natural halloysite nanoclay, an abundantly available tubule form of kaolin, was studied as an efficient adsorbent both for cationic and anionic aqueous dyes. Tubular morphology of halloysite favors dye removal, as compared with chemically similar platy kaolin multilayer stacks. The influence of temperature, pH, ionic strength and adsorbent amount of cationic Rhodamine 6G and anionic Chrome

IOP Conf. Series: Materials Science and Engineering **64** (2014) 012043 doi:10.1088/1757-899X/64/1/012043

azurol S revealed molecular details involved in the adsorption process. Ionic interaction is the major force in Rhodamine6G adsorption on halloysite, while hydrogen bonding was the most essential for Chrome azurol S adsorption.

Halloysite is novel nano-material which can be used in water filter system. The removal efficiency of halloysite is higher than most of conventional adsorbents. Besides, halloysite can be regenerated by burning after adsorption and be reused. Dye removal efficiency exceeded 99.9% after 5 reuse cycles for negative Chrome azurol S and 95% for positive Rhodamine 6G. Optimal pH for Rhodamine 6G adsorption was in the range of 8 and 9 while acidic solutions were favorable for Chrome azurol S adsorption. Higher temperatures favor Rhodamine 6G adsorption, indicating the endothermic nature of the adsorption.

Acknowledgments

Support with NSF-1029147 and NSF-EPS1003897 grants are acknowledged.Any opinions, findings, and conclusions or recommendations expressed in this report are those of authors and do not necessarily reflect the view of National Science Foundation. Authors are also thankful to Applied Minerals Inc., USA for providing halloysite.

References

[1] Gupta V K and Suhas 2009 *J. Environ. Manage.***90** 2313
[2] Sandeman S R *et al* 2011 *J. Colloid Interface Sci.* **358** 582
[3] Zhuang X *et al* 2009 *Chem. Mater.* **21** 706
[4] Crini G 2006 *Bioresour. Technol.* **97** 1061
[5] Saha B *et al* 2011 *J. Phys. Chem. C* **115** 8024
[6] Luo P *et al* 2011 *Korean J. Chem. Eng.* **28** 800
[7] Benkli Y E *et al* 2005 *Water Res.* **39** 487
[8] Teng M and Lin S 2006 *Desalination* **194** 156
[9] Tehrani-Bagha A R *et al* 2011 *Desalination* **266** 274
[10] Annadurai G *et al* 2002 *J. Hazard. Mater.* **92** 263
[11] Li L *et al* 2006 *J. Colloid Interface Sci.* **300** 52
[12] Pasbakhsh P *et al* 2013 *Appl. Clay Sci.***74** 47
[13] Abdullayev E *et al* 2013 *ACS Appl. Mater. Interf.* DOI: 10.1021/am400936m
[14] Veerabadran N G *et al* 2007 *NANO* **2** 115
[15] Abdullayev E *et al* 2011 *ACS Appl. Mater. Interface* **3** 4040
[16] Zhao Y *et al* 2010 *Sep. Sci. Technol.* **45** 1066
[17] Suh Y *et al* 2011 *J. Nanosci. Nanotech.* **11** 661
[18] Wei W *et al* 2012 *Macromol. Mater. Eng.* **301** 645
[19] Zhao Y *et al* 2010 *Water Sci. Technol.* **62** 937
[20] Abdullayev E and Lvov Y 2010 *J. Mater. Chem.* **20** 6681
[21] Luo P *et al* 2010 *Water Res.* **44** 1489
[22] Zhao Y *et al* 2010 *J. Hazard. Mater.***178** 658
[23] Luo P *et al* 2011 *Ind. Eng. Chem. Res.* **50** 10246
[24] Xie Y *et al* 2011 *Chem. Eng. J.* **168** 959
[25] Abdullayev E *et al* 2008 *Polymer Mater. Sci. Eng.* **99** 121
[26] Abdullayev E *et al* 2009 *ACS Appl. Mater. Interface* **1** 1437
[27] Joshi A *et al* 2013 *Langmuir* DOI: 10.1021/la3044973
[28] Yelleswarapu C S *et al* 2010 *OpticsComm.* 283 (2010), 438

IOP Conf. Series: Materials Science and Engineering **64** (2014) 012044 doi:10.1088/1757-899X/64/1/012044

Numerical Investigation of Nanofluid-based Solar Collectors

M Karami[1],M Raisee[1] and S Delfani[2]
[1] School of Mechanical Engineering, College of Engineering, University of Tehran, PO Box 14395-515, Tehran, Iran
[2] Department of Installations, Building and Housing Research Center (BHRC), Tehran, PO Box 13145-1696, Iran

E-mail: karamimaryam@ut.ac.ir

Abstract. Solar thermal collectors are applicable in the water heating or space conditioning systems. Due to the low efficiency of the conventional collectors, some suggestions have been presented for improvement in the collector efficiency. Adding nanoparticles to the working fluid in direct absorption solar collector, which has been recently proposed, leads to improvement in the working fluid thermal and optical properties such as thermal conductivity and absorption coefficient. This results certainly in collector efficiency enhancement. In this paper, the radiative transfer and energy equations are numerically solved. Due to laminar and fully developed flow in the collector, the velocity profile is assumed to be parabolic. As can be observed from the results, outlet temperature of collector is lower than that obtained using uniform velocity profile. Furthermore, a suspension of carbon nanohorns in the water is used as the working fluid in the model and its effect on the collector efficiency is investigated. It was found that the presence of carbon nanohorns increases the collector efficiency by about 17% compared to a conventional flat-plate collector. In comparison with the mixture of water and aluminium nanoparticles, a quite similar efficiency is obtained using very lower concentration of carbon nanohorns in the water.

1. Introduction

A drastic increase in energy consumption and global problems caused by fossil fuels such as climate change and greenhouse effect, are the main reasons of larger attention to renewable energies in the world. On the other hand, lack of fossil fuel sources and increased energy demand due to population growth, building and industries development leads to move faster toward alternative energies for more environmental friendly options.

One of the best alternative energy sources which can be used instead of fossil fuels is solar energy considering economical, environmental and safety concerns. Thus, in the last two decades, solar energy technologies have been continually developed for efficient collection and conversion of solar radiation to useful energy.

Conventional solar collectors suffer from many limitations, such as surface-based absorption of incident solar flux causing high heat losses and the low thermal properties of the working fluid. Therefore, an enhancement in the efficiency of solar harvesting devices is a basic challenge which requires great effort.

Direct Absorption Solar Collectors (DASC), initially proposed in 1970s by Minardi and Chuang [1], utilize the concept of volumetric-based absorption of solar radiation to improve the collector efficiency. They developed a direct collector that absorbed solar radiation by the black fluid (water with 3.0 g/l. India ink), which flows in transparent plastic tubing formed in an oval shape.

An approach to conquer other limitations of solar collectors (i.e. poor absorption and thermal properties of the working fluid) is to add small sized solid particles to the working fluid as an

[1] Corresponding author. Tel.: +98 21 88255942
E-mail address: karamimaryam@ut.ac.ir (M. Karami)

IOP Conf. Series: Materials Science and Engineering **64** (2014) 012044 doi:10.1088/1757-899X/64/1/012044

absorbing medium, as proposed and studied originally by Hunt et al. [2] and Abdelrahman et al. [3]. It is believed that incident radiation scattering which is caused by small particles, leads to an increase in radiation absorption within the fluid and thus improvement in collector efficiency. On the other hand, other thermophysical properties of nanoparticles in colloidal suspensions such as thermal conductivity and convective heat and mass transfer (Prasher et al. [4] and Krishnamurthy et al. [5]) is enhanced; therefore, direct absorption receivers which apply particle-laden-fluid as working medium have a great potential for better harnessing solar radiation.

Another configuration for DASC was proposed by Arai et al. [6] as a "volume heat-trap" solar collector in which a "fine-particle semitransparent liquid suspension" (FPSS) is used as a heat storage fluid. Based on this concept, the complete model was developed by Kumar and Tien [7] which considered the interaction of radiative flux with convective transport in an absorbing, emitting, and scattering particulate-laden falling film without neglecting the flow field or scattering in the film, not assuming the properties to be spectrally constant.

Recently many researchers have investigated the effect of using nanofluids (a liquid-nanoparticle suspension) as the working fluid on the efficiency improvement of DASC. Tyagi et al. [8] have numerically evaluated the performance of a "nanofluid-based direct absorption collector" using aluminum nanoparticle suspensions in water. The efficiency enhancement of 10% was obtained using DASC in comparison to a conventional flat-plate collector. A schematic of direct absorption solar collectors is shown in Figure 1.

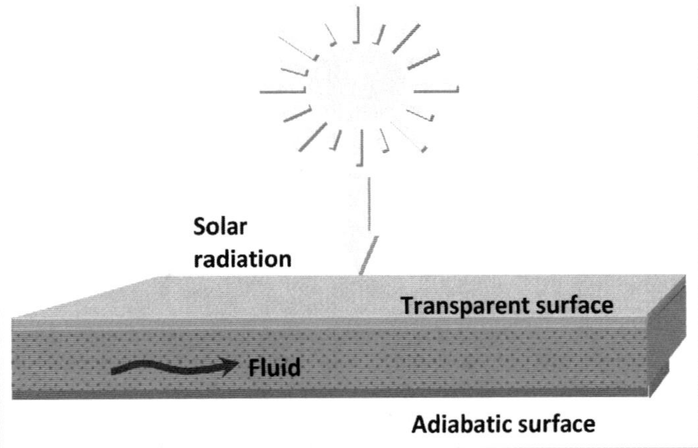

Figure 1. Schematic of a direct absorption solar collector

The radiation absorption characteristics of Ni nanoparticles are studied by Kameya and Hanamura [9]. Using spectroscopic transmission measurement and prediction process based on Mie theory, they found that the absorption coefficient of Ni-based nanofluid is much higher than that of base liquid.

Because of very high thermal conductivity, the family of carbon-based nanostructured materials like carbon nanoparticles, nanotubes and nanohorns can be one of the best candidates for solar energy harvesting in collectors. Sani et al. [10] used the nanofluids based on the carbon nanohorns as direct sunlight absorbers in their work. With regard to promising thermal and optical properties of nanohorns, a considerably higher sunlight absorption and thermal conductivity than pure water is obtained.

A numerical investigation similar to Tyagi et al. [8] was conducted by Otanicar et al. [11] and the numerical results are validated by comparisons made with experimental data obtained from a micro solar collector. They have reported an increase in collector efficiency for nanofluids made from carbon nanotubes, graphite and silver of nanoparticles.

Veeraragavan et al. [12] presents an analytical model that studies the effect of heat loss, particle loading, solar concentration and channel height on collector efficiency. They have also determined an optimum receiver length where the total efficiency is maximized.
In this paper, a numerical investigation is performed to calculate the DASC efficiency. First, the results of model using variable velocity are compared with those of a model using uniform velocity. Then, the extinction coefficient of carbon nanohorns is calculated according to Lambert-Beer law for efficiency calculation of collector based on single wall carbon nanohorns (SWCNH) suspensions in water. As another part of this study, the effect of volume fraction and collector height on the efficiency is considered. Finally, the efficiency improvement by carbon nanohorns-based nanofluids in direct absorption solar collector is compared with a flat-plate collector and aluminium nanofluid-baesd DASC.

2. Mathematical model

The incident solar intensity neglecting atmospheric absorption is calculated using the blackbody relation given by Eq. (1):

$$I_{b\lambda}(\lambda,T)=\frac{2hc_0^2}{\lambda^5\left[\exp\left(\frac{hc_0}{\lambda k_B T}\right)-1\right]} \quad (1)$$

where h is Planck's constant, k_B is the Boltzmann constant, and c_0 is the speed of light in a vacuum.

The radiative transfer equation (RTE), which accounts for the change of intensity as the radiation passes through an absorbing, emitting and scattering medium is written as [13]:

$$\frac{\partial i_\lambda}{\partial y}=-K_\lambda i_\lambda+a_\lambda i_{\lambda b}+\frac{\sigma_{s\lambda}}{4\pi}\int_{\omega_i=0}^{4\pi} i_\lambda(y,\omega_i)\Phi\Phi(\lambda,\omega,{}_i)d\omega_i \quad (2)$$

where i is the incoming radiation after subtracting the reflected component, λ is the wavelength, ω is solid angle, K_λ ($=a_\lambda+\sigma_{s\lambda}$) is the extinction coefficient composed of the absorption coefficient a_λ and the scattering coefficient $\sigma_{s\lambda}$.

In this study, we utilize the Lambert-Beer law for calculation of carbon nanohorns-based nanofluid extinction coefficient K(λ), which utilizes the spectral transmittance T(λ):

$$T(\lambda)=e^{-K(\lambda).l} \quad (3)$$

where l is the propagation length within the medium. In the current study, the spectral transmittance of carbon-nanohorns is obtained from the experimental work of Sani et al. [10].

In the current study, the following assumptions are made for numerical solution of RTE: the gain by emission is neglected because of low temperatures in the collector and the gain by incoming scattering is very small compare to the absorbed radiation due to the small size particle approximation (Rayleigh scattering). Under the assumptions, the second and third terms (augmentation by emission and incoming scattering) can be eliminated and thus, the equation of radiative transfer can be simplified as:

$$\frac{\partial i_\lambda}{\partial y}=-K_\lambda i_\lambda \quad (4)$$

The temperature distribution within the collector working fluid is obtained by solving steady state two-dimensional energy equation shown below:

$$\rho c_p u(y)\frac{\partial T}{\partial x} = k_{eff}\frac{\partial^2 T}{\partial y^2} - \frac{\partial q_r}{\partial y} \tag{5}$$

where q_r is the radiative heat flux obtained from:

$$q_r = 2\pi\int i_\lambda d\lambda \tag{6}$$

and u(y) is the flow velocity as a function of collector depth.

A schematic of computational domain of the solar collector is shown in Fig. 2. The boundary condition at the top surface of the collector was considered as convection heat transfer, whilst the bottom was considered as adiabatic.

The collector efficiency can be defined as:

$$\eta = \frac{\int_0^h \rho u(y)\left[T_{out}(y) - T_{in}\right]dy}{G_T} \tag{7}$$

where $T_{in}(=35^\circ C)$ is the fluid inlet temperature and $G_T(=1000\,W/m^2)$ is the incident solar flux on the collector.

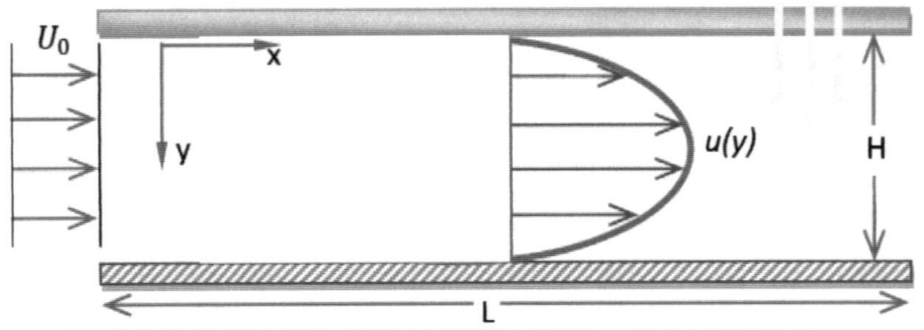

Figure 2. The schematic of computational domain

Equations (2) and (4) were solved numerically using the forward difference implicit method and the results are compared with those provided by Tyagi et al. (2009) in Figure 3 for the same working conditions. As can be seen, the results of two models are in acceptable agreement.

3. Results and discussion

CNHs, which discovered in 1998 by Dr. Sumio Iijima's research group [17], have many advantages such as carbon nanotubes. For inatance, it has huge surface areas and therefore can be fabricated with high purity. Also, unsimilar to carbon nanotubes (CNT), carbon nanohorns can be made in a simple way without using any catalyst. Another advantage of CNH is the better thermal property in absorbing radiation comprised with CNT [18].

In the earlier works [10, 18 and 19], the researchers studied the absorption and scattering properties of aqueous suspensions of single wall carbon nanohorns and calculated stored energy fraction for different concentrations. In this paper, the intention of the authors is to investigate the effect of carbon nanohorn-based nanofluid on the performance of direct absorption solar collector.

Fig. 3 shows the temperature distribution as a function of the normalized depth of the collector (y/H) and at various SWCNH concentrations. This plot shows that, as expected, the temperature of the fluid increases by increasing SWCNH concentration. Moreover, the change in temperature is not uniform. The temperature increase is the highest for the top layers and decreases with depth, because the maximum attenuation in the spectral intensity occurs in the top layers, and decreases along the depth of the collector.

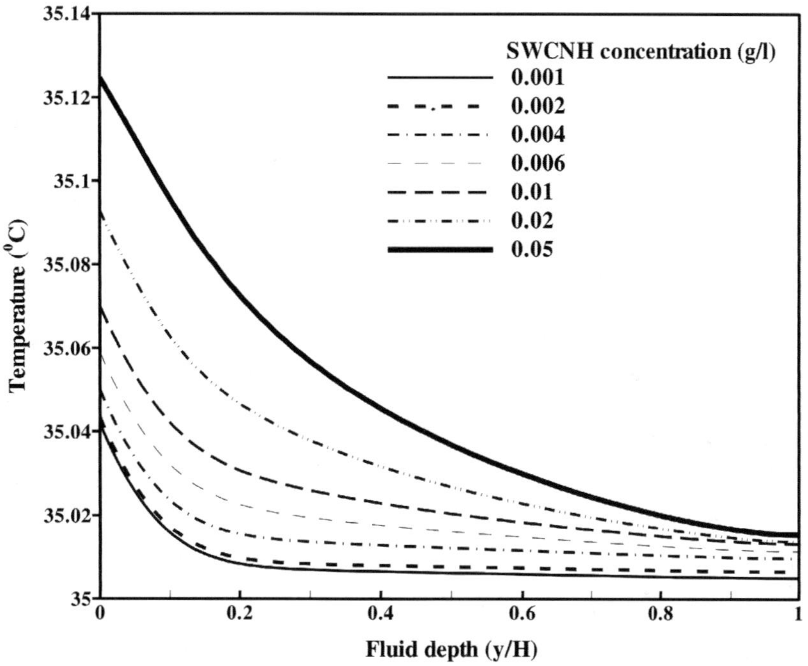

Figure 3. Temperature distribution within the solar collector for the various concentration of SWCNH (g/l)

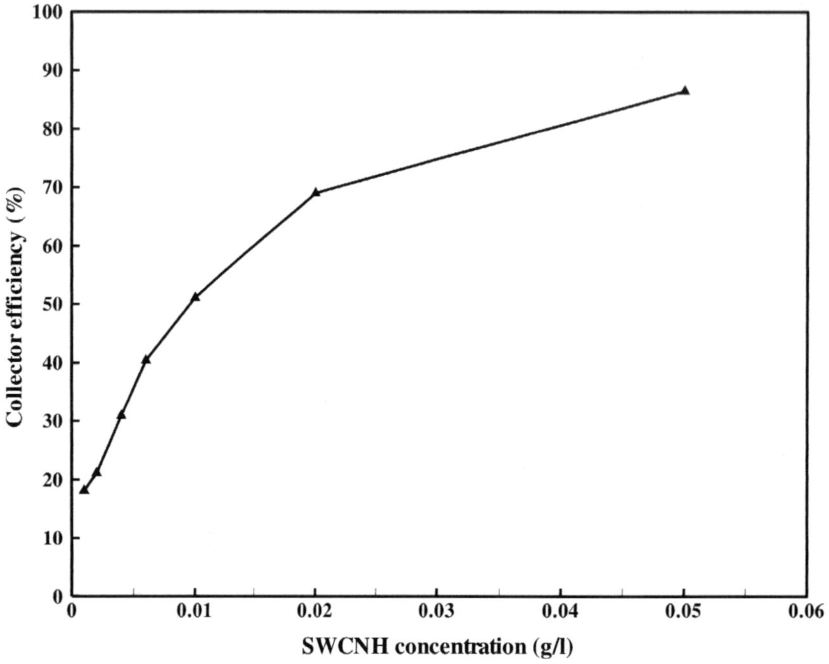

Figure 4. Collector efficiency for the various concentration of SWCNH (g/l)

The influence of SWCNH concentration on the collector efficiency is presented in Fig. 4. With the addition of more nanoparticles, more solar flux incident is expected to absorb and hence, the collector efficiency increases. This increase is considerably rapid at low concentrations and then, becomes slow at higher concentrations.

The variation of the collector efficiency versus height (H) is shown in Fig. 5. It is seen that by increasing collector height, its efficiency increases and finally reaches an asymptotic value of about 90%. This can be explained by the fact that more incident solar flux is attenuated across the greater depth and is absorbed by nanofluid which leads to higher temperature and thus higher solar collector efficiency.

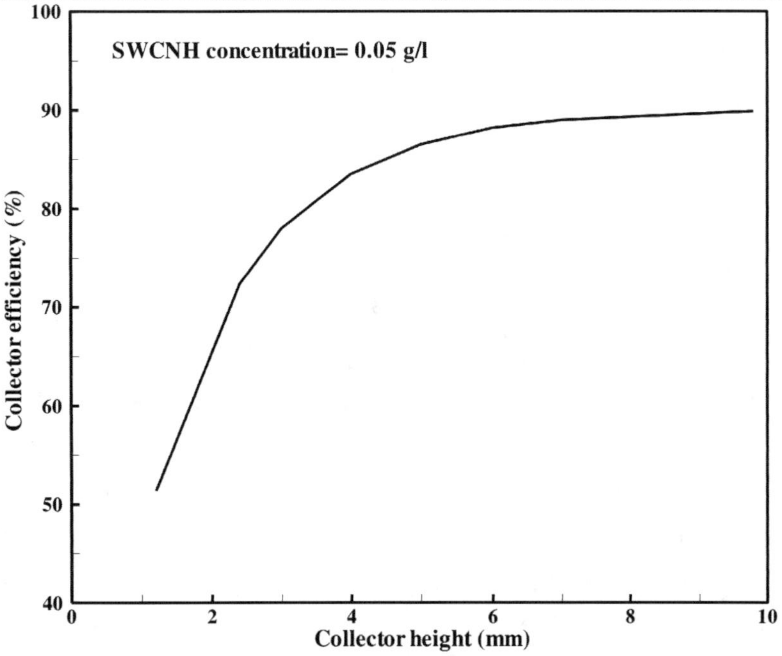

Figure 5. Collector efficiency as a function of collector height (H)

To better understand the effect of height on the collector efficiency, the temperature contours for three heights are shown in Fig. 6. These contours indicate that more solar radiation is absorbed in bottom levels by increasing collector height; hence, the outlet temperatures are rised. In collector height of 5mm, about 95% of incoming solar radiation is absorbed. This result indicates that increasing collector height will be valuable only up to a certain limit, beyond which the increase in efficiency would be miniscule

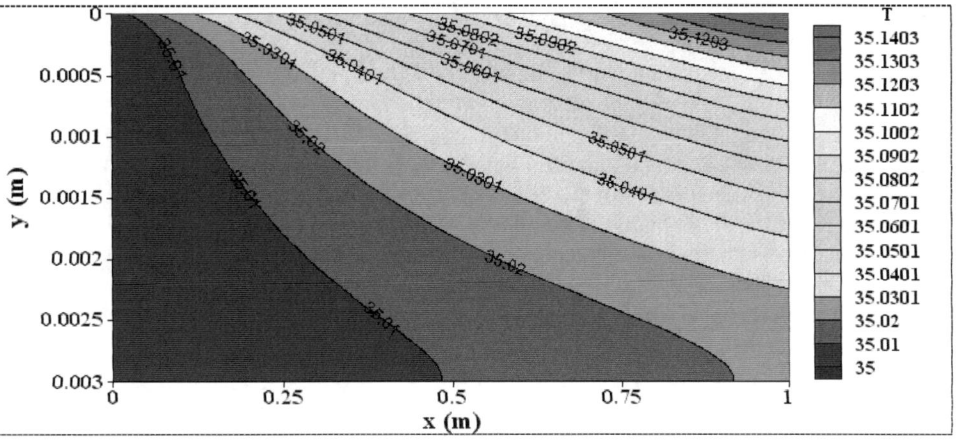

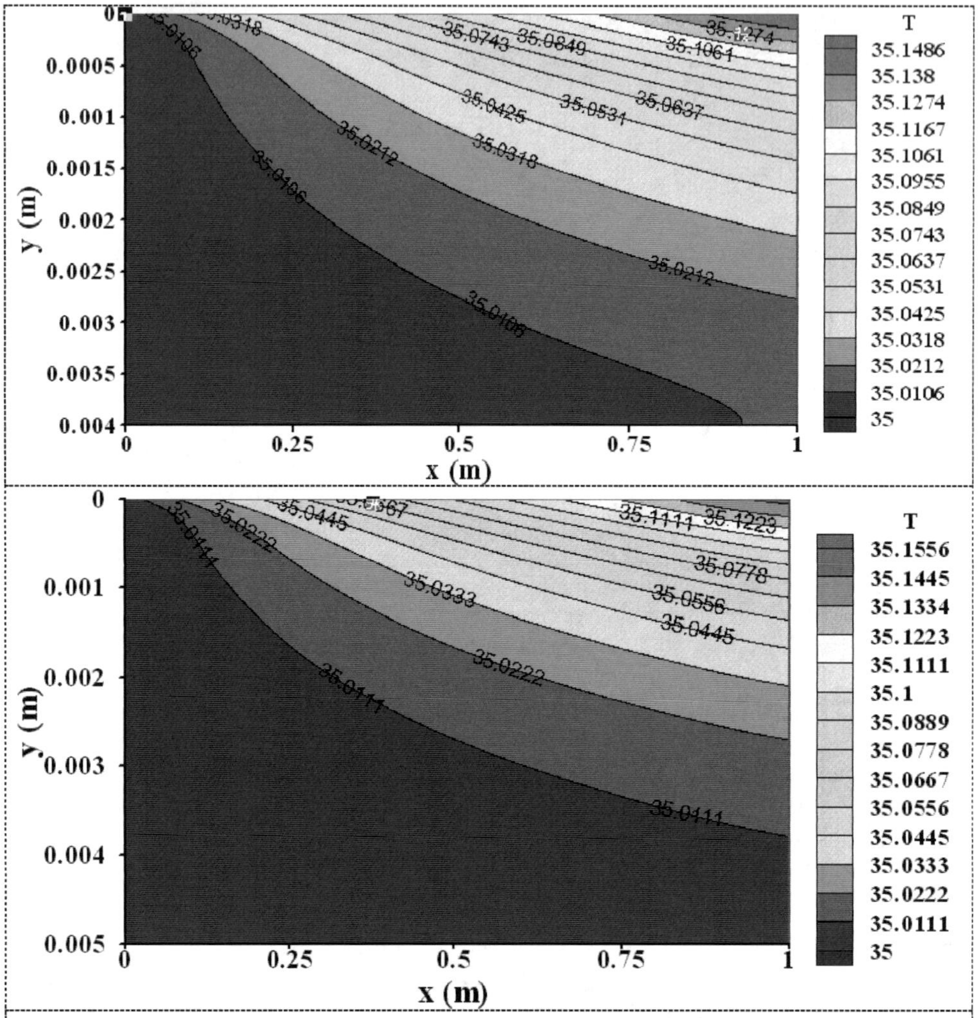

Figure 6. Outlet temperature contours for three different collector heights at 0.05 g/l concentration of SWCNH

To highlight the comparison between our nanofluid-based collector model and the conventional one, the efficiency of these collectors is compared and the results are shown in Fig. 7. The efficiency improvement of about 0.8% is obtained by adding 0.02 g/l carbon nanohorns (h=5mm) to the base fluid (water) compare to a typical flat-plate collector. This increase is 9% and 17% for 0.05 g/l carbon nanohorns and collector heights 3 and 5 mm, respectively. This increase in the efficiency using carbon nanohorns shows the potential use of this nanofluid-based collector in solar heating systems

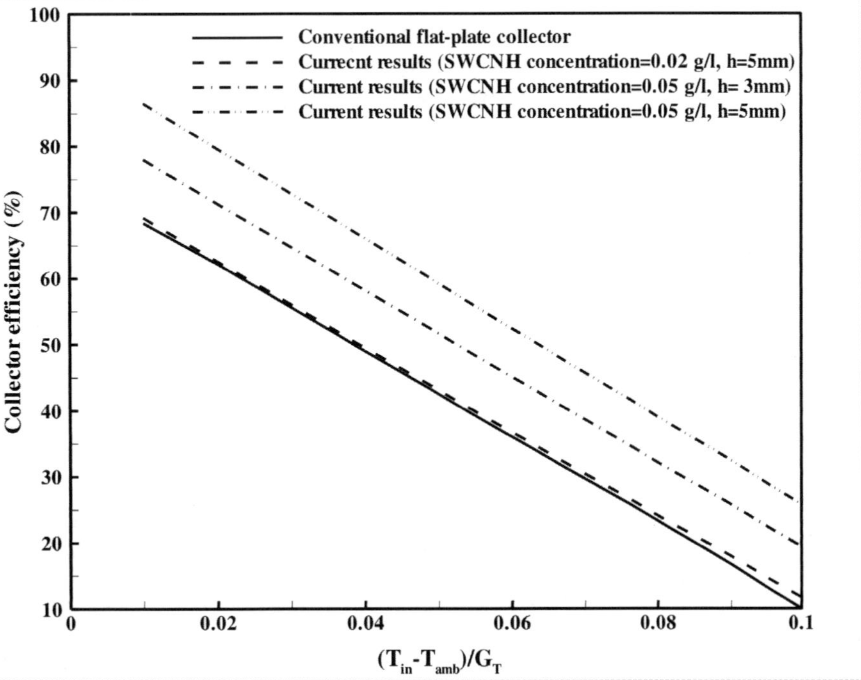

Figure 7. Variation of the collector efficiency versus the normalized fluid inlet temperature ($T_{in} - T_{amb}/G_T$)

For making comparison, the collector efficiency is plotted using carbon nanohorns- and aluminium nanoparticles-based nanofluid as the working fluid of DASC (results of Tyagi's study) (Fig. 8). At the same collector height, a little difference (about 1%) is shown in the calculated efficiencies, whereas aliminium nanoparticles concentration in the water is very upper than to carbon nanohorns concentration. This represents that carbon nanohorns, the most recent in the family of carbon nanostructures, can play a very promising role in enhancing the DASC efficiency.

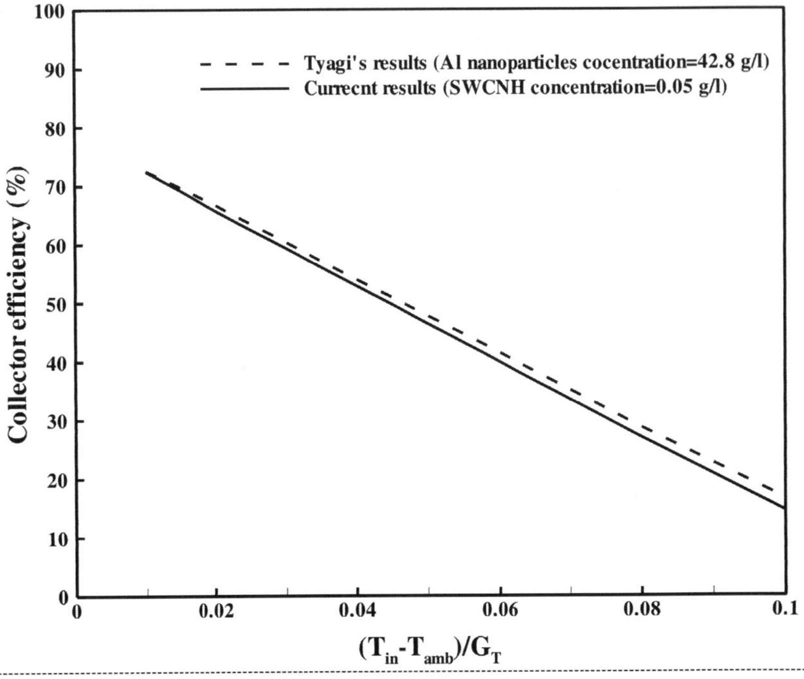

Figure 8. Comparison of the collector efficiency for aqueous suspension of Al nanoparticles and SWCNH (H=2.4 mm)

4. Conclusion

In this paper, a two-dimensional heat transfer model for analysis of nanofluid-based direct absorption solar collector was developed. The velocity profile between two plates of collector is assumed parabolic and its effect on the results is considered. It was found that the calculated outlet temperatures and efficiencies assuming constant velocity is higher than those obtained based on variable velocity. The performance of direct absorption solar collector is considered using the model with a mixture of carbon nanohorn and water as the working fluid. The effect of various parameters such as nanoparticle concentration and collector height on the efficiency was evaluated. The results show that the collector efficiency can be increased significantly by increasing height of the collector and the carbon nanohorns concentration, but only up to a certain limit, beyond which the increase in efficiency would be miniscule. Finally, the efficiency of the solar collectors utilizing aqueous suspension of carbon nanohorns (h=5mm, 0.05 g/l SWCNH concentration) is higher about 17% than that of a conventional flat-plate collector. At the same conditions, the collector performance can be better by adding nanohorns in proportion to aluminium nanparticles to the base fluid.

Nomenclature

A	Area (m^2)
a	Absorption coefficient
c_0	Speed of light ($m\ s^{-1}$)
c_p	Specific heat ($J\ kg^{-1}\ K^{-1}$)
D	Mean particle diameter (m)
f_v	Particle volume fraction (%)
G_T	Incident solar flux ($W\ m^{-2}$)
h	Planck's constant (J s)
H	Thickness of solar collector (m)

i	Intensity (W m^{-2} str^{-1} μm^{-1})
k	Thermal conductivity (W m^{-1} K^{-1})
k_B	Boltzmann constant (J K^{-1})
K	Extinction coefficient (m^{-1})
L	Length of the collector (m)
m	Complex refractive index
n	Index of refraction
q	Heat flux (W m^{-2})
T	Temperature ($^{\circ}C$)
u	Fluid velocity (m s^{-1})
U	Inlet fluid velocity (m s^{-1})
x,y	Coordinates

Greek symbols

α	Size parameter
η	Collector efficiency (%)
κ	Index of absorption
λ	Wavelength (m)
ρ	Density (kg/m^3)
ω	Solid angle (str)
σ	Scattering coefficient
Φ	Phase function

References

[1] Minardi J.E. and Chunag H.N. Performance of a Black Liquid Flat-Plate Solar Collector. Solar Energy 17, 179-183,1975.

[2] Hunt A.J. Small Particle Heat Exchanger. Lawrence Berkeley Laboratory, LBL-7841,1978.

[3] Abdelrahman M., Fumeaux M. and Suter P. Study of solid-gas-suspensions-used for direct absorption of concentrated solar radiation. Solar Energy 22, 45-48,1979.

[4] Prasher R., Phelan P. E. and Bhattacharya P. Effect of Aggregation Kinetics on the Thermal Conductivity of Nanoscale Colloidal Solutions (Nanofluid). Nano Letters, 6, No.7, 1529-1534,2006.

[5] Krishnamurthy S., Bhattacharya P., and Phelan P. E. Enhanced Mass Transport in Nanofluid. Nano Letters, 6, No. 3, 419-423, 2006.

[6] Arai N., Italya Y.and Hastani M. Development of a Volume heat-trap type solar collector using a fine-particle semitransparent liquid suspension (FPSS) as a heat vehicle and heat storage medium. Solar Energy, 32, 49-18563,1984.

[7] Kumar S. and Tien C.L. Analysis of combined radiation and convection in a particulate laden falling film. Heat transfer Phenomena in Radiation, Combustion and Fires, National Heat Transfer Conference, 1989.

[8] Tyagi H., Phelan P.and Prasher R.S. Predicted Efficiency of Nanofluid- Based Direct Absorption Solar Receiver. Journal of Solar Energy Engineering, 131, 2009.

[9] Kameya Y. and Hanamura K. Enhancement of solar radiation absorption using nanoparticle suspension. Solar Energy, 85, 299–307, 2011.

[10] Sani E., Barison S., Pagura C., Mercatelli L., Sansoni P., Fontani D. and Jafrancesco D. and Francini F. Carbon nanohorns-based nanofluids as direct sunlight absorbers. Optics Express180, 18, No.5, 2010.

[11] Otanicar T.P., Phelan P.E., Prasher R.S., Rosengarten G. and Taylor A.A. Nanofluid-based direct absorption solar collector. Journal of Renewable and Sustainablw Energy 2, 2010.

[12] Veeraragavan A., Lenert A, Yilbas B., Al-Dini S., Wang E. N. Analytical model for the design of volumetric solar flow receivers. International Journal of Heat and Mass Transfer 55, 556–564, 2012.
[13] Siegel R. and Howell J.R. Thermal radiation heat transfer. Third ed., Hemisphere publishing corporation, 1993.
[14] White F.M. Fluid Mechanics. Fifth ed., McGraw-Hill Co., Boston, 2003

Phase Change Insulation for Energy Efficiency Based on Wax-Halloysite Composites

Yafei Zhao, Suvhashis Thapa, Leland Weiss, Yuri Lvov*

Institute for Micromanufacturing, Louisiana Tech University, Ruston, LA 71270

E-mail: ylvov@latech.edu

Abstract. Phase change materials (PCMs) have gained extensive attention in thermal energy storage. Wax can be used as a PCM in solar storage but it has low thermal conductivity. Introducing 10% halloysite admixed into wax yields a novel composite (wax-halloysite) which has a thermal conductivity of 0.5 W/mK. To increase the base conductivity, graphite and carbon nanotubes were added into the PCM composite improving its thermal energy storage. Thermal conductivity of wax-halloysite-graphite (45/45/10%) composite showed increased conductivity of 1.4 W/mK (3 times higher than the base wax-halloysite composite). Wax-halloysite-graphite-carbon nanotubes (45/45/5/5%) composite showed conductivity of 0.85 W/mK while maintaining the original shape perfectly until 91 °C (above the original wax melting point). Thermal conductivity can be further increased with higher doping of carbon nanotubes. This new composites are promising heat storage material due to good thermal stability, high thermal/electricity conductivity and ability to preserve its shape during phase transitions.

1. Introduction

Due to the increasing demand for energy and energy sustainable development, organic solid–liquid phase change materials (PCMs) have attracted considerable attention for storing thermal energy and controlling temperature by storing and releasing significant latent heat during phase change process [1-3]. Among them, paraffin wax was widely employed for thermal energy storage in many applications due to high heat of fusion, appropriate melting temperature, high specific heat capacity, chemically inert, stable, noncorrosive and nontoxic properties [2, 4-6]. However, low thermal conductivity (0.25 W/mK) and shape instability above the wax melting point limit its applications. To overcome these problems, a wide range of research has been undertaken to enhance wax thermal conductivity and form shape stabilized material. The commonly used method is to disperse in-wax carbon nanotubes, expanded graphite, copper and other metal nano/micro particles to form the composites [2-5, 7-12]. Carbon nanotubes and metal particles essentially increase the composite cost while decreasing the total volume of latent heat thermal storage. Moreover, these composites often have poor stability [3].

Halloysite nanoclay, a natural environmentally friendly low-cost material, attracted a great interest due to its hollow tubule structure, high surface area and good polymer mixability without need for exfoliation. Halloysite nanotubes are formed by rolling kaolinite alumosilicate sheets [13-16]. The inner lumen of halloysite tubes can be loaded with various chemicals, such as metals, chemical inhibitors, dyes, antiseptics, drugs and proteins [17-21]. Zhang et al. found that halloysite maintains wax shape during heat absorption and phase change and can improve thermal conductivity [22].

Compared with the aforementioned materials, halloysite nanotubes as additives are well mixable with wax and yield a pseudo ceramic "skeleton" that maintains composite structure even above the wax melting temperature.

In this paper, a shape stabilized phase change material of wax and halloysite was developed. It has form-stable shape even when the temperature exceeds the melting point and has a higher thermal conductivity than wax. Heat shape stabilization and thermal conductivity of the composite was optimized by through the additional of graphite or carbon nanotubes.

2. Experimental and Methods

2.1. Materials and preparation of wax/halloysite composites

Halloysite from Applied Minerals Inc (USA) was used without further treatment. The inner and outer diameter of halloysite is in the range 10-20 nm and 40-60 nm, while the wall thickness is 15-20 nm. The average surface area of halloysite is 54 m^2/g. Technical grade wax (melting point Tm=50-53 °C), isooctane, graphite, and carbon nanotubes were purchased from Sigma-Aldrich. Halloysite clay and wax were mixed in the isooctane solution. The mixture was stirred and refluxed under 80 °C for 3 h. Then the mixture was vacuum dried overnight. The wax/HNT composites were prepared with the different weight ratios. In order to increase thermal storage, graphite and carbon nanotubes were added in the mixture at the beginning. The following steps were consistent with the same method mentioned above.

2.2. Thermal conductivity measurement

Composites were tested in a specially designed chamber. It was fabricated from 5.8 mm thick polyacrylate with dimensions of 35 x 30 x 15 mm. Ambient temperature around the setup was maintained at 25 °C. Data was recorded and analyzed via Lab-View setup on a connected computer. The four k-type thermocouples provided temperature data which was used to determine the average thermal conductivity k of each mixture tested. Equation 1 provided the basis for this calculation where q was the heat transfer from the resistance heater (measured via heat flux sensor), A was the area of heat transfer, ΔT was the temperature gradient between thermocouple readings and Δz was the vertical distance between them.

$$k = \frac{q/A}{\Delta T/\Delta z} \qquad (1)$$

3. Results and Discussion

3.1. Thermal characteristics of PCM composites

Figure 1 shows the thermal properties of wax and wax composites by using DSC, including transition temperature (T_t), melting temperature (T_m) and latent heat (ΔH). A sharp principal peak which signifies a solid-to-liquid transition and a minor peak which represents a solid-to-solid transition appear in pure wax [23]. In shape-stabilized phase changing materials, these two peaks of paraffin wax are still exist and obvious. This could be explained as there were no chemical reactions between paraffin wax, graphite and carbon nanotubes in the preparation of the shape-stabilized PCM. The shape stabilized composites exhibit almost the same peaks as that of paraffin wax, meaning the transition temperature (T_t) and melting temperature (T_m) are constant within a standard deviation of ±1 °C, although halloysite, graphite and carbon nanotubes were dispersed throughout the paraffin wax. The solid-to-liquid phase change is discussed here because it is the main phase change of paraffin wax. One sharp phase change peak at 52.3 °C is observed for pure wax which shows the melting point is 52.3 °C. In addition, the latent heat is 171.0 J/g. The lower strength of the peaks and smaller area below the peaks indicate that all the composites have lower latent heat than pure wax due to the reduced wax content.

The thermal properties data obtained from these characterizations are also shown in Table 1 and compared with the theoretical latent heat calculated according to Equation (2).

$$\Delta H_{composite} = \omega \Delta H_{wax} \quad (2)$$

where $\Delta H_{composite}$, ω and ΔH_{wax} are the calculated latent heat of the shape stabilized PCM, mass ratio of wax in the composite and latent heat of wax, respectively. The measured latent heat fits well with the calculated values, yielding differences less than -14 %.

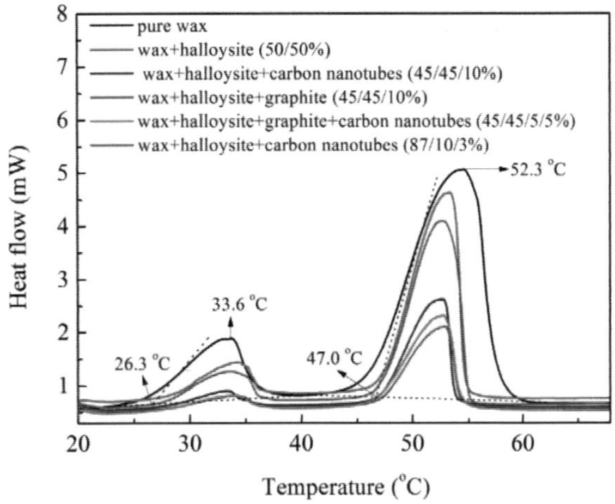

Figure 1. DSC curves of pure wax (a), wax/halloysite (50/50%) (b), wax/halloysite/carbon nanotubes (45/45/10%) (c), wax/halloysite/graphite (45/45/10%) (d), wax/halloysite/graphite/carbon nanotubes (45/45/5/5%) (e) and wax/halloysite/carbon nanotubes (87/10/3%) (f).

Table 1. Thermal properties of pure wax and shape stabilized PCM composites

Wax:halloysite:graphite :carbon nanotubes (w%)	Melting				
	Transition T (°C)	Melting point T (°C)	Total latent heat ΔH (J/g)	Calculated ΔH (J/g)	Difference (%)
100:0:0:0	47.0	52.3	171.0	—	—
50:50:0:0	46.8	53.5	87.1	85.5	-1.9
45:45:10:0	47.1	52.9	68.4	76.9	11
45:45:5:5	46.8	52.7	88.1	76.9	-14
87:10:0:3	47.4	52.7	145.7	148.7	2.1
45:45:0:10	47.1	52.7	75.7	76.9	1.6

3.2. Thermal conductivity of PCM composite

Thermal conductivity is one of the most important properties of PCMs. The thermal conductivity of different PCM composite was summarized in Table 2. Adding 10% of halloysite can increase the thermal conductivity of wax twice and the composite could keep shape above wax melting temperature 50 °C. Further increasing the ratio of hallosite to 50% resulted in the decrease of PCM composite's thermal conductivity but increase the shape maintenance temperature to 70 °C. In order to improve both the thermal conductivity and shape-stabilized temperature, carbon nanotubes and graphite were introduced to the PCM composite. 5% carbon nanotubes can increase the thermal conductivity whereas the composite would deform its shape at 50 °C. However, 10% graphite can enhance both the thermal conductivity as high as 1.4 W/mK and shape deformation temperature up to 81 °C. Graphite alone dispersed into paraffin wax can increase the thermal conductivity to 0.65 W/mK, but it cannot enhance the shape stable temperature. When wax/halloysite composite contains both graphite and carbon nanotubes, shape deformation temperature can be increased to 91 °C.

Table 2. Thermal conductivity of PCM composites

Wax:halloysite:graphite:carbon nanotubes (wt %)	Thermal conducivity (W/mK)	Shape deformation T (°C)
100:0:0:0	0.25	—
0:100:0:0	0.10	53
90:10:0:0	0.50	60
50:50:0:0	0.36	70
85:10:0:5	0.57	53
45:45:10:0	1.40	81
90:0:10:0	0.65	53
45:45:5:5	0.85	91

3.3. Stable performance of wax/composites

The leakage tests above the melting temperature of wax were executed to compare the form-stable performance of the as-prepared composite PCMs. Figure 3 shows the photographs of pure wax, and different mixtures of wax/halloysite/graphite kept at 90 °C for 20 min. As seen from the photographs of the 3 cm diameter cylindrical samples, all the pure wax and wax/carbon nanotubes (90/10%) melted to liquid and wax/halloysite (50/50%) had minimal liquid leakage while wax/halloysite/graphite (45/45/10 %) and wax/halloysite/graphite/carbon nanotubes (45/45/5/5 %) maintained form without liquid release from the matrix.

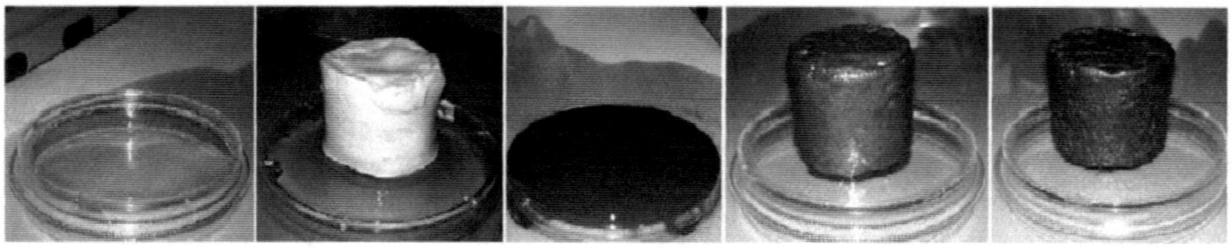

a b c d e

Figure 3. Photographs of pure wax (a), wax/halloysite (50/50 %) (b), wax/graphite (90/10 %) (c), wax/halloysite/graphite (45/45/10 %) (d) and wax/halloysite/graphite/carbon nanotubes (45/45/5/5 %) (e) at 90 °C.

The leakage test confirms that halloysite provides mechanical strength for the phase changing composites and prevents the seepage of the melted paraffin wax. In the shape stabilized composites, halloysite, graphite and carbon nanotubes serve as supporting materials and thermal conductivity promoter of wax. However, graphite or carbon nanotubes alone cannot improve the stability of wax significantly, and must be combined with halloysite to produce shape-stabilized material (Figure 3c). The wax undergoes phase transition in the network of halloysite, graphite and carbon nanotubes. Because of the uniform dispersion of halloysite into wax and the interfacial adhesion between wax, halloysite, and graphite, these composites could maintain their shapes even heated above the phase change temperature. There is no leakage of the wax even above melting point. These composite samples could maintain shape perfectly even after subjection to 50 melting–freezing cycles.

4. Conclusion

Doping wax with 50-nm diameter halloysite clay nanotubes provide a skeleton to maintain wax shape during heat absorption and phase transition even when the temperature of the phase change material exceeds the melting point. Therefore, such composite containing of 40-50% halloysite nanotubes allows wax melting without its leakage (while melting temperature is preserved). Halloysite also improves the thermal conductivity of wax composite. The thermal conductivity can be further adjusted by adding graphite or carbon nanotubes to the wax/halloysite composite. Due to good thermal stability, high thermal conductivity, ability to preserve shape during wax melting, and abundant availability of this natural tubule clay, the phase change insulation composite based on halloysite can be used to conserve energy in large scale practical applications.

Acknowledgments

NSF CMMI-1029147, NSF ECCS-1053729 and EP-1003897 grants are acknowledged for support. Any opinions, findings, and conclusions or recommendations expressed in this report are those of authors and do not necessarily reflect the view of National science foundation. Authors are also thankful to Applied Minerals Inc., USA for providing halloysite.

References

[1] Li B X *et al* 2013 *Chem. Engineer. J.* **215–216** 819
[2] Sharma A *et al* 2009 *Renewable Sustainable Energy Reviews* **13** 318
[3] Wang J F *et al* 2010 *Solar Energy* **84** 339
[4] Ehid R *et al* 2012 *Energy Conver. Manag.* **57** 60
[5] Zhang J S *et al* 2012 *Solar Energy* **86** 1142
[6] Li B X *et al* 2013 *ACS Sustainable Chem. Engineer.* **1** 374
[7] Sarı A *et al* 2009 *Chem. Engineer.* J.**155** 899
[8] Wu S Y *et al* 2010 *Energy Fuels* **24** 1894
[9] Yavari F *et al* 2011 *J. Phys. Chem. C* **115** 8753
[10] Cui Y B *et al* 2011 *Solar Energy Mater. Solar Cells* **95** 1208
[11] Ji P J *et al* 2012 *Chem. Engineer. Sci.* **81** 140
[12] Zhang H Z *et al* 2012 *Solar Energy Mater. Solar Cells* **102** 93
[13] Weng O Y *et al* 2012 *J. Am. Chem. Soc.* **134** 1853
[14] Lvov Y and Abdullayev E 2013 *Progress in Polymer Sci.* accepted, published online
[15] Luo P *et al* 2010 *Water Research* **44** 1489
[16] Zhao Y F *et al* 2010 *J. Hazard. Mater.* **178** 658
[17] Lvov Y *et al* 2008 *ACS Nano* **2** 814
[18] Abdullayev E *et al* 2011 *ACS Appl. Mater. Interfaces* **3** 4040
[19] Zhao Y F *et al* 2010 *Sepera. Sci. Technol.* **45** 1066
[20] Zhao Y F *et al* 2010 *Water Sci. Technol.* **62** 937
[21] Luo P *et al* 2011 *Industrial Engineer. Chem. Research* **50** 10246
[22] Mei D *et al* 2011 *Solar Energy Mater. Solar Cells* **95** 2772
[23] Sun Z M *et al* 2013 *Thermochimica Acta* **558** 16

IOP Conf. Series: Materials Science and Engineering **64** (2014) 012046 doi:10.1088/1757-899X/64/1/012046

Phosphonium modified clay/polyimide nanocomposites

Hatice Ceylan, Emrah Çakmakçı, Aslı Beyler-Çiğil, Memet Vezir Kahraman

Marmara University, Department of Chemistry, 34722 Goztepe-Istanbul, Turkey
E-mail: hatice.ceylan@marmara.edu.tr

Abstract. In this study, octyltriphenylphosphonium bromide [OTPP-Br] was prepared from the reaction of triphenylphosphine and 1-bromooctane. The modification of clay was done by ion exchange reaction using OTPP-Br in water medium. Poly(amic acid) was prepared from the reaction of 3,3',4,4'-Benzophenonetetracarboxylic dianhydride (BTDA) and 4,4'-Oxydianiline (ODA). Polyimide(PI)/clay hybrids were prepared by blending of poly(amic acid) and organically modified clay as a type of layered clays. The morphology of the Polyimide/ phosphonium modified clay hybrids was characterized by scanning electron microscopy (SEM). Chemical structures of polyimide and Polyimide/ phosphonium modified clay hybrids were characterized by FTIR. SEM and FTIR results showed that the Polyimide/ phosphonium modified clay hybrids were successfully prepared. Thermal properties of the Polyimide/ phosphonium modified clay hybrids were characterized by thermogravimetric analysis (TGA).

1. Introduction

Polyimide (PI) is one of the most important super-engineering materials because of its mechanical properties and its thermal stability [1-5]. Electronic industry is the typical extensive applications of polyimides including interlayer insulation films, buffer coating, films, alpha-ray shielding films, and alignment films for liquid crystal displays [1, 6-7].

The mechanical and thermal properties of polymers are generally improved by the addition of inorganic additives. The challenges in this area of high performance organic/inorganic hybrid materials are to obtain significant improvements in the interfacial adhesion between the polymer matrix and the reinforcing material since the organic matrix is relatively incompatible with the inorganic phase [8-9].

Clays are widely used in various areas of science. The use of clay in polyimide (PI) nanocomposites has attracted considerable attention in the field of materials science and technology [10]. Polyimide/organically modified clay nanocomposites appear as a material possessing improved tensile properties, thermal stability [11].

In order to disperse hydrophilic clay silicate layers into the hydrophobic polymer matrix, either the clay or the polymer need to be modified to enhance their compatibility. In practical purpose the first one is chosen, for its simplicity and better dispersion of clay layers into the polymer matrix. To make the clay compatible to the polymer matrix, bulky ions such as quaternary ammonium, quaternary phosphonium, and imidazolium are generally incorporated into the clay galleries. The choice of the modifier is based on the processing temperature of the nanocomposites and desired property in the final nanocomposites [11].

In this study, the modification of clay was performed by the incorporation of octyl triphenylphosphonium bromide into clay's interlayer spaces. We prepared poly(amic acid) from the reaction of 3,3',4,4'-Benzophenonetetracarboxylic dianhydride (BTDA) and 4,4'-Oxydianiline (ODA). Polyimide-clay hybrids (nanocomposites) were prepared by blending poly(amide acid) and organically modified clay.

2. Materials and methods

2.1. Materials

3,3',4,4'-Benzophenonetetracarboxylic dianhydride (BTDA) and 4,4'-Oxydianiline (ODA) were purchased from Sigma-Aldrich and were desiccated under vacuum overnight at 120°C. N.N. Dimethylformamide (DMF) was also purchased. Clay was obtained from a local supplier. The composition of the clay is given in Table I.

2.1. Methods

2.1.1. Preparation of octyl triphenylphosphonium bromide

Octyl triphenylphosphonium bromide were prepared from poly(amic acid), according to reference [10]. A solution of 1-bromooctane(0.02 moles) and triphenylphosphine(0.04 moles) was heated under reflux for 24 hours and continuous stirring. The mixture was then cooled and the yellow product was filtered, washed with diethyl ether and then dried at room temperature for 24 hours. Preparation route to octyl triphenylphosphonium bromide is given in Figure 1.

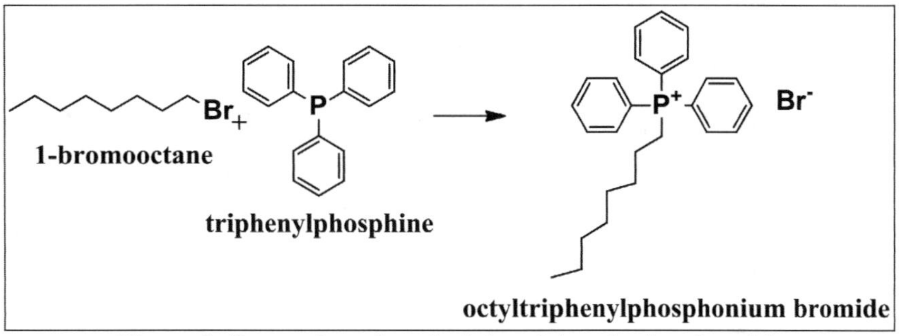

Figure 1. Preparation of octyl triphenylphosphonium bromide

2.1.2. Modification of the clay

Modification of clay was performed according to [12]. The clay was dispersed in 100 mL distilled water at 80 °C and stirred for 1h. A solution containing phosphonium salt with 70 mL distilled water was then added drop wise into suspension over a period of 10 min. The stirring was continued at the same temperature for 2h. The phosphonium modified clay was then filtered by using vacuum pump and washed distilled water to remove impurities and then dried at 100 °C for 24 h. The composition of clay can be seen in Table 1.

Table 1. Compositon of unmodified Clay

Metal Oxides	*Percent weight (%)*
SiO_2	62.01
Al_2O_3	15.21
Fe_2O_3	6.58
CaO	1.19
MgO	2.28
K_2O	1.13
Na_2O	1.59

2.1.3. Preparation of BTDA-ODA polyimide films

BTDA-ODA films were prepared from poly (amic acid), according to [13]. In order to synthesis poly(amic acid) (PAA) solutions, ODA(0.01 moles) was firstly put into a flame dried three-neck flask containing 16 mL of DMF under a nitrogen purge. After ODA was fully dissolved in DMF,

BTDA(0.01 moles) was added to flask batch by batch. The mixture was then stirred at room temperature for 24 h to take a viscous PAA solution. Polyimide films were prepared by casting viscous PAA solutions on clean dust-free glass plates. Then, thermal imidization was performed stepwise at 100, 200 and 300 °C for 1h at each temperature. Polyimide films were removed from glass plates by immersing in distilled water at 80 °C. Preparation route to polyimide films is given in Figure 2.

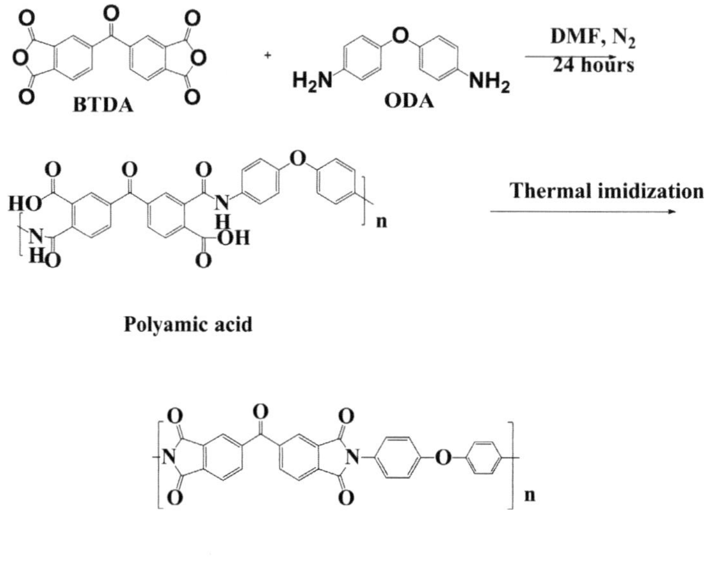

Figure 2. Preparation of polyimide film

2.1.4. Synthesis of BTDA/ODA-modified clay polyimide composites
BTDA/ODA-modified clay polyimide films were prepared, according to [13]. 3 wt % modified clay containing composite film was prepared by adding appropriate amount of modified clay to the PAA which was prepared as described above. Briefly, PI/modified clay composites films were prepared by dispersing 0.02 g of modified clay in 10 g of PAA solution. The mixture was stirred for 24 h to give a homogenous solution. Prior to thermal imidization this solution was sonicated for 30 min. Then cast films were cured as described above.

3. Characterization
FT-IR spectrum was recorded on Perkin Elmer Spectrum100 ATR-FTIR spectrophotometer. Thermogravimetric analyses (TGA) of the Polyimides were performed using a Perkin-Elmer Thermogravimetric analyzer Pyris 1 TGA model. Samples were run from 30 to 750 °C with heating rate of 10 °C / min under nitrogen atmosphere. SEM imaging of the films were performed on Philips XL30 ESEM-FEG/EDAX. The specimens were prepared for SEM by freeze-fracturing in liquid nitrogen.

4. Result and Discussion

4.1. Modification of Clay
In this study bentonite clay was modified with a phosphonium salt. The modification of clay was proved via FTIR analysis. ATR-FTIR spectra of neat clay and the phosphonium salt modified clay are given in Figure 3. In the spectrum of neat clay, the peaks at around 3600 cm^{-1} are attributed to the -OH stretching. The peaks at 997 cm^{-1} and 511 cm^{-1} were due to characteristic Si-O ether bonds and Al-O

bonds, respectively. After modification of clay, new peaks were observed in the FTIR spectrum. As can be seen in Fig. 3, the peaks at 2975 and 2855 cm^{-1} belong to the $-CH_2$ and $-CH_3$ of the groups of the octyl moiety. Moreover in this spectrum the peak at 1629 cm^{-1} can be attributed to the aromatic double bonds. These newly formed peaks prove that octyl triphenylphosphonium bromide was successfully intercalated between the layers of clay.

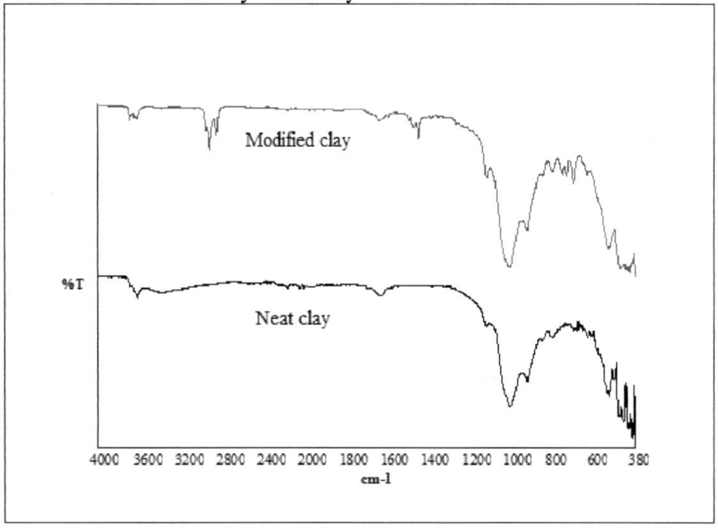

Figure 3. The FTIR spectra of neat clay and modified clay

4.2. Polyimide/PI-composite Synthesis

The FTIR spectra of PAA, PI and PI/clay composite are given in Figure 4. In the spectrum of PAA, the characteristic amide bonds stretching were found at 1660 cm^{-1}. After imidization this amide carbonyl peak was completely disappeared, indicating the full imidization. Moreover the spectrum of PI displays the characteristic carbonyl groups at around 1780 and 1715 cm^{-1}. When modified clay was incorporated into PI matrix, these carbonyl groups were slightly shifted. This can be attributed to the interactions between clay and PI. Also it can be interpreted that this situation is a result of the good dispersion of clay owing to the organophilic octyl groups.

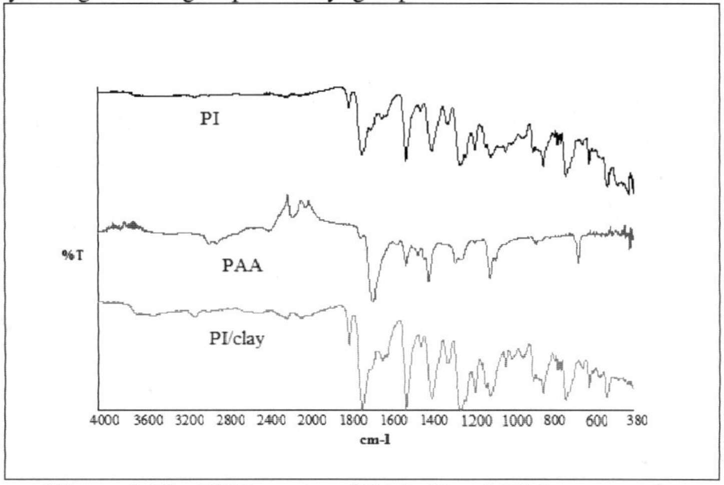

Figure 4. The FTIR spectra of PAA, PI and PI/clay composite

4.3. Thermo-gravimetric analysis

Thermal stability of the clay-PI composite was characterized by TGA under nitrogen atmosphere. The TGA spectra of PI and PI-clay composite are given in Figure 5. As can be seen the addition of phosphonium modified clay was found to improve the thermal stability of neat PI. The onset degradation temperature of neat clay was 480 degrees while it was shifted to 500 degrees when 3 wt. % clay was added. Moreover the maximum weight loss temperature of the composite PI was also higher than that of pure PI. The char yields of the PI and PI-Clay composite films were also collected at 750 degrees. As expected the char yield of the composite films produced more char than BTDA-ODA PI films. Thus it can be said that the addition of octyl triphenylphosphonium salt modified clay not only dispersed well in the organic PI matrix, it also improved the thermal properties of PI.

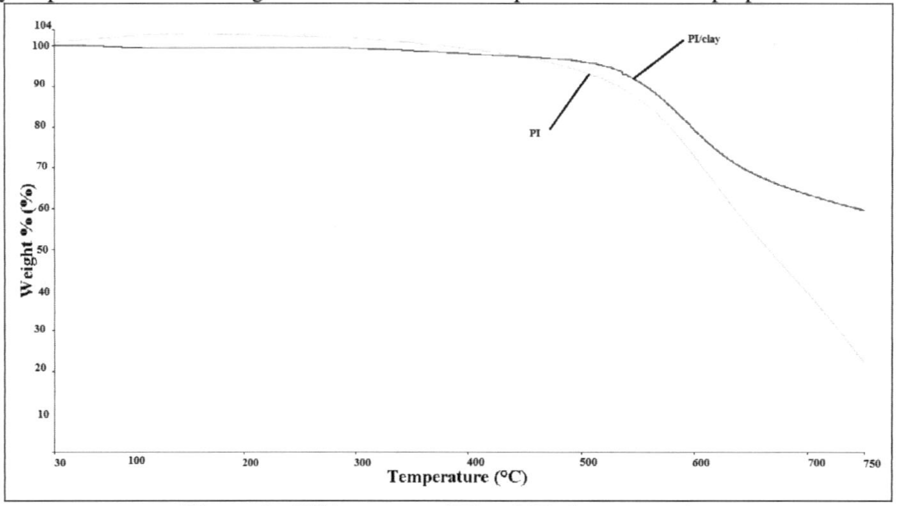

Figure 5. TGA spectra of PI and PI-clay composite

4.4. SEM analysis

The surfaces of PI and polymer film are normally investigated by means of the SEM technique. SEM micrographs of PI and PI/Clay films were given in Figure 6 from which obvious differences can be observed. Figure 6 show that the pure PI film has a very smooth surface. Surface roughness in nanocomposite increased with respect to pristine polyimide. It can be clearly seen that the particles (clay) are distributed uniformly in the polymer matrix for the hybrid film containing 3 wt % of clay.

Figure 6. SEM micrographs of a) pristine PI and b) the PI/clay film

5. Conclusions

Clay has been modified using octyl triphenylphosphonium bromide. The modified clay was characterized by using FTIR spectroscopy and TG analysis. FTIR spectra show that phosphonium salt is successfully intercalated into the interlayer of clay. Thermal stability of PI/modified clay film with appropriate clay content was better than this of original PI film. Morphology study with scanning electron microscopy showed that the surface roughness in nanocomposite increased with respect to pristine polyimide.

References

[1] Baiqing Tang, Dongdan Cai, Jun Sun, Jianjun Wang, Lixing Dai, 2013, *Polymer Composites* DOI 10.1002/pc.22616.
[2] Yi-He Zhang, Zhi-Min Dang, Shao-Yun Fu, John H. Xin, Jian-Guo Deng, Juntao Wu, Shiyong Yang, Lai-Feng Li, Qing Yan, 2005, *Chemical Physics Letters*, 401, 553-557.
[3] Emrah Çakmakçı, Aslı Beyler Çiğil, Özkan Danış, Serap Demir, Memet Vezir Kahraman, 2013, *Starch*, 65, 1-7.
[4] Yi-He Zhang, Jun-Tao Wu, Shao-Yun Fu, Shi-Yong Yang, Yan Li, Lin Fan, Robert K.-Y. Li, Lai-Feng Li , Qing Yan, 2004, *Polymer*, 45, 7579-7587.
[5] T. Agag, T. Koga, T. Takeichi, 2001, *Polymer*, 42, 3399-3408.
[6] Soane D, Martynenko Z. Polymers in microelectronics: *Fundamentals and applications*. Amsterdam: Elsevier, 1989.
[7] A.S. Mathews, I. Kim, and C.S. Ha, 2007 *Macromol. Res.*, 15, 114
[8] Jang BZ. 1992, *Compos Sci Technol*; 44:333.
[9] Sohn JE. 1985 *J Adhes*; 19:15.
[10] Saadon A. Aowda, Emad A.Jaffar Al-Mulla, Sadiq . J. Baqir, 2011 *Journal of al-padisiyah for pure science*, 16, 1-10.
[11] Supratim Suin, Sandip Maiti, Nilesh K. Shrivastava, B.B. Khatua, 2014, *Materials and Design* 54, 553-563.
[12] Supratim Suin, Nilesh K. Shrivastava, Sandip Maiti, B.B. Khatua, 2013, *European Polymer Journal* 49, 49-60.
[13] Çakmakci,E., Gungor,A., 2013, *Polym.Degrad.Stab.*, 98, 927-933.

IOP Conf. Series: Materials Science and Engineering **64** (2014) 012047 doi:10.1088/1757-899X/64/1/012047

Aging Resistant Elastomer Reinforcement by Antioxidant Loaded Clay Nanotubes

Ye Fu [1,3], Detao Zhao[2], Wencai Wang[1,2], Yuri Lvov [3,*]
1. State Key Laboratory of Organic Inorganic Composites, Beijing University of Chemical Technology, China.
2. Key Laboratory of Beijing City on Preparation and Processing of Novel Polymer Materials, Beijing University of Chemical Technology, China.
3. Institute for Micromanufacturing, Louisiana Tech University, USA.

Abstract. An aging resistant styrene-butadiene rubber (SBR) composite was prepared by using N-isopropyl-N'-phenyl-p-phenylenediamine (antioxidant 4010NA) loaded halloysite nanotube (HNT) as filler. The antioxidant loaded in the tube lumens of HNTs allowed for sustained released through HNT openings, provided longer supply of antioxidant in rubber composite and enhanced the aging resistance of rubber. Surface modifying of HNTs with silane coupling agent was used to improve the performance of SBR/HNTs nanocomposites. The aging resistance of SBR/HNTs was studied by oxygen adsorption and heat aging method. The antioxidant loaded HNTs not only worked as fillers for better mechanical properties but also can be used for improving aging resistant.

1. Introduction

Halloysite nanotube (HNT) is an economically viable two-layered aluminosilicate with a predominantly hollow tubular structure in the submicron range. The size of halloysite particles varies within 1-1.5 μm in length and 10-15 nm of inner diameter. Due to its high aspect ratio (L/D ~ 30), it gives a large amount of filler-polymer interactions compared to spherical nanoparticles improving mechanical and thermal properties of the polymeric composites [1-4]. Loading these tube lumens with N-isopropyl-N'-phenyl-p-phenylenediamine (antioxidant 4010NA) and doping them at 30-50 % into styrene-butadiene rubber (SBR) allows for controlled sustained release halloysite nanotube openings, increasing the amount of antioxidant used in the rubber composite [5-8]. The sustained release of antioxidant can avoid the blooming caused by the high antioxidant content in rubber. Tight interfacial bonding and fine dispersion of the clay are two crucial factors to determine the performance of the SBR/HNTs nanocomposites, which can be achieved by surface modification of HNT with silane coupling agent. The aging resistance of SBR/HNTs was studied by oxygen adsorption and heat aging method [9-11].

* To whom any correspondence should be addressed.

2. Results and Discussion

2.1 Loading procedures of Antioxidant

HNT (two times weight of antioxidant) was mixed thoroughly with antioxidant acetone solution (30 g/L) and placed in a vacuum chamber at 100 torr for 30 minutes then at standard atmosphere for 15 minutes. This vacuum cycle was repeated three times followed by washing and drying.

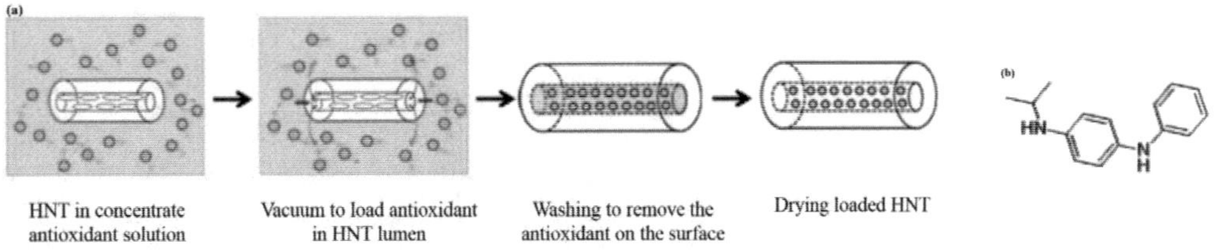

Fig. 1 (a) Loading procedure of active agents into halloysite nanotube and (b) the formula of antioxidant 4010NA.

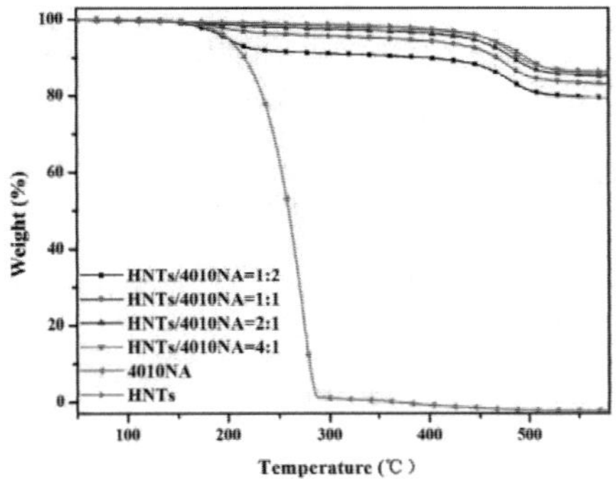

Fig. 2 TGA curves of antioxidant 4010NA loaded HNTs

The TGA curves of samples with different HNTs/antioxidant weight ratios are shown in Fig. 2. The loading efficiency calculated with the formula (I) is listed in Table 1. The loading efficiency increases with the decreasing of weight ratio of HNTs/4010NA until its maximum value.

$$c_A = (1 - w_{H\text{-}A}/w_H) * 100\% \qquad \text{(I)}$$

c_A- the loading efficiency of antioxidant
$w_{H\text{-}A}$- the weight of antioxidant loaded HNTs at 600°C
w_H- the weight of HNTs at 600°C

Table 1 Loading efficiency of different weight ratios of HNTs/4010NA

HNTs/4010NA	**4:1**	**2:1**	**1:1**	**1:2**
Loading efficiency	0.8%	1.5%	3.8%	8.1%

2.2 Release of Antioxidant from HNT lumens

The release of antioxidant in cyclohexane is much faster than in water (Fig. 3), attributed to the higher solubility of antioxidant in cyclohexane. The large release amount at the initial stage is acceptable. For one reason is the release in rubber will be slowed down as the HNTs are coated with the cross-linked polymer, another is the aging resistant system of elastomer needs an initial concentration of antioxidant.

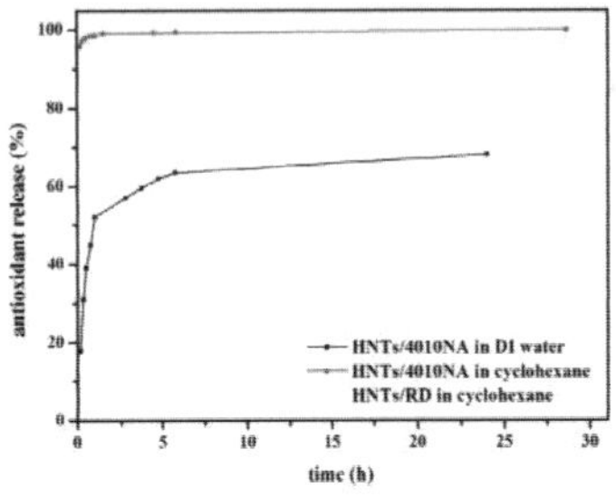

Fig. 3 Antioxidant release from loaded HNTs in water and cyclohexane

2.3 Preparation of the SBR/HNTs nanocomposites

SBR and HNTs were compounded with rubber additives with a two-roll mill at room temperature. The compound was press-cured to a 1-mm-thickness sheet at 170°C×Tc90 and then cut into specimens for measurements.

Table 2 Composition of SBR/HNTs nanocomposites[a]

Sample code	SBR	HNTs (phr)	HNTs/4010NA (phr)	4010NA (phr)
SBR#1	100	-	-	-
SBR#2	100	40	-	-
SBR#3	100	-	-	1.5
SBR#4	100	-	40	-
SBR#5	100	-	40	1.5

[a] Rubber intergradients: zinc oxide, 5 phr; stearic acid, 1 phr; dicumyl peroxide (DCP), 1 phr.

2.4 Morphology and mechanical properties of SBR/HNTs nanocomposites

The morphology of SBR/HNTs nanocomposites is shown in Fig. 4. The dispersion of HNTs with silane coupling agent is much better and the majority of HNTs are embedded in the matrix.

Table 3 shows the mechanical properties of SBR/HNTs nanocomposites before and after aging at 120°C for 24h and 72h. It is revealed that the overall mechanical properties are increased with silane coupling agent. That is because the organic layer brought on the HNT surface by silane decreases the surface energy and increases the disperation of HNTs in SBR while enhancing the interaction between HNTs and rubber. After aging, less decrease of mechanical properties was performed by the antioxidant loaded HNTs reinforced SBR than the HNTs reinforced SBR with adding antioxidant directly while mixing. The sustained release of antioxidant makes it long-acting and avoids the blooming caused by the high antioxidant content in the nanocomposites.

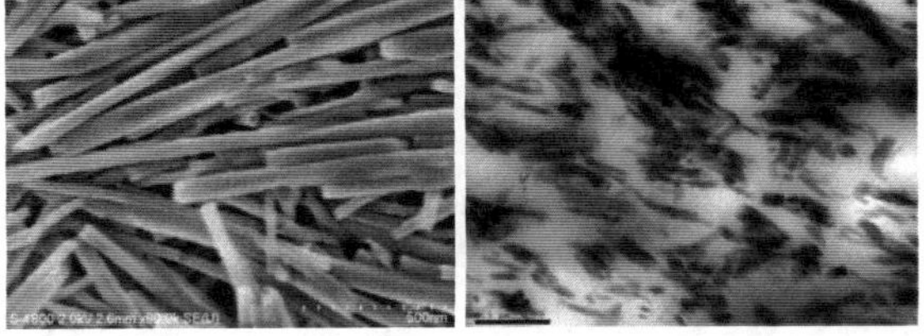

Fig. 4 SEM image of halloysite nanotubes and TEM image of SBR/HNTs nanocomposites SBR#5.

Table 3 the mechanical properties of SBR/HNTs nanocomposites before and after aging

Sample code		Ts/MPa	Eb/%	Hardness (shore A)
Before aging	SBR#1	2.36	563	42
	SBR#2	14.97	523	64
	SBR#3	2.41	557	43
	SBR#4	15.11	517	64
	SBR#5	15.20	515	65
Changes after aging at 120°C×24h	SBR#1	-76%	-48%	+12
	SBR#2	-70%	-51%	+12
	SBR#3	-57%	-31	+8
	SBR#4	-60%	-34%	+10
	SBR#5	-53%	-27%	+8
Changes after aging at 120°C×72h	SBR#1	-	-55%	+14
	SBR#2	-	-57%	+14
	SBR#3	-	-36%	+9
	SBR#4	-	-39%	+11
	SBR#5	-	-31%	+9

3. Conclusion

In this work, an aging resistant rubber nanocomposite was prepared by using antioxidant loaded HNTs as not only controlled sustained release of antioxidant but also fillers improved the mechanical properties of rubber composite. After 72 hrs aging at 120° C in oxygen atmosphere, halloysite-4010NA antioxidant rubber composites have shown ca. 31 % elasticity drop as compared with 55 % drop for unloaded rubber accomplished.

4. References

[1] Abdullayev E and Lvov Y 2011, Halloysite clay nanotubes for controlled release of protective agents *Journal of nanoscience and nanotechnology*. **11** 10007-26.

[2] Du M, Guo B, and Jia D 2010, Newly emerging applications of halloysite nanotubes: a review *Polymer International*. **59** 574-82.

[3] Lvov Y, Shchukin G, Möhwald H 2008, Price R Halloysite clay nanotubes for controlled release of protective agents *ACS Nano*. **2** 814-20.

[4] Veerabadran N, Price R and Lvov Y 2007, Clay nanotubes for encapsulation and sustained release of drugs *Nano*. **2** 115-20.

[5] Raman V, Rooj S, Das A, Stöckelhuber K, Simon F, Nando G, Heinrich G 2013, Reinforcement of Solution Styrene Butadiene Rubber by Silane Functionalized Halloysite Nanotubes *Journal of Macromolecular Science, Part A*. **50** 1091-106.

[6] Baldwin J andBauer D 2008, Rubber oxidation and tire aging-a review *Rubber Chemistry and Technology*. **81** 338-58.

[7] Ismail M. and Wazzan A 2006, Evaluation of new thermal stabilizers and antifatigue agents for rubber vulcanizates *Polymer-Plastics Technology and Engineering*. **45** 751-58.

[8] Zhang J, Wang J, Wu Y, Sun W and Wang Y 2009, Thermal Behaviour and Improved Properties of SBR and SBR/Natural Bitumen Modified Bitumens *Iranian Polymer Journal*. **18** 465-78.

[9] Wang Y, Zhang H, Wu Y, Yang J and Zhang L 2005, Structure and properties of strain - induced crystallization rubber - clay nanocomposites by co-coagulating the rubber latex and clay aqueous suspension *Journal of applied polymer science*. **96** 318-23.

[10] Tian M, Cheng L, LiangW, Zhang L 2006, Overall properties of fibrillar silicate/styrene–butadiene rubber nanocomposites *Journal of applied polymer science*. **101** 2725-31.

[11] Zhang L, Wang Y, Wang Y, Sui Y, Yu D 2000, Morphology and mechanical properties of clay/styrene - butadiene rubber nanocomposites *Journal of Applied Polymer Science.* **78** 1873-78.

IOP Conf. Series: Materials Science and Engineering **64** (2014) 012048 doi:10.1088/1757-899X/64/1/012048

Phosphorylated nano-diamond/ Polyimide Nanocomposites

Aslı Beyler-Çiğil, Emrah Çakmakçı, Memet Vezir Kahraman

Marmara University, Department of Chemistry, 34722 Goztepe-Istanbul, Turkey
E-mail: asli.beyler@marmara.edu.tr

Abstract. In this study, a novel route to synthesize polyimide (PI)/phosphorylated nano-diamond films with improved thermal and mechanical properties was developed. Surface phosphorylation of nano-diamond was performed in dichloromethane. Phosphorylation dramatically enhanced the thermal stability of nano-diamond. Poly(amic acid) (PAA), which is the precursor of PI, was successfully synthesized with 3,3',4,4'-Benzophenonetetracarboxylic dianhydride (BTDA) and 4,4'-oxydianiline (4,4'-ODA) in the solution of N,N-dimethylformamide (DMF). Pure BTDA-ODA polyimide films and phosphorylated nano-diamond containing BTDA-ODA PI films were prepared. The PAA displayed good compatibility with phosphorylated nano-diamond. The morphology of the polyimide (PI)/phosphorylated nano-diamond was characterized by scanning electron microscopy (SEM). Chemical structure of polyimide and polyimide (PI)/phosphorylated nano-diamond was characterized by FTIR. SEM and FTIR results showed that the phosphorylated nano-diamond was successfully prepared. Thermal properties of the polyimide (PI)/phosphorylated nano-diamond was characterized by thermogravimetric analysis (TGA). TGA results showed that the thermal stability of (PI)/phosphorylated nano-diamond film was increased.

1. Introduction

Polyimide (PI) is known to be a high-performance material. Polyimides rank among the most heat-resistant polymers and are widely used in high temperature plastics, adhesives, dielectrics, photoresists, nonlinear optical materials, membrane materials for separation etc. Additionally, polyimides are used in a diverse range of applications, coating, including the fields of aerospace; they are also used in liquid crystal alignments, composites, electroluminescent devices, electrochromic materials, polymer electrolyte fuel cells, polymer memories, gas separation membranes, Nanofiltration membranes, chemical sensor, fiber optics, etc [1-3].

Organic/inorganic nano-hybrid materials have rapidly received worldwide attention and become a new field on materials and membrane research because of their outstanding characteristics, such as excellent tensile strength, low thermal expansion coefficient and dielectric constant, better penetrant discrimination, and resistance to organic solvents. To prepare surface bound metal nanoparticles on polyimide surface would have great potential applications. Various nano-scale additives, such as Ag nanoparticles, TiO2 nanoparticles, Au nanoparticles and platin nanoparticles have been incorporated into the polyimide matrix [4-6].

Over the last 10 years, nano-sized diamond particles known as nano-diamonds (NDs) have attracted much attention because of their outstanding physical properties and low toxicity, as well as the possibility of modifying their surface chemistry [7].

In this work, we report the preparation of phosphorylated nano-diamonds/ polyimide (PI) nanocomposites.

2. Materials and methods

2.1. Materials

4,4′-Oxydianiline (ODA) and 3,3′,4,4′-benzophenonetetracarboxylicdianhydride (BTDA) were purchased from Sigma Aldrich and used without further purification. Nano-diamond (ND), dimethyl formamide (DMF), methylphosphonic dichloride and dichloromethane were also purchased from Sigma Aldrich.

2.2. Methods

2.2.1. Functionalization of nano-diamonds by phosphorylation

The procedure for the synthesis of nanodiamonds was adopted from reference [7]. Methylphosphonic dichloride (0.4 g) was added to a dispersion of nano-diamonds (1 g) in dry dichloromethane. The mixture was stirred for 3 days at 40 °C. The functionalized nano-diamonds were obtained after centrifugation (10.000 rpm for 10 minutes for 3 times) under ambient conditions, followed by a drying step at 150 °C under reduced pressure for 24 h. Phosphorylated nano-diamonds are shown in Scheme 1.

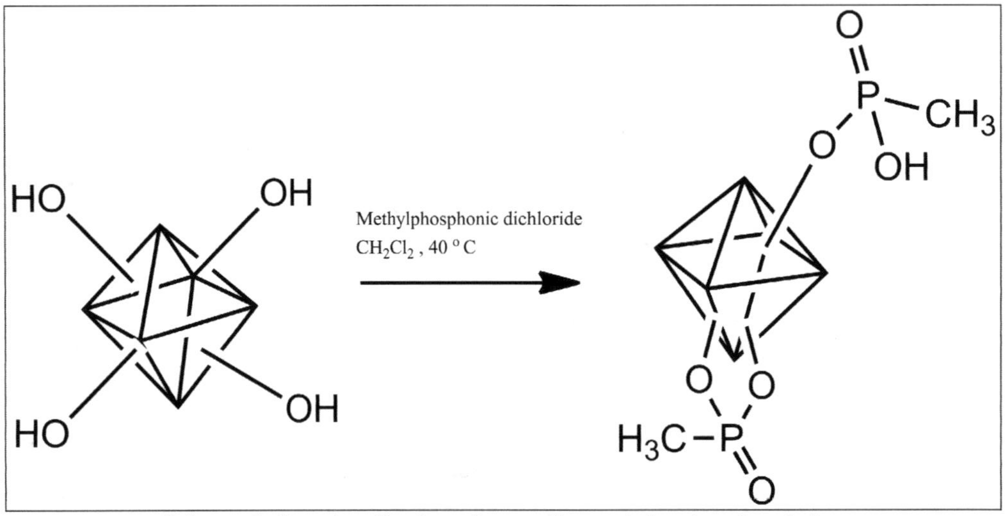

Scheme 1. Functionalization of nano-diamonds by phosphorylation

2.2.2. Synthesis of polyamic acids (PAA)

To synthesize polyamic acid (PAA) solutions, ODA (4 g) was first put into a flame dried three-neck flask containing (6.5 g) ml of DMF under a nitrogen purge. After ODA was completely dissolved in DMF, BTDA (42 g) was added to the flask. Procedure of polyamic acid synthesis is shown in Scheme 2. The mixture was stirred overnight at room temperature to obtain a viscous PAA solution. The resulting PAA solution was ≈ 20 wt %.

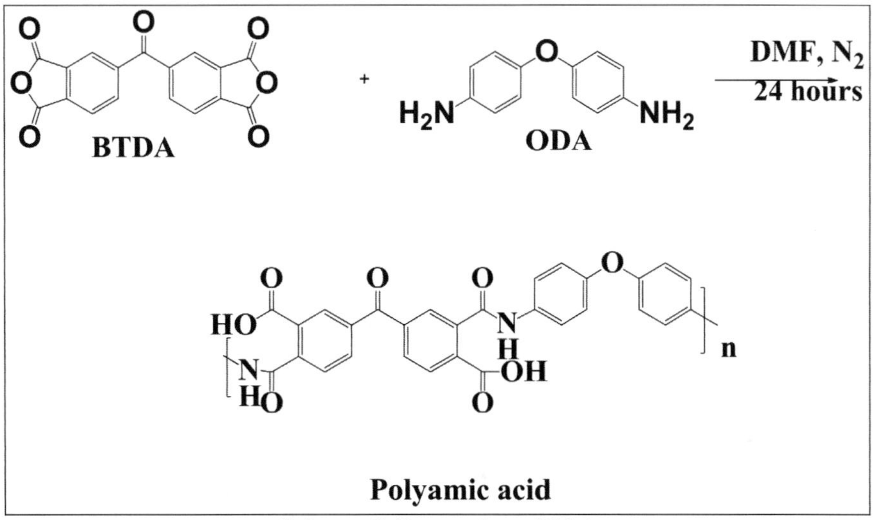

Scheme 2. Preparation of PAA

2.2.3. *Synthesis of BTDA/ODA-phosphorylated nano-diamond*

Composite membranes were prepared by adding appropriate amounts of phosphorylated nano-diamonds (0 wt% and 1 wt%) into PAA solutions. Membranes were prepared by casting viscous mixture solutions on clean dust-free glass plates. Then thermal imidization was performed stepwise at 100, 200, and 300 °C for 1 h at each temperature. The preparation of 1 wt% phosphorylated nano-diamond containing polyimide films were prepared by dispersing 0.02 g phosphorylated nano-diamond of in 10 g of PAA solution. The mixture was stirred for 24 h and these solutions were sonicated for 30 min. Then cast films were cured as described above. Hybrid films were removed from glass plates by immersing in distilled water at 80 °C.

2.2.4. *Characterization methods*

FTIR spectrum was recorded on Perkin Elmer Spectrum100 ATR-FTIR spectrophotometer. Thermogravimetric analyses (TGA) of polyimide films were performed using a Perkin-Elmer Thermogravimetric analyzer Pyris 1 TGA model. Samples were run from 30 to 750 °C with heating rate of 10 °C/min under air atmosphere.
DSC measurements were performed using Pyris Diamond DSC. Samples were run from 30 to 300 °C with a heating rate of 10 °C/min. Glass transition temperatures were obtained from the second heating scan.
SEM imaging of the films were performed on Philips XL30 ESEM-FEG/EDAX. The specimens were prepared for SEM by freze-fracturing in liquid nitrogen and applying a platinum coating.

3.Result and Discussion

In this study BTDA/ODA polyimide film and phosphorylated nano-diamonds containing polyimide film were prepared and characterized. PNDs particles were successfully dispersed in polyimides up to 1 wt%.

3.1. *FTIR studies*

Fig. 1 shows the ATR-FTIR spectra of neat BTDA/ODA PAA, pure BTDA/ODA polyimide film and composite polyimide films. Pure polyimide film shows characteristic polyimide absorption bands at 1776 cm^{-1} and 1713 cm^{-1} which correspond to asymmetric and symmetric imide carbonyl stretchings, respectively. Also the peak at around 1370 cm^{-1} is attributed to C-N bond stretchings.
Moreover, disappearance of –NHCO- peak at 1546 cm^{-1} (which can be seen in the spectrum of PAA)

indicates that polyimides were synthesized successfully. In the ATR-FTIR spectra of composite film, peaks related with phosphorylated nano-diamonds can also be seen in addition to typical polyimide absorption bands.

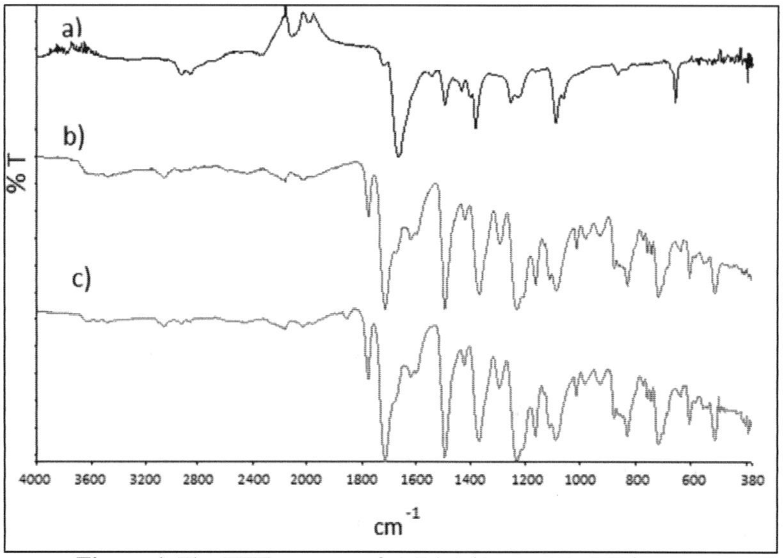

Figure 1. The FTIR spectra of a) PAA b) pure PI c) PI/PNDs

3.2. *Thermal properties*

Figure 2 show the thermal degradation behavior of hybrid film and pure polyimide film under air atmospheres. Under air atmosphere the onset weight loss temperature and the maximum weight loss temperature of pure BTDA/ODA polyimide were found as 546 and 619 °C respectively. From the results it can be seen that the addition of phosphorylated nano-diamonds significantly increases both the onset degradation and the maximum weight loss temperatures. As the amount of PNDs content was increased in the composite films, thermal properties were also increased. The addition of phosphorylated nano-diamonds greatly improves the thermostability of polyimide composites.

We found the glass transition temperature of pure BTDA/ODA polyimide film as 297 °C. When PNDs was incorporated into polyimide matrix, glass transition temperatures of hybrid composites decreased. The glass transition temperature of PI/PNDs was found as 268 °C.

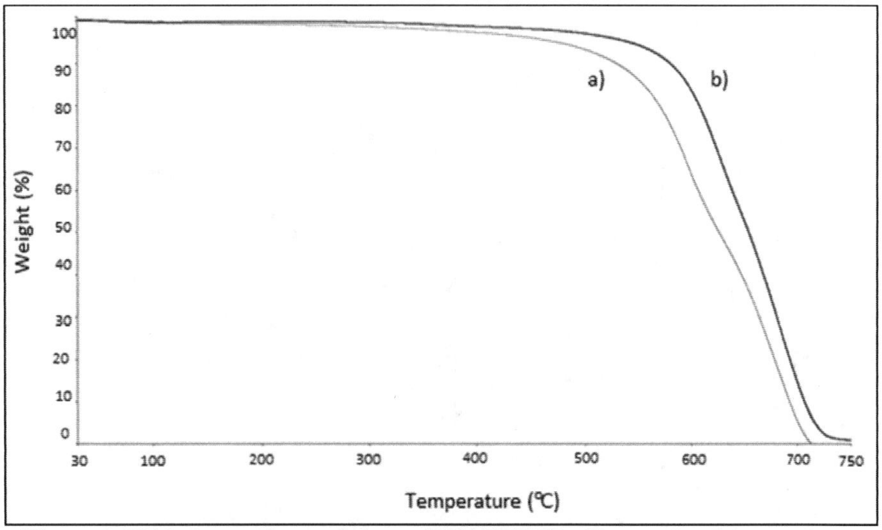

Figure 2. TGA thermograms of a) pure PI b) PI/PNDs films under air atmosphere

3.3. Morphology

Fig. 3a-b shows the fractured surface morphology of polyimide composites. We can see from SEM micrographs that pure polyimide has a homogenous, crack free, nonporous and a uniform surface (Fig. 3a). In the SEM images of phosphorylated nano-diamonds containing sample (Fig. 3b). Fig. 3b shows of the surface of PI/PNDs film and in this image it can be clearly seen that some phosphorylated nano-diamond particles were well entrapped in the polymer matrix.

4. Conclusions

In this study, pure BTDA-ODA polyimide films and phosphorylated nano-diamond containing BTDA-ODA PI films were prepared via thermal imidization of polyamic acid precursors. Surface phosphorylation of nano-diamond was performed in dichloromethane. Phosphorylation dramatically enhanced the thermal stability of nano-diamond. The effect of phosphorylated nano-diamonds of the composites was investigated.

IOP Conf. Series: Materials Science and Engineering **64** (2014) 012048 doi:10.1088/1757-899X/64/1/012048

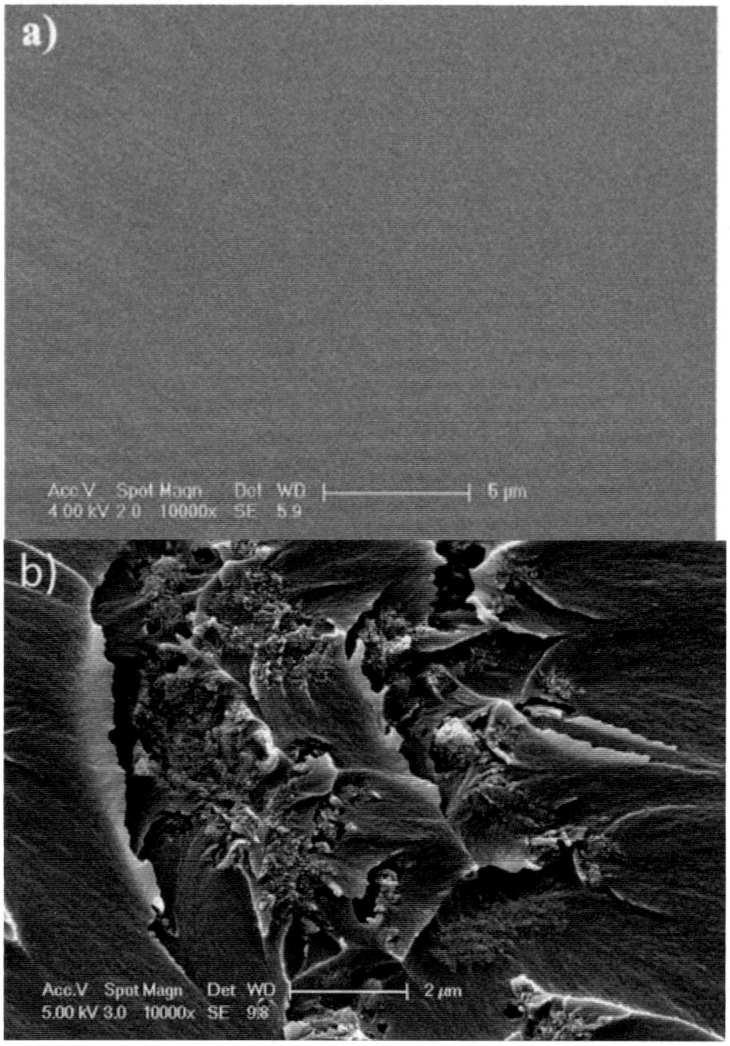

Figure 3. SEM micrographs a) pure PI x10.000 b) PI/PNDs x10.000

References

[1] Emrah Çakmakçı, Aslı Beyler Çiğil, Özkan Danış, Serap Demir, Memet Vezir Kahraman, 2013, Starch, 65, 1-7.

[2] Sankar C. Khatua, Sukumar Maiti, 2002, European Polymer Journal, 38, 537-543.

[3] Sobhan Niyogi, Sukumar Maiti, Basudam Adhikari, 2000, 459-464.

[4] Youchang Xiao, Kai Yu Wang, Tai-Shung Chung, Juina Tan, 2006, Chemical Engineering Science, 61, 6228-6233.

[5] Ying Kong, Hongwei Du, Jinrong Yang, Deqing Shi, Yunfang Wang, Yuanyuan Zhang, Wei Xin, 2002, Desalination, 146, 49-55.

[6]Emrah Çakmakçı, Atilla Güngör, 2013, Polymer Degradation and Stability, 98, 927-933.

[7]Charlene Presti, Johan G. Alauzun, Danielle Laurencin, and P. Hubert Mutin, 2013, Chemistry of Materials, 25, 2051-2055.

IOP Conf. Series: Materials Science and Engineering **64** (2014) 012049 doi:10.1088/1757-899X/64/1/012049

Influence of carbon nanotubes on the properties of epoxy based composites reinforced with a semicrystalline thermoplastic

A Díez-Pascual, P Shuttleworth, E Gónzalez-Castillo, C Marco, M Gómez-Fatou and G Ellis
Institute of Polymer Science and Technology (ICTP-CSIC), Juan de la Cierva 3, 28006 Madrid, Spain
E-mail: adiez@ictp.csic.es

Abstract. Novel ternary nanocomposites based on a thermoset (TS) system composed of triglycidyl p-aminophenol (TGAP) epoxy resin and 4,4'-diaminodiphenylsulfone (DDS) curing agent incorporating 5 wt% of a semicrystalline thermoplastic (TP), an ethylene/1-octene copolymer, and 0.5 or 1.0 wt% multi-walled carbon nanotubes (MWCNTs) have been prepared via physical blending and curing. The influence of the TP and the MWCNTs on the curing process, morphology, thermal and mechanical properties of the hybrid nanocomposites has been analyzed. Different morphologies evolved depending on the CNT content: the material with 0.5 wt% MWCNTs showed a matrix-dispersed droplet-like morphology with well-dispersed nanofiller that selectively located at the TS/TP interphase, while that with 1.0 wt% MWCNTs exhibited coarse dendritic TP areas containing agglomerated MWCNTs. Although the cure reaction was accelerated in its early stage by the nanofillers, curing occurred at a lower rate since these obstructed chain crosslinking. The nanocomposite with lower nanotube content displayed two crystallization peaks at lower temperature than that of pure TP, while a single peak appearing at similar temperature to that of TP was observed for the blend with higher nanotube loading. The highest thermal stability was found for TS/TP (5.0 wt%)/MWCNTs (0.5 wt%), due to a synergistic barrier effect of both TP and the nanofiller. Moreover, this nanocomposite displayed the best mechanical properties, with an optimal combination of stiffness, strength and toughness. However, poorer performance was found for TS/TP (5.0 wt%)/MWCNTs (1.0 wt%) due to the less effective reinforcement of the agglomerated nanotubes and the coalescence of the TP particles into large areas. Therefore, finely tuned morphologies and properties can be obtained by adjusting the nanotube content in the TS/TP blends, leading to high-performance hybrid nanocomposites suitable for structural and high-temperature applications.

1. Introduction

Epoxy resins are versatile thermosetting polymers widely used in structural applications because of their high stiffness, chemical and heat resistance derived from the cross-linked structure, and easy processability. However, they are inherently brittle and have low wear resistance, which hinder their use for certain structural, adhesive, or coating applications. To address this problem, thermoplastic modifiers [1], rubber particles [2] or hyperbranched polymers [3] have been blended with epoxy resins. On the other hand, multi-walled carbon nanotubes (MWNTs) have outstanding mechanical, thermal and electrical properties and low density, thus are ideal candidates for reinforcing polymer matrices. Poor nanofiller-matrix interfacial adhesion and MWCNT agglomeration are generally the main challenges when dealing with nanotube-reinforced polymer composites. However, it has been recently demonstrated [4,5] that the uneven or asymmetric distribution of nanofillers within the matrix can lead to significant improvements in key properties. In the present work, we have followed this approach in order to achieve property improvements in an aerospace-grade high performance epoxy system. To attain this goal, the resin has been modified with a semicrystalline thermoplastic, an ethylene/1-octene copolymer and MWCNTs. An extensive characterization has been carried out in order to analyze the effect of MWCNT content, location and level of dispersion on the morphology, thermal and mechanical properties of the resulting hybrid composites.

IOP Conf. Series: Materials Science and Engineering **64** (2014) 012049 doi:10.1088/1757-899X/64/1/012049

2. Experimental

The epoxy (TS) system consisted of a high-performance trifunctional epoxy, triglycidyl p-aminophenol (TGAP) with a 4,4′-diaminodiphenyl sulfone (DDS) curing agent, supplied by Huntsman Advanced Materials, LLC. CVD MWCNTs were obtained from Cheap Tubes Inc (VT, USA). The thermoplastic (TP), an ethylene/1-octene copolymer with commercial name DOWLEX SC2107, was provided by Dow Chemical. The TP particle size was reduced using a cryogenic impact mill.

The neat TS was prepared by blending TGAP and DDS (100/49 resin/hardener wt% ratio) in a planetary centrifugal mixer for 5 min at 300 rpm and room temperature. The mixture was then poured into silicone moulds and cured in an oven at 130 °C for 1.5 h, followed by postcuring at 177 °C for 2 h. Blends of TS/TP (5 wt%) were prepared in a two-stage process: The TP powder was initially mixed with the TGAP in the planetary centrifuge for 10 min at 500 rpm. After a few minutes of stirring at 130 °C, DDS was added stepwise under continuous stirring until a homogeneous mixture was obtained, and the blend was then subjected to the above indicated curing cycle. Regarding TS/TP/MWCNT composites, the TP powder was initially blended with TGAP in the planetary mixer for 10 min at 500 rpm; afterward, the MWCNTs were added and the mixing continued for another 10 min. Subsequently, the mixture was stirred at 130 °C and the curing agent was added, and finally it was cured under the same conditions.

3. Results and discussion

The effect of the TP and MWCNTs on the curing behaviour of TS was investigated by differential scanning calorimetry (DSC), and the calorimetric parameters derived from the non-isothermal thermograms are collected in Table 1: curing enthalpy (ΔH), curing onset (T_o) and the slope of the curve in the region of the curing peak (*m*). The neat resin begins the curing at 130 °C, and shows a single step process with a slope of -0.55 W/g °C. Similar behaviour is observed for the TS/TP binary mixture, with the curing occurring at a slightly lower rate. However, composites incorporating MWCNTs start the curing at a lower temperature, and exhibit a two-stage process, the first step taking place at a significantly lower rate and the second at similar rate to that of the neat TS. These results hint that the MWCNTs have a dual effect: firstly, due to their higher thermal conductivity, they accelerate the onset of the curing process, which is initiated in those regions where CNTs are preferentially located. Secondly, they appear to impose a certain hindrance in the epoxy crosslinking that takes place at a lower rate. Afterwards, since curing is a dynamic process, it is extended to the entire composite, occurring at similar rate and the same temperature for all the samples compared to the neat matrix. Interestingly, as the CNT content increases, the aforementioned effects are stronger: T_o shifts to lower temperature and the slope of the first stage decreases. This is related to the morphology of the composites, as discussed later. Further, ΔH decreases in the presence of the MWCNTs, also pointing towards a hampering of the epoxy crosslinking. In fact, it has been reported [6] that well-dispersed CNTs can act as a physical hindrance to the monomer mobility and thus lower the heat of reaction.

Table 1. Thermal data derived from DSC experiments

SAMPLE	ΔH^a (J/g)	T_o^a (°C)	m^a (W/g °C)	T_m (°C)	T_c (°C)	X_c (%)
TS	203	130	-0.55			
TP				121	99	39
TS/TP (5 wt%)	177	130	-0.46	115	72/16	41
TS/TP (5wt%)/MWCNT (0.5 wt%)	175	128	-0.023/-0.45	123	82/23	48
TS/TP (5wt%)/MWCNT (1.0 wt%)	189	124	-0.015/-0.48	112	93	36

[a]Determined from non-isothermal curing thermograms. ΔH: curing enthalpy; T_o: curing onset; *m*: slope of the non-isothermal curve in the region of the curing peak. T_c and T_m: crystallization and melting temperatures, respectively. X_c: degree of crystallinity.

The morphology of the samples was analyzed by scanning electron microscope (SEM), and typical *images* of ternary composites with 0.5 and 1.0 wt% are shown in Fig. 1. In TS/TP (5.0 wt%)/MWCNT (0.5 wt%) nanocomposite (Fig. 1a and 1b), the TP particles are randomly and homogenously dispersed within the TS, showing matrix-dispersed droplet type morphology. The MWCNTs appear to be preferentially located at the TS/TP interphase (see Fig. 1b), probably due to their higher affinity with the TP owing to their hydrophobic nature. This phenomenon has been regarded as a mechanism to reduce the size of the minor polymer phase. Further, no MWCNT aggregates were observed in the whole examined regions. The well-dispersed nanofillers act as a solid barrier around the TP particles, thereby preventing the coalescence. However, SEM micrographs of TS/TP (5.0 wt%)/MWCNTs (1.0 wt%) show a very different morphology (Fig. 1c and 1d). The size of the TP domains is considerably larger, and they appear interconnected due to the coalescence of several particles, leading to a gross dendritic morphology. The MWCNTs are located within the TP phase (Fig. 1d), and appear quite aggregated. A plausible explanation could be that the higher CNT concentration resulted in higher viscosity of the phases, hindering their diffusion and migration over the composite.

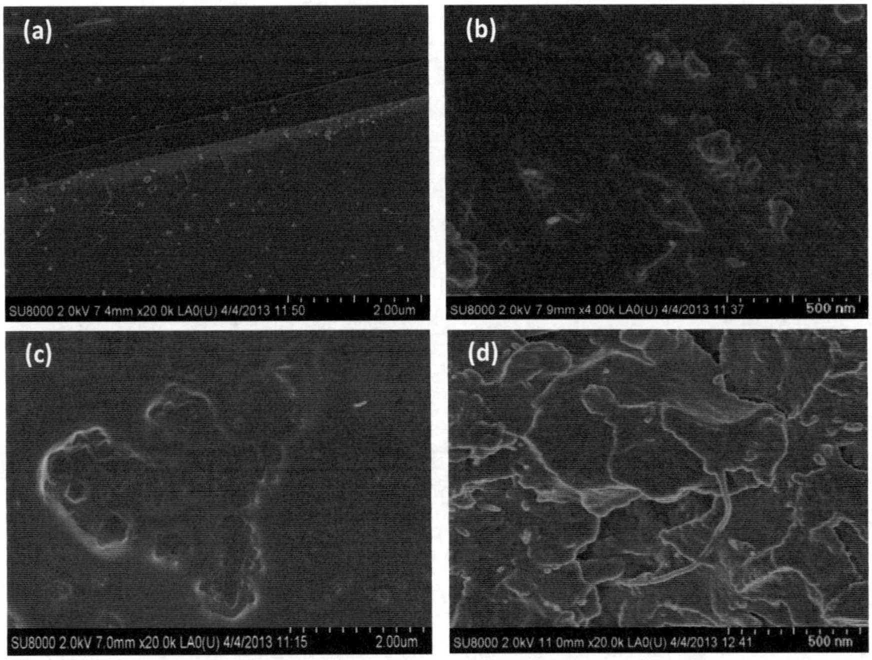

Fig. 1 SEM micrographs at different magnifications of: (a) and (b) TS/TP (5 wt%)/MWCNT (0.5 wt%); (c) and (d) TS/TP (5 wt%)/MWCNT (1.0 wt%).

The crystallization behaviour of the composites was investigated by DSC (Fig. 2). Neat TP shows a sharp crystallization peak T_c at about 100 ºC, with a slight tail corresponding to smaller crystal sizes at the lower temperature side. On the contrary, TS/TP (5.0 wt%) displays two small, broad exothermic peaks at much lower temperatures than that of the neat TP, around 72 and 16 ºC (Table 1). These are probably related to two populations of crystals with different average crystal sizes. According to the SEM images, the TP phase forms isolated and well-dispersed micro domains confined within the epoxy phase. Each domain can have a different nucleating activity depending on the size, degree of dispersion, presence of impurities, etc. An analogous behaviour has been reported for immiscible polymer blends where the crystalline polymer is finely dispersed in small domains (at the micro- or nanoscale), and is denominated "confined or fractionated crystallization" [7]. TS/TP (5 wt%)/MWCNT (0.5 wt%) also shows two broad crystallization peaks related with TP particles with different crystal sizes, hence different degrees of confinement. Besides, the T_c is shifted to higher values, probably due to the nucleating effect of the nanofillers. However, TS/TP (5 wt%)/MWCNT (1.0 wt%) only exhibits a single crystallization. The disappearance of the lower temperature exotherms on increasing the MWCNT content is indicative of the elimination of the confinement effect in this sample. From the SEM images, both TP and the MWCNTs are agglomerated, with considerably large TP domains, hence the polymer crystallizes in a similar manner to neat TP. Further, the T_c is slightly higher than that of TS/TP(5.0 wt%)/MWCNTs(0.5 wt%), which could be related to the dispersion of the nanofillers inside the TP domains.

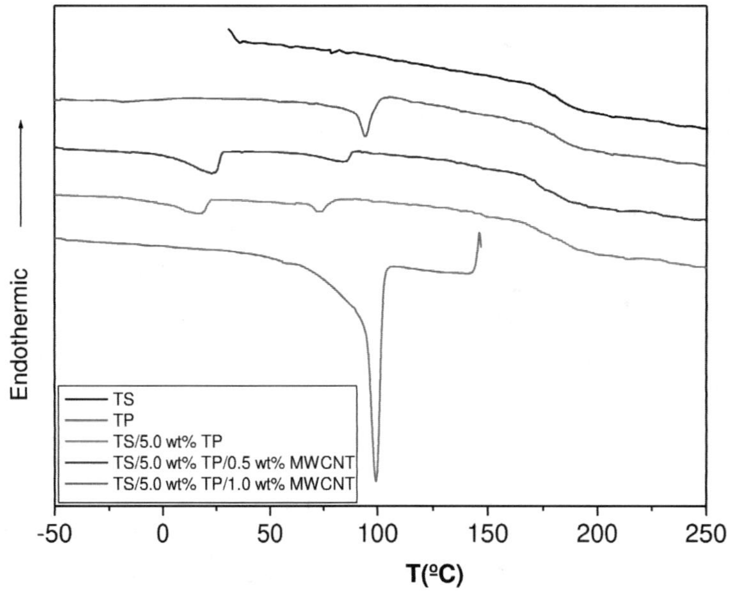

Fig. 2

Dynamic DSC cooling (bottom) scans for the different samples

The thermal stability of the samples was investigated under nitrogen atmosphere and the results are presented in Fig. 3. The initial degradation temperature of TS increases upon addition of both TP and MWCNTs, due to the higher thermal stability of these materials compared to the neat resin. The hybrid containing 0.5 wt% MWCNTs shows around 25 C increment compared to that of neat TS. This outstanding increase is attributed to a synergetic stabilizing effect of the isolated TP domains and the well-dipersed MWCNTs that can act as a barrier and effectively hinder the diffusion of the degradation products from the bulk of the matrix to the gas phase. Further, the increase in thermal conductivity due to the presence of the MWCNTs facilitates heat dissipation within the nanocomposite. However, for the hybrid with 1.0 wt% MWCNTs the improvement is smaller (~15 °C), since both the TP and the MWCNTs are quite agglomerated, hence exert a less efficient barrier effect. Qualitatively similar trends are found for the temperatures corresponding to 10% weight loss and the maximum rate of weight loss.

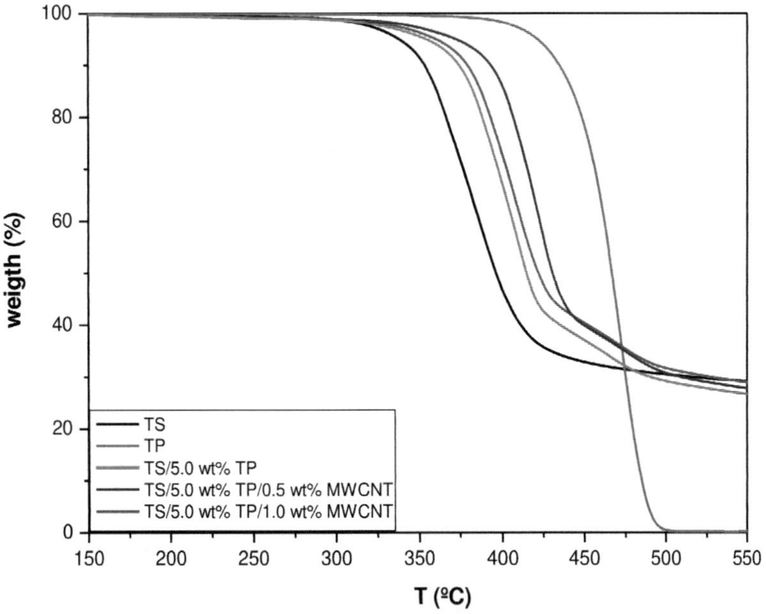

Fig. 3 TGA thermograms under nitrogen atmosphere at a rate of 10 ºC/min

The mechanical properties of the hybrids were characterized by tensile tests (Fig. 4). The Young´s modulus (E) of TS is ~2.8 GPa. The TP acts as a plasticizer and causes a 16% decrease in the matrix modulus. In contrast, the incorporation of MWCNTs imparts the high stiffness of the fillers to the matrix, and the TS/TP (5 wt%)/MWCNT (0.5 wt%) has ~12% higher E than TS. In this case, the nanotubes are predominantly located at the TS/TP interphase and might improve the matrix-TP interfacial adhesion. However, for the nanocomposite with 1.0 wt% MWCNTs, E is slightly lower than that of the matrix due to the less effective reinforcement of the agglomerated nanotubes that are mainly located inside the TP domains. Regarding the tensile strength (σ_y), the trends observed are similar to those found for the modulus, albeit the percentage increments attained are higher, i.e. ~24% for TS/TP (0.5 wt%)/MWCNT (0.5 wt%) compared to neat TS, suggesting that the nanotubes are more efficient in enhancing the matrix strength.

Focusing on the elongation at break (ε_b), a noticeable improvement is attained upon addition of the TP to TS due to its higher ductility. Nevertheless, TS/TP/MWCNT samples display lower ε_b than TS/TP (0.5 wt%), the reduction being more pronounced for that with higher nanotube loading since agglomerates strongly obstruct the plastic deformation of the matrix. The area under the tensile curve is a measure of the toughness of the material. As expected, TS/TP (0.5 wt%) shows higher toughness than the neat matrix (about 45%); the incorporation of 1.0 wt% MWCNTs to TS or TS/TP results in moderate toughness reductions, since the presence of agglomerates nucleates cracks, resulting in premature failure hence lower impact resistance. However, TS/TP (5.0 wt%)/MWCNT (0.5 wt%) shows ~52% higher value than that of TS. It seems that the toughening effect is boosted in this sample due to a synergistic action between the TP and the MWCNTs. Moreover, this superior performance could be related to the matrix-dispersed droplet type morphology observed for this composite, where small TP domains were found to be homogeneously dispersed within the matrix.

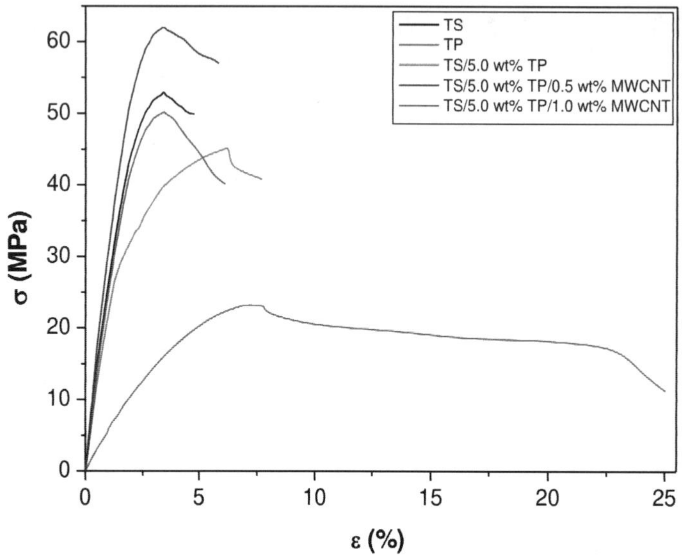

Fig. 4 Room temperature tensile curves for the different samples.

4. Conclusions

Novel ternary nanocomposites based on a thermoset epoxy matrix toughened with a semicrystalline thermoplastic copolymer and reinforced with MWCNTs have been developed. A significant change in the phase morphology of TS/TP blends from droplet-dispersed type to agglomerated TP domains with dendritic structure was found when the CNT concentration was increased from 0.5 to 1.0 wt%. The MWCNTs have a dual effect on the epoxy curing process: they accelerated the cure onset albeit hindered its progress since they impeded chain crosslinking. The nanocomposite with the lowest nanotube content showed two crystallization peaks at lower temperature than that of neat TP, related to confined crystallization of isolated domains with different average crystal sizes. However, a single peak appearing at similar temperature to that of TP was observed for that with 1.0 wt% MWCNTs, since the coalescence of the TP domains eliminated the confinement effect. The highest thermal stability and best mechanical properties were found for TS/TP (5.0 wt%)/MWCNTs (0.5 wt%) nanocomposite, due to synergistic effects of both TP and the nanofiller.

Acknowledgments

Financial support from the Ministerio de Ciencia e Innovaci6n (MICINN, Project MAT2010-21070-C02-01) is gratefully acknowledged. AD thanks to the Consejo Superior de Investigaciones Cientifícas (CSIC) for a JAE-doc postdoctoral contract cofinanced by the EU. PS thanks to the MICINN for a Juan de la Cierva postdoctoral contract (JCI-2011-10836). EIGC acknowledges CONACYT of the Mexican Government for a postgraduate fellowship.

References

[1] Hourston D J and Lane J M 1992 *Polymer* **33** 1379
[2] Frigione M, Acierno D and Mascia L 1999 *J. Appl. Polym. Sci.* **73** 1457
[3] Ratna D, Varley R, and Simon G P 2003 *J. Appl. Polym. Sci.* **89** 2339
[4] Li Y and Shimizu H 2008 *Macromolecules* **41** 5339
[5] Liu L *et al* 2009 *Polymer* **50** 3072
[6] G6nzalez-Dominguez J M *et al* 2011 *J. Phys. Chem. C* **115** 7238
[7] Salmeron-Sanchez M *et al* 2006 *J. Polym. Sci. Part B: Polym. Phys* **44** 815

Investigation of reaction conditions on morphology and optical properties of Zinc Oxide Nanorods

H A Almodarresiyeh[1]**, A V Abakshonok**[2]**, V E Agabekov**[1]**, A N Eryomin**[2]**, S N Shahab**[1]

[1] Institute of physical organic chemistry of NAS of Belarus, Surganova 13, 220072, Minsk, Belarus

[2] Institute of chemistry of new materials of NAS of Belarus, F. Skarina 36, 220141, Minsk, Belarus

E-mail: thr3059@Yahoo.com

Abstract. Zinc oxide nanoparticles (ZnO NPs) were synthesized by precipitation method in the presence of various polymers. Rod shaped ZnO NPs (length ~ 1 micron) were obtained at 70 °C in a reaction medium containing 10-20 mM of zinc nitrate hexahydrate ($Zn(NO_3)_2 \cdot 6H_2O$), 0.05-0.1 mg/ml of polyethylenimine (PEI) and 20 mM of hexamethylenetetramine (HMT). Properties of ZnO NPs were characterized by fluorescence, UV-visible spectroscopy, atomic force and transmission electron microscopy.

1. Introduction

Due to the fact that the properties of nanoparticles (NPs) are strongly dependent on their size and shape the synthesis of rodlike metal and semiconductor NPs is of interest [1-3]. Spherical metal NPs are characterized by only one plasmon resonance, whereas by extending a particle in one dimension a second lower energy resonance band in the longitudinal direction takes place. In general, the number of surface plasmon resonance peaks increases for particles having aspect ratios (length/diameter, L/D) larger than 1 [4, 5]. The crystal grows faster in only one direction or anisotropic growth is required for the formation of nanorods and nanowires. So, in the synthesis of metal and semiconducting NPs, control over the shape and size is one of the important and challenging tasks. Nanocomposite films containing rod-like NPs with unique optical properties are suitable for use in optoelectronic devices [6-9].

In connection with these findings, the synthesis of rod-like NPs of ZnO, which is a wide band gap semiconductor with large excitation energy and an important functional material with near electric conductivity, optical transparency and piezoelectricity have attracted considerable attention. ZnO is a promising material for electronic and optoelectronic applications such as solar cells, liquid crystal displays, gas sensors, surface acoustic wave devices etc. The emission properties of ZnO NPs in the visible region depend on their synthetic method. Now various techniques have been developed for the preparation of ZnO nanorods such as ZnO vapor phase deposition methods, metalorganic chemical vapor deposition, hydrothermal method and so on [10-14]. But these methods require severe reaction conditions, such as high temperature or accurate gas concentration, flow rate or scarce raw materials or complex process, and so on. Therefore, the synthesis of ZnO NPs by deposition method (precipitation) from an aqueous solution is simpler and more accessible.

2. Experimental procedure

All the chemical materials used in these experiments were purchased from Sigma-Aldrich and used without any further purification. To obtain ZnO NPs, an aqueous solution of $Zn(NO_3)_2.6H_2O$ and a polymer were mixed and the resulting solution was cooled for 5 minutes in an ice bath. A solution of hexamethylenetetramine (HMT, $C_6H_{12}N_4$, 20 mM) was added dropwise to the cooled aqueous solution of 0.1 mg/mL polymer and 20 mM $Zn(NO_3)_2.6H_2O$ with vigorous stirring by a magnetic stirrer. Then, the resulting ZnO gel was placed in a water bath heated to 70 °C and maintained under these conditions with stirring for 4 hours. The obtained ZnO particles were stored at room temperature in the light.

Samples containing rod or spherical ZnO particles were separated by centrifugation. The resulting ZnO precipitates were resuspended in ethanol and then were washed 3 times with the same solvent while being centrifuged for 10 min at 5000 rev/min. Samples were dried at room temperature to remove excess ethanol.

Size and shape of ZnO particles were evaluated using optical microscope Planar MB (Belarus), atomic force microscope Nanoscope-3D («Veeca», USA) and transmission electron microscope JEM-100CX II (JEOL, Japan). The absorption and fluorescence spectra of ZnO NPs were recorded by spectrofluorimeter CM 2203 («SOLAR», Belarus).

3. Results and discussion

The ZnO nanorods were synthesized by reaction of $Zn(NO_3)_2$ and HMT in the presence of a polymer as a structure-modifying agent. The formation of ZnO nanorods in the condition of HMT is depicted by reactions 1–4.

$$Zn(NO_3)_2 \rightarrow Zn^{2+} + 2(NO_3)^- \quad (1)$$

$$C_6H_{12}N_4 + 6H_2O \rightarrow 6HCHO + 4NH_3 \quad (2)$$

$$NH_3 + H_2O \rightarrow NH_4^+ + OH \quad (3)$$

$$Zn^{2+} + 2OH^- \rightarrow ZnO\downarrow + H_2O \quad (4)$$

The effects of various polymers on the synthesis of ZnO were investigated. The most effective additive for the synthesis of rod-like ZnO particles is PEI and the least is PVP-40 (Table 1 and Figure).

Table 1. Characteristics of absorption and fluorescence spectra (1 cm quartz cuvette) of ZnO suspension, obtained in the presence of different polymers

polymers	λ_{max}, nm	D_{max}	λ_{exc}, nm	λ_{fl}, nm	I_{max}, rel.unit
Polyethylenimine (PEI), ~60 kDa	301 367	1.51 1.30	353	630	0.40
Poly(diallyldimethylammonium chloride), 100-200 kDa	302	0.78	352	614	1.45
Poly(allylamine hydrochloride), ~17 kDa	304 359	1.45 1.22	354	616	0.41
Poly(allylamine hydrochloride), ~70 kDa	302 360	1.51 1.33	350	629	0.45
Polyvinylpyrrolidone (PVP-40), 40 kDa	302 370	0.81 0.56	359	631	1.09
Polyvinylpyrrolidone, 24 kDa	302 368	0.91 0.68	358	634	1.11

Note: the reaction medium containing 0.1 mg/mL polymer, 20 mM $Zn(NO_3)_2.6H_2O$ and 20 mM HMT. λ_{exc} and λ_{fl}: Maximum excitation and fluorescence, respectively.

In the reaction medium, containing 10 and 20 mM $Zn(NO_3)_2$, when the magnification is 400 times rod-like ZnO particles can be revealed by using optical microscope (Planar MB). However, their number decreases at concentrations $Zn(NO_3)_2$ 30 and 50 mM, but the length of the particles increases. In the presence of 70 and 100 mM $Zn(NO_3)_2$ mainly spherical particles are formed. Fluorescent properties of the particles depend on the concentration of $Zn(NO_3)_2$ (Table 2). With increasing concentration from 10 to 30 mM, the fluorescence intensity of the particles decreases. However, at higher concentrations of $Zn(NO_3)_2$ fluorescence samples are practically unchanged (Table 2). Thus, the presence of more than 20 mM zinc nitrate in the medium of synthesis is inexpedient, since it does not increase the content of the rod-like particles, but only affects their dimensions. In a medium containing more than 70 mM of $Zn(NO_3)_2$ substantially spherical particles are formed.

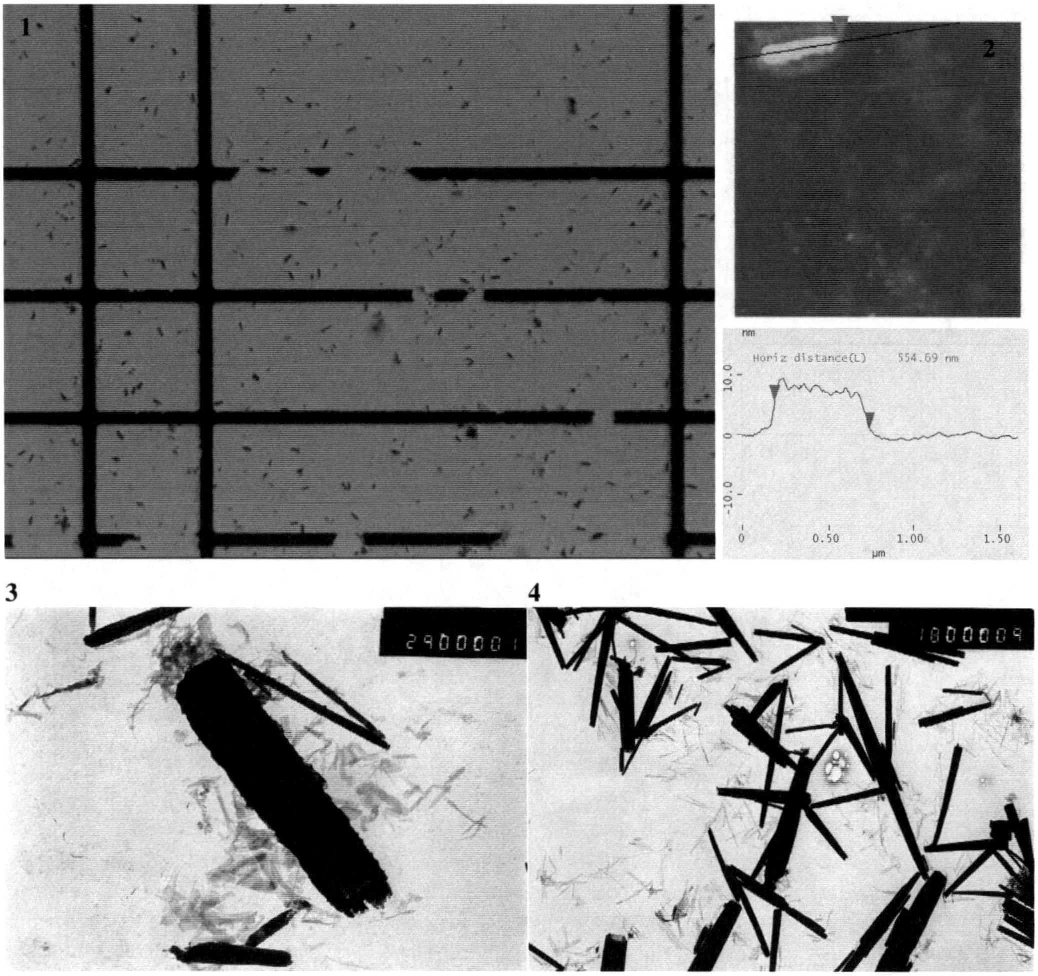

Figure. Images of ZnO particles: microscope field Planar MB (magnification 400 times) (1), the AFM Nanoscope-3D (2) and TEM JEM-100CX II (3 and 4). Scan Window 3.0×2.1 (1) and 4.8×3.3 (2) microns).

Table 2. Characteristics of absorption and fluorescence spectra ZnO suspension, obtained at different concentrations of $Zn(NO_3)_2$.

$Zn(NO_3)_2$, mM	λ_{max}, nm	D_{max}	λ_{exc}, nm	λ_{fl}, nm	I_{max}, rel. units
10	282	1.92	356	641	1.36
	369	1.25			
20	294	2.81	355	632	0.92
	368	2.11			
30	285	2.21	358	621	0.75
	369	1.15			
50	303	1.96	357	637	0.70
	369	1.15			
70	315	3.34	358	634	0.70
	370	2.06			
100	318	3.36	354	631	0.76
	369	2.34			

Note: except $Zn(NO_3)_2$, the reaction medium contained 0.1 mg/ml PEI and 20 mM HMT.
λ_{exc} and λ_{fl}: Maximum excitation and fluorescence, respectively.

Table 3. Characteristics of absorption and fluorescence spectra (1 cm quartz cuvette) ZnO suspension, obtained at different concentrations of PEI.

PEI, mg/ml	λ_{max}, nm	D_{max}	λ_{exc}, nm	λ_{fl}, nm	I_{max}, rel. unit
0	303	1.22	355	627	0.77
	362	1.03			
0.05	304	1.91	359	618	0.42
	368	1.72			
0.10	303	2.38	355	626	0.36
	366	2.13			
0.15	302	1.22	361	619	0.40
	363	0.94			
0.20	303	1.98	357	622	0.34
	366	1.73			
0.30	302	1.67	359	614	0.58
	367	1.41			

Note: except to PEI, the reaction medium contained 20 mM Zn $(NO_3)_2.6H_2O$, and 20 mM HMT.

λ_{exc} and λ_{fl}: Maximum excitation and fluorescence, respectively.

If the medium containing 0.05 and 0.10 mg/ml of PEI, the maximum amount of particles are nanorod. These particles have intense fluoresce less than those obtained without PEI. In the presence of 0.15 mg/ml PEI less ZnO rod particles are formed than in the case of 0.05 and 0.10 mg/ml of resin, but they are longer. In the case of 0.2 and 0.3 mg/ml PEI rod-like particles are formed, but their number is minimal. A sample of the particles formed in the presence of 0.3 mg/ml PEI showed a maximum fluorescence at 514 nm when it is excited by light with a wavelength of 325 nm. When λ_{exc} = 355-367 nm, all the particles fluoresce at 614-626 nm.

4. Conclusion

Rod-like zinc oxide was synthesized at 5-6 °C in an aqueous solution containing 0.05-0.10 mg/ml of PEI (~ 60 kDa), 10-20 mM $Zn(NO_3)_2$.6H2O and 20 mM HMT, followed by aging at 70 °C for 4-6 hours.

5. References

[1] Golovin A B, Stromer J and Kreminska L 2012 Aligned Layers of Silver Nano-Fibers *Materials.* **5** 239-47

[2] Agabekov V, Ivanova N, Dlugunovich V and Vostchula I 2012 Optical Properties of Polyvinyl Alcohol Films Modified with Silver Nanoparticles *Journal of Nanomaterials.* 1-5

[3] Dirix Y, Bastiaansen C, Caseri W and Smith P 1999 Preparation, structure and properties of uniaxially oriented polyethylene-silver nanocomposites *J. Mater. Sci.* **34** 3859-66

[4] Bernabò M, Pucci A, Ramanitra H and Ruggeri G 2010 Polymer Nanocomposites Containing Anisotropic Metal Nanostructures as Internal Strain Indicators *Materials.* **3** 1461-77

[5] Hao E, Schatz G and Hupp J 2004 Synthesis and Optical Properties of Anisotropic Metal Nanoparticles *Journal of Fluorescence.* **14** 331-41

[6] Cascaval C N, Cristea M, Rosu D, Ciobanu C, Paduraru O and Cotofana 2007 Preparation and characterization of polyvinyl alcohol colloidal silver nanocomposites *Journal of Optoelectronics and Advanced Materials.* **9** 2116-20

[7] Dabbousi B O, Bawendi M G, Onitsuka O and Rubner M F 1995 Electroluminescence from CdSe quantum-dot polymer composites *Appl. Phys. Lett.* **66** 1316-18

[8] Fernandes D M, Hechenleitner A W, Lima S M, Andrade L H C, Caires A R L and Gómez Pineda E A 2011 Preparation, characterization, and photoluminescence study of PVA/ZnO nanocomposite films *J.Materials Chemistry and Physics* **128** 371–76

[9] Saini I, Rozra J, Sharma A, Sharma P and Dhiman R 2012 Cu nanoparticles induced modification in optical and structural properties of PVA *Solid State Physics, AIP Conf.* **1447** 219-20

[10] Zhang H, Feng J, Wang J and Zhang M 2007 Preparation of ZnO nanorods through wet chemical method *Materials Letters* **61** 5202–5

[11] Verges M A , Mifsud A and Serna C J 1990 Formation of Rod-like Zinc Oxide Microcrystals in Homogeneous Solutions *J. CHEM. SOC. FARADAY TRANS,* **86** 959-63

[12] Sua W, Huanga J and Lina C 2008 Improving the property of ZnO nanorods using hydrogen peroxide solution *Journal of Crystal Growth* **310** 2806–9

[13] Pauporte Th and Lincot D 2001 Hydrogen Peroxide Oxygen Precursor for Zinc Oxide Electrodeposition *Journal of The Electrochemical Society* **148** 310-14

[14] Chen Y, Jiang F, Wang L, Mo C, Pu Y and Fang W 2004 Influence of hydrogen peroxide solution on the properties of ZnO thin films *Journal of Crystal Growth* **268** 71–5

IOP Conf. Series: Materials Science and Engineering **64** (2014) 012051 doi:10.1088/1757-899X/64/1/012051

Effect of annealing temperature on electrical properties of poly (methyl methacrylate): titanium dioxide nanocomposite films using spin coating deposition technique

L N Ismail[1], Z Habibah[1], S H Herman[1] and M Rusop[2]
[1]NANO-ElecTronic Centre (NET), Faculty of Electrical Engineering, Universiti Teknologi MARA (UiTM) ,40450 Shah Alam, Selangor, Malaysia
[2]NANO-SciTech Centre (NST), Institute of Applied Sciences, Universiti Teknologi MARA (UiTM), 40450 Shah Alam, Selangor, Malaysia

Email: lyly2909@gmail.com

Abstract. Nanocomposite poly (methyl methacrylate) :titanium dioxide (PMMA :TiO_2) film were deposited on glass substrate. The effect of annealing temperature, especially on electrical, dielectric and the morphological properties of the thin films were investigated by current-voltage (I-V) measurement, impedance spectroscopy, and FESEM. The annealing temperature is varies from 120°C, 140°C, 160°C, 180°C and 200°C. The electrical properties results showing when nanocomposite film annealed at 120°C produce the lowest current. Meanwhile, when the annealing temperature increased, the current increased drastically and this indicates the PMMA:TiO_2 nanocomposite film are no longer having insulating properties. The dielectric properties also indicate that film annealed at 120°C has the best dielectric properties compared to other temperature. The FESEM results show that as the temperature increased, the PMMA:TiO_2 nanocomposite film started to create a phase separation between the PMMA matrix and TiO_2 nanoparticles.

1. Introduction

Presently, research to develop composite materials between polymers with inorganic materials has been steadily growing to improve the properties of polymer to be used in multiples field. The composition between two materials usually identifies as nanocomposite or hybrids materials. Nanocomposite material is defined a composite between two material which polymer with inorganic nanoparticles materials [1]. By combining these two materials, novel functions can be generated which leads to a wide range of interesting applications.

Basically, the properties of the nanocomposite materials solely depend on the preparation process. For examples the ratio of concentration between the polymers and inorganic material, the temperature use, and types of deposition techniques. There are two general approaches to prepare the nanocomposite materials. First, by using top-down technique, where the inorganic particles in nanometer (nm) sized were synthesized to form nanostructures then added into polymer. Second approached by using bottom up techniques, where the inorganic particles were directly dispersed into polymer matrix from a molecular

IOP Conf. Series: Materials Science and Engineering **64** (2014) 012051 doi:10.1088/1757-899X/64/1/012051

level [2]. This integration of polymer and inorganic components is achieved by the conventional sol-gel process. Meanwhile, spin coating technique are the most convenience method to obtain a good nanocomposite thin films [3].

Poly (methyl methacrylate) (PMMA) is one of the most versatile polymeric materials. It has been widely used for applications such as solar cells, optic [4, 5], and electro-optics [5] and as resist for photolithography. PMMA offer low-temperature process, it strong and lightweight and it was reported that PMMA can transmit up to 92% of visible light [6, 7]. PMMA also has been assessed as to be used as gate dielectric due to its high resistivity [3, 8]. Apart of then numerous advantages, PMMA also has disadvantage for example its low dielectric properties [9] and pinholes are formed when several thicknesses is reduced. Titanium dioxide (TiO_2) is one of the promising candidate materials in organic-inorganic applications. Due to its excellent dielectric and optical properties, it is extensively being used as thin film in optical devices [10] and also as a filler to enhance the polymer or organic capabilities. Some researchers have demonstrated by incorporation the TiO_2 into PMMA, they obtain an improvement in the optical properties for example capabilities in absorbing light in UV region [11-13].

To our knowledge, there is no report on the dielectric and electrical characterizations behavior of PMMA:TiO_2 nanocomposite film as dielectric layer. The study of electrical and dielectric properties of metal-insulator- metal (MIM) structure using PMMA:TiO_2 nanocomposite film as dielectric would gives useful information when to be fabricate organic capacitor or organic field effect transistor (OFET). In this work, the PMMA:TiO_2 nanocomposite thin film were prepared by spin coating method deposited on Al/Ti/glass substrates and annealed at temperature ranging from 120 to 200 ^{0}C. We report on the electrical, dielectric and surface morphology properties of the PMMA:TiO_2 nanocomposite films.

2. Experimental method

2.1 Metal-Insulator-Metal (MIM) Fabrication

The preparation of nanocomposite PMMA:TiO_2 thin film were prepared as reported in previous publication [14]. Before coating, the glass substrate was clean by sonication method in acetone, methanol and de-ionized (DI) water for 10 minute respectively. Titanium (Ti) and aluminium (Al) with the thickness of 60 and 40 nm were deposited on glass substrate acting as bottom electrode. Nanocomposite PMMA:TiO_2 were formed on Al/Ti/glass by spin coating and dry at 50 ^{0}C for 5 min. These steps were repeated several times to obtain a certain thickness. The final films were annealed only at once at temperature ranging from 120 to 200 ^{0}C for 30 min. For measurements of electrical and dielectric properties, Al electrodes were evaporated on the surface of the film through a mask as top electrodes to form metal-insulator-metal (MIM) configuration. The device structure in this work was then, Al/PMMA:TiO_2/Al/Ti/glass.

2.2 Device Characterization

The current-voltage (*I-V*) and measurement was performed at room temperature by the two point probe method using Keithley 4200 semiconductor characterization system connected to a probe station. Meanwhile, the dielectric properties were measured using impedance spectroscopy (Solatron S1260). To test the electrical properties, the bias voltage is swept from 0V to 5. The surface morphology and cross-section image of the thin film was measured using Field Emission Scanning Electron Microscope (FESEM, JEOL JSM 7600F).

3. Result and discussion

3.1 Surface Morphology

In order to study the annealing process, the FESEM images of the samples were measured, before and after annealing. The FESEM images of the samples before and after annealing at 120 and 180°C, are shown in Fig. 1 (a, b, c) respectively. The thickness of the thin films was found to be in the 987, 890, 793, 150 and 36 nm for 120°, 140°, 160°, 180° and 200°C respectively. Although, the film thickness for sample that being anneal at temperature 120° to 140° C is greater than 500 nm, the optimization during the deposition process on the spin speed and time can be varied in order to have film thickness below than 500 nm. Drastic changes in the film thickness for the sample anneal at 180 and 200°C is because PMMA can withstand the maximum temperature is 160°C before it melt which is consistent with the temperature that is being reported by other researchers [15, 16]. The confirmation of existing of TiO_2 nanoparticles on the surface of the PMMA:TiO_2 nanocomposite films have been published in our previous work [17]. The density of TiO_2 nanoparticles on the surface increased noticeably and porous surface occur after annealed the PMMA:TiO_2 nanocomposite film above 120°C for 30 min. When anneal the nanocomposite film at 180°C and above cause the phase separation between the PMMA and TiO_2 nanoparticles. This is because the PMMA start to degrade at temperature of 160°C [15].

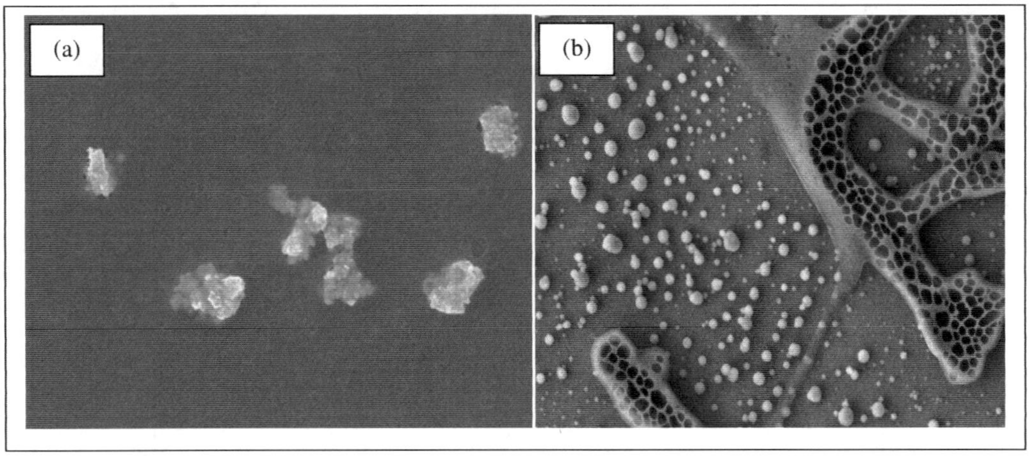

Figure 1. FESEM images of the nanocomposite PMMA:TiO_2 (a) anneal at 120°C (b) anneal at 180°C

3.2 Dielectric Properties

In order to investigate the effect of annealing temperature on the dielectric properties of nancomposite PMMA:TiO_2 thin films, the dielectric studies were performed using impedance spectrometer. Fig. 2 shows the real permittivity, ε_r of PMMA:TiO_2 nanocomposite film with five different annealing temperatures. All PMMA:TiO_2 nanocomposite thin films reveal that the real permittivity, ε_r are dependent to the annealing temperature. The values of real permittivity, ε_r at frequency of 1 kHz and 1 MHz are reducing when the annealing temperature is increased. The nanocomposite film anneal at 120°C gives the highest real permittivity, ε_r (12 at 1kHz and 10 at 1MHz) is due to an increased of total polarization arising from dipoles and trapped charge carriers. Films annealed at 180°C and 200°C give the lowest real permittivity, $\varepsilon_r \sim 0$ at same frequency. This indicates that weak polar nature of the PMMA as the matrix in the nanocomposite films. Hence, from the view point of dielectric constant, only three temperatures are suitable for obtaining a good insulation for metal-insulator-semiconductor (MIS) capacitor and organic field effect transistor (OFET) fabrication. Fig. 2 also indicates that the variation of real permittivity, ε_r with frequency from 10Hz to 10 MHz for nanocomposite PMMA:TiO_2 films. The observed decreased of dielectric constant with increasing frequency is because the induced dipoles in the film tried to rearrange themselves in the direction of applied field.

Fig. 3 shows the imaginary permittivity, ε'' for PMMA:TiO_2 nanocomposite film anneal at various temperature in the range of 120 to 200°C as a function of measuring frequency. The dependence of imaginary permittivity, ε'' on temperature measured at two selected frequencies (1 kHz and 1 MHz) shows that, as temperature increased, the imaginary permittivity, ε'' is decreased. Large imaginary permittivity, ε'' at low frequencies for all samples indicate that the films showing that the nanocomposite film severe high leakage current problem [18]. The imaginary permittivity, ε'' are nearly 0 when the nanocomposite PMMA:TiO_2 films were anneal at temperature above 180°C. Table 1 show the dielectric properties measurement results for the nanocomposite PMMA:TiO_2 at different annealing temperature.

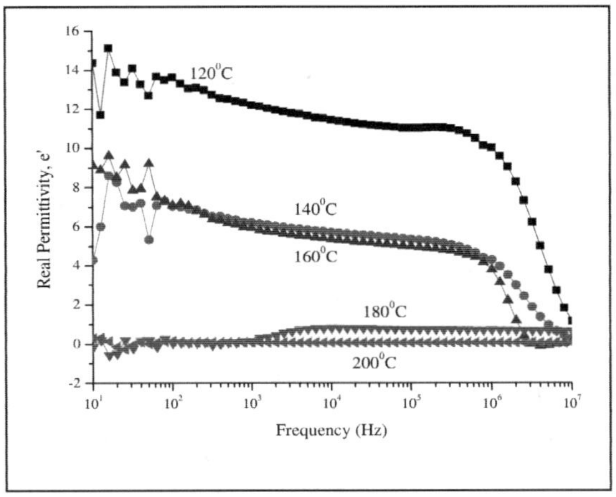

Figure 2. Real permittivity of PMMA:TiO_2 nanocomposite film anneal at different temperature.

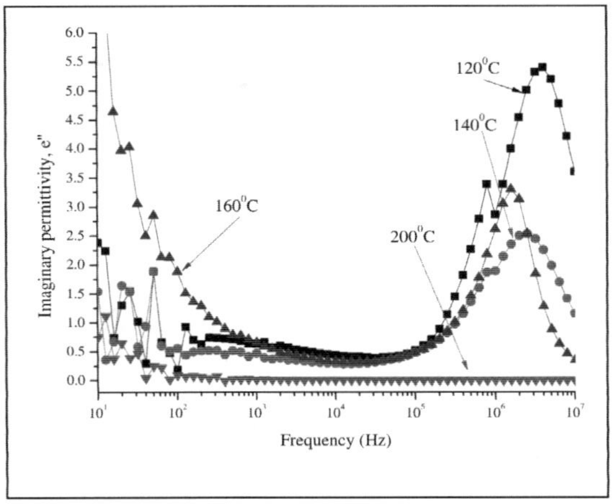

Figure 3. Imaginary permittivity of PMMA:TiO_2 nanocomposite film anneal at different temperature.

Table 1. Summary of dielectric properties of $PMMA:TiO_2$ nanocomposite films

Temperature (°C)	Film thickness (nm)	Real Permittivity, ε_r		Imaginary Permittivity at 1 kHz ε''
		1 kHz	1 MHz	
120	520	12	10	2.84
140	490	6	4.3	1.88
160	390	5.6	3.7	2.56
180	150	0.2	0.7	0
200	36	0.1	0.1	0

3.3 Electrical properties

I-V of the nanocomposite film annealed at 120, 140 and 160°C has the lowest current in the range of nanoampere (nA). On the contrary, the film annealed at 180 and 200°C has the highest current which is in the miliampere (mA). I-V characteristics for film annealed at 200°C showing the high rectifying curve because the film thicknesses are approximately 36nm compare to other samples. Film thickness of the insulator are responsible in determine the device performance because a variation in film thickness will cause a large fluctuation in its electrical behaviour of the dielectric layer [19-21].

The leakage current density-voltage characteristics were also examined because the insulating property is critical design consideration for the insulator to be used as insulator or dielectric in metal-insulator – semiconductor (MIS) capacitor or organic field effect transistor (OFET). Fig. 4 shows the J-V properties of MIM structure nanocomposite film with different anneal temperature. The leakage current density for nanocomposite film annealed at 120°C is $3.14e^{-9}$ A/cm^2 at 5V, while it increased to $4.26e^{-9}$ and $7.41e^{-9}$ A/cm^2 at the same voltage applied as the annealing temperature increased to 140 and 160°C respectively. The insulating properties are reduced when the annealing temperature is increased from 160 to 200°C. The nanocomposite film annealed at 200°C (3.3^{e-6} A/cm^2) has larger leakage current that anneal at 180°C (1.02^{e-6} A/cm^2), is due to the perforated surface morphology of the nanocomposite film that cause the Al top contact is diffuse into nanocomposite film.

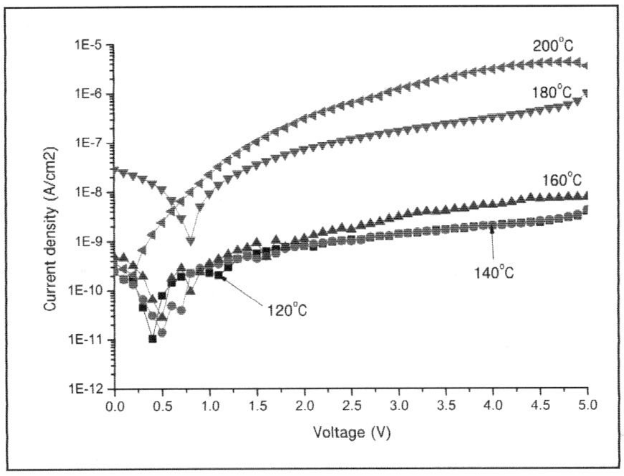

Figure 4. Leakage current density characteristics of $PMMA:TiO_2$ nanocomposite film anneal at different temperature

4. Conclusion

In conclusion, we have successfully prepared good nanocomposite PMMA:TiO_2 films on glass substrates by using the simple deposition technique that is spin coating. The effect of annealing temperature on the dielectric, electrical and morphology of the nanocomposite PMMA:TiO_2 are examined. The current-voltage and dielectric characteristics indicate that the optimum annealing temperature for nanocomposite film to be used as insulator in OFET or organic capacitor 120°C. Anneal at 180°C and above, makes the interaction between polymer chains and TiO_2 nanoparticles become weakened and at the same time the mobility of TiO_2 nanoparticles is augmented. Under these conditions, the nanocomposites films are no longer having insulation properties.

Acknowledgement

The authors would like to thank all members of NANO-ElecTronic Centre (NET) Universiti Teknologi MARA Malaysia for their cooperation and support. Also thanks to Ministry of Higher Education (MOHE) Malaysia for supporting the research under SLAB and Excellence Fund Research Management Institute (RMI) Universiti Teknologi Mara, Shah Alam (Project Code: FRGS/1/2012/TK02/UITM/03/8) for their financial support.

References

[1] W. Zheng and S.-C. Wong: Composites Science and Technology 63 (2003) 225.
[2] L. Matejka: Hybrid Nanocomposites or Nanotechnology: Electronic, Optical, Magnetic and Biomedical Applications, L. Merhari, Ed.: Springer, 2009.
[3] M. Morales-Acosta, M. Quevedo-López, B. Gnade, and R. Ramírez-Bon: Journal of Sol-Gel Science and Technology 58 (2010) 218.
[4] S. Li, M. M. Lin, M. S. Toprak, D. K. Kim, and M. Muhammed: NANO REVIEWS, 2010.
[5] X. Sun, X. Chen, X. Liu, and S. Qu: Applied Physics B: Lasers and Optics (2010) 1.
[6] N. G. Semaltianos: Microelectronic Journal 38 (2007) 754.
[7] T. S. Huang, Y. K. Su, and P. C. Wang: Japanese Journal of Applied Physics 47 (2008) 3185.
[8] J. Puigdollers, C. Voz, I. Martín, M. Vetter, A. Orpella, and R. Alcubilla: Synthetic Metals 146 (2004) 355.
[9] M. Na, and S.-W. Rhee: Organic Electronics 7 (2006) 205.
[10] M. Inkyo, Y. Tokunaga, T. Tahara, T. Iwaki, F. Iskandar, C. J. Hogan, and K. Okuyama: Industrial & Engineering Chemistry Research 47 (2008) 2597.
[11] M. M.Demir, K. Koynov, U. Akbey, C. Bubeck, I. Park, I. Lieberwirth, and G. Wegner: American Chemical Society 40 (2007) 1089.
[12] A. H. Yuwono, Y. Zhang, J. Wang, X. H. Zhang, H. Fan, and W. Ji: Chemistry of Materials 18 (2006) 5876.
[13] L. N. Ismail, N. N. H. N. Mohamad, M. S. Shamsudin, H. Zukefle, M. H. Abdullah, S. H. Herman, and M. Rusop: Japanese Journal of Applied Physics 51 (2012) 06FG09.
[14] R. R. Prakash, S. Pandiarajan, P. Venkatesh, and N. Kamaraj: International Conference in Emerging Trends in Electrical and Computer Technology (ICETECT) (2011) 46.
[15] A. T. Sellinger, A. H. MArtin, and J. M. Fitz-Gerald: Thin Solid Films 516 (2008) 6033.
[16] L. N. Ismail, S. A. Farahiyah, Z. Habibah, S. H. Herman, and M. Rusop: IEEE Symposium in Humanities, Science and Engineering Research (SHUSER) (2012) 259.
[17] C. Yang, G. Hu, and J. Wang: Microelectronics Journal 91 (2012) 132.
[18] S. Krishnan, E. Stefanakos, and S. Bhansali: Thin Solid Films 516 (2008) 2244.
[19] V. D. Costa, M. Romeo, and F. Bardou: Journal of Magnetism and Magnetic Material 258 (2003) 90.
[20] S. M. Sze, Physics of Semiconductor Devices: Wiley Inter-Science Publication, 1981

Applications of web produced by hot air assisted melt differential electrospinning method

Mahmoud M Bubakir[1], Haoyi Li[1, 2], Weifeng Wu[1], Xiaohu Li[1], Shuai Ma[1],Weimin Yang[1, 2]

[1]College of Mechanical and Electrical Engineering, Beijing University of Chemical Technology, Beijing 100029, China
[2]State Key Laboratory of Organic–Inorganic Composites, Beijing University of Chemical Technology, Beijing 100029, China

Abstract: Melt electrospinning, a technique that has gained increasing attention since it easily can generate continuous ultrafine fibers directly from polymer melts without the use of any solvent. Therefore, it is considered as a safe, cost effective, and environmental friendly technique. However, with all those great advantages, the technique still suffers some drawbacks such as: large fiber diameter and low throughput. The hot air assisted melt differential electrospinning (MDES) is a new technique invented by our research team that can solve or eliminate those drawbacks. The most important features of our used apparatus are: Needleless nozzle that could generate multiple Taylor cones around the bottom edge of the nozzle, which can result in a high throughput. The stretching force acting on the jets can be further strengthened by an air current provided by an air pressure gun. Interference between the high voltage supply and temperature sensors could be prevented through the grounding of the nozzle. The ultrafine pp webs produced using the same apparatus was in the micro/nano scale with a diameter of 600nm-6um and a smooth surface. Porosity of the webs ranges from 86.5%-99.4% when different collecting devices are used. The resultant ultrafine webs were applied in three areas: oil sorption, water treatment, and hydrophilic PP membrane. The results were very promising as for oil the sorption capacity was 129.0g/g; for water treatment, the rejection rate for 3um particles was 95%. And for the hydrophilic PP membrane, the water sorption capacity was 12.3 g/g.

Keywords: Melt differential electrospinning (MDES), fiber diameter, oil sorption, water treatment, hydrophilic PP membrane

1. Introduction

With the increasing advancement in science and technology research, electrospinning has been attracting increasing attention, probably due in part to a surging interest in nanotechnology, or interest in its ability for the fabrication of continuous polymeric ultrafine fibers or fibrous structures, it is a high efficient method for producing nano-fibers, which can be classified into two types: solution electrospinning (SES) and melt electrospinning (MES). However, the MES is believed to be an alternative method of SES, because it is more environmental-friendly, low-cost, and high efficient in spinning than the latter[1].When the diameters of the polymeric fibers decreased from micrometers to nanometers, it provided several amazing characteristics such as high surface area to volume ratio, high porosity, flexibility in surface functionalities, and superior mechanical properties. These outstanding properties of electrospun nanofibers make them to be the optimal candidates for a broad range of important applications in widely different areas such as drug carriers, tissue scaffolds, wound dressing, reinforcement materials, filters, protective clothing, electrodes, sensors, catalysts, etc. [2].The origin of electrospinning as a viable fiber spinning technique can be traced back to a hundreds of years ago. The process of electrospinning, also termed as electrostatic spinning, was first conceived by Lord Rayleigh [3] in the late 19th century. In 1902, the first patent for the production of fibers from a solution jet by using electric field was issued to Cooley [4] in the USA. From 1934 to 1944, Formhals published a series of patents [5–9], describing an experimental setup for the production of polymer filaments using an electrostatic force. Since then, the electrospinning truly surfaced as a feasible technique for spinning fibers with small diameters. Following the work of Formhals, the focus changed to developing a better understanding of the electrospinning process. In 1969, Taylor published his work[10] on the shape of the polymer droplet produced at the tip of the needle when an electric field was applied. Taylor found that the pendant droplet developed into a cone when the surface tension was balanced by electrostatic force and the jets were ejected from the vertices of the cone with diameters significantly smaller than the diameter of the needle when the electrostatic force exceeded its surface tension. This conical shape of the jet was later referred as the “Taylor Cone” in subsequent work. In the subsequent years, interest shifted away from the fundamental understanding of the electrospinning process to studying the structural morphology of the electrospun nanofibers and the relationship between the structural features and individual processing parameters. In 1971, Baumgarten[11] reported to make an apparatus to electrospun acrylic fibers with diameters in the range of 500 to 1100 nm. In 1981, LarrondoandMandley[12, 13]successfully produced polyethylene and polypropylene fibers by electrospinning their melts. They found that the fibers electrospun from a melt were relatively larger in diameter than fibers electrospun from a solution. In the early 1990s, several research groups, especially the Reneker group, revived the electrospinning by demonstrating the fabrication of ultrathin fibers from various polymers [14].Since then, the electro-spinning regained an upsurge attention in the research due to the potential applications of these nanofibers in various areas. Up to date, over a hundred synthetic and natural polymers have been successfully electrospun into fibers with diameter ranging from a few nanometers to several micrometers[15].In recent years, the research interest on electrospinning has been quickly shifted from the fabrication of nanofibers to the application of these resultant electrospun nanofibers with various functionalities. Some of these applications have reached the industrial level. Moreover, the results on the application areas showed that electrospinning is a versatile technique with broad applications in biomedical applications, as tissue engineering scaffolds, in wound healing, drug delivery, filtration, as affinity membrane, in immobilization of enzymes, small diameter vascular graft implants, healthcare, biotechnology, environmental engineering, defense and security, and energy storage and generation and in various researches that are ongoing[16-20]

2. Melt electrospinning

Melt electrospinning is a promising approach which has several advantages over solution electrospinning. Even though much of the research ,industrlization, and commercial development to date has been in solution electrospinning, but once the complications associated with Melt electrospinning method such as : relative thick fiber diameter, high viscosity and device complexity for high temperatures and high voltage were solved, this solvent-free method will open the doors to new horizons in modeling electrospinning without any risk of solvent evaporation, and complicated process for the solvent recycling as well as mass production of micro-nano fibers .It can be seen from literatures that many researchers focused on improvement of electrospinning equipment, fibers made from special materials and structure, and on nano scale fibers based on test results or microscopic images. In 1981 Larrondo and Manley[21-23] confirmed the feasibility of polymer MES according to their experiments when they established a simple device and electrospun PP fiber with diameter of about 50μm successfully by extruding the melt with a piston. Moreover, the formation and properties of fibers, the flow field during spinning process and the deformation of melt droplets were analyzed in detail. In 2004, Lyons[24] summarized the theoretical model of MES in his doctoral dissertation, In the same year, he published an article and pointed out that the electric field strength and polymer molecular weight have an important effect on the morphology and diameter of MES fibers.[25] Eduard [26]presented a gas-assisted polymer melt electrospinning process where under the investigation condition in electrospinning of polylactic acid (PLA) melt, air drag produced an additional 10% thinning compared to the un-assisted melt electrospinning process. Christopher[27] made Poly (butylene terephthalate), polypropylene, and polystyrene fiber by a single orifice melt blowing apparatus. They produced nanofibers with average diameters less than 500 nm. Ogata N, et al from university of Fukui in Japan designed a melt electrospinning setup that could produce 1μm fiber with laser heating source. Although the indirect flash heating system had some advantages, but laser equipment was of high energy consumption and risk. Shimada N, et al followed to heat a membrane with customized linear laser source, by which a line of Taylor cones was obtained and a row of jets were produced. This method increased the productivity comparing with the point laser source, but the yield was relatively low. Michal Komarek and Lenka Martinova from technical university of Liberec, Czech Republic invented two melt electrospinning devices, One is a “rod” style spinning head; another is a slot-shaped spinning head. The later one gave inspiration about the design of the melt electrospinning apparatus, yet it didn’t combine the screw continuous extrusion device, and the melt distribution at the slot was poor. In addition, the number of the fibers was limited when using a slot only. W. M.Yang research group from Beijing university of chemical technology, China, utilized typical materials commonly used in fiber manufacturing industry such as PP, PE, PA, PET, PLA,PCL, PES etc., to make fibers through self-designed MES equipment. The effect of voltage and temperature on fiber morphology was studied, and a few micron-grade fibers were achieved.[28, 29]

3. Design and Features of the melt differential electrospinning Setup

The hot air assisted melt differential electrospinning is a new technique invented by our research team that can solve or eliminate the drawbacks of melt electrospinning technique. It is a technique that provides a differential electrostatic melt spinning nozzle, through which a shunt within the nozzle cone can produce a plurality of individual fibers， functioning to resolve the key difficulties of device using single orifices, the self-designed apparatus consists of six major parts as shown in (Figure 1) : power supply of high-voltage, heating system, an air pressure gun, a copper circular ring functioning as electrode A, a needleless cone-shaped nozzle, and a collecting device. The power supply device secures a maximum output of 100 kV and a maximum current output of 2 mA. The electric heating ring with a power of 300 W was custom-built. The function of hot air can be summarized into three main points: Preserving heat of runner, accelerating thinning of flying jets, and keeping the environmental temperature high, that's higher than the softening point which will increase the action time of electric force and reduce fiber diameter (Figure 2 and Figure 3).

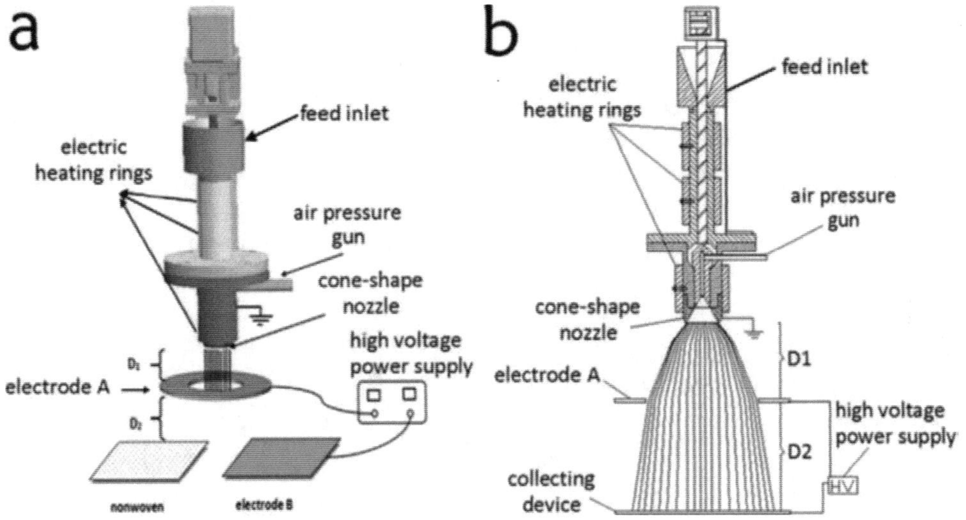

Figure 1. Schematics of a self-designed needleless MDES apparatus.

The unique features of the apparatus are: the generation of multiple Taylor cones around the bottom edge of the nozzle, resulting in a high throughput; the air current driven by an air pressure gun further strengthened the stretching force acting on the jets. More importantly, it could contract the flying jets with the pressure difference between the air current and the atmosphere to prevent the polymer jets from being attracted by the inner circle of electrode A; and two kinds of collecting devices were applied, namely, the nonwoven and electrode B. The nozzle was grounded to prevent any kind of interference between the high-voltage supply and temperature sensors;

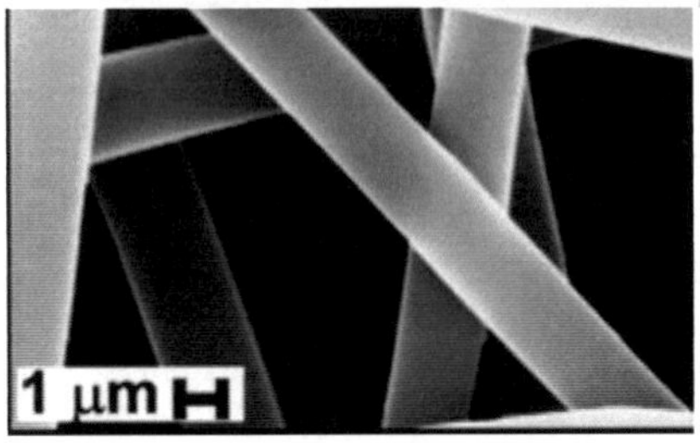

Figure 2. Un-assisted: 1-2 µm

Figure 3. Hot air assisted: 0.2-0.8 µm

With all those features, the device has the advantages of simple structure, capability of spinning thin fibers, high spinning efficiency, and the ultrafine PP webs produced are in the micro/nano scale with a diameter of 6um-600nm and a smooth surface. Porosity of the webs ranges from 86.5%-99.4% when different collecting devices are used.

4. Applications of web produced by hot air assisted MDES

Since the early 1980's, electrospun polymer nanofibers have already been proposed for many different applications. Reviewing the number of patents, we can see that approximately two-thirds of all electrospinning applications are in the medical field. Of the remaining patents, one-half deals with filtration applications, and all other applications share the remaining half [30]. But most of applications were using solution electrospun fibers, and the use of melt electrospun fibers in many applications is still limited, some groups started to research the feasibility of using melt electrospun fibers in different applications. The web fabricated by hot air assisted melt differential electrospinning method was applied experimentally in three areas of research, and gave promising results. The areas were:

4.1. *Sea Oil-Spill Cleanup applications*

Polypropylene (PP), a synthetic polymer and general purpose plastic, has great mechanical behavior, together with other excellent properties, like nontoxicity, processibility, and low cost, PP has been widely used in many applications, but have not been used in oil sorption as melt electrospun produced fiber before. Ultrafine polypropylene (PP) webs as oil sorbents used in this application were produced via the melt-differential-electrospinning self– designed apparatus and characterized by SEM and contact-angle analysis. PP webs of various diameters and porosities were obtained by the manipulation of the applied electrical field. The effects of the fiber diameter and porosity on the oil-sorption capacity and oil retention behavior were investigated. The experimental results showed that for web diameter on the microscale, the porosity played an important role in determining the oil-sorption capacities. The maximum oil-sorption capacity of the resulting PP webs with regard to the kinds of oil used (motor – peanut) were 129 and 80 g/g, respectively; these values were approximately six to seven times higher than that of the commercial PP nonwoven for the corresponding oils. The high oil-sorption capacity of the melt-electrospun PP webs was attributed to its small fiber diameter, high porosity, olephilicity, and hydrophobicity. Additionally, the great mechanical performance of the resulting webs ensured their reusability and recoverability. They could be reused up to seven times while still maintaining a sorption capacity of around 80 g/g, and above 97% of the sorbed oil could be recovered, which indicated excellent reusability and recoverability.(Figure 4)[31]

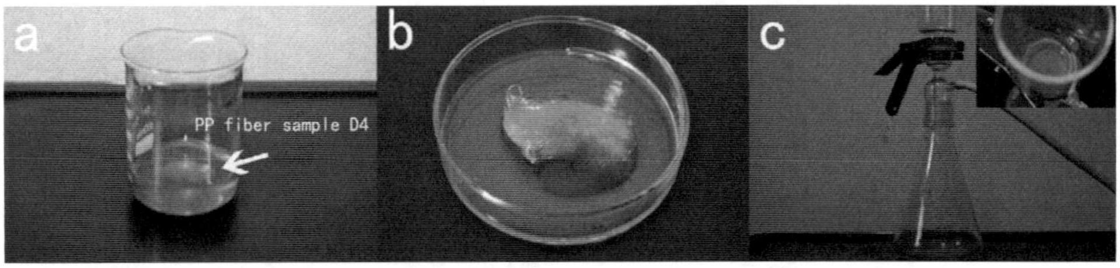

Figure 4. Oil sorption / desorption process

4.2. Fabricated oriented fiber membranes for microfiltration applications
The main aim of this study was to explore the influence of arrangement for oriented fibers on filtration efficiency. The melt-differential-electrospinning self–designed technique was used to fabricate oriented fiber membranes for this application. During the experiment, Polypropylene (PP) with a melt flow rate of 2000 g/10min was used. Electrospinning was carried out in a humidity of 25% at 24 C^0, and the nozzle temperature was set at 240C^0. The PP melt were electrospun at the condition of single-electrode with a feed rate of 15 g/h controlled by a pump. A cylinder collector (diameter: 100mm, maximum rotating speed: 1400 rpm) was used to get the oriented fibers and was placed 260mm off the effects of the nozzle (Figure 5).

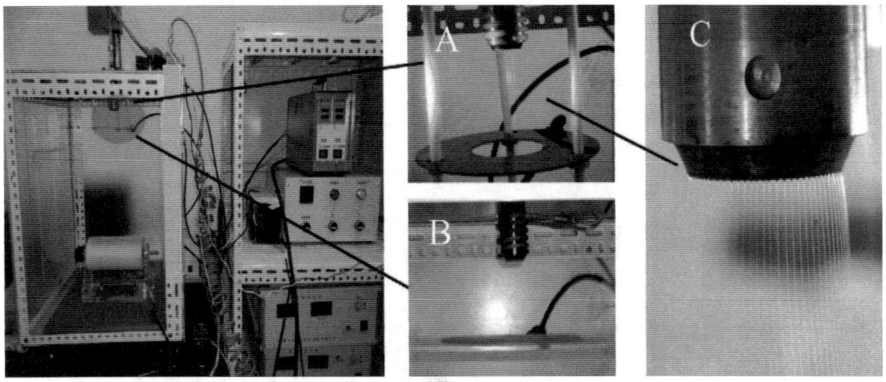

Figure 5. Images of melt electrospinning apparatus: (A) single electrode as receiver, (B) ring electrode with a cylinder collector, (C) melt jet.

The oriented PP fibers were stacked under different azimuthal angles based on the horizontal datum of 0 degrees to obtain a series of filtration membranes (Figure 6), and used these samples to do orthogonal experiments. Then, six prepared PP membranes were treated by hot-pressing at 90C^0, along with the one untreated original membrane. The oriented fiber membrane were 100 mm in diameter and Characterized by scanning electron microscopy (SEM, HITACHI S4700, Japan) after gold coating to examine the morphology of electrospun PP fiber filtration membrane, and the fiber diameter was measured by the image analysis software (Image J 2X). A micrometer was used to measure the thickness of the membranes.

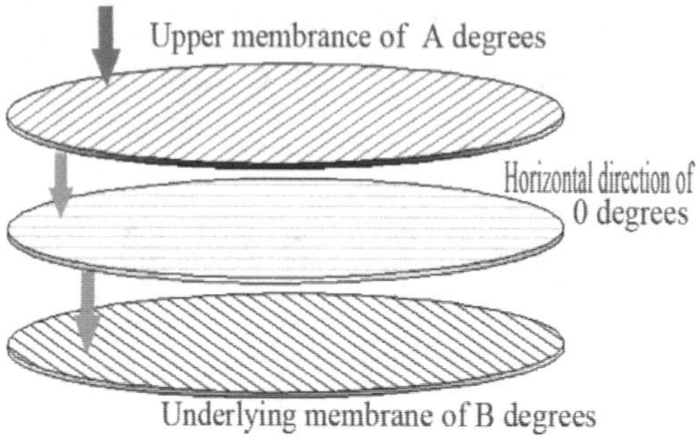

Figure 6. Schematic diagram of sample under different azimuthal angles

The mechanical properties of membranes were assessed by a material testing system (Instron5565, Instron Co.) at a loading velocity of 10 mm/min. And the thermal properties of membranes were evaluated by differential scanning calorimetry(DSC, Q2000, TA Instruments) at a heating rate of $10C^0$/min under nitrogen atmosphere from 30 to $250C^0$. The contact angle between the electrospun PP membrane and water was measured by using a contact angle analysis system (OCA20, Dataphysics). Two microliters of deionized water were dropped on the surfaces of samples. Image of solid/liquid interface was captured by a charge-coupled device camera. The contact angle data of PP membrane (Figure 7) illustrate that the oriented PP membrane showed a high water contact angle of (127. 9°) than the random PP membrane of (124.4°), which may be due to the smaller pore size of oriented membrane than that of the original membrane.

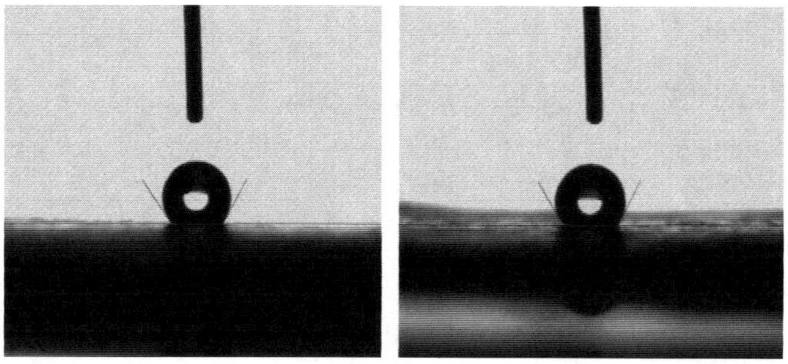

Figure 7. Contact angle of (a) oriented PP membrane (127. 9°) and (b) random PP membrane (124.4°)

The particle rejection test used 300 nm-3μm carbonate particles to prepare 1% wt of the original solution. The concentration of the solution was examined by UV spectrophotometer (UV-9000). The rejection efficiency indicates the rejection performance of one particle which is a very important indicator during the actual filtration. The results of the rejection efficiency of different membranes to the particles showed that rejection of all membranes after hot-pressing treatment could reach more than 90%, while the original membrane exhibited only 75% rejection. This conclusion shows that hot-pressing process made the film more dense and fiber surface much more attached. Also, when the interval angle of adjacent membranes was smaller, the filtration efficiency was greater. And the results demonstrated that electrospun oriented membranes showed more efficiency in rejecting the 300 nm

IOP Conf. Series: Materials Science and Engineering **64** (2014) 012052 doi:10.1088/1757-899X/64/1/012052

-3μm particle, and still maintained the same permeate flux as that of the original membrane. The electrospun oriented fiber membranes also showed good overall mechanical performance.

4.3. Hydrophilic PP membrane

The hydrophilic modification of PP and the transfer mechanism of water have been studied by many researchers, but few research works about the production of hydrophilic modified PP ultra-thin fibers by melt electrospinning and hydrophilic performance of these fibers have been reported. In this study, pure ultra-fine PP fibers and PP blending fibers have been fabricated by the melt-differential-electrospinning self– designed apparatus In order to study the modification effect of the different hydrophilic modification methods. Different kinds of hydrophilic agents (Nano Calcium Carbonate， SDBS， Arstar-7008)as additives and (emulsifier OP-10， TF-629,TF-629C) as coating respectively were mixed with Polypropylene (PP) (melt flow rate of 2000 g/10min),and the mixtures spun under specific spinning conditions(spinning temperature: 230°C； top plate voltage : 30kv； bottom plate voltage : 65kv； distance between sprayer and top plate: 50mm； distance between top and bottom plates: 120mm).Experiments were done to determine Water absorption capacity, and Water delivery rate where mathematical equations have been applied to calculate the average for both. Fiber diameter and surface morphology of elecrospun fibers are examined by scanning electron microscopy (SEM). Image processing software (Image J 2X) is applied to calculate the average fiber diameters. The water contact angle of fibers is measured by a contact angle meter. Images of solid/liquid interface are captured by a charge-coupled device camera, and the average contact angle value is obtained by measuring the same sample at five different positions. Our results showed that the method of melt-differential-electrospinning can produce mass blending ultra-fine fibers with considerable production efficiency (effective output is about 9g/h) , and diameters of which mainly range from 2 to 6μm. In addition, after adding or coating with hydrophilic agents, the contact angles decrease in different extent and hydrophilicity enhanced. Adding additives alone Arstar-7008 has better effect in hydrophilic modification of PP than that of nano-Calcium Carbonate and analytic pure SDBS. Arstar-7008, composed of 60% PP and 40% additives, can dramatically enhance the hydrophilic performance of PP so that PP fibers can absorb and deliver water repeatedly and constantly. The average water absorption capacity was 12.30g/g, and the average water delivery rate was 1.69(g/min*g). The fibers produced by melt electrospinning have better texture. With comparison among several hydrophilic agents as coatings, TF-629 enhances the hydrophilic performance of PP fibers most. Coatings can easily infiltrate into water and wick fast and high, whereas this kind of modifier has poor sustainability, which means they can easily fall off or be soluble in water. When using additives and coatings.

5. Conclusions

From the experimental work done, we can conclude that the fibers or webs produced by the melt-differential-electrospinning, could be efficiently used in many environmental engineering applications, such as oil-spill cleanup , water treatment, and many other applications .

6. References

[1] X. Wang, Y. Liu, C. Zhang, Y.An , X.He, W.M.Yang Simulation on electrical field distribution and fiber falls in melt electrospinning, J. Nanosci. Nanotechnol. 13 (2013) 4680–4685.

[2] Ramakrishna S, ebrary Inc. An introduction to electrospinning, and nanofibers, World Scientific 2005; 398.

[3] Barnes CP, Sell SA, Boland ED, Simpson DG, BowlinGL. Nanofiber technology: Designing the next generation of tissue engineering scaffolds. Adv Drug DelivRev 2007; 59(14): 1413-1433.

[4] Cooley, J.F.: US692631 (1902).

[5] Formhals A. US patent 1,975,504, 1934.

[6] Formhals A. US patent 2,160,962, 1939.

[7] Formhals A. US patent, 2,187,306, 1940.

[8] Formhals A. US patent, 2,323,025, 1943.

[9] Formhals A. US patent, 2,349,950, 1944.

[10] Taylor G. Electrically Driven Jets. In Proceedings of the Royal Society of London Series a-Mathematical and Physical Sciences1969; 313(1515): 453-475.

[11] BaumgartenPK. Electrostatic spinning of acrylic microfibers. J Colloid Interface Sci 1971; 36(1): 71-79.

[12] Larrondo L, St. John M R. Electrostatic fiber spinning from polymer melts. I. Experimental observations on fiber formation and properties. J PolymSciPolymPhys Ed 1981; 19(6): 909-920.

[13]Larrondo L, St. John MR. Electrostatic fiber spinning from polymer melts. II. Examination of the flow field in an electrically driven jet. J PolymSci PolymPhys Ed 1981; 19(6): 921-932.

[14] Doshi J, Reneker DH. Electrospinning process and applications of electrospun fibers. J Electrostat 1995; 35(2&3): 151-160.

[15] Subbiah T, Bhat GS, Tock RW, Parameswaran S, RamkumarSS. Electrospinning of nanofibers. J ApplPolymSci 2005; 96(2): 557-569.

[16]How TV .Synthetic Vascular Grafts and Methods of Manufacturing Such Grafts.US Patent 4552707, 1985.

[17]Reneker DH, Yarin AL, Fong H, Koombhongse S. Bending instability of electrically charged liquid jets of polymer solutions in electrospinning. J ApplPhys 2000; 87:4531–47

[18] Stitzel J, Pawlowski K, Wnek G, Simpson D, Bowlin G. Arterial smooth muscle cell proliferation on a novel biomimicking,
biodegradable vascular graft scaffold. J BiomaterAppl 2001; 16:22–33.

[19] Kim JS, Reneker DH. Mechanical properties of composites using ultrafine electrospunfibers.Polym Compos 1999; 20:124–31.

[20] Ramakrishna S, Fujihara K, Teo WE, Yong T, Ma Z, Ramaseshan R. Electrospun nanofibers: solving global issues. Mater
Today 2006; 9:40–50.

[21]L. Larrondoand R. S. J. Manley, *J. Polym Sci.* 19, 909 (1981).

[22] L. Larrondo and R. S. J. Manley, *J. Polym. Sci.* 19, 921 (1981).

[23] L. Larrondo and R. S. J. Manley, *J. Polym. Sci.* 19, 933 (1981).

[24] J. Lyons, Melt-electrospinning of thermoplastic polymers: An experimental and theoretical analysis, Ph.D. Dissertation,
Drexel University, Philadelphia (2004).

[25] J. Lyons, C. Li, and F. Ko, *Polymer* 45, 7597 (2004).

[26]E. Zhmayev,D. Cho, and Y. L. Joo, *Polymer* 51, 4140 (2010).

[27] C. J. Ellison, A. Phatak, D. W. Giles, C. W. Macosko, and F. S.Bates, *Polymer* 48, 306 (2007).

[28] Y. Liu, R. J. Deng, M. F. Hao, H. Yan, and W. M. Yang, *Polym. Eng. Sci.* 50, 2074 (2010).

[29] M. F. Hao, Y. Liu, X.T. He, Y. M. Ding, and W. M. Yang, *Adv.Mater. Res.* 221, 111 (2011).

[30]Burger C, Hsiao BS, Chu B. Nanofibrous materials and their applications. AnnuRevMater Res 2006; 36:333–68.

[31] H. Li, W. Wu, M. Bubakir, H. Chen, X. Zhong, Z. Liu, Y. Ding, W.M.Yang, " Polypropylene fibers fabricated via a needleless melt-electrospinning device for marine Oil-spill cleanup" J applPolymSci,2014,131,40080 .

IOP Conf. Series: Materials Science and Engineering **64** (2014) 012053 doi:10.1088/1757-899X/64/1/012053

Electrical Characterization of Metal-Ferroelectric-Insulator-Semiconductor having Double Layered Insulator for Memory Applications

L N Ismail[1], M. H. Wahid[3], Z Habibah[1], S H Herman[1], M. D. Rozana[3] and M Rusop[2]
[1]NANO-ElecTronic Centre (NET), Faculty of Electrical Engineering, Universiti Teknologi MARA (UiTM) ,40450 Shah Alam, Selangor, Malaysia
[2]NANO-SciTech Centre (NST), Institute of Applied Sciences, Universiti Teknologi MARA (UiTM), 40450 Shah Alam, Selangor, Malaysia
[3]Faculty of Applied Sciences, Universiti Teknologi MARA (UiTM), 40450 Shah Alam, Selangor, Malaysia

Email: lyly2909@gmail.com

Abstract. Metal-ferroelectric-insulator-semiconductor (MFIS) devices were successfully fabricated using poly(vinylidene fluoride-trifluoroethylene) (PVDF-TrFE) and poly (methyl methacrylate): titanium dioxide (PMMA:TiO_2) nanocomposite as ferroelectric and insulator films, respectively on n-type silicon (*n*-Si) substrate. Both ferroelectric and insulator films were prepared by sol-gel spin coating method. The electrical behaviour of metal-ferroelectric-metal (MFM) structure with PVDF-TrFE film and metal-insulator-metal (MIM) structure PMMA:TiO_2 film exhibited different current characteristics. The capacitance of the MFIS devices was found to be 0.42 and 0.29 nF at frequency of 1kHz and 1 MHz respectively. Meanwhile, the dielectric loss values are constant (~60 x 10^{-3}) in the frequency range from 100 Hz to 100 kHz. *I-V* results for MFIS are much higher than MIM and MFM is due to there is a trapped holes/electron located at the semiconductor-insulator interface which contributes to high leakage current in the MFIS device. We conclude, although interposing the PMMA :TiO_2 nanocomposite insulator layer between the semiconductor and Al electrodes degrades the MFIS performance, nevertheless, they remain sufficiently good for use in organic electronic devices.

1. Introduction

An organic material has shown potential to be used in the memory applications. Their attractive features such as simpler fabrication process compared that being employed in the conventional silicon (Si) technology, low-power operation and large capacity for data storage. Metal-ferroelectric-insulator-silicon (MFIS) is a promising candidate for non-volatile random access memory because of its high speed, low power consumption and non-destructive read-out operation [1-2]. The main characteristic of MFIS is capacitance-voltage (*C-V*) measurement can affect the information storage capability and stability of memory devices. Poly(vinylidenefluoride-trifluoroethylene) (PVDF-TrFE) is an organic ferroelectric material that has been studied widely for memory application owing its unique characteristics which can be process at low temperature processed, low polarization density and larger coercive field [3]. Poly (methyl methacrylate) (PMMA) is one of the most versatile polymeric materials. It has been widely used for applications such as solar cells, optic [4, 5], and electro-optics [5] and as resist for photolithography. PMMA offer low-temperature process, it strong and lightweight and it was reported that PMMA can transmit up to 92% of visible light [6, 7]. PMMA also has been assessed as to be used as gate dielectric due to its high resistivity [3, 8]. Apart of then numerous advantages, PMMA also has disadvantage for example its low dielectric properties [9] and pinholes are formed when several thicknesses is reduced. Titanium dioxide (TiO_2) is one of the promising candidate materials in organic-inorganic applications. Due to its excellent dielectric and optical properties, it is extensively being used as thin film in optical devices [10] and also as a filler to enhance the polymer or organic capabilities. Some researchers have demonstrated by incorporation the TiO_2 into PMMA, they obtain an improvement in the optical properties for example capabilities in absorbing light in UV region [11-13]. Bae and Kang et.al has fabricated ferroelectric field-effect-transistor (FeFET) by blending film of PVDF-TrFE and poly (methyl methacrylate) (PMMA) and the result show considerable memory performance with long data retention and low operation voltage [4-6]. Nevertheless, there were less studies on memory application having PMMA:TiO_2 nanocomposite in the MFIS structure.

In this paper, we report the fabrication and characterization of metal-ferroelectric-insulator-semiconductor (MFIS) devices using PMMA:TiO2 nanocomposite and PVDF-TrFE (70:30 mol%) as insulator and ferroelectric layer, respectively. The obtained nanocomposites and ferroelectric films are deposited by spin coating onto glass substrates and characterized by current-voltage (*I-V*) and capacitance-voltage (*CV*) for electrical properties, polarization-field (*P-E*) for the ferroelectric properties, and field emission scanning electron microscopy (FESEM) for the morphology. In this work, we concentrate on *I-V* and *P-E* characteristics for the PMMA:TiO2 and PVDF-TrFE films and the effect of different structure of MFIS device.

2. Experimental method

2.1 Metal-Ferroelectric-Insulator-Semiconductor (MFIS) Fabrication

The metal-ferroelectric- insulator-semiconductor (MFIS) structures were assembled on p-type silicon (*p*-Si) substrate. Aluminum (Al) top electrodes were thermally evaporated on polymer films using shadow mask with the holes size of 1um. The PMMA:TiO_2 nanocomposite solutions were prepared as reported in earlier publication [7]. PVDF-TrFE (70:30 mol%) with concentration at 30g/L was prepared by dissolving in 2-butanone for 24 hours. Spin coating process was utilized to produce ~250nm thick PVDF-TrFE thin film. The PVDF-TrFE thin films were then annealed at 120°C for 2 hours to improve it crystallinity. The aluminum (Al) was patterned through shadow mask by thermal evaporation to form top contact electrode. The current-voltage (*I-V*) and measurement was performed at room temperature by the two point probe method using Keithley 4200 semiconductor characterization system connected to a probe station. Meanwhile, the dielectric properties were measured using impedance spectroscopy (Solatron S1260). To test the electrical properties, the bias voltage is swept from 0V to 10V. The surface morphology and cross-section image of the thin film was measured using Field Emission Scanning Electron Microscope (FESEM, JEOL JSM 7600F).

3. Summary

Figure 1 illustrate the current-voltage *(I-V)* curves of PMMA:TiO_2 nanocomposite film.The *I-V* curve indicates that PMMA:TiO_2 nanocomposite film still maintaining its insulator behavior where the current are small in the nanoampere (nA) range. The leakage current for PMMA:TiO_2 is 6.43 x 10^{-8} A/cm_2 measured at -5V. This value is comparable and slightly higher compared to other organic polymer dielectric. The large leakage current obtained in sample using PMMA:TiO_2 as insulator could be explained by the amount of TiO_2 in the PMMA matrix and it change the bonding configuration of PMMA and PMMA:TiO_2. These charges behave as conduction paths for electrons and holes and therefore lead to an increase in the leakage current [8]. Figure 2 shows the ferroelectric characteristics of PVDF-TrFE thin film without PMMA:TiO_2 nanocomposite film measured from 30 to 100V drive voltage at frequency of 100Hz . The saturation of polarization reversal was archived at the highest drive voltage which results the remnant polarization (P_r) of 93 mC/m2 with coercive field (E_c) of 72 MV/m. These are the indication of stable polarization reversal and polarization retention as the thin film able to produce high P_r at low E_c [9]. FE-SEM images on Fig. 3 (a) and (b) show the PMMA:TiO_2 nanocomposite and PVDF-TrFE film deposited on glass substrate. It can be seen that, the nanocomposite PMMA:TiO_2 film is like plastic and there are particles that accumulate in the nanocomposite film. Meanwhile, PVDF-TrFE thin film consisted of elongated crystallites which indicating the ferroelectric crystal and enhanced the crystallinity of the thin film [10].

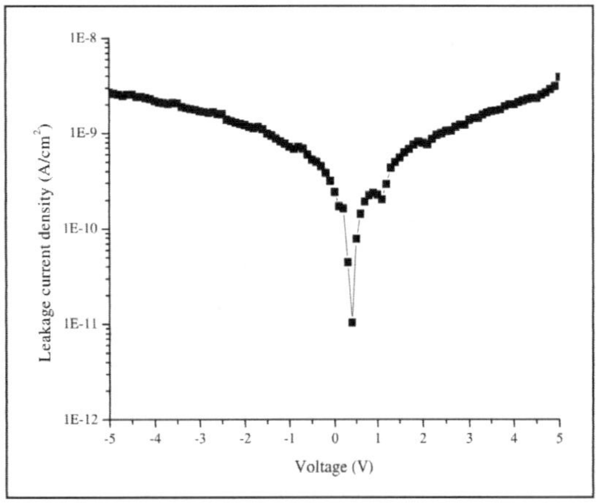

Figure 1. I-V curves for PMMA:TiO_2 nanocomposite film.

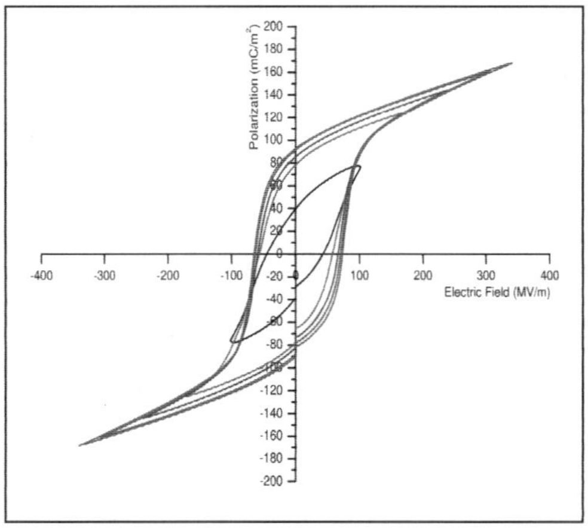

Figure 2. P-E curve for PVDF-TrFE film.

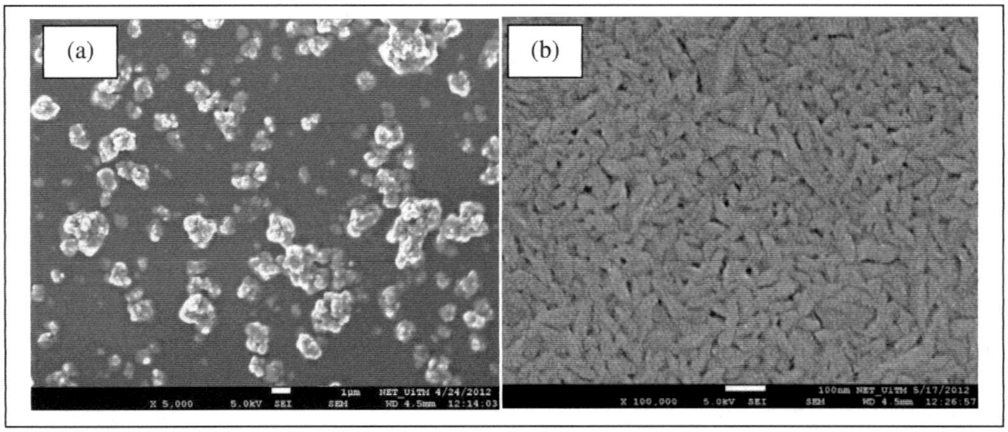

Figure 1. FESEM images of the (a) PMMA:TiO_2 nanocomposite (b) PVDF-TrFE film

References

[1] J. Sun, *et al.*, "A model for the electrical characteristics of metal-ferroelectric-insulator-semiconductor field-effect transistor," *Current Applied*

Physics, vol. 12, pp. 760-764, 2009.
[2] Y. G. Xiao, *et al.*, "Factors for the polarization lifetime in metal-ferroelectric-insulator-semiconductor capacitors," *Solid-State Electronics,* vol. 73, pp. 84-88, 2012.
[3] R. Kalbitz, *et al.*, "Fixed negative interface charges compromise organic ferroelectric field-effect transistors," *Organic Electronics,* vol. 13, pp. 875-884, 2012.
[4] I. Bae, *et al.*, "Organic ferroelectric field-effect transistor with P(VDF-TrFE)/PMMA blend thin films for non-volatile memory applications," *Current Applied Physics,* vol. 10, pp. e54-e57, 2010.
[5] S. J. Kang, *et al.*, "Printable Ferroelectric PVDF/PMMA Blend Films with Ultralow Roughness for Low Voltage Nonâ€ Volatile Polymer Memory," *Advanced Functional Materials,* vol. 19, pp. 2812-2818, 2009.
[6] J.-M. Kim, *et al.*, "Effect of CdSe nanoparticles in poly methyl methacrylate tunneling layer on the performance of nonvolatile organic memory device," *Microelectronic Engineering,* vol. 98, pp. 305-308, 2012.
[7] L. N. Ismail, *et al.*, "Effect of Solvent on the Dielectric Properties of Nanocomposite Poly(methyl methacrylate)-Doped Titanium Dioxide Dielectric Films," *Japanese Journal of Applied Physics,* vol. 51, p. 06FG09, 2012.
[8] E. H. Oulachgar, *et al.*, "Fabrication and Characterization of GaAs MOS Capacitor With CVD Grown Polymer-Based Thin Film as a Gate Dielectric," *Electron Devices, IEEE Transactions on,* vol. 57, pp. 1942-1947, 2010.
[9] T. Furukawa, Takahashi, Y. and Nakajima, T.. "Recent advances in ferroelectric polymer thin films for memory applications," *Current Applied Physics,* vol. 10, pp. e62-e67, 2010.
[10] J.S. Lee, K.J. Kim and A. Prabu. "Ferroelectric P(VDF/TrFE) Ultrathin Film for SPM-based Data Storage Devices " Solid State Phenomena, vol. 124-126, pp. 303-306, 2007.
[11] J.S. Lee, K.J. Kim and A. Prabu, "Ferroelectric P(VDF/TrFE) Ultrathin Film for SPM-based Data Storage Devices " Solid State Phenomena, vol. 124-126, pp. 303-306, 2007

IOP Conf. Series: Materials Science and Engineering **64** (2014) 012054 doi:10.1088/1757-899X/64/1/012054

Memristive behaviour of spin coated titania thin film

N S Kamarozaman[1,*], S H Herman[1,] and M A Mahmudin[1]**
[1]NANO-ElecTronic Centre (NET), Faculty of Electrical Engineering, Universiti Teknologi MARA (UiTM) ,40450 Shah Alam, Selangor, Malaysia

E-mail: sheeraee@yahoo.com*, hana1617@salam.uitm.edu.my**

Abstract. This paper presents the memristive behaviour of spin coated titania thin films. The precursor molarity of titania thin film was varied from 0.05 to 0.4 M to study the effect of precursor molarity on the memristive behaviour of the thin films. From the observation, although the film thickness increased with the precursor molarity, the resistance ratios of the best switching loop for all samples showed no significant differences. However, it was found that the sample with less precursor molarity (device that having thinner film) required lesser time to produce the stable switching loop compared to the sample with higher precursor molarity (device that having thicker film).

1. Introduction

For over 150 years, there were only resistor, capacitor and inductor known as the fundamentals passive circuit element, until 1971, Dr Leon Chua argued that there should be the fourth element to complete the symmetry of the basic fundamental passive circuit element. The first physical modeled of a memristor was realized by HP Labs only in 2008 due to the advancement of nanotechnology [1]. The term 'memristor' is basically the simplified term of 'memory resistor' which indicates its ability to memorize the recent dynamic resistance even when the bias is turned off. The memristor exhibits I-V characteristics of bow-ties or known as pinched hysteresis loop that crosses at zero point [1, 2]. Basically, the memristor is based on metal-insulator-metal (MIM) structure. This simpler structure compared to transistors and its ability to be used as nonvolatile memory, switching devices and smart interconnects has attracted much attention from researchers recently [3, 4]. It has been reported that the memristive behaviour becomes noticeable at nano scale thickness [2-4], but then there are few studies reported that the switching behaviour can also be observed even in micro or millimeter scale thickness recently [5].

Titania is a promising active layer to be used in memristor due to its simple structure and compatibility with standard complementary metal-oxide semiconductor technology [6, 7]. It is easy to grow the films on various substrates using different deposition methods such as sputtering, atomic layer deposition (ALD), dip-coating, anodization and spin-coating method [7-10]. Furthermore, Debashis Panda et al., has stated that the TiO_2 shows different types of switching depending on the structures and stoichiometry of the films [11]. Due to that, TiO_2 is chosen as an active layer for the device in this work. Although sputtering and atomic layer deposition (ALD) are well-known methods used to fabricate the device due to the better control of the film thickness and oxygen content during deposition process, both methods require high cost [10, 12, 13]. Thus, a simpler and inexpensive method, sol-gel spin coating is chosen. Titania thin film was spin-coated on conductive

IOP Conf. Series: Materials Science and Engineering **64** (2014) 012054 doi:10.1088/1757-899X/64/1/012054

ITO-coated glass substrate followed by deposition of Pt metal contact to form a memristive device. The effect of precursor molarity differences on the switching behaviour of the memristive device was studied. The ITO was chosen as the bottom electrode because it was reported that it improves the reproducibility of I-V curve and shows non-volatile characteristics [12], while Pt as top electrode shows better memristive behaviour with titania layer compared to Au and Ti [14].

2. Experimental method

2.1. Memristive device fabrication

The ITO-coated substrate as a bottom electrode for the device was subjected to a standard cleaning procedure using methanol and deionized water (DI) for 10 minutes each in an ultrasonic bath. The titania solution was prepared using titanium (IV) isopropoxide, $Ti[OCH(CH_3)_2]_4$ (Sigma Aldrich) as a precursor, glacial acetic acid (GAA) as the stabilizer, Triton X-100 as the surfactant and ethanol absolute as the solvent. The precursor molarity was varied from 0.5 to 0.4M. The solutions were stirred at 2 rpm for 3 hour in room temperature. The solution was then deposited onto the ITO substrate by using spin coating method. The spin speed was set to 3000 rpm for 60 seconds. During this process, 10 drops of titania sol-gel were dropped onto the spinning ITO substrate. Next, the titania thin film was annealed for 20 minutes in the furnace with temperature of 250 °C. Lastly, Pt as a top electrode for the device was sputtered with 60-nm thickness.

2.2. Device characterization

The current-voltage (I-V) measurement was performed by the two point probe method using Keithley 2400 semiconductor characterization system connected to a probe station at room temperature. To test the memristive behaviour, the bias voltage is swept from 0V to 5V, then 5V to -5V and -5V back to 0V while simultaneously measuring the current. From the result, resistance (R_{OFF}/R_{ON}) ratio can be obtained as shown in figure 1. The surface morphology of the samples were observed by using Field Emission Scanning Electrode Microscope (FESEM, JEOL JSM 7600F) and the thickness of the TiO_2 thin film was measured by using Surface Profiler (VEECO DEKTAK 750 Profile System). The average thicknesses of the films were calculated by taking measurements at 5 different points on one sample.

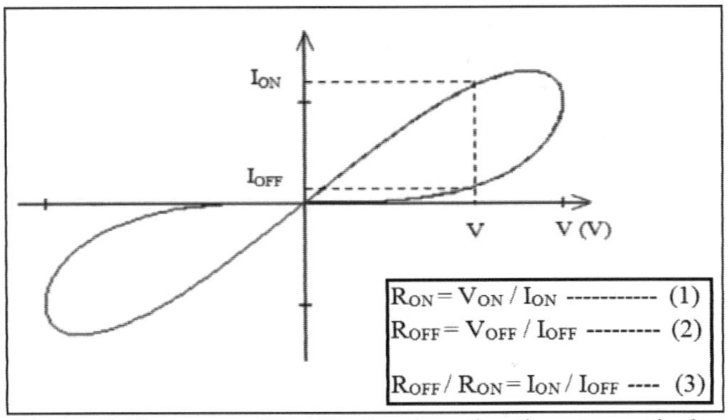

Figure 1. Resistance (R_{OFF}/R_{ON}) ratio estimation at a particular voltage

3. Result and discussion

3.1. Memristive behaviour

IOP Conf. Series: Materials Science and Engineering **64** (2014) 012054 doi:10.1088/1757-899X/64/1/012054

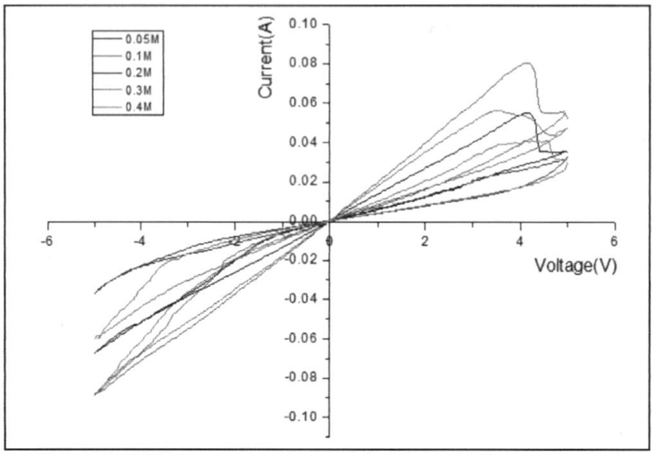

Figure 2. I-V characteristics of samples which respect to the different precursor molarity for 0.05M, 0.1M, 0.2M, 0.3M, and 0.4M

Figure 2 shows the I-V characteristics for 5 samples which have different precursor molarity of 0.05M, 0.1M, 0.2M, 0.3M, and 0.4M, respectively. From figure 2, all samples show the bias-dependant switching characteristic that match the electrical behaviour reported for memristor [1, 3]. From the result, it is observed that all samples have almost similar resistance (R_{OFF}/R_{ON}) ratio ranging from 1.80 to 2.29 as shown in table 1, despite having deposited using different precursor molarity. The sample 0.4M gave the highest resistance ratio, while the sample 0.2M gave the lowest resistance ratio.

It has been reported that the resistance ratio is higher for thinner film thickness [1, 15]. However, in this work, it is observed that the sample with 0.4M with thin film of 120nm thickness still shows the memristive behaviour. This result is agreed with work reported by T. Prodromakis et al. which stated that the memristive behaviour still can be observed even in microscale memristor [5]. The reason why film thickness did not affect the memristive behaviour in this work will be investigated further.

Table 1. Resistance ratio and maximum current at low resistance state for all samples with different precursor molarity

Precursor (TTIP) molarity (M)	Resistance ratio (R_{OFF}/R_{ON})	Maximum current, I_{max} at LRS (A)	Thickness (nm)
0.05	1.96	0.054	21.5
0.1	2.03	0.079	33.4
0.2	1.80	0.023	62.2
0.3	2.00	0.056	85.3
0.4	2.29	0.040	120.0

3.2. Device performance

Figure 3 (a) to (e) shows the I-V characteristics of the samples that have different precursor molarity and the insets correspond to the FESEM surface morphology images. To test the repeatability of the results, the measurement was done three times for each sample.

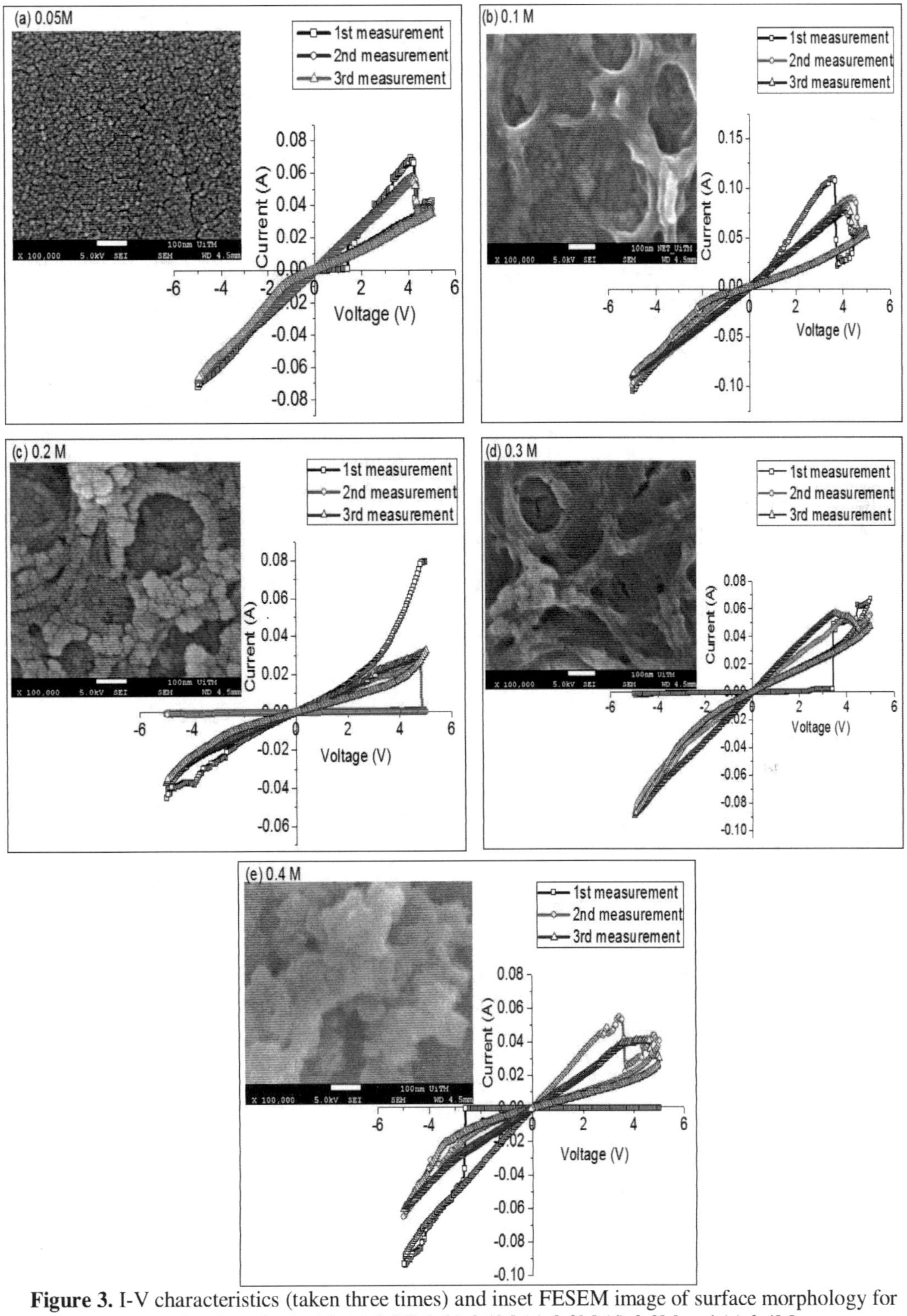

Figure 3. I-V characteristics (taken three times) and inset FESEM image of surface morphology for samples deposited with (a) 0.05M (b) 0.1M (c) 0.2M (d) 0.3M and (e) 0.4M

From figure 3, it was observed for samples of 0.05M and 0.1M, the same switching loop was reproduced during the second measurement. However, for sample 0.2M, 0.3M, and 0.4M, the stable switching loop is only obtained during third measurement. From the FESEM images of the surface morphology, it was observed that, sample 0.05M gave the smoothest surface morphology, while sample with precursor molarity of 0.4M, the grains that seen on the surface morphology are dense with agglomerate particles. It can be seen that the higher the precursor molarity, the thickness of titania thin film is increased as shown in table 1. Thus, we hypothesize that the thicker the titania thin film, the surface became more agglomerate, and the more time was taken to the memristor to achieve the stable switching loop. Note that the switching behaviour is dealing with the movement of positive charges, oxygen vacancies. Due to that, most have reported that the path creation of the oxygen vacancies in the active layer can be created by electroforming process [16-18]. However, this electroforming process requires high voltage to the virgin sample and might affect the device performance [17-19]. Thus, it is desirable to eliminate this process by controlling the oxide layer thickness and switching cycling [16, 20] . We found that the device with thinner film thickness takes a shorter time interval to get a stable switching loop compared to the device with thicker film thickness. Thus, from this work, it can be said that the conductive path is easier to be formed within device that having thinner film thickness without more measurement cycles compared to the device that having thicker film thickness.

4. Conclusion

We have successfully deposited titania thin film by applying sol-gel spin coating method. The titania was deposited onto several samples with variation of precursor molarity ranging from 0.05M to 0.4M. The variation of precursor molarity gave a little impact to the resistance ratio which can be conclude that, the memristive behaviour of all the samples were not affected despite of the precursor molarity differences. Nevertheless, it was observed that the sample with less precursor molarity (thinner film) required lesser time to obtain stable switching loop compared to the sample with high precursor molarity (thicker film).

Acknowledgments
This work is partially supported by Research Management Institute (RMI), UiTM and Malaysian Ministry of Higher Education (MOHE) under the Exploratory Research Grant Scheme (ERGS) (Project code :600-RMI/ERGS 5/3 (27/2013) / ERGS/1/2013/TK04/UITM/03/05).

References

[1] D. B. Strukov, G. S. Snider, D.R. Stewart, R.S. Williams 2008 *Nature* 80–83.
[2] B. Hayes 2011 *American Scientist* **99** 106-110.
[3] R. Williams 2008 *IEEE Spectrum* **45** 28-35.
[4] A. Shadaram, S. Mirzakuchaki, F. Zakerian 2011, *Canadian Journal on Electrical and Electronics Engineering* **2** 346-352.
[5] T. Prodromakis, K. Michelakis, C. Toumazou 2010 *Electronics Letters* **46**.
[6] Nadine Gergel-Hackett, Behrang Hamadani, Barbara Dunlap, John Suehle, Curt Richter, Christina Hacker, D. Gundlach 2009 *IEEE Electron Device Letters* **30** 706-708.
[7] K. Miller, K.S. Nalwa, A. Bergerud, N.M. Neihart, S. Chaudhary 2010 *IEEE Electron Device Lett.* 737-739.
[8] T. Prodromakis, K. Michelakisy, C. Toumazou 2010 *Proceedings of 2010 IEEE International Symposium on*, , pp. 1520-1522.
[9] E. Gale, R. Mayne, A. Adamatzky, B. de Lacy Costello 2014 *Materials Chemistry and Physics* **143** 524-529.
[10] J. Joshua Yang, Nobuhiko P. Kobayashi, John Paul Strachan, M.-X.Zhang, Douglas A. A.Ohlberg, Matthew D. Pickett, Zhiyong Li, Gilberto Medeiros-Ribeiro, R.S. Williams 2011 *Chem. Mater.* **23** 123–125.

[11] D. Panda, A. Dhar, S.K. Ray 2012 *IEEE Trans. on Nanotechnology* **11** 51-54.
[12] O. Kavehei, C. Kyoungrok, L. Sangjin, K. Sung-Jin, S. Al-Sarawi, D. Abbott, K. Eshraghian, *IEEE 54th International Midwest Symposium on Circuits and Systems (MWSCAS)*, 2011, pp. 1-4.
[13] C. Yakopcic, A. Sarangan, J. Gao, T.M. Taha, G. Subramanyam, S. Rogers 2011 *Proceedings of the IEEE National Aerospace and Electronics Conference (NAECON)* 101-104.
[14] Z. Aznilinda, S.H. Herman, R.A. Bakar, M. Rusop 2013 *Applied Mechanics and Materials* **393** 74-78.
[15] T.H. Kim, E.Y. Jang, N.J. Lee, D.J. Choi, K.-J. Lee, J.-t. Jang, J.-s. Choi, S.H. Moon, J. Cheon 2009 *Nano Lett.* **9** 2229-2233.
[16] J. Joshua Yang, Feng Miao, Matthew D Pickett, Douglas A A Ohlberg, Duncan R Stewart, Chun Ning Lau, R.S. Williams 2009 *Nanotechnology* **20**.
[17] S.B. Lee, H.K. Yoo, K. Kim, J.S. Lee, Y.S. Kim, S. Sinn, D. Lee, B.S. Kang, B. Kahng, T.W. Noh 2012 *Nanotechnology* **23** 315202(315209pp).
[18] J.P. Strachan, J.J. Yang, L.A. Montoro, C.A. Ospina, A.J. Ramirez, A.L.D. Kilcoyne, G. Medeiros-Ribeiro, R.S. Williams 2013 *Beilstein J. Nanotechnol.* **4** 467–473.
[19] D.S. Jeong, R. Thomas, R.S. Katiyar, J.F. Scott, H. Kohlstedt, A. Petraru, C.S. Hwang 2012 *Rep. Prog. Phys.* **75** 076502 (076531pp).
[20] Q. Mao, Z. Ji, J. Xi, J. 2010 *Phys. D: Appl. Phys.* **43** 395104 (395105pp).

The potential role of amorphous lead titanate thin films as nanodielectric layer for capacitor applications

Z Nurbaya[1, 3]* and MRusop[1, 2]

[1]NANO-ElecTronic Centre (NET), Faculty of Electrical Engineering, Universiti Teknologi MARA, 40450 Shah Alam, Selangor, Malaysia
[2]NANO-Sci Tech Centre (NST), Institute of Science, Universiti Teknologi MARA, 40450 Shah Alam, Selangor, Malaysia
[3]Razak School of Engineering and Advanced Technology, Universiti Teknologi Malaysia, 54100 Kuala Lumpur, Malaysia

*E-mail: nurbayazainal@gmail.com

Abstract. A Study of thin films preparation faces many challenges due to the environmentally sensitivity that could initiate effect towards device performance. Requirement to fabricate the passive device like capacitor at such nanoscaled thickness is quite challenges that would initiate the effect on the device performance. Nowadays, capacitor not just only existed by conventional dielectric material, it can become a complex device when ferroelectric materials get involved. This material promotes useful knowledge of their polarization behavior which contains switching element that premier contribution for memory storage applications. The present study demonstrates the potential of amorphous $PbTiO_3$ thin films as nanodielectric layer for high performance capacitor at low voltage applications. This unique characteristic of amorphous structure performs such incredible high dielectric constant value about ~100 and tangent loss of 4-5%, measured at 1 kHz. The $PbTiO_3$ films also involve other electrical measurement like P-E hysteresis loop.

Keywords: Lead titanate thin films; dielectric property, ferroelectric behaviour, metal-ferroelectric-metal capacitor

1. Introduction

Lead titanate ($PbTiO_3$) is a perovskite structured ceramic family that has ferroelectric behaviour. This comes across to the attention that the energy storage elementary in memory device has been influenced by this factor. Thus, as pointed out by Haertling et al in her extensively literature on ferroelectric ceramic overview; ferroelectric behaviour is the reason to gain high dielectric constant of thin films [1]. This highlights to the significant of $PbTiO_3$ ferroelectric ceramic material to be used in a broad range of device application, for example metal-ferroelectric-metal (MFM) capacitor [2], metal ferroelectric semiconductor (MFS) transistor [3], [4], and non-volatile random access memories (NVRAMs) [5]. Various processing methods have been used to prepare $PbTiO_3$ thin films, such as metal organic chemical vapour deposition [6]–[8], radio frequency magnetron sputtering [9]–[11], and sol-gel processing [12]–[16].

The films preparation by sol-gel processing has attracted many researchers because of its ability to deposit large area as well as being cost effective production in line with industrial needs. It also produces such a various structural property which depends on the effect during deposition process. Crystalline $PbTiO_3$ films with purely perovskite structure were obtained in most studies preferred using platinised silicon substrate with titanium adhesion, with contribution of highly rapid annealing treatment up to 650~800°C [17]–[19]. This however gives impact on mass production cost, where low synthesis temperature could be the best alternative to be implemented for the preparation of ceramic films on soda lime glass substrate. In [20], Cheng et al had successfully prepared nanocrystalline $PbTiO_3$ films on glass substrate and spent most of his time to value the films potentialities through electrical measurement. In this study, we purposely prepare amorphous films of $PbTiO_3$ in order to investigate its property mainly for high performance capacitor applications.

2. Experimental Method

Preparation of thin films $PbTiO_3$ using indium doped tin oxide; ITO glass substrate is most preferable for metal-ferroelectric-metal, MFM capacitor due to high conductivity behaviour of ITO itself. This study presents an experimental investigation on the potentiality of purely amorphous $PbTiO_3$ films under sub-100nm thickness at low processing temperature. Among various techniques in chemical solution deposition, spin coating technique shows interesting way in approaching sub-100nm scaled film thickness with high films densification controlled. Basically the technique promotes homogeneity of multilayer coating which also covering large and very complex area. Thus zero defect films could yield to high performance thin films capacitor devices.

2.1. $PbTiO_3$ solutions

T The $PbTiO_3$ precursor solutions were made through polymerization at open ambient temperature. The starting materials involved alkoxide material, Titanium tetraisopropoxide, TTiP (Sigma Aldrich, 97%) and lead acetate trihydrate (Sigma Aldrich, 99.9%) with the solvent, 2-methoxyethanol (Sigma Aldrich, 99.9%). A mixture of lead powders and solvent were mixed in 50ml beaker glass and stirred at 100°C for about 1hour. The stirring heat then were reduced to 60°C, while quickly added the titanium isopropoxide into the solutions and continuously stirred for the next 30minutes. The process at this stage was handled in very quick movement so that the alkoxide solution will not vaporized quickly with humidity at room temperature. Then, the prepared solutions were added with an appropriate amount of lead powder that sufficient to avoid lead compensation during annealing treatment. After that, the solutions were gone through ageing process about 24~36 hours with tight sealed beaker.

2.2. $PbTiO_3$ films processing

Sol-gel transformations will be completed by deposition of light yellowish color of $PbTiO_3$ solutions onto cleaned ITO layer of glass substrate. First and for most, the cleaned and clear substrates were dried at 150°C in hot furnace before proceed to the solutions deposition. The reason behind this is to promote such homogenous films developed on top of ITO surface, thus reduced out the existence of interfacial layer problem. About 20minutes before deposition, $PbTiO_3$ solutions were sonicated moderately in ultrasonic bath at 50°C of sonication temperature. The spin coater was set at 3000rpm, 30s of deposition time and the films was coated in five repetitions layer. The heat treatment involved repetition of drying process at 200°C and annealing process at 550°C for 10minutes and 1hour respectively.

2.3. Films Characterizations

The prepared sol-gel derived spin coated $PbTiO_3$/ITO glass thin films at sub-100nm thickness were characterized in both structural and electrical properties. Surface morphological study of the films were observed through field emission scanning electron microscopy (FESEM-JEOL) while to investigate the films uniformity by topological study were performed by atomic force microscopy,

AFM (Park System XE-100). Electrical characterization of $PbTiO_3$ thin films were carried out for ferroelectric and dielectric property which is measured by P-E hysteresis loop measurement (Radiant Technology, Precision LC) and impedance spectroscopy (Agilent 4294A). For this purposes, aluminium (Al) was deposited on the films as top metal contact by thermal evaporation based on Al/$PbTiO_3$/ITO/glass configuration.

3. Results and Discussions

3.1. Nanostructural $PbTiO_3$thin films

$PbTiO_3$ films surface morphology were observed at x50,000 magnification, 5.0kV and 10.4μA of emission power as shown in figure 1. The films demonstrate noncrystalline structure, provided by the formation of multisize flakes and grains formation. Even though the films were in nonuniform structure, thus having such high densification films without existence of microcrack surface films would be the premier contribution for ceramic thin films capacitor. This is in agreement with Bao et al that mentioned denser films gives the best impact on the dielectric properties [21]. Furthermore, EDX analysis as shown in figure 1b indicates that Pb element facilitate high binding energy performance due to the fact that the films were at influenced of 10mol% Pb excess.

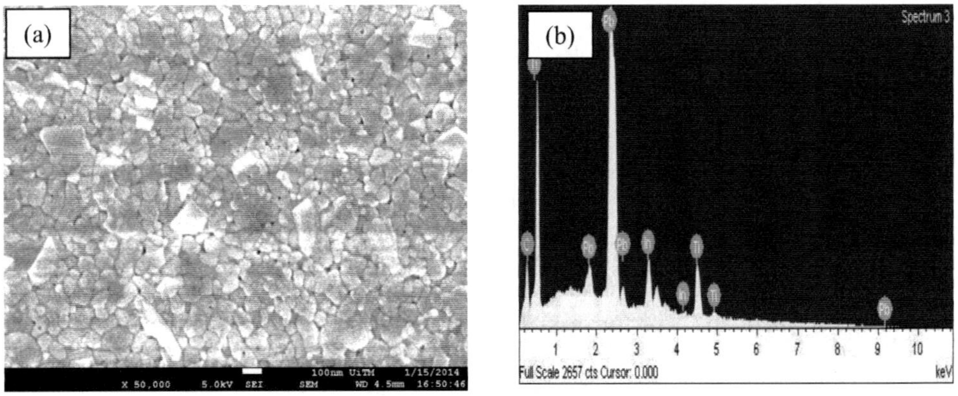

Figure 1. FESEM result of $PbTiO_3$ thin films, (a) surface morphology and (b) EDX analysis

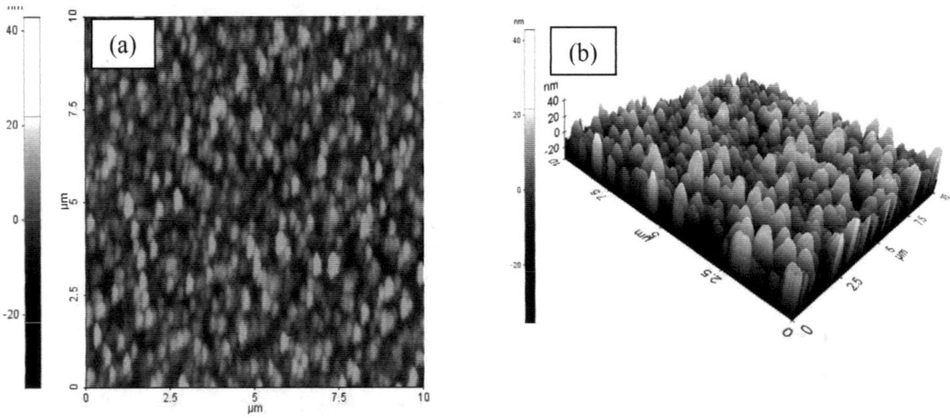

Figure 2. AFM result of $PbTiO_3$ thin films, (a) planar view and (b) 3-D view

Study on the films surface roughness that is topographical source is important to acknowledge at, thus, ushering the smooth electronic conduction when applying source. The films surface topology were observed by using AFM as shown in figure 2, set by planar view and 3-D view of the films. The details of the scanning parameters were set at localized $10\mu m^2$ scanning area, 2.50 servo gain and 1.50Hz scan rate. Films surface topology indicates the average surface roughness and average grain volumes were achieved about 8.905nm and $1.2x10^{-3}\mu m^3$ respectively.

3.2. Electrical Measurement

3.2.1. Ferroelectric behaviour. The ferroelectric behaviour of $PbTiO_3$ films are demonstrated by the plot of P-E hysteresis loop at various dc bias which are 30, 75, 100, and 150V as shown in figure 3. Plus, the details of spontaneous polarization (P_s), remanent polarization (P_r), and coercive field (E_c) value of are tabulated in table 1. The area of the films according to sandwich-type MFM configurations that measured by using surface profiler is about $1.08x10^{-5}cm^2$ and the thickness is 139.47nm.

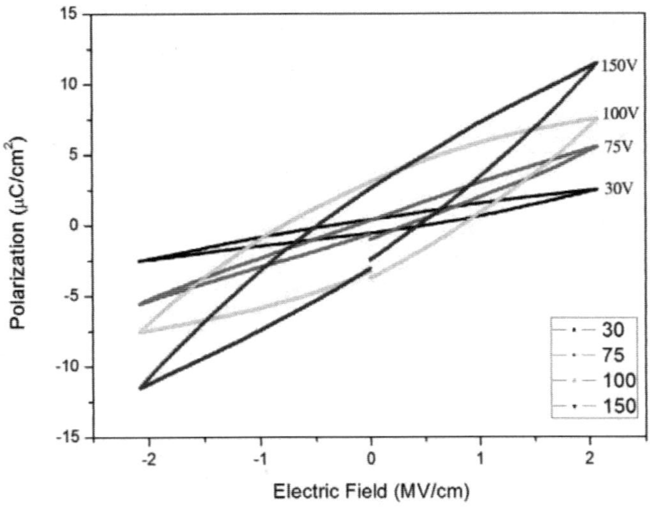

Figure 3. Plot of P-E hysteresis loop of $PbTiO_3$ thin films measured at various dc bias

Table 1. Ferroelectric properties at various dc biases obtained at 500Hz

DC value (V)	Spontaneous Polarization, P_s ($\mu C/cm^2$)	Remanent Polarization, P_r ($\mu C/cm^2$)	*Coercive Field, E_c (kV/cm)*
30	0.13	0.58	450
75	4.95	0.32	500
100	7.51	3.05	800
150	11.25	2.62	500

Ferroelectricity behaviour in this study shows the P-E loops were meeting the agreements where the spontaneous polarization and electric field source are approximate to each other. This is due to the fact that the ferroelectric polarization happened between two conductive plates and creates the hysteresis which is caused by the application of an external voltage on the internal electric dipoles. The P-E result in this study is about 12.5μC/cm^2 at 2MV/cm obtained at very high electric field applications due to the performance of amorphous structured $PbTiO_3$ films. It was found that Many studies had impressively prepared the films on crystalline Pt/Ti/Si substrate as reported by Wang et al in his tremendous literature review on ferroelectric behaviour of lead based titanate films [22]. Closest reference made by Cheng et al, claimed that highest spontaneous polarization could be obtained with such ITO glass substrate as indicated in his study [20].

3.2.2. Dielectric measurement. The measurement was conducted as same parameter (area and thickness) as done in ferroelectric measurement based on MFM configuration. The analyzer was set up with 500mV ac signal and performed in range of 40Hz to 100 kHz frequencies as shown in figure 4. The k value is having constantly space charge moment about ~99 with tolerance of ±7% measured at 1kHz frequencies complimented with low dissipation energy about 4 to 5%. Whereas, above than 10kHz, the dielectric polarizations were dropped drastically to a specific value indicated dipole moment polarizations. Hence, the films are believed in well suited for DC power application when there highlight significant *k* value approached to the lowest frequency. High densification films brought by the addition of Pb excess as shown in Figure 1 are in addition to demonstrate such high dielectric constant which is in agreement with studies reported by Bao et al in [21].

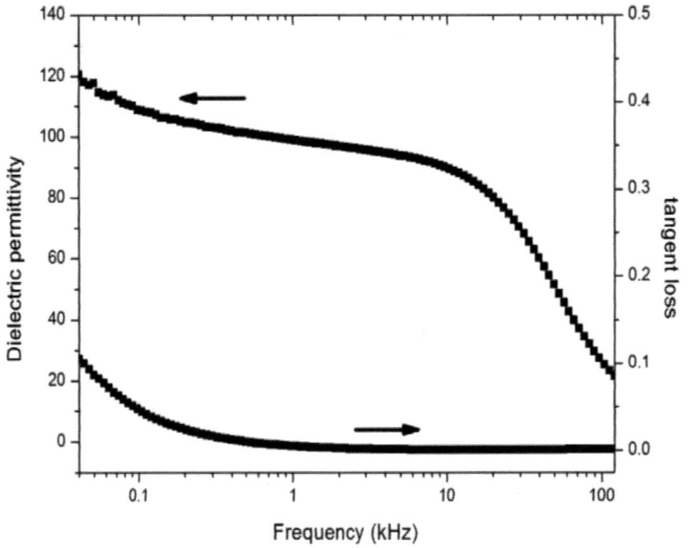

Figure 4. Plot of dielectric versus frequency of $PbTiO_3$ thin films

The measured capacitance density, (F/cm^2) is related to the dielectric permittivity according to following express:

$$\frac{C}{A} = \frac{\varepsilon_r \varepsilon_o}{d} \tag{1}$$

Where, A, ε_r, ε_o, and *d* comprises of films area, relative permittivity of dielectric material, free space permittivity, films thickness respectively. Thus, the capacitance density is about ~0.63μF/cm^2 with ±7% tolerance that is quite interesting to acknowledge at sub-100 nm thickness films.

4. Conclusions

Using sol-gel spin coating method with low synthesis temperature, amorphous nanostructured $PbTiO_3$ thin films were obtained. The plots of P-E hysteresis loop and dielectric versus frequency were successfully performed on Al/$PbTiO_3$/ITO glass at remarkable ~140nm thickness. For various dc bias applications on ferroelectric measurement, it demonstrates approximate relation between spontaneous polarization and dc source. In this study, the ferroelectric behaviour of the films were not affected by the increase of electric field that generated by dense films. Furthermore, high dielectric constant with low dissipation energy presented in this study shows that the $PbTiO_3$ thin films are potential as nanodielectric films for future multilayered nanoscaled ferroelectric capacitor device applications.

Acknowledgement

The author would like to thank for the financial support of Ministry of High Education, Malaysia (MOHE).

References

[1] Haertling G H 1999 *J. Am. Ceram.* **82** 797.
[2] Lee K, Rhee B R, and Lee C 2001 *J. Korean Phys. Soc.* **38** 723.
[3] Tang M H, Zhou Y C, Zheng X J, Yan Z, Cheng C P, Ye Z and Hu Z S 2007 *Solid. State. Electron.* **51**, 371.
[4] Wu S Y 1974 *IEEE Trans. Electron Devices* **21** 499.
[5] Chen H, Tsaur S, and Lee J Y 1998 *Jpn. J. Appl. Phys.* **37**, 4056.
[6] Wang C H and Choi D J 2001 *J. Am. Ceram. Soc.* **13** 207.
[7] Vellaikal M and Kingon A I 1996 *Thin Solid Films* **287** 139.
[8] Chen H Y, Lin J, Tan K L 1996 and Feng Z C *Thin Solid Films* **289** 59.
[9] Chang W L and He J L 2004 *J. Electroceramics* **13** 35.
[10] Pignolet A, Schmid P E, Wang L and Levy F 1991 *J. Phys. D Appl. Phys* **619** 58.
[11] Iijima K, Tomita Y, Takayama R and Ueda I 1986 *J. Appl. Phys.* **60** 361.
[12] Sidorkin A, Nesterenko L, Sidorkin A, Ryabtsev S and Bulavina G 2010 *Solid State Sci.* **12** 302.
[13] Tahan D M and Safari A 1992 *In Applications of Ferroelectrics, 1992. ISAF '92* 420.
[14] Morita T and Cho Y 2005 *J. Korean Phys. Soc.* **46** 10.
[15] Moon J, Kerchner J A, Lebleu J, Morrone A A and. Adair J H 1997 *J. Am. Ceram. Soc.* **80** 2613.
[16] Bao D, Yao X, Wakiya N, Shinozaki K and Mizutani N 2002 *Mater. Sci. Eng. B*, **94** 269.
[17] Bakar R A, Bakar M S A and Rusop M 2011 *Int. Conf. Electron. Devices, Syst. Appl.* 237.
[18] Lee C Y, Tai N H, Sheu H S, Chiu H T and Hsieh S H 2006 *Mater. Chem. Phys.* **97** 468.
[19] Yadav H O 2004 *Ceram. Int.* **30** 1493.
[20] Cheng S D, Kam C H, Zhou Y, Que W X, Lam Y L, Chan Y C and Gan W S 2000 *Thin Solid Films* **375** 109.
[21] Bao D, Yao X, Wakiya N, Shinozaki K and Mizutani N 2002 *Mater. Sci. Eng. B* **94** 1.
[22] Wang D G, Chen C Z, Ma J and Liu T H 2008 *Appl. Surf. Sci.* **255** 1637.

IOP Conf. Series: Materials Science and Engineering **64** (2014) 012056 doi:10.1088/1757-899X/64/1/012056

Effects of shear during the cooling on the rheology and morphology of immiscible polymer blends

S Hammani [1-3], N Moulai-Mostefa[2], L Benyahia[3] and J F Tassin[3]
[1]LAFPC, University Saad Dahlab of Blida, Route de Soumaa, 09000 Blida, Algeria
[2]LME, University of Medea, Ain d'Heb, 26001 Medea, Algeria
[3]Polymères, Colloïdes, Interfaces, Université du Maine-CNRS, 72085 Le Mans, France

E-mail: Hammani71@yahoo.fr

Abstract. The aim of this work was the generation of a microfibrillar structure in immiscible polymer blends using a new technique. The blend polymer model is the emulsion formed by a mixture of polypropylene (PP) with polystyrene (PS) in the proportion of PP10/PS90. In the first case the pellets of polystyrene and polypropylene were blended on the twin-screw mini extruder in the classical manner with different shear rates. In the second case, the same blend was prepared in the same way followed by a dynamic cooling at different shear rates. The phase morphologies of PP in the blend were determined by Scanning Electron Microscopy on two directions (transversal and longitudinal direction to the flow). In the two cases, the dispersed phase size decreased with the increase of the shear rate in the extruder. An anomaly was registered in the classical method at 200 rpm, where the size of the dispersed phase increases with the increase of the shear rate. The dynamic cooling technique recorded smaller diameters (4 to 5 times) of the dispersed phase compared to the conventional technique. In addition, the reappearance of the microfilaments at 200rpm was observed. The rheological properties were determined by RS100 (Thermo Scientific Haake). Using this new technique, it was noticed that he elastic modulus increases with one decade compared to the classical method and the complex viscosity decreases with the increase of the shear rate. An anomaly was registered in the classical technique, where the dynamic viscosity at 200rpm increases with increasing the shear rate in the extruder.

1. Introduction

Blending of polymers is an effective way to obtain new materials with improved properties [1]. Since most of the polymers are immiscible, thus a decrease of specific properties is usually observed. The extrusion of an immiscible polymer blend in which the dispersed phase forms in situ reinforced nodules or fibers is the preferable way to achieve the highest mechanical properties [2]. In order to obtain such structure, so called microfibrillar composites were developed [3-5].

The morphology of the dispersed phase is the key parameter which governs the all properties of the materials. It is directly related to the processing conditions and concentration of the dispersed phase [6]. Various phenomena were related to the dynamics of droplet such as deformation, rupture, time-dependent behavior and coalescence. Several studies have been carried out in order to control the morphology of the blend polymer, such as the rapid cooling of the extruded in iced water technique [7] and, quenching the extrudate [8-10]. The major studies were conducted on specific devices [11, 12]. Most properties of immiscible polymer blends depend essentially on the morphology development during the phase preparation [13]. The principal goal to structuring the polymer blends was to attain a possible finest morphology and good distribution of the dispersed phase in the matrix.

During the preparation of the blend polymer, a great diversity of morphologies (droplets, filament and co-continuous phases) can be obtained [1]. The final size and shape of the dispersed phase are determined by the competition between coalescence and breakup. Coalescence is a process in which two or more particles collide and physically merge into one particle. Two classes of coalescence play critical roles in the development of morphology during the processing of immiscible polymer blends [14, 15]. In the first class, the coalescence is caused by the thermal Brownian motion and, in the second class; the collisions are caused by hydrodynamic forces [16]. On the other hand the breakup of the dispersed phase during the extrusion of the polymer blends is mainly controlled by the interfacial tension, rheological properties and complex strain field of the extruder [17]. However, deformation and rupture of the droplets are governed by two factors: the viscosity ratio between dispersed phase and the matrix phase, and the capillary number which presents the ratio between the viscous and interfacial forces [18].

This investigation was performed on the extruder in order to approximate real conditions for the preparation of the polymer blend. The main goal of this study was to introduce a new technique for the preparation of the polymer blend. The classical technique consists in the blending at different shear rate followed by the solidification of the molten blend under air flow. However in this early stage, the static coalescence plays an important role. In order to suppress the static coalescence of the dispersed phase, the dynamic cooling technique was adopted to produce the finest system morphology and smallest droplet sizes. This technique was based on two parameters during the cooling stage (rotation speed and temperature). Due to the difference in sensitivity of the blend compounds, the temperature variation during extrusion can produce different ratios of viscosity in the blend and that gives an opportunity for manipulation finest droplet and fibrillar behavior in blends. The dynamic solidification of the dispersed phase due to the crystallization was shown to be a good way to improve the stability of the fibrillar structures with respect to the shear out during extrusion.

2. Experimental part

2.1. Materials

The materials used in this study were commercial grades. Standard polystyrene was supplied by BASF S.A (Germany). Its average molar mass was $\bar{M}_w$= 215160g/mol with polydispersity coefficient I=2.35. Isotactic polypropylene (100-GA01) was produced by INEOS Company (Switzerland), with a melt flow index (MFI) of 0.9g/10min and density of 0.923 g/cm^3.

2.2. Sample preparation

The principal factor to take into account was the difference in temperatures between the dispersed phase (melting temperature) and the matrix phase (glass transition temperature). All polymers were dried at least 18 h in a vacuum oven at 80°C. The polymer granulate was pre-mixed and injected in the twin screw mini extruder (Thermo Haake) at 200°C. In the conventional technique the samples were extruded at (30, 60, 90, 120, 150 and 200) rpm and 200°C, followed by cooling stage on air in the extruder. The dynamic cooling technique used for structuring the blend polymers is described as follows. The sample was blended for 3mn at 200°C and 60rpm, followed by a rapid change of the extrusion speed while keeping the same temperature. After 3min, the cooling phase was started, maintaining the speed extrusion to the point of crystallization of the dispersed phase (maximum torque of the extruder), in order to freeze the generated microstructures (suppress of the static coalescence).

2.3. Scanning Electron Microscopy

A Hitachi S-3340 Scanning Electron Microscopy (SEM), operating at 15KV accelerating voltage, was used to observe the blend morphology. The surfaces taken from cryofractured samples were observed by SEM in two directions to the flow. After drying procedure, the samples were coated with gold-palladium thick film. The adopted technique to detect the morphological evolution was based on an image of the sample on two-dimensional (2D) object, because the fibrillar morphology can be viewed as droplets when the fibrils are observed perpendicularly to the flow direction. For this reason, it was suggested [19, 20] to examine the collection of images from two orthogonal planes [21]. An average of 50 to 100 particles per sample was examined.

2.4. Thermal properties

The thermal properties were carried out by Differential Scanning Calorimetry using a Q100 DSC (TA Instruments, USA), with samples of about 5 mg in aluminum pans under nitrogen atmosphere. The temperature and heat flow calibrations were performed using Indium (T_m=156.6°C) at a heating and cooling rate of 10°C/min. The matrix (PS) presents a glace transition temperature of about 96°C and a melting and crystallization temperatures of the dispersed phase (PP) as 165°C and 112°C respectively. These parameters are valuable information, which enables us to get a large difference between the crystallization of the dispersed phase and the glass transition of the matrix temperatures.

2.5. Rheological properties

The rheological characterization of the pure and blends polymers was carried out using a stress controlled rheometer (RS 100) from Thermo Haake Rheostress, using parallel plate geometry (diameter 20 mm and a gap of 1 mm). Samples with 20 mm and a thickness of 1 mm were compression molded from the test material using hydraulic press (Carven-USA) at 200°C. For all measurements, the stress amplitude was fixed at 10Pa to insure the viscoelastic linearity of the response. All materials functions were determined in the frequency range from 0.001 to $10s^{-1}$.

3. Results and discussions

3.1. Morphological observation

3.1. 1. Static cooling technique blends morphology

Figure 1 shows the micrographs of PP10/PS90 blends prepared with different shear rates using static cooling technique. The micrographs on the left side were obtained on the perpendicular facet to the flow, on the right side the micrographs were photographed on the longitudinal facet. The sample sheared at 30rpm presents deformed droplets with a shape factor slightly greater than 1.

The sample sheared at 60rpm presents a mixture of filaments and droplets oriented in the flow direction. The blend polymer sheared at high speed presents a circular shape of the dispersed phase without orientation; this is probably due to a static cooling stage. The dispersed phase has the necessary time to relax before the crystallization. It was found that the major reduction of the particle size of the dispersed phase occurs at an initial stage of less than 1 or 2min, during blending [13]. However, an anomaly was recorded at 200rpm, where the size of the dispersed phase increases with the increase of the extrusion speed. These observations are in good agreement with those obtained by [22, 23]. It is may be due to the intensity of the extrusion which promotes dynamic calescence of the dispersed phase.

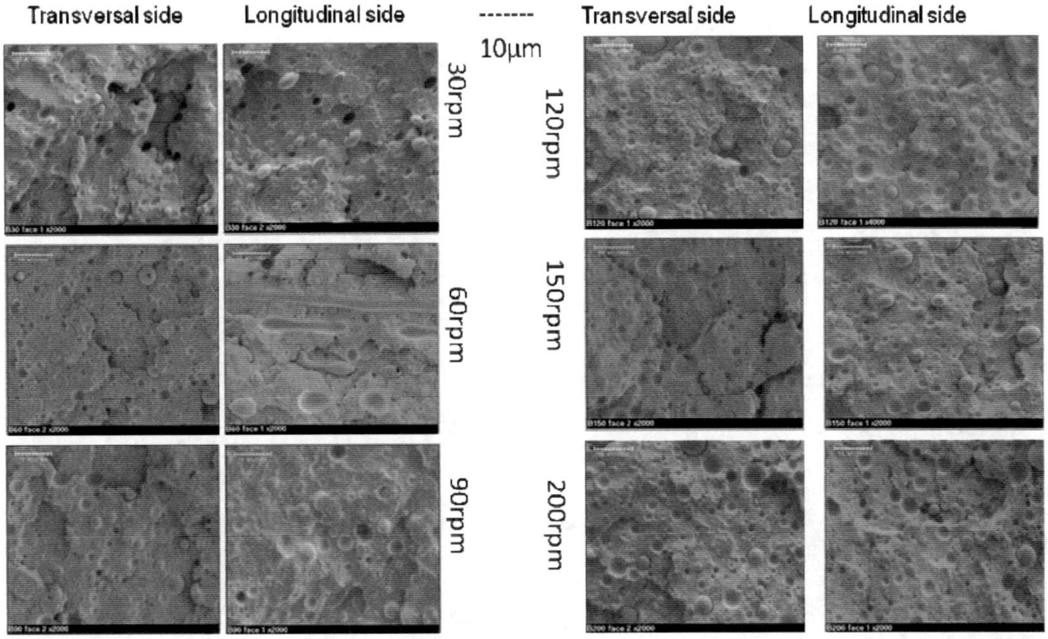

Figure 1. Micrographs of the blends PP/PS (90/10) prepared with static cooling technique: extruded at different speeds (30, 60, 90, 120, 150, and 200rpm). All blends were photographed in both directions (transversal and longitudinal to the flow).

3.1.2. Dynamic cooling technique blends morphology

The morphology development obtained by dynamic cooling technique is shown Figure 2. When the sample was not cooled by dynamic technique, the dispersed phase was presented in the transversal side by circular coarseness form. In the longitudinal side, a fibrillar form of the dispersed phase was developed with length that exceeds 100µm. When the dynamic cooling technique was applied, the size of the dispersed phase decreases sharply. Reducing the size of the dispersed phase is inversely proportional to the increase of the extrusion speed. Fibrillar microstructure reappeared again in the speed extrusion of 200rpm with a shape factor of 20; this may be due to the extrusion intensity.

The treatment of the micrographs (Image J software) of the blends prepared with the two techniques was realized in terms of the average diameter of the dispersed phase (Table 1). The average diameter of the dispersed phase was decreased with increasing the extrusion speed in the two cases.

Table 1. Average diameters of the dispersed phase prepared by static and dynamic cooling techniques

Sample	PP0.9/PS (10/90)	
Extrusion speed (rpm)	d (conventional technique)	d (dynamic technique)
30, (60/30)	2.5	0.75
60, (60/60)	1.8	/
90, (60/90)	1.5	0.35
120, (60/120)	1.2	0.31
150, (60/150)	1.1	0.27
200, (60/200)	1.15	0.20

The dynamic cooling technique develops diameters five times lower than in the classical technique. This may due to the suppression of the coalescence in the cooling stage.

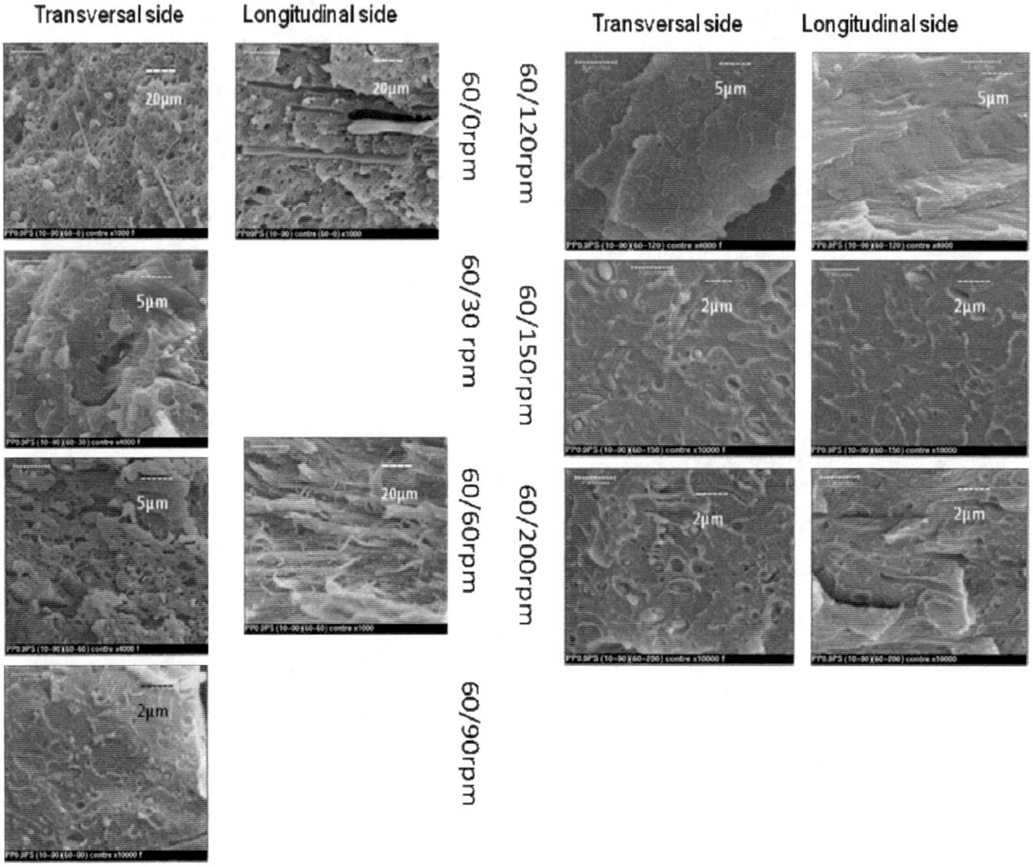

Figure 2. Micrographs of the blends PP/PS (90/10) prepared with dynamic cooling technique: extruded at 60rpm flowed by jump of speeds at (30, 60, 90, 120, 150, and 200rpm) and cooled at the same speed. All blends were photographed in both directions (transversal and longitudinal to the flow).

3.2. Rheological properties

3.2.1. Static cooling technique

Frequency sweeps were performed at the same conditions (F=0.001 to 10 s^{-1} and 200°C). Figure 3-a and Figure 3-b present respectively loss modulus and complex viscosity in function of the shear frequency for the blends prepared with the static cooling technique. All values of storage modulus were situated in between of those of the components of the blends Figure 3-a. A less decrease was noticed when the extrusion speed increases. An enhanced elasticity at low frequency is obviously shown. The shoulder in G' of the blend with droplet represents an additional relaxation due to the deformation of the dispersed phase [24]. Molten polymers show constant viscosity (Newtonian behavior) only in the low frequency region. At higher frequency, the viscosity generally decreases with the increase of the shear rate, indicating a pseudoplastic behavior. The rheological properties of the blends were predominant by the matrix phase. An anomaly was observed in the evolution of the complex viscosity of the blend prepared at 200rpm, where the complex viscosity increases with the increasing of the extrusion speed. This may be explained by the increase of the size morphology of the dispersed phase at the same extrusion speed (Table 1) [22-23]. At low frequencies a negative deviation behavior of the complex viscosity was noticed, which probably originates in slip at the interface [25-26] due to poor or insufficient adhesion between the matrix and the dispersed phase.

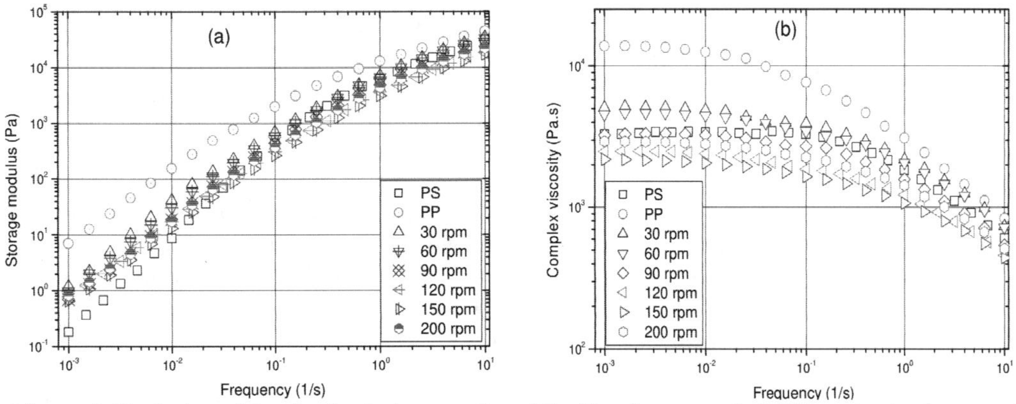

Figure 3. Evolution of the rheological properties of the blends prepared at different extrusion speeds by static cooling technique.

3.2.2. Dynamic cooling technique

Figure (4-a) shows the evolution of storage modulus of PP10/PS90 blends prepared with dynamic cooling technique. All the storage moduli are situated in between the moduli of the components. The elasticity of the blends was predominant by the elasticity of the dispersed phase. At high frequencies the storage moduli of the blends were superposed on the moduli of both components. The anomaly shown in the first case (static cooling) was suppressed when the cooling dynamic technique is used. All values of complex viscosities Figure (4-b) were situated in between the components of the blends (PP and PS). Increasing the rotation speed in the extruder decreases slightly Newonian viscosity. The rheological properties of the polymer blends are predominated by those of the disperse phase, thus increasing the elasticity and viscosity of the blends. At low frequencies, a pronounced deviation on the elastic modulus was observed. It presents the additional relaxation of the dispersed phase [27]. Obviously, these rheological parameters deviating from linear rheological model seems a useful approach to assessing the critical point of phase separation in polymer blend [28].

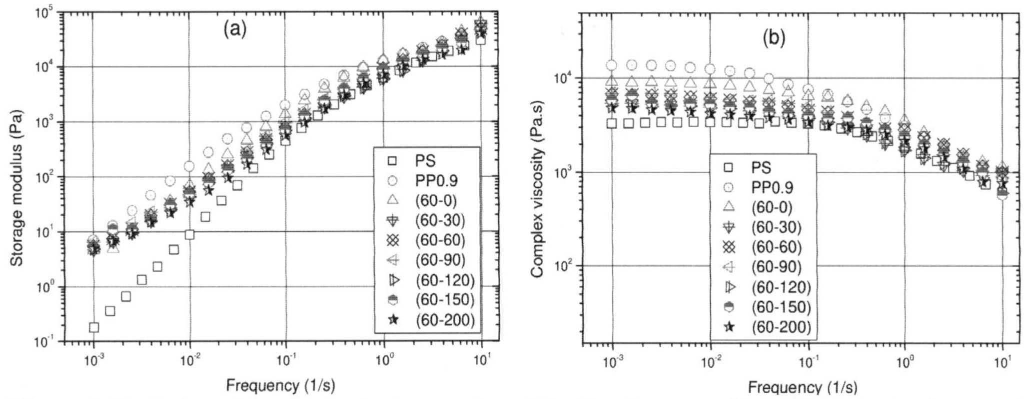

Figure 4. Evolution of the rheological properties of the blends prepared by different extrusion speeds by dynamic cooling technique.

4. Conclusion

In this study the morphology evolution in polymer blend composed by polypropylene (10%) and polystyrene (90%) was evaluated by two techniques, static and dynamic cooling. The new technique was based on the crystallization of the dispersed phase during the extrusion. The size of the droplets generated by the dynamic cooling technique is 4 to 5 times less than that generated by the conventional technique. The new technique minimizes the coalescence of the dispersed phase during the blend solidification. At high shearing, the cooling dynamic technique generated microfibrillar morphology with a high shape factor (20). In the two techniques, the reduction in domain size by the increase of the shear rate in the molten blends originates from the breakup of the droplets caused by the coalescence suppression effect. The storage modulus developed by dynamic cooling technique was higher compared to the conventional technique by one decade. At static cooling technique the elastic moduli and the complex viscosity were predominant by the matrix phase, at dynamic cooling technique; the rheological properties of the blends were predominant by the dispersed phase. The storage modulus in low frequencies region developed by the new technique present a pronounced deviation; this corresponds to an increase in elasticity and longer relaxation time compared to the matrix. The main result in this study is the important role played by the coalescence in the determining of the morphology of the blends.

References

[1] Utracki LA. *Polymer alloys and blends*. 1989 New York: Hanser Publishers.
[2] Mei Y, Huang Y, He Y, Yang Q, 2012 *J. Appl. Poly. Sci.* **124** 6 4838.
[3] Karingamanna J, Sabu T and Kuruvilla J 2011 *J. Poly. Res.* **18** 1
[4] Scott, C E and Macosko, C W 1994 *Polymer.* **35** 5422
[5] Panamoottil S M, Potschke P, Lin RJT, Bhattacharyya D, Fakirov S 2013 *eXP. Poly. Lett.* **7** 607
[6] Son Y and Yoon J T 2001 *Polymer.* **42** 7209
[7] Hua. C, Wang. D, Xiang. B, Chiou, B S and Sun. G. 2010, *J. Poly. Sci: Poly. Phy.* **48** 921
[8] Favis B D, Chalifou J P, 1988, *Polymer* **29** 1761
[9] Bordereau V, Shi Z H, Utracki L A, Sammut P and Carrega M 1992 *Poly. Eng. Sci.* **32** 24 1846.
[10] Shi Z H and Utracki L A 1992 *Poly. Eng. Sci.* **32** 24 1834.
[11] Cassagnau P, Michel A 2001, *Polymer.* **42** 3139.
[12] Deyrail Y, Cassagnau P 2004 *J. Rheo.* **43** 3 505.
[13] Utracki L A, *Polymer Blends Handbook* 2002 Kluwer Academic Publishers.
[14] Lyu S P, Bates F S, Macosko C W, 2000 *J. AIChE.* **46** 2 229.
[15] Dimzoski B, Fortelný I, Šlouf M, Nevoralová M, Michálková D, Mikešová. 2011 *J e-Poly* **1** 1 1.
[16] Hong. J S, Kim. Y K, Ahn. K H, Lee. S J and Kim. C, 2007 *Rheol. Acta.* **46** 469.
[17] Wu S 1987 *Poly. Eng. Sci.* **27** 335.
[18] Li. W, Schlarb. A K, Weber. M and Evstatiev. M. 2010 *Poly. Bull.* **64** 483.
[19] Steinmann S, Gronski W and Friedrich C. 2001 *Polymer* **42** 15 6619.
[20] Heeschen WA. 1995 *Polymer.* **36** 9 1835.
[21] Omonov T S, Harrats C, Moldenaers P and Groeninckx G 2007 *Polymer.* **48** 5917.
[22] Ghodgaonkar P G, Sundararaj U 1996 *Poly. Eng. Sci.* **36** 12 1656.
[23] Sundararaj U, Macosko CW 1995 *Macromolecules.* **28** 2647.
[24] Yu W, Zhou W and Zhou C 2010 *Polymer.* **51** 2091.
[25] Van Puyvelde P, Velankar S, Moldenaers P 2001 *Curr. Opin. Coll. Int. Sci.* **6** 457.
[26] Hammani S, Moulai-Mostefa N, Benyahia L and Tassin J F 2012 *J. poly. Res.* **19** 9940.
[27] Fan X, Wang Z, Wang K, Deng H, Chen F, Fu Q 2012 *J. Appl. Poly. Sci.* **125** S1 E292.
[28] Min Z and Qiang Z 2008 *Sci. China. Ser. B-Chem.* **51** 1 1.

Surface modification-a novel way of attaching cocatalysts on CdS semiconductors for photocatalytic hydrogen evolution

Weili Yu,[1,2] Tayirjan Isimjan,*[3] Bin Lin,[1] and Kazuhiro Takanabe*[1]

[1] Division of Physical Sciences and Engineering, King Abdullah University of Science and Technology (KAUST), KAUST Catalysis Center, Thuwal, 23955-6900, Saudi Arabia

[2] School of Materials Science and Engineering, Changchun University of Science and Technology, 7989 Weixing Road, Changchun 130022, P. R. China

[3] Solar and Photovoltaics Engineering Research Center, KAUST, Thuwal, 23955-6900, Saudi Arabia

*Email: Kazuhiro.Takanabe@kaust.edu.sa; Tayirjan.Isimjan@kaust.edu.sa

Abstract. Noble metals as cocatalysts for hydrogen evolution are widely investigated for semiconductor photocatalytic water splitting. In this paper, we present a novel way to attach not only noble metals, but also transitional metals onto CdS nanocrystals as cocatalysts for hydrogen evolution. The hydrogen evolution performances for each metal were compared and result shows that Pd attached CdS gives the highest hydrogen evolution rate of 250 μmol/h. The amounts of metal ions attached on the surface were measured by inductively coupled plasma optical emission spectrometry (ICP-OES). This work confirms that surface modification is a promising way of attaching cocatalysts onto semiconductor photocatalysts.

1. Introduction

Photocatalytic hydrogen evolution using semiconductor powder materials has gained considerable attention because of the importance of utilizing solar energy as renewal energy to form green fuel [1]. Nanocrystals (NCs) in particular have seen significant growth as the photocatalyst to generate hydrogen due to superior photostability, larger absorption spectral range and properties that deliver multiple electrons with minimal structural perturbations compared to organic or organometallic molecules. For semiconductor photocatalyst system, cocatalysts play an extremely important role as it helps to promote charge separation, which in return reduces bulk and surface electron/hole recombination [2]. In general, there are two main ways to deposit cocatalyst on semiconductors, i.e. photodeposition and impregnation. However, both processes inherit the following disadvantages. For example, both ways always result in nonuniform distribution of the cocatalysts on the surface. Second, the size distribution of cocatalyst is hard to be controlled hence results are difficult to be reproduced. Third, noble metals are still widely used which hindered the wide application due to the high cost of these materials. Therefore, alternative methods and low-cost cocatalyst materials are still much needed for the synthesis of low-cost, highly stable and effective water splitting photocatalyst systems.

Semiconductor CdS has been reported as effective photocatalyst for H_2 evolution in literatures, and effective cocatalys are always needed for this photocatalyst to produce hydrogen, thus CdS nanocrystals was chosen as photocatalyst for testing our surface modification approach [3,4] and materials. Here, by modifying the CdS NCs' surface with organic molecules, i.e. 1,2-Ethanedithiol

(EDT), we achieved a stable, reproducible and efficient way of depositing various of cocatalysts on CdS surface.

2. Materials and methods

2.1. Materials
Sodium sulfide nonahydrate ($Na_2S{\cdot}9H_2O$, ≥98%) and cadmium(II) acetate nonahydrate ($Cd(OAc)_2{\cdot}9H_2O$, ≥98%) were purchased from Acros Organics. 1,2-Ethanedithiol (EDT), triethanolamine (TEOA), ascorbic acid (AA), copper(II) sulfate pentahydrate, ferric nitric acid nonahydrate, cobalt chloride, gold(III) chloride trihydrate, palladium(II) chloride, trisodium rhodium(III) hexachloride were all purchased from Sigma-Aldrich.

2.2. Methods

2.2.1. Synthesis of CdS. CdS nanocrystals was synthesis according to a previous report [5]. In detail, 400 ml of a Na_2S (0.14 M) aqueous solution and 500 ml of a $Cd(OAc)_2$ (0.14 M) aqueous solution were mixed together under vigorous stirring. After being continuously stirred for 24 h, the solution was filtered and the solid was washed several times with deionized water. The yellow solid was then dispersed in 100 ml of deionized water and transferred to a Teflon-lined stainless autoclave for hydrothermal processing. After being heated at 453 K for 3 days, the yellow solid was filtered again and kept under vacuum until dry.

2.2.2. Attaching metals ions as cocatalysts on CdS. A solution of CdS (100 mg) in toluene (50 ml, >99.9%, Sigma-Aldrich) was sonicated at room temperature, 1 ml of 1% (v/v) EDT solution in ethanol (absolute grade, Sigma-Aldrich) were added. The resulting mixtures were stirred for 6 h and centrifuged. The precipitates were washed with ethanol (20 ml) 3 times and redispersed in ethanol (40 ml). Then, 0.1 mol/l M (M=Cu, Fe, Mn, Co, Au, Pd, Rh) ions in ethanol (10 ml) was added to this suspension. The suspension was stirred for 1 h and then centrifuged. The resulting precipitates were washed with ethanol (20 ml) 3 times, centrifuged, and dried under vacuum at 50°C to obtain a greenish-yellow solid in quantitative yield.

2.2.3 Characterization. Photocatalytic testing was conducted using a recirculating batch reactor and top-irradiated photocatalytic reactor. The accumulated gaseous products were analyzed with a Bruker 450 gas chromatograph equipped with a TCD and a molecular sieve 5A column. The light source was an Asahi Spectra MAX-303 (300 W Xe lamp) with a cold mirror and various band-pass filters. The photon irradiance was measured using an EKO LS-100 spectroradiometer. 50 mg of photocatalyst was immersed by sonication in 25 ml of a solution of 0.5 M Na_2S and 0.5 M Na_2SO_3. The reactor was connected to a recirculation system, and the dissolved air in the solution was evacuated before the photocatalytic test.

3. Results and discussion
We used 1,2-ethanedithiol (EDT) as a molecular linker to attach various metal ions to CdS NCs surfaces through exceptionally stable metal-thiol bond. Since the metal ions are only attached to the surface with SH group, the position of the metals can be confirmed and The amounts (0.1~0.3 wt%, ICP results) that required to obtain an optimal H_2 evaluation rate are significantly lower than that of the metal deposited by photo-deposition (normally around 1 wt%), as shown in Table 1. Not only noble metals, but also transitional metals can be attached to the surface of CdS NCs (Figure 1). For the transitional metals, Ni shows the highest H_2 evolution rate of 110 μmol/h. Mn and Co are also effective cocatalysts for H_2 evolution. Figure 2 shows the H_2 evolution rate when noble metals are attached. All the noble metals improved H_2 evolution rate compare to bare CdS NCs, among which,

the highest hydrogen evaluation rate of 255 (μmol/h) was achieved by Pd which is around 30% higher than that of photo deposited Pd. These results confirm that surface modification is an effective way of attaching various metals as cocatalysts on semiconductor surface. Interestingly, some metals (such as Cu^{2+}, Fe^{3+}) had negative effect, while others improved the hydrogen evolution rate. One has to see that almost any metal ions can be attached to semiconductor surface through this surface modification approach while photo deposition only works for certain metals.

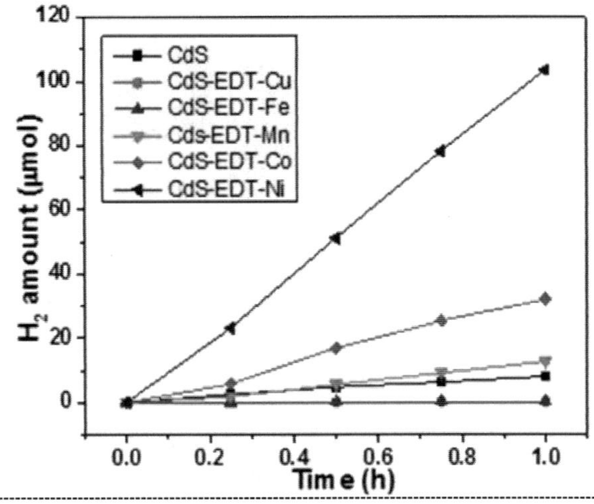

Figure 1. The H_2 evolution amount vs. time for transitional metals. Amount of catalysts: 50 mg; Electrolyte solution: 25 ml, Na_2S (0.5 M) / Na_2SO_3 (0.5 M). Light source: ASAHI Xe lamp, 300 W.

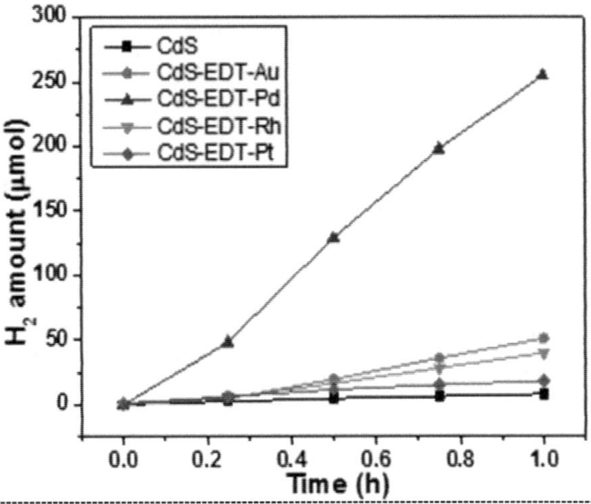

Figure 2. The H_2 evolution amount vs. time for noble metals. Amount of catalysts: 50 mg; Electrolyte solution: 25 ml, Na_2S (0.5 M) / Na_2SO_3 (0.5 M). Light source: ASAHI Xe lamp, 300 W.

In our recent research work [6], we have proved that only 0.01 ml of 1% EDT solution (in ethanol) was enough to fully cover 100 mg CdS NCs surface. Here, we used 1 ml 1% EDT solution in order to make sure that the complete surface of CdS NCs (100 mg) was saturated with EDT molecules. The metals ions were then added into the solution, which will be linked onto the CdS surface via coordination bond between the metal ions and EDT molecules. The exact amounts of metal ions that attached onto the surface of CdS NCs were measured by ICP, as shown in Table 1.

Table 1. ICP results for metals and the H_2 evolution rates on the samples prepared with different metal ions.

Metal ions	**Metal (wt%)**	**H_2 (μmol/h)[a]**
Fe	0.07	0.2
Mn	0.28	25
Co	0.10	32
Ni	0.14	110
Rh	0.24	40
Pd	0.36	255
Pt	0.33	18

[a] Conditions: 50 mg-cat, 20°C, 0.5 M Na_2S, 0.5 M Na_2SO_3, 25 ml, 300 W Xe lamp.

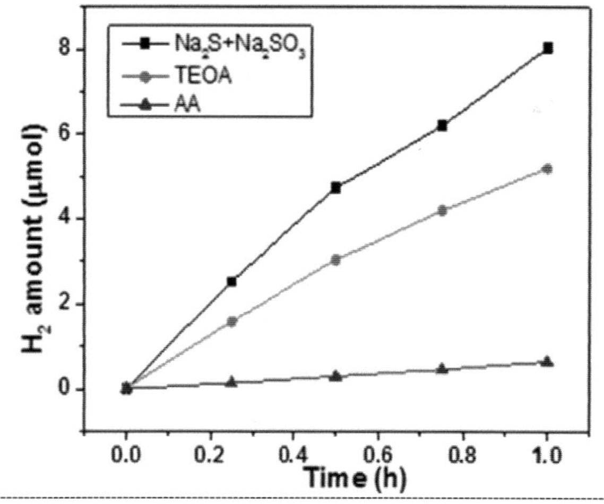

Figure 3. The effect of sacrificial reagents on H_2 evolution. Amount of catalysts: 50 mg; Cocatalyst: Ni; Electrolyte solution: 25 ml, Na_2S (0.5 M) / Na_2SO_3 (0.5M) (Black square); triethanolamine (TEOA, red cycle); ascorbic acid (AA, blue triangle). Light source: ASAHI Xe lamp, 300 W.

The effect of various sacrificial reagents such as Na_2S/Na_2SO_3, TEOA and AA was also studied and the corresponding hydrogen evaluation rates are shown in Figure 3. The results show that the highest H_2 evolution rate was obtained when Na_2S/Na_2SO_3 was used as sacrificial reagent. The enhancement of H_2 evolution rate was attributed to improved S oxidation rate assisted by the sacrificial reagents [7].

4. Conclusion

We demonstrated EDT as a highly efficient molecular linker to attach various metal ions onto CdS NCs surfaces through stable metal-thiol bonds. The performance of different metals as cocatalyst are fully investigated, among which Ni and Pd show H_2 evolution rate 110 and 255 μmol/h, respectively. The effect of sacrificial reagents was also investigated and Na_2S/Na_2SO_3 sacrificial reagent show

IOP Conf. Series: Materials Science and Engineering **64** (2014) 012057 doi:10.1088/1757-899X/64/1/012057

better performance than TEOA and AA. This approach also can be used as a universal method to load metal ions or semiconductor nanocrystals onto photocatalysts for hydrogen evolution.

References

[1] Takanabe K and Domen K 2011 Toward visible light response: Overall water splitting using heterogeneous photocatalysts. *Green* **1** 313-322.
[2] Kudo A and Miseki Y 2009 Heterogeneous photocatalyst materials for water splitting. *Chem. Soc. Rev.* **38** 253-278.
[3] Zong X, Yan H, Wu G, Ma G, Wen F, Wang L and Li C 2008 Enhancement of photocatalytic H_2 evolution on CdS by loading MoS_2 as cocatalyst under visible light irradiation. *J. Am. Chem. Soc.* **130** 7176–7177.
[4] Li Q, Guo B, Yu J, Ran J, Zhang B, Yan H and Gong J 2011 Highly efficient visible-light-driven photocatalytic hydrogen production of CdS-cluster decorated graphene nanosheets. *J. Am. Chem. Soc.* **133** 10878–10884.
[5] Yan H, Yang J, Ma G, Wu G, Zong X, Lei Z, Shi J and Li C 2009 Visible-light-driven hydrogen production with extremely high quantum efficiency on Pt–PdS/CdS photocatalyst. *J. Catal.* **266** 165-168.
[6] Yu W, Isimjan T, Gobbo S D, Anjum D H, Abdel-Azeim S, Cavallo L, Garcia-Esparza A T, Domen K, Xu W and Takanabe K 2014 Tethering metal ions to photocatalyst particulate surfaces via bifunctional molecular linkers for efficient hydrogen evolution. *ChemSusChem*, in press.
[7] Matsumura M, Saho Y and Tsubomura H 1983 Photocatalytic hydrogen production from solutions of sulfite using platinized cadmium sulfide powder. *J. Phys. Chem.* **87** 3807-3808.

IOP Conf. Series: Materials Science and Engineering **64** (2014) 012058 doi:10.1088/1757-899X/64/1/012058

New approach on the catalytic oxidation of methanol to formaldehyde over MoO_3 supported on nano hydroxyapatite catalysts

A A Said*, M M Abd El-Wahab and A M Alian
Chemistry Department, Faculty of Science, Assiut, University, Assiut, Egypt

Abstract: Molybdenum oxide (20 wt. %) supported on nano hydroxyapatite mixed was prepared by impregnation method and calcinated at 400°, 500°, 600° and 700°C in static air atmosphere. The catalysts were characterized by thermogravimetry (TG), differential thermal analysis (DTA), X-ray diffraction (XRD), Transmission Electron Microscope (TEM) and nitrogen sorption measurements. The gas–phase oxidation of methanol to formaldehyde was carried out in a conventional fixed flow bed reactor. The obtained results clearly revealed that the formation of $CaMoO_4$ spinel nano particles was active and selective catalyst towards the formation of formaldehyde. The maximum yield of formaldehyde was 97% on the catalyst calcined at 400 °C. Moreover, the yield of formaldehyde was found unaffected by increasing the calcination temperature up to 700° C.
Keywords: MoO_3; nano hydroxyapatite support; structure; surface area; oxidation; methanol.

1. Introduction

Hydroxyapatite (HAP) has attracted the attention of researchers to biomaterials field in recent years [1]. HAP is a highly non-stoichiometric calcium phosphate compound with a Ca/P molar ratio ranging from 1.5 to 1.67. Moreover, HAP also is known [2] to have the character of both acidic and a basic catalyst when its Ca/P ratio is between 1.5 and 1.67. On the other hand, formaldehyde is obtained from oxidative dehydrogenation of methanol by two commercial processes: (i) On iron-molybdate based catalysts and (ii) On silver based catalysts. In both processes, a conversion of about 99 % and 92 % selectivity respectively are obtained the temperature of reaction is about 400° C in the first and 650° C in the second process. The silver process has high operating costs due to steam consumption. Moreover, the selective oxidation of methanol to formaldehyde takes place on MoO_3 mixed oxide catalysts [3,4]. However, the catalytic gas oxidation of methanol to formaldehyde over MoO_3 supported on nano hydroxyapatite to our knowledge has not yet been reported. Therefore, the main objective of the present work is synthesis of formaldehyde by oxidative dehydrogenation of methanol over MoO_3 (20 wt %) supported on n-HAP as a competitive catalyst with previous applied industrial catalysts.

2. Experimental

2. 1. Materials

Methyl alcohol, citric acid, calcium nitrate, diammonium hydrogen phosphate and ammonium hydroxide were obtained as pure reagents and were used without further purification. Ammonium heptamolybdate (AHM) was obtained from Merck, (Darmstadt, Germany).

* **Corresponding author**: e-mail: aasaid55@yahoo.com, Tel/Fax: 00201227433186 - 0020 88 2342708

2.2. Synthesis of nano hydroxyapatite by sol-gel method

The citric acid sol–gel method [5] has been used for the synthesis of nanocrystalline hydroxyapatite (nHAP) powder from calcium nitrate, diammonium hydrogen phosphate and citric acid. The pH of solution was adjusted by ammonia solution.

2.3. Preparation of nHAP-supported 20 wt. % MoO_3 catalyst

Nano hydroxyapatite supported with 20 % MoO_3 (wt./wt.) was prepared by the impregnation method. The sample produced was dried in an oven at 100°C for 24 h before being calcined at 400°, 500°, 600° and 700° C for 4 h in static air atmosphere, then quenched to room temperature. The prepared catalyst is donated by nHAP20Mo.

2.4. Apparatus and techniques

Thermal analysis

Thermogravimetry (TG) and differential thermal analysis (DTA) thermogravimetry for pure nHAP and AHM supported on HAP were recorded upon heating up to 600° C at 10 ° C min^{-1} and a 40 ml min^{-1} flow of air atmosphere, using a Shimadzu Thermal Analyzer TA60 Apparatus (Japan).

X-ray diffraction (XRD)

XRD analysis of the catalysts was performed with a Philips diffractometer (Model PW 2103), (The Netherlands) λ = 1.5418 Å, 35 kV and 20 mA) with a source of CuKα radiation (Ni filtered). Average particle sizes were estimated from X-ray line width broadening using the Scherrer equation [6].

Nitrogen gas adsorption

Nitrogen gas adsorption-desorption isotherms were measured at -196°C using a Nova 3200 (Quantachrom Instrument Corporation, USA).

Catalytic activity measurements

The catalytic activity of the catalysts for the gas-phase oxidation of methanol was carried out at the reaction temperature of 400° C in a conventional fixed–bed flow type reactor at atmospheric pressure using dry nitrogen as carrier gas. A mixture of methanol and nitrogen was introduced into the reactor after nitrogen was bubbled through methyl alcohol saturator. The total flow rate was fixed at 100 ml min^{-1} in the gas feed. The gases after reaction were chromatographically analyzed by FID with a Unicam ProGC using a DNP glass column (2m) for analysis of the reaction products of methanol on the tested catalysts.

3. Results and discussion

3.1. Thermal analysis

TG and DTA curves of original catalyst are shown in Fig. (1).The results show that the weight loss can be differentiated into three regions such as (1) 30°–150° C, (2) 150°–250° C and (3) 400°–500° C. In the first region which accompanied with endothermic peak at 120° C, the weight loss is attributed to the removal of physically adsorbed water. The second stage of the weight loss from 150°-250° C which associated with sharp exothermic peak maximized at 190° C is due to the removal of structural water or decomposition of organic constituents. The third stage of the weight loss from 400°-500° C which accompanied with sharp exothermic peak maximized at 475° C, may be corresponded to the $CaMoO_4$ formation [7].

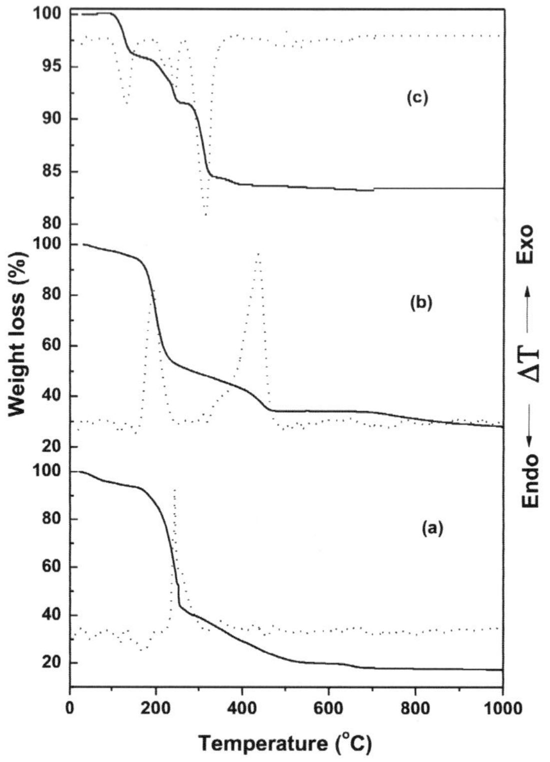

Fig. (1): TG and DTA curves of pure nHAP (a), nHAP20Mo (b) and AHM (c).

3.2. X-ray diffraction (XRD)

X-ray diffraction patterns of the powder calcined at 400°C in air atmosphere for 4h were carried out and presented in Fig. (2). Phase analysis was done using PDF card no. 85-0585 for the catalyst n-HAP20Mo, which provide information in the 2θ range of 4° – 80°. The diffractogram of the sample calcined at 400° C shows new peaks are appeared. These peaks may be attributed to the formation of new phase nanocrystalline $CaMoO_4$. The intensity of peaks of $CaMoO_4$ is due to planes around (101), (112), (004), (200), (211), (114), (123), (220), (312) and (224).

3.3. Surface area measurements

The specific surface area $\mathbf{S_{BET}}$ was obtained by applying the BET equation in its normal range of applicability (0.05-0.30 p/p°) and the values are cited in Table (1). From Table (1), it seems that, raising the calcination temperature from 400° up to 700° C led to a continuous decrease in S_{BET} values of pure HAP whereas, a sharp decrease in S_{BET} value of n-HAP20Mo on calcination at 500° C. This behavior may due to the complete formation of $CaMoO_4$ which exhibits low surface area. Moreover, from the above, Tables it important to mention here that the sharp reduction in S_{BET} value of the n-HAP20Mo calcined at 500°C may be attributed to a remarkable decrease in the total pore volume by 460 % compared to the sample calcined at 400°C. These results are in agreement with the results of thermal analysis in which the formation of $CaMoO_4$ was found between 400°-500° C.

3.4. Crystallite size determination

The average grain sizes of the catalysts calcined at 400° C were calculated from XRD line broadening method and specific surface area (S_{BET}) and TEM.
Table (2) shows the average crystallite sizes for the calcination of n-HAP20Mo powders calculated by the above three methods.

Table (1): Texture data variation of S_{BET} and S_t of n-HAP and n-HAP20Mo catalysts calcined at different temperatures in air atmosphere for 4h.

Temperature (°C)	S_{BET} (m^2g^{-1})		S_t (m^2g^{-1})	
	n-HAP	n-HAP20Mo	n-HAP	n-HAP20Mo
400	56.1	32.1	55.2	31.9
500	47.3	5.71	47.2	5.61
600	22.5	4.61	22.4	4.51
700	19.4	3.22	0.21	3.22

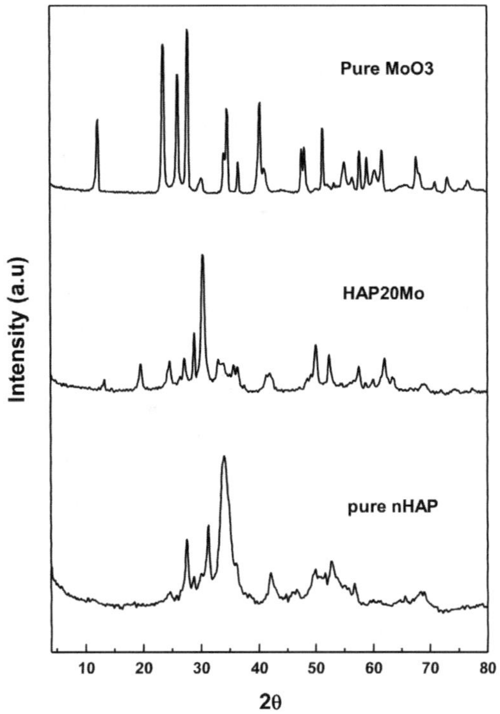

Fig. (2): X-ray diffraction patterns of pure nHAP, nHAP20Mo and pure MoO_3 calcined at 400 °C in a static air atmosphere for 4 h.

Table (2): Calculated average crystallite size of nHAP20Mo calcined at 400°C.

Temperature (°C)	Average crystallite size (nm)		
	XRD	S_{BET}	TEM
400	17.05	18.41	17.06

3.5. Catalytic activity

The catalytic oxidation of methyl alcohol on n-HAP20Mo calcined at 400°, 500°, 600° and 700° C was carried out at the reaction temperature of 400° C and the results are shown in Fig. (3). It shows that nHAP20Mo is highly active and selective catalyst towards the formation of formaldehyde (major ≈ 97%) and formic acid (minor ≈ 3%) for all calcined catalysts. These results are very interesting not only for the high selectivity of the catalyst towards the formation of formaldehyde but also the capacity of the catalyst unaffected by increasing the calcination temperature, up to 700° C. In addition the selectivity of the catalyst is also, unaffected by the sharp decrease in the surface area as indicated in Table (1). Moreover, it is important to mention here that the formation of $CaMoO_4$ spinel resulted from the interaction between MoO_3 and nHAP is accompanying with existence of high oxidation state of molybdenum Mo^{6+}. These results are value added to the advantages of our prepared catalyst.

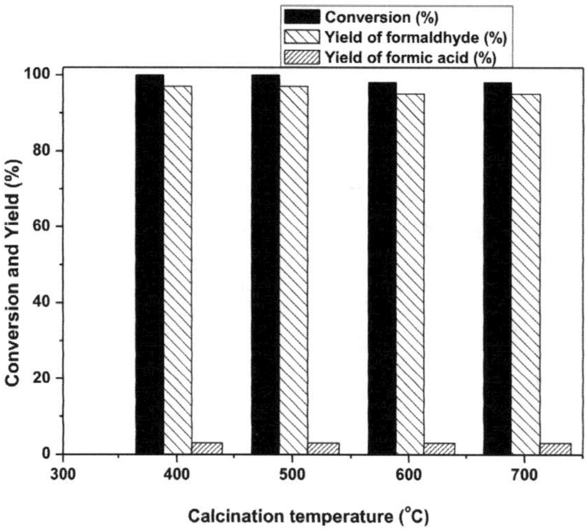

Fig. (3): Catalytic oxidation of methanol over nHAP20Mo at 400°C at different calcination temperatures.

Conclusions

From the results presented and the discussion above, it can be concluded that:

- nHAP20Mo calcined from 400° up to 700 °C acts as active and selective catalysts towards the formation of formaldehyde. The maximum yield of formaldehyde (≈97%) with low yield of formic acid (≈ 3%) as a side reaction was obtained.
- The activity and selectivity of nHAP20Mo are unaffected by increasing the calcination temperature up to 700° C.

References

[1] R. Murugan and S. Ramakrishna, Comp. Sci, Technol. 65(2005) 2385.
[2] L. C. Kebby and K. W. Hall. J. Catal. 31 (1973) 65.
[3] K. Routray, W. Zhou, C.J. Kiely, W. Grünert, I. E. Wachs, Journal of Catalysis 275 (2010) 84.
[4] G. Jin, W. Weng, Z. Lin, N F. Dummer, S H. Taylor, C J. Kiely, J K. Bartley, G J. Hutchings, Journal of Catalysis 296 (2012) 55.
[5] A. Bigi, E. Boanini and K. Rubini, J. Solid State Chem. 177 (2004) 3092.
[6] B. Imelik, J. C. Vedrine (Eds.) Catalyst Characterization: Physical techniques for solid materials, Plenum, New York, 1994.
[7] J. W. Yoon, J. H. Ryu, K. B. Shim. Materials Science and Engineering B 127 (2006) 154.

AUTHOR INDEX

AUTHOR INDEX